Disaster Resilience and Green Growth

Series Editors

Anil Kumar Gupta, National Institute of Disaster Management, New Delhi, Delhi, India

SVRK Prabhakar, Climate Change Adaptation, Institute of Global Environment Strategies, Kanagawa, Japan

Akhilesh Surjan, College of Indigenous Futures, Arts and Society, Charles Darwin University, Darwin, Australia

Over the years, the relationship between environment and disasters has received significant attention. This is largely due to the emerging recognition that environmental changes - climate change, land-use and natural resource degradation make communities more vulnerable to disaster impacts. There is a need to break this nexus through environment based and sustainability inclusive interventions. Science – technology and economic measures for disaster risk management, hence, need to adapt more integrated approaches for infrastructure and social resilience. Environmental and anthropogenic factors are key contributors to hazard, risk, and vulnerability and, therefore, should be an important part of determining risk-management solutions.

Green growth approaches have been developed by emphasizing sustainability inclusion and utilizing the benefits of science-technology interventions along policy-practice linkages with circular economy and resource efficiency. Such approaches recognize the perils of traditional material-oriented economy growth models that tend to exploit natural resources, contribute to climate change, and exacerbate disaster vulnerabilities, Green growth integrated approaches are rapidly becoming as preferred investment avenue for mitigating climate change and disaster risks and for enhancing resilience. This includes ecosystem-based and nature-based solutions with potential to contribute to the resilience of infrastructure, urban, rural and peri-urban systems, livelihoods, water, and health. They can lead to food security and can further promote people-centric approaches.

Some of the synergistic outcomes of green growth approaches include disaster risk reduction, climate change mitigation and adaptation, resilient livelihoods, cities, businesses and industry. The disaster risk reduction and resilience outcome of green growth approaches deserve special attention, both for the academic and policy communities. Scholars and professionals across the domains of DRR, CCA, and green growth are in need of publications that fulfill their knowledge needs concerning the disaster resilience outcomes of green growth approaches. Keeping the above background in view, the book series offers comprehensive coverage combining the domains of environment, natural resources, engineering, management and policy studies for addressing disaster risk and resilience in the green growth context in an integrated and holistic manner. The book series covers a range of themes that highlight the synergistic outcomes of green growth approaches.

The book series aims to bring out the latest research, approaches, and perspectives for disaster risk reduction along with highlighting the outcomes of green growth approaches and including Science-technology-research-policy-practice interface, from both developed and developing parts of the world under one umbrella. The series aims to involve renowned experts and academicians as volume-editors and authors from all the regions of the world. It is curated and developed by authoritative institutions and experts to serve global readership on this theme.

Anil Kumar Gupta • Akhilesh Gupta •
Pritha Acharya
Editors

Disaster Risk and Management Under Climate Change

Springer

Editors
Anil Kumar Gupta
ECDRM Division, International Cooperation, CAP-RES DST Project
National Institute of Disaster Management (Ministry of Home Affairs)
New Delhi, India

Akhilesh Gupta
Department of Science and Technology
Science and Engineering Research Board, S&TI Policy Coordination, Climate Change Programme, Government of India
New Delhi, India

Pritha Acharya
ECDRM Division, CAP-RES DST Project
Centre for Excellence on Climate Resilience, National Institute of Disaster Management (Ministry of Home Affairs)
New Delhi, India

ISSN 2662-4885 ISSN 2662-4893 (electronic)
Disaster Resilience and Green Growth
ISBN 978-981-99-4104-9 ISBN 978-981-99-4105-6 (eBook)
https://doi.org/10.1007/978-981-99-4105-6

This Springer imprint is published by the registered company Springer Nature Singapore Pte Ltd.
The registered company address is: 152 Beach Road, #21-01/04 Gateway East, Singapore 189721, Singapore

Message

MESSAGE

I am happy to present the Policy Document on Climate Change and Disaster Risk Management. The aim of this publication is to showcase India's knowledge bank on variety of climatic issue and related disasters to the world. It is one of a kind endeavour by the Government of India to document the researches and case studies done by imminent scientists and researchers which not only focuses the climate change issues but also talk about scientifically and technologically advanced solutions for it.

India is party to the Sendai Framework for Disaster Risk Reduction (SFDRR) to implement disaster risk reduction strategies at national and sub-national levels. For this global treaty on disaster risk reduction under the United Nations International Strategy for Disaster Risk Reduction, Government of India formulated the National Disaster Management Plan in 2016 for achieving substantial reduction in disaster risk and losses in lives, livelihoods, health and in the economic, physical, social, cultural and environmental assists of persons, businesses and communities.

Government of India takes into account the global and regional trends in making policies for climate change and disaster resilience. It incorporates the approach enunciated in the SFDRR and SDGs (Sustainable Developmental Goals) for disaster risk reduction, which are global frameworks under the support of United Nations to which India is signatory.

In its pursuit of excellence, the Government of India endeavours to emulate and even improve upon the best practices in various types of resilience building measures to disaster risk reduction and climate change impacts. This can be achieved firstly by documenting the knowledge of best practices and understanding of climate change and disasters to provide a competitive advantage for organizations and departments to make India disaster and climate resilient. I am sure that this publication would be useful to a range of readers from academia, research, policy and practice.

Date:03.03.2020

(Prakash Javadekar)

Foreword

FOREWORD

Fight against natural disasters and climate change is never ending as the complexities continue to grow with the pace of economic development and exploitation of natural resources. What was revolutionized as a pathway for human development has become the reason of possible extinction of human race along with other species. Modernization and industrialization is not wrong when its essence and foundation of functions is based on sustainable development of ecosystem and as long as human respects the nature. India is determined to strive for the balance between human development and prosperous nature. This book is one example of many endeavours that the government of India has taken to move towards this target.

Ministry of Environment, Forest and Climate Change is taking an initiative of presenting a book, at COP-26, on Publication on Climate Change and Disaster Risk Management. The intention behind presenting this book is to present India's knowledge repository on the issues of Climate Change and Natural Disaster.

This book consists of variant knowledge of scientists and researchers of the country as well as of the global scientists and researchers whose authentic and years long research and case studies are of imminent importance for the risk management of disasters and the gruesome impacts of climate change. For the first time ever, Government of India decided to present this repository of knowledge on disaster management and climate change under the supervision of Ministry of Environment, Forest and Climate Change (MoEF&CC). For this novel work MoEF&CC coordinated with National Institute of Disaster Management (NIDM), the leading institute of the country for capacity development in India and the region.

I congratulate the project team and all the authors involved in this prestigious book for their hard work and commitment. This book has been prepared taking into account the vast geography and variant climate and geophysiology of the country; issues related to climate change and natural disasters; and the best possible and technologically advanced solutions.

[R P Gupta]

New Delhi, the 14th September, 2021

Preface

Climate change is a reality beyond doubts or conflicts and so are the disasters. As a known contributor to hydro-climatic hazards, climate change impacts are also associated with increasing socio-cultural, political and economic challenges relating to livelihoods, food or nutrition security, migration, conflicts. Therefore, the framework for addressing climate change and its impacts, the central call is for understanding and addressing disaster risks and vulnerabilities. Synergies of green growth, circular economy and other climate change mitigation means with adaptation and DRR are crucial for holistic perspectives of "Resilience and Sustainability" in the present and future. India is vulnerable to multiple natural hazards due to its diverse geological variations, posing a risk to not only the people and infrastructure but also multiple sectors, industries and businesses. The Government of India has enacted a comprehensive Disaster Management Act in 2005 and adopted the National Policy on Disaster Management in 2009, which signalled a paradigm shift from relief-centric approach to a comprehensive, pro-active approach focused on disaster risk reduction.

Preparing a disaster management plan for India, taking into account global experience, and incorporating the new perspectives on disaster management prove more challenging than we had initially envisaged. India's vision 2047, under the *Azadi ka Amrit Mahotsav*, recognizes the contexts of changing horizons of disaster risks along with weather extremities and likely black swan events including health epidemics and human-induced disasters. There are many new developments, particularly, the adoption of three landmark International Agreements in the year 2015 having significant bearing on India's policies for disaster management and climate change, i.e., Sendai Framework for Disaster Risk Reduction in March 2015, Sustainable Developmental Goals 2015–2030 in September 2015, Paris Agreement on Climate Change UNFCCC, December 2015 and its latest 27th COP meeting in November 2022, and the United Nations Convention to Combat Desertification (UNCCD), 14th Conference of Parties in September 2019.

National Institute of Disaster Management (NIDM) and Department of Science and Technology (DST) under the CAP-RES project, under Climate Change

Programme (NKMCC) organized "Resilience and Sustainability Summit: Vision 2047 (RESSUMMIT2047)" to review and deliberate upon the key aspects of present and futuristic disaster risks and climate contexts and their changing scenarios. UNDP, GIZ, and a large number of international and national organizations joined as organizing collaborators. Ministry of Environment Forest and Climate Change and NIDM has adopted a novel, concise and action oriented approach towards structuring the document on Disaster Risk and Management under Climate Change. The chapters of this book are focused on thematic areas for planning and implementation with clear defined roles and responsibilities. The book has attempted to highlight the issues (past, present and futuristic) related to climate change and their solutions, which are innovative, technologically advanced, sustainable and systematic. This book is equipped with injunctions and suggestions for stakeholders, central and state agencies about their roles and responsibilities for mainstreaming climate resilient and disaster management plans into developmental planning.

The publication has been prepared to showcase our knowledge bank on assessing and managing the disasters on national and sub-national levels and borrowed advanced technology information from outside the India which is relevant to India's diverse geography and climate. This knowledge compendium also has a sync with the recently held 1st RESSUMMIT2047 at New Delhi (January 2023) focused on integrated resilience and sustainability. The editors of the book are hopeful that the contents would be very useful for policy planners, researchers, academicians, practitioners, scholars and civil society working on related issues in developing and transition economies in the perspective of climate and disasters.

New Delhi, Delhi, India — Anil Kumar Gupta
New Delhi, Delhi, India — Akhilesh Gupta
New Delhi, Delhi, India — Pritha Acharya

Acknowledgement

On the recommendation of the Ministry of Environment, Forests and Climate Change (MoEFCC) of Government of India, this strategically and academically important exercise to document on "Disaster Risk and Management under Climate Change; Context and Lessons from India" has been undertaken. In order to arrive at the present form, the endeavour has been supported by several institutions and experts in related subjects.

Then Minister of Environment Forests and Climate Change Mr Prakash Javadkar has given a foreword to the book and I am grateful for the encouragement received from him. I am grateful to Mr. R. P. Gupta, then Secretary, MoEFCC, Dr. J R Bhatt, then Advisor, MoEFCC, for their kind support to this endeavour. Support of Mr. B. H. Anil Kumar, Mr. Sanjeev Jindal and Major General Manoj Bindal, then Executive Director(s), NIDM, has been important in facilitating the initial coordination. Thanks are also due to Mr. Rajendra Ratnoo, present ED NIDM for his positive thoughts. Dr. Akhilesh Gupta, Secretary, SERB, Sr. Advisor DST-GOI who is a Senior Editor has been source of immense inspiration and guidance in this entire process.

I acknowledge the support and cooperation of the authors who could manage to contribute and update their chapters based on reviews in tight timelines, without which the development of the book would have not been possible. Thanks are due to all the reviewers. I extend my gratitude to Mr. Sanjiv Jindal, Joint Secretary, DM, Government of India and Shri Kamal Kishore, Member Secretary NDMA for his support. I extend my heartfelt thanks to Dr. Nisha Mendiratta, Head, CCP, DST, Mr. Shantanu Goel of Bombay Natural History Society, Dr. Susheela Negi, Scientist E, DST, Dr. Santosh Kumar, Professor, NIDM, Dr. Chandan Ghosh, Professor, NIDM, Dr. Ajinder Walia, Associate Professors and Dr. Sushma Guleria, Assistant Professor of NIDM for their support and insights. I also would like to thank Ms Pritha Acharya who has coordinated for preparing the manuscript. I also extend my sincere thanks to Ms. Nainika Singh and Ms. Jasmine Biba, MoEF &CC, Ms. Shruti Bansal and Ms. Taniya Gupta, Research Assistant, CCDRM Project, NIDM, Ms. Pritha Acharya, Ms. Shweta Bhardwaj, Ms. Atisha Sood, Dr. Sweta

Baidya, Dr. Sanayanbi Hodam (Research Fellows, NIDM), Dr Anjli Barwal, Ms. Richa Srivastava (Consultants, NIDM) and Ms. Thinles Chondol, Ms.Fatima Amin and Mr. Harshit Sharma (YP, NIDM) for the help in developing the book. Thanks to the Senior Editor Springer Nature Ms Aakanksha Tyagi and Mr Muthuneela Muthukumar for processing the printing and publishing support.

(Prof. Anil K. Gupta)
Head, ECDRM Division
Director of Projects & CoE, International Cooperation,
NIDM New Delhi

Contents

Part III Tools and Strategies

About the Editors

Anil Kumar Gupta is a noted resilience and sustainability strategist, full professor of Policy-planning and Strategies, Integrated Resilience; head of the ECDRM Division, International Cooperation, Director of projects and Centre for Excellence on Climate Resilience; and nodal officer for Central Ministries at National Institute of Disaster Management (Govt. of India), New Delhi. He is known for national, subnational, and sectoral planning, pioneering initiatives in adaptation, S&T innovations and NBS strategies, operational research and capacity building. He is also a member of the expert team for WMO Climate Statement and core group member of IUCN-Council of Ecosystem Management. He is elected Fellow of Society for Earth Scientists, Fellow of International Society of Environmental Botany (ISEB), and Vice-chair of Association of Occupational & Environmental Health (AOEH). He is founder Chair of Resilience Forum (ResFor) and General Secretary of the Alliance for Integrated Sustainability-Resilience-DRR (Allsure).

Akhilesh Gupta is the Secretary of Science and Engineering Research Board (SERB) and Senior Advisor at the Department of Science and Technology (Govt. of India). Dr. Gupta currently heads the Policy Coordination and Programme Management (PCPM) division and is the overall in charge of five National Missions at DST—National Mission on Interdisciplinary Cyber Physical System, National Mission on Quantum Technology and Applications, National Super-computing Mission, National Mission on Strategic Knowledge for Climate Change, and National Mission for Sustaining the Himalayan Ecosystem. He is a Fellow of Indian National Academy of Engineering (FNAE), Indian Meteorological Society (FIMS), and Association of Agro-meteorologists (FAAM). Dr Gupta has been one of the authors of India's National Action Plan on Climate Change (NAPCC) and head of the Secretariat which drafted India's new Science, Technology and Innovation Policy, which is under finalization. He was Secretary of the University Grants Commission (UGC) and President of Indian Meteorological Society, and an internationally known climate science promoter and diplomat.

Pritha Acharya is a research associate at National Institute of Disaster Management (NIDM). She is an environmental sustainability and biodiversity professional. She holds expertise in the field of disaster risk reduction, climate change adaptation, nature-based solutions and policy planning for disasters, climate resilience, and sustainability. She has worked with the International Union for Conservation for Nature (IUCN) for their Business and Biodiversity Programme and is currently looking after the Climate Adaptive Planning for Resilience and sustainability in Multi-hazard Project (CAP-RES) (DST supported) at NIDM. She is a founder of the Resilience Forum.

Part I
Overview and Major Climatic Disasters

Chapter 1
Climate Change: Extremes, Disasters and Call for Resilient Development

Anil K. Gupta, Pritha Acharya, and Akhilesh Gupta

Abstract World has observed significant rise in global temperature which is clearly evident from the increased atmospheric temperature, oceanic temperature, melting of glaciers, rising sea level, shifting of local vegetation and agricultural practices, intense heat waves, migration of birds and animals, endangered corals species, flora and fauna etc. Future scenario is gruesome and expected to get worsen. Science fraternity, worldwide claims that, the developing countries are expected to suffer more at the hand of Climate Change. The future predicted for the developing countries, and world at large is the ubiquitous presence of stresses such as scarcity of water resources, food insecurity, non-existence of many island countries and non-existence of many animal and plant species. In the light of keeping the ecology and environment from getting distressed to worse, Climate Change Adaptation (CCA) and Mitigation has the real potential to overcome this worse condition by building Climate Resilience in the community and ecology. This chapter discusses about the interrelatedness between increased disaster's frequency with Climate Change, need for CCA and mitigation strategies for building resilience, present condition of the country's vulnerability to Climate Change, associated disasters, and initiatives taken by India to address Climate Change.

Keywords Climate change · Disasters · Mitigation · Adaptation · Resilience

A. K. Gupta (✉)
ECDRM Division, International Cooperation, CAP-RES DST Project, National Institute of Disaster Management (Ministry of Home Affairs), New Delhi, India

P. Acharya
ECDRM Division, CAP-RES DST Project, Centre for Excellence on Climate Resilience, National Institute of Disaster Management (Ministry of Home Affairs), New Delhi, India

A. Gupta
Department of Science and Technology, Science and Engineering Research Board, S&TI Policy Coordination, Climate Change Programme, Government of India, New Delhi, India

A. K. Gupta et al. (eds.), *Disaster Risk and Management Under Climate Change*, Disaster Resilience and Green Growth,
https://doi.org/10.1007/978-981-99-4105-6_1

1.1 Introduction

The world has witnessed the calamities due to Climate Change umpteenth of time and the future is no different. Over the last few decades, rise in global temperature has been observed with adequate evidences of increased atmospheric and oceanic temperature causing melting of glaciers, rising sea levels, shifting of tree lines, intense heat waves, migration of animals etc. (ISDR 2009). Climate Change is caused by increased Greenhouse Gases (GHGs) emissions (globally), which is being accelerated by human activities such as burning of fossil fuel, land use changes, raising of livestock, decaying of landfill wastes, natural gas production and transportation etc. (NRC). As per the IPCC Fourth Assessment Report, this increasing atmospheric and oceanic temperature is higher than it used to be during at least the past five centuries (IPCC 2007a, b). The human interventions and activities have increased CO_2 emissions of 278 ppm during pre-industrial time to 405 ppm reported in September 2018 (UNFCCC 2007; NOAA-ESRL 2018). Reports have suggested this as the fastest warming trend of the earth (UNFCCC 2007). World-wide, human population has increased the concentration of emitted GHGs including Methane (by almost two-and-a-half times than the pre-industrial areas), and Nitrous oxide (risen by almost 15% since 1750).

Future conditions are expected to get worse, especially for the developing countries in the form of water scarcity, food insecurity for about billions of population (UNFCCC 2007). India being a developing country is one of the most vulnerable countries to the Climate Change impacts, the impacts of which can now easily be seen in the form of increased water stress, heat waves, drought, storms and flooding, and greater than before risk on health and livelihoods (GCRI, Battelle,, and PNWD 2009). The following table provide a gist of some of the future global climatic projections and uncertainties for India (Table 1.1):

1.2 Present Scenario

1.2.1 Climate Change and Disasters

Climate Change amplifies the risk of disasters, by (i) increasing the frequency and intensity of extreme/disastrous events; and (ii) decreasing the resilience of surroundings. Reports suggest that annually 26 million people globally are forced to displace due to disastrous events like floods, storms etc. (NRC, n.d.). Worldwide increase in intense floods, drought, forest fire, cyclones, storm and heat waves has a strong and menacing link with Climate Change. Profound numbers of literature provide evidence of linkage of anthropogenic Climate Change with natural disasters (Asian Development Bank 2015). According to EM-DAT, the frequency of natural disasters has increased threefold in the last four decades, from 1300 events in 1975–1984 to over 3900 in the period of 2005–2014 (EM-DAT 2015). Number of hydrological

Table 1.1 Global climatic observation and expected impact (GCRI, Battelle, and PNWD 2009; GoI M 2010)

Observations	Projected impacts
Glacier melting with an average rate of 10–15 m/year	Flooding in river valleys, followed by decreased flow, causing water scarcity for drinking and irrigation Sea level rise as a result of thermal expansion and exchange of water between ocean and other forms of fresh water may also contribute to coastal inundation
Warming of 0.5 °C is expected by 2030 and a warming of 2–4 °C by the end of this century	Higher levels of tropospheric ozone pollution and other forms of air pollution in major cities Increase instances of forest fires in the mountainous regions Increased intensity and frequency of cyclonic activity and storm surges will lead to inundations of low lying areas along the coast-lines
Increased precipitation, expected to occur in the form of lesser monsoon days with extreme rainfall events	Summer Intensification, leading to more warm summers Fast and earlier snowmelt, which could adversely impact agricultural production Dry vegetation burning will increase atmospheric dust and smoke Drought like situations are likely to increase due to decrease in number of rainy days Increase in flood and drought events will influence the ground water recharge

and meteorological event has gone up sharply during 2005–2014, with the annual number of category 5 storms tripled up between 1980 and 2008 (Independent Evaluation Department (IED) 2013). One another aspect of damage from Climate Change is economic loss of a country. From 2005–2014, $142 billion worth capital is lost due to increased number of natural disasters.

IPCC report has summarized the observed climatic trends and their possible impacts. Table 1.2 gives a brief list of observed trends with their associated impacts.

1.2.2 Concept of Climate Change Adaptation and Mitigation

We cannot prevent Climate Change impeccably, therefore we need to adapt and/or mitigate to this varying climate and possibly predicted its outcomes. Climate Change Mitigation is the process of identifying the factors causing it, followed by controlling the Climate Change phenomenon by eliminating/or reducing the identified root causes. For e.g. GHG emissions are the driving cause for Climate Change (Global Warming), so mitigation here would be reducing GHG emissions. Whereas Climate Change Adaptation is the process of acclimatizing such as building climate resilient environment to minimize the negative outcomes of Climate Change. The eventual

Table 1.2 Climatic trends and their possible impacts (ISDR 2009)

Phenomenon and direction of trend	Observable impacts
Fewer cold days and night, frequent hot days and night	Rise in agricultural productivity in colder region and reduction in warmer regions Increased insect outbreaks and linked diseases Increased snow melting that may impact water resources Reduced mortality from cold exposure Declining air quality in cities
Frequent heat waves	Reduced yields in warmer regions due to heat stress Increased risk of bushfires Increased water demand, water-quality problems This heat stress will increase heat-related mortality. Old people, chronically sick and very young people will be impacted the most
Rise in precipitation frequency	Soil erosion and crop damage Deterioration of surface and ground water Water scarcity may be relieved Count and vulnerability related to death, injuries, skin diseases will go up Loss of life and property due to flooding causing disruption of settlements, commerce, transport and societies Pressures on urban and rural infrastructure
Increased drought affected area	Land degradation, lower crop yields, soil erosion and crop damage Increased livestock deaths Increased risk of wildfire Scarcity of food and water shortage, causing increased risk of malnutrition, water and food-borne diseases Migration of humans and animals
Increasing intensity of tropical cyclones	Damage to crops and trees High risk of diseases related to water Post-traumatic stress disorder Disruption by flood and high winds Migration, loss of property
Rise in sea levels	Salinization of irrigation water and freshwater systems, which will eventually decrease freshwater availability Increase in flood mortality rated Migration-related health effects High costs of coastal protection and relocation Relocation of people and infrastructure

intent of either mitigation or adaptation is to reduce the vulnerability or risk associated with Climate Change. There is a rising need of Climate Change Adaptation, so that communities can adjust with the projected future climatic scenario.

Several studies and research has shown the linkage between Disaster Risk Reduction (DRR) and Climate Change Adaptation (CCA) (ISDR 2009). Disaster Risk and Climate Change share the similar problem of changing weather and climate. Disaster Risk is considered to be an ongoing issue while climate change

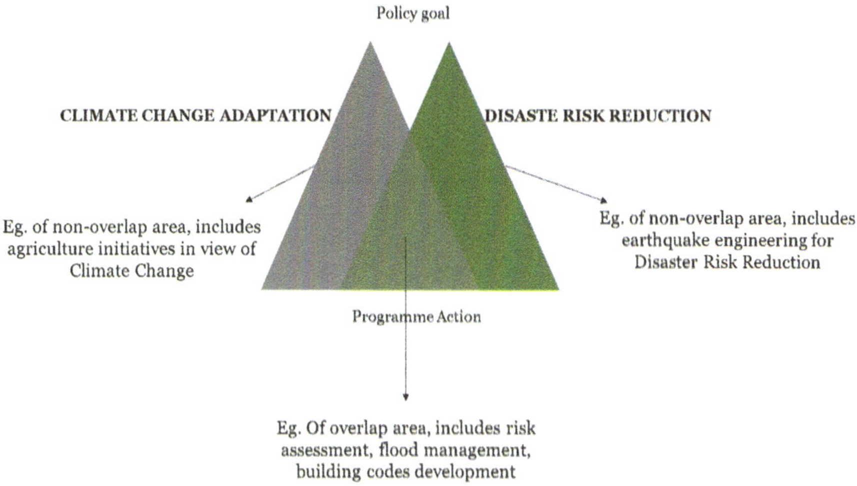

Fig. 1.1 Overlaps and non-overlapping area of CCA and DRR

is understood as an emerging issue. Similar methodologies for risk assessment, prediction models are used for both to understand their behaviour and consequences. Therefore it is wise to implement CCA and DRR in an integrated manner (ISDR 2009). Figure 1.1 explains the area of convergence and divergence for CCA and DRR. This can be explained by taking flood management which is a concern for both DRR and CCA. Another example is risk assessment where both share similar methodology. On the other hand disaster risk engineering is an area of concern of Disaster Management and share no similarity with CCA. Figure 1.1 also justifies the fact that, CCA at many cases (e.g. Flood Management, Drought Management) will eventually result in reducing disaster risks.

1.2.3 Early Warning Systems and Risk Communication

Preparedness for disasters also needs effective early warning and risk communication systems. Early warning systems help in reducing the impact of hazards by reducing loss of life and cutting the economic losses. Thus early warning is also a major element of disaster risk reduction. To be effective, early warning systems need to actively involve the communities at risk, facilitate public education and awareness of risks, effectively disseminate messages and warnings and ensure there is constant state of preparedness (UNISDR 2006). The topic is further covered in details in Chap. 3.

1.2.3.1 Call for Resilience

Disasters are the events of environmental extremes which are inevitable entities of this living world (Gupta and Nair 2011). Environmental changes induced climatic variations in terms of extreme weather for example and act as the cause whereas the disasters can be seen as the impacts. Environmental changes act as drivers of disasters primarily in three ways viz. (i) climate change; (ii) natural resource and; (iii) land-use changes (Gupta and Nair 2011). This inextricable linkage between environment and disasters has been understood in the following three interfaces:

1. Environmental degradation leading to water and climate related disasters due to degradation of natural resources.
2. Environmental degradation causes vulnerability by causing new hazards and aggravating precursors of other disaster events.
3. Disasters impact the environment by effecting natural resources, processes and ecosystems, creating future emergency conditions and situations.

Climate Change has now become a constant phenomenon and it is evident from the past decadal trends, that the conditions will not be much different, rather will tend to intensify in the upcoming times. Adverse impacts of Climate Change can be minimised through effective adaptation strategies. Despite this, Climate Change may trigger emergency scenarios even after adaptation measures are at place. Thus adaptation strategies should be flexible enough to eventually achieve sustainability in an integrative and holistic manner—making the systems climate resilient.

Climate Resilience development refers to the development where the system components (including social, economic and environmental) has the capacity:

- To cope up with the additional stresses induced due to climate change,
- To adapt, adjust, acclimatize in a desirable manner, in order to address the predicted climatic variability.
- So that the evolved system will no longer be severely impacted by foreseen climatic condition

India has a diverse climatic profile, and topology; because of which climatic risk is non-uniformly distributed. The impact of Climate Change can be seen all across the country, for example, in the form of heat spell in Ahmedabad caused temperature of around 45 °C during summer daytime; floods of 2018 in Kerala has caused economic loss of about $3 billion to the country; poor air quality of Delhi; Rajasthan saw record breaking summer in May, 2016, during which temperature of about 51 °C was recorded; and cyclone Vardah of Dec 2016 in Odisha which is considered to be one of the deadliest (Sarkar 2017; Anand 2018).

These incidences and projected variability in climatic parameters demands for developing climate resilient nation.

Several initiatives have been taken by Government of India (GoI) under this; India has always been an active participant of United Nation Framework for Convention on Climate Change (UNFCCC). National Action Plan on Climate

Change (NAPCC) has also been released by GoI on 30th June 2008. This led to the establishment of the Executive Committee on Climate Change. It is being handled by Prime Minister's Council on Climate Change and coordinated by the Ministry of Environment, Forest and Climate Change (MoEF&CC) National Disaster Management Plan had been prepared with the objective of mainstreaming Disaster Risk Reduction into developmental agenda (NDMA 2016). National Institute of Disaster Management is mandated for training, research, developing documentation and database related to Disaster Management.

1.3 International Policy Framework

Disaster management (DM) saw a paradigm shift from relief and response centric approaches to mitigation and preparedness centric approaches and now with the variability of climate change, DM is undergoing a second paradigm shift towards environmental hazard and vulnerability (Gupta et al. 2014). NDMP Gupta 2019 has acknowledged that sustainable development needs to be disaster resilient and be adaptive to climate change impacts (NDMA 2019). Below are the international frameworks that call upon integrating CCA and DRR to achieve sustainable development.

1.3.1 SDGs (Sustainable Development Goals)

DRR, CCA and Resilience cut across different sectors and aspects of development. 25 targets of SDGs framework are directly or indirectly related to DRR in ten of the 17 SDGs. SDGs' targets identify the immediate need to reduce climate and disaster risks and emphasizes resilience building of communities and nations to achieve SDG. Explicit references for DRR, CCA and Resilience can be observed in goals and targets especially meant for poverty, hunger, healthy lives, building resilient infrastructure, education, sustainable management of water, Climate Change, resilient cities, and marine and terrestrial ecosystem.

1.3.2 Paris Agreement on Climate Change at COP21

The Paris Agreement is a legally binding agreement within the United Nations Framework Convention on Climate Change (UNFCCC) which focused on combating climate change through reducing the GHG emissions and accelerating and intensifying actions and investments needed for a sustainable low carbon future

(UNFCCC 2011). It was negotiated by 196 members and was adopted on 12th December 2015 during the 21st Conference of Parties (COP 21).[1]

1.3.3 Sendai Framework for Disaster Risk Reduction (SFDRR)

The Sendai Framework for DRR (Disaster Risk Reduction) was adopted at the third UN World Conference in Sendai, Japan, on March 18, 2015. It is the successor instrument of HFA (Hyogo Framework for Action) 2005–2015 and a non-binding voluntary agreement for 15 years which recognizes the responsibility of States for DRR and the shared responsibility with local governments, private sector and other stakeholders. Key features of SFDRR are:

- Shifting focus from Disaster management to disaster risk management
- Emphasis on the need of strengthening the disaster risk governance by placing governments at the center of disaster risk reduction
- wider approach of disaster risk reduction, focusing on both natural and man-made hazards and related environmental, technological and biological hazards and risks
- recognition to implementation of integrated environmental and natural resource management techniques

1.3.4 United Nations Conventions to Combat Desertification (UNCCD) 2018–2030 Strategic Framework

The United Nations Convention to Combat Desertification (UNCCD),[2] established in 1994 is a legally binding agreement of 197 parties which links sustainable land management to environment and development. The strategic framework under the convention commits to achieve Land Degradation Neutrality (LDN) by building a future that avoids, minimizes, and reverses desertification, land degradation and mitigates the effects of drought in the affected areas at all levels to achieve a land degradation-neutral world, consistent with the 2030 Agenda for Sustainable Development (COP-14 2019).

[1]More information on the UNFCCC COP21 can be retrieved form https://unfccc.int/process-and-meetings/the-paris-agreement/what-is-the-paris-agreement

[2]More information on the UNCCD COP14 can be retrieved form http://unccdcop14india.gov.in/conference-of-parties

Table 1.3 Incorporation of four priorities for action under the SFDRR in NDMP

Sendai framework for DRR (2015–2030) priority	Chapters with the priority as its dominant theme
Understanding disaster risk	Chapters 2 and 3
Strengthening disaster risk governance to manage disaster risk	Chapters 3, 4, 5, 6, 8 and 9
Investing in disaster risk reduction for resilience	Chapters 3, 4, 5, 6, 7 and 8
Enhancing disaster preparedness for effective response and to 'Build Back Better' in recovery, rehabilitation and reconstruction	Chapters 4, 6, 7, 8, 9 and 10

1.4 National Policy Frameworks

1.4.1 Implementing the International Climate Change and Disaster Frameworks in India

Year 2015 provided tremendous opportunity for the world to take well thought, holistic and integrated actions towards sustainable development with emphasis on disaster risk reduction with the three global landmark agreements—SFDRR, SDGs and Paris Agreement on Climate Change. India adopted these agreements as a signatory member and strives to comply with them. The country is making efforts to achieve the global targets by making advancement in the entire disaster management cycle by following the recommendations in these agreements and by adopting internationally accepted best practices. National Disaster Management Plan (NDMP), 2016 is the first ever plan at national level for disaster management and was revised in 2019 in alignment with the guidelines of SFDRR, SDGs and UNFCCC COP 21 focusing on actions for disaster risk reduction.

Moreover, India being the signatory to the UNCCD COP, adopted the 'Delhi Declaration' which encourages a proactive approach to reduce risks and impacts of desertification/land degradation and drought through the implementation of drought preparedness plans and increased risk mitigation for drought and sand and dust storms (COP-14 2019) (Table 1.3).

1.4.2 National Policy Framework on CCA and DRR

1.4.2.1 National Policy on Disaster Management

National Policy on Disaster Management (NPDM) was approved on October 22nd, 2009 with the vision of building a safe and disaster resilient India by developing a holistic, proactive, multi-disaster oriented and technology driven strategy through a culture of prevention, mitigation, preparedness and response" (Rijiju 2015) in pursuance with Disaster Management Act, 2005. The objectives of NPDM are listed in the following box.

Box 1.1 NPDM Objectives (NDMA 2009)

- Promoting a culture of prevention, preparedness and resilience at all levels through knowledge, innovation and education.
- Encouraging mitigation measures based on technology, traditional wisdom and environmental sustainability.
- Mainstreaming disaster mitigation into the developmental planning process.
- Establishing institutional and techno-legal frameworks to create an enabling regulatory environmental and aa compliance regime.
- Developing contemporary forecasting and early warning systems backed by responsive and fail-safe communication with information technology support.
- Ensuring efficient response and relief with a caring approach towards the needs of the vulnerable sections of the society.
- Undertaking a reconstruction as an opportunity to build disaster resilient structures and habitat for ensuring safer living.
- Prompting a productive and proactive partnership with the media for disaster management.

1.4.2.2 National Action Plan on Climate Change

NAPCC was released in 30th June 2008 and has eight missions on water, renewable energy, energy efficiency agriculture etc. The focus area of NAPCC is to promote and create awareness on Climate Change, need of Climate Change Adaptation and Mitigation, promoting Energy Efficiency, and Natural Resource Management (NAPCC 2008). However, four additional missions were added under the NAPCC in 2018–2019, acknowledging the emergent thrust areas viz. National Wind Mission, Health Mission, Waste to Energy Mission and Coastal Mission (The Indian Express 2020). Apart from the major missions under the NAPCC, various important and innovative initiatives on mitigation and adaption were also introduced. The report released by GoI on NAPCC in 2018 highlights all these initiatives taken up by the centre. The missions are described in brief in the paragraphs below;

National Solar Mission

The mission works to reduce fossil fuel consumption for energy sources by promoting significant use of solar energy and also recognizes the utilization of other renewable energy sources. It also encourages various R&D programmes to make the solar energy economically affordable in comparison to fossil fuels based energy options (GoI 2018).

National Mission for Enhanced Energy Efficiency (NMEEE)

The mission aims to conserve and promote efficient use of energy by enabling policies and incorporating innovative energy efficient models. Under this mission several initiatives had been taken like distributing 23.39 crore LED light, providing free clean cooking gas to women below/poverty line (GoI 2018; Rattani 2018).

National Mission on Sustainable Habitat

The mission aims to promote sustainable urban planning by incorporating relevant policies, encouraging research in the areas of buildings, waste management, water resources and transportation. In addition the mission also focuses on CCA through infrastructure resilience, community based disaster management and measures for improving the warning systems for extreme events (GoI 2018; Rattani 2018).

National Water Mission

The mission aims to promote sustainable supply of water by incorporating water conservation practices, minimizing wastage of water, increasing water use efficiency to 20%, ensuring uniform water distribution across the country (GoI 2018; Rattani 2018).

National Mission for Green India

The mission aims to enhance, increase, and protect country's green cover (with a target of 33% tree cover) and encourages the direct participation of communities through JFMs to manage land degradation. It also focuses on enhancing ecosystem services including biodiversity, wetlands, habitat restoration also realizing the outcomes with Carbon Sequestration as added benefit (GoI 2018).

National Mission for Sustaining the Himalayan Ecosystem

The mission aims at incorporating management practices to protect and safeguard Himalayan ecosystem including conservation of forest cover, biodiversity, prevention of glacier melting etc. The mission also aims to maintain two-third of the area under forest cover in the hilly regions, incentivising community based management of the ecosystems along with knowledge sharing and raising awareness on Climate Change Impacts and Adaptation in the Himalayan ecosystems (GoI 2018).

National Mission for Sustainable Agriculture

The mission aims to improve sustainability and to encourage climate resilient agriculture through the development of insurance policies, promoting climate resilient crops, capacity building etc. (GoI 2018; Rattani 2018).

National Mission on Strategic Knowledge for Climate Change

The mission aims at developing thorough knowledge warehouse related to Climate Change related aspects to gain better understanding about impacts, and challenges of Climate Change (GoI 2018).

National Health Mission (NHM)

The mission aims at achieving universal access to equitable, affordable and quality health care services that are accountable and responsive to people's needs through strengthening of Health System, Reproductive-Maternal-Neonatal-Child and Adolescent Health (RMNCH+A), and Communicable and Non-Communicable Diseases. The mission has two sub missions; the National Rural Health Mission (NRHM) and the National Urban Health Mission (NUHM) (MHFW 2020).

National Wind Energy Mission

The mission aims to enhance the use of renewable source of energy by enhancing the wind energy generation capacity, efficiency and infrastructure as well as identify high wind power potential zones within the country. Through this mission India attempts to generate 60 GW of wind energy by 2022 (MNRE 2020; TheEconomicTimes 2020).

1.4.2.3 Waste to Energy Mission

This mission aims to increase the share of renewable energy in India's total energy budget. It explores the availability of technological options to set up projects primarily to recover energy in the form of Biogas/Bio-CNG/Electricity from agricultural, Industrial and urban wastes of renewable nature such as municipal solid wastes, vegetable and other market wastes, slaughterhouse waste, agricultural residues and industrial/STP wastes and effluents (NMRE 2019).

1.4.2.4 National Coastal Mission

The mission aims to primarily conserve the coastal environment but at the same time promotes livelihood through development, revenue generation and providing employment (MoEF&CC 2018).

1.4.3 State Action Plan on Climate Change

India has varied geographical and topographical features and in order to mainstream climate resilient development, both Center and Sates needs to take joint responsibility. NAPCC also understand that to address Climate Change adaptation and mitigation challenges active participation of state and local government is required and therefore made mandatory the formation and active preparation of SAPCCs for all Indian state and Union Territories with an aim to mainstream climate change action into local-level planning (GIZ 2014). According to MoEF&CC's data, 33 SAPCCs for states and UTs has been prepared (MoEF&CC 2020). Table 1.4 explains the suggested criteria that need to be considered to develop SAPCC.

1.5 Way Towards Sustainable Development

Considering the discussions in the earlier sections of this chapter, adapting to the changing climatic scenarios demands for an integrated approach promoting combined efforts of policy makers, planners, scientific fraternity and communities to design suitable strategies to mainstream Climate Change Adaptation (CCA) and Disaster Risk Reduction (DRR) into the development planning process (Gupta et al. 2016). Given the history of past weather events, it is evident that Climate Change would be a permanent phenomenon. Howsoever, the lessons learned from successes and failures in tackling and managing these events, it is evident that promoting a safe and a sustainable future would require systems and sectors to be climate risk proof.

Table 1.4 Common Framework for formulating SAPCCs (Rattani 2018)

Criteria	Consideration
State Climatic Profile	Existing climatic profile and projected trends of the state
Climate Change vulnerability assessment	Examination of present and projected vulnerability temporally and spatially
GHG Emissions	Current sector-wise emissions and estimation of future needs
Climate Change Strategy	Identification of existing programs, gaps and proposed schemes considering cost-benefit analysis
Climate Change Action Plans	Identification of existing funding agencies, check for additional funds requirement

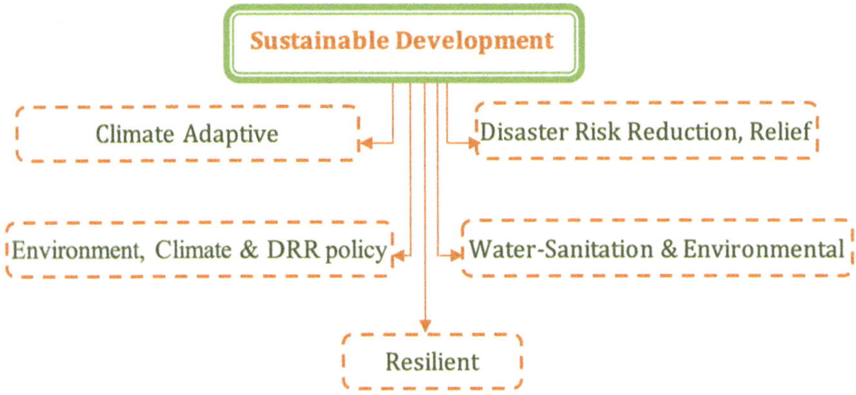

Fig. 1.2 Illustration depicting the pillars of Sustainable Development

This section identifies five important pillars for promoting safe and a sustainable future (Fig. 1.2):'

1.5.1 Climate Adaptive Development and Mitigation

It is proposed to include Disaster Risk Reduction along with adaptation and mitigation at national/sectorial, sub-national and local levels. The focus area here should be:

1.5.2 GHG Mitigation

India has made commitment to reduce emissions intensity of its GDP to 20–25% from 2005 levels by 2020 (Pahuja et al. 2014). Consequently NAPCC with initial 8 missions were launched to control Climate Change, of which two missions are specifically to cut down fossil fuel usage and reduce energy consumption. This shows that policies and mitigation measures are already in place. Adequate monitoring and tracking of the mitigation action, has the potential to strengthen the implemented policies.

1.5.3 DRR and Resilient Development

Development is a continuous process and many developing countries like India where infrastructure development is the primary focus, disasters impede the growth.

Disaster Management has experienced a second paradigm shift from relief to risk reduction in response to the fact that building resilient communities enhances individual safety. Thus resilient development has the potential to reduce disaster risk.

1.5.4 Land-Use Planning

Land use planning is very powerful in mitigating disaster risk which can improve the resilience of the people. Spatial distribution is considered to promote sustainable land use development and also reduces vulnerability of hazards.

1.5.5 Environment, Climate and Disaster Risk Reduction Policy Tools

Integrative approach for taking climatic and disaster risk/vulnerability assessment into decision tools like Environmental Impact Assessment (EIA) and Environmental Management Plan (EMP). The promising approach includes:

- Disaster Impact Assessment (DIA) integration with EIA is one good approach to regulate/eliminate development projects with very high risk and vulnerability to disasters
- Rapid Environmental Impact Assessment (REIA) w.r.t disasters, is a powerful tool to identify and prioritize the harmful disaster impacts on environment.
- Post-disaster needs assessment (PDNA) is essential to understand essential requirements (e.g. food, clothing, housing, first aid etc.) after disasters.
- Natural Resource Management (NRM) to ensure optimum utilization of resources (including forests, biodiversity, land, livestock etc.) which are the basic requirements for human survival.

1.5.6 Resilient Agriculture

Agriculture sector as a whole is worst hit by disasters and climate Change presents severe challenges to the productivity of agricultural crops. According to FAO, almost 60% of the global population is dependent in agriculture, at the same time it also shares linkages with other sectors. As the global population is increasing at an unprecedented rate, the concerns are to ensure food and nutritional security for the world. Planned agriculture should be adopted to increase resilience of agricultural crop production. Management practices that support crop production under adverse climatic conditions should be adopted. Practices such as in-situ moisture conservation, residue incorporation, instead of burning, water harvesting, and location specific agronomic and nutrient management practices (NICRA 2012) and disaster resilient agriculture practices are proposed to promote climate resilient agriculture.

1.5.7 Water-Sanitation and Environmental Health

Lack of sanitation measures and systems can cause severe health issues especially with young children. United Nation General Assembly has also considered the access to safe water and sanitation as human rights. Reports claim that just 62% of the worldwide population has access to improved sanitation facility (that means human waste is not in contact with human) (Unite For Sight 2015). Howsoever, disaster and climate change related challenges are often neglected when environmental health is concerned. Thus there is a need to focus and research on 'Environmental Health in Disaster and Emergency', 'Climate Change and Disaster Epidemiology', 'Climate Change Adaptation Plan for Health Sector', 'Health Risk Analysis' and 'Industrial Chemical Accidents and Pollution'.

1.5.8 Disaster Risk Reduction, Relief and Recovery

Prior preparation to disastrous events is very important to reduce the risk associated. The focus and research areas as well as policy actions should include

- Integrating climate and disaster resilience into relief and reconstruction, to promote resilient development post disaster.
- Sustainable recovery and green growth strategies, prevents environmental degradation, promotes sustainability and inhibits inefficient natural resources utilization

1.5.9 Mainstreaming CCA-DRR into development

Integration of Climate Change Adaptation (CCA) and Disaster Risk Reduction (DRR) within planning and implementation framework at different levels is needed to ensure the journey for sustainable development (Gupta et al. 2014). India launched the National Action Mission on Climate Change (NAPCC) in 2008 as part of its Nationally Determined Contributions (NDCs).

The approach should rather focus on sharing of co-benefits between the sectors. This has also been highlighted in the Prime Minister Agenda 10, under its various points. Point 1 and 2 of PM agenda 10 says 'All development sectors to imbibe the principles of disaster risk management' and 'Working towards risk coverage towards all'. Thus mainstreaming CCA-DRR into the development process calls for an integrated approach where all the sectors should work in tandem to address disasters through risk reduction.

There are several ways and methods through which DRR and CCA measures have been integrated to reduce the risks of natural disasters. Approaches and pathways for inclusion of CCA and DRR into development plans are (Gupta et al. 2014):

- Establishing appropriate framework-technological, legal, financial and institutional-for planning and implementation.
- Promoting multi-sectoral developmental process and mitigation measures addressing disaster management concerns.
- Integration of disaster risk reduction planning and policies in a holistic, participatory, inclusive and sustainable manner.

1.5.10 Legal Policy Framework for Disaster Management

1.5.10.1 Governance Structure

Disaster Management Act, 2005 provides for institutional framework for disaster management.

Ministry of Home Affairs (MHA) is the nodal Ministry to deal with all the issues concerning with the disaster management. The Central Relief Commission (CRC) is constituted in Ministry of Home Affairs for coordinating with the relief operations.

National Disaster Management Authority (NDMA) has been constituted for laying down policies and guidelines for disaster management in India. Disaster Management Authority at State and District Level are being constituted for laying down guidelines at state and district level.

National Institute of Disaster Management (NIDM) is responsible for human resource development through planning and research development, capacity building, and documentation and policy advocacy. NIDM collaborates with MHA, NDMA, Central and State, and local government as well as various other stakeholders to build capacities towards climate resilience.

Natural Disaster Response Force (NDRF) has been constituted under section 44 of DM Act, 2005 for strengthening the response system of the country. NDRF team is trained in handling various types of natural, man-made and non-natural disasters.

Environment and Natural Resources Laws in DRR

Regulatory provisions related to environment and natural resources—water, land, agriculture, forests, wildlife, habitats, ecosystems; procedures and planning—Environmental clearance, EIA, audit, risk analysis, land-use and zoning, emergency preparedness; and environmental services-drinking water, sanitation, waste management, preventive-health, including climate mitigation and adaptation etc. play an important role in addressing hazards, reducing underlying causes of vulnerability, enhancing capacity and thereby, relate to Disaster Risk Reduction (Gupta et al. 2016). For mitigating climatic hazards and minimizing the impacts of natural disasters and for improving livelihoods and overall well-being of the people, central and state Governments have implemented a number of schemes, whose activities are facilitated further by the involvement of PRIs, NGOs etc. These plans and projects (given in Table 1.5) are the best examples of implementing climate resilient development plans at national and sub national level.

Box 1.2 Environment Protection Act 1986
Environment Protection Act (EPA) 1986:

It is an umbrella legislation which tries to cover the gap areas of major environmental challenges lacking in other laws. The Environmental Impact Assessment (EIA) also comes under EPA. The Coastal Regulation Zone (CRZ) Notification 1991 also came under EPA.

Various other constitutional provisions for DRR came out and have been implemented under the EPA. Gupta et al. 2013 have listed in detail all the regulatory provisions supporting CCA and DRR under EPA.

1.6 Conclusion and Way Forward

Resilient development is needed to cope up with the adverse impacts of disaster. Impacts of Climate Change is noticeable worldwide, the increased frequency and intensity of disasters can be correlated with the changing climatic scenario. India has already taken several initiatives to support climate resilient infrastructure including announcement of NAPCC, formulation of 32 SAPCCs, preparation of NDMP, contributions towards the NDCs, agreement with Paris Climate Change Agreement, partner with SFDRR etc. To promote these initiatives, mitigation efforts should be included at all stages of disasters; e.g. at pre-disaster stage, DIA should be included in EIA to support disaster safe developmental activities; during-disaster stage ensuring Natural Resource Management; at post-disaster stage, sustainable recovery and green growth strategies should be adopted. For climate resilience, integration of CCA and DRR is required to enhance efficiencies and effectiveness, promote synergies, and cohesiveness in policies, plans and projects. The best way to integrate is to mainstream CCA and DRR in all existing and future plans, programmes and

Table 1.5 Brief description of some of the Schemes and Programs where climate and disaster risk reduction can be integrated

Schemes	Details
Swacch Bharat Mission	It is a national campaign of the Government of India, to clean the streets, roads and infrastructure of the country. It has two arms: SBM (urban) and SBM (rural)
Atal Mission for Rejuvenation of Urban Transformation	It was launched in 2015 to provide basic services to household i.e. water supply, sewerage, urban transport
Smart Cities Mission	The initiative aims to drive the economic growth and quality of life in cities by enabling local area development through the use of technology, information, infrastructure development and services
Urban and Regional Development plans Formulation and Implementation	It has provisions for rainwater harvesting, conservation of urban water bodies, water supply system, waste water management system, energy efficiency, strategic plan for new and renewable energy efficiency, strategic plan for new and renewable energy, alternate sources of energy to meet the city demand
Affordable Housing in Partnership (AHP)	GoI approved the scheme of Affordable Housing in Partnership (AHP) as part of Rajiv Awas Yojana (RAY) on 3rd September 2013 to increase affordable housing stock, as part of the preventive strategy
National Rural Health Mission (NRHM)	It was launched by Ministry of Health and Family Welfare in 2005 with an objective to provide support to health care systems in rural areas of 18 states through provision of physical infrastructure, human resources, equipment, emergent transport, drugs, diagnostics and other support
Sarva Shiksha Abhiyan (SSA)	GoI's flagship program—implemented in partnership with states—to achieve Universalization of Elementary Education (UEE) in a time bound manner, making free and compulsory education to children of 6–14 years age group
Pradhan Mantri Gram Sadak Yojana (PMGSY)	PMSGY is launched by GoI in December 2000. It is a fully funded Centrally sponsored Scheme to provide all weather road connectivity in rural areas of the country
National Rural Livelihood Project (NRLP)	The NRLP is implemented in 13 high poverty states accounting for about 90% of the rural poor in the country. Intensive livelihood investments would be made by the NRLP in 107 districts, 422 blocks and 13 states (Assam, Bihar, Chattisgarh, Jharkhand, Gujarat, Maharashtra, Madhya Pradesh, Orissa, Rajasthan, Uttar Pradesh, West Bengal, Karnataka and Tamil Nadu)
Special Package for Drought Mitigation Strategies	The project objectives are to restore ecological balance by harnessing, conserving and developing natural resources like soil, water and forest and improve the ecosystem by checking soil erosion and deforestation. GoI in 2009 approved a special package for implementing drought mitigation strategies in Bundelkhand region Uttar Pradesh and Madhya Pradesh, to be implemented over a period of 3 years starting 2009–10

(continued)

Table 1.5 (continued)

Schemes	Details
National Watershed Development Project for Rain fed Areas (NWDPRA)	The project aims at in-situ moisture conservation primarily through vegetative measures to conserve rainwater, control soil erosion and generate the green cover both on arable and non-arable lands. The scheme is implemented at the field level by an inter-disciplinary team of members from line departments of state government and the beneficiaries of the watershed
Pradhan Mantri Krishi Sinchai Yojana	PMKSY is an amalgamation of ongoing schemes viz. Accelerated Irrigation Benefit Program (AIBP) of the Ministry of Water Resources, River Development and Ganga Rejuvenation (MoWR, RD&GR), Integrated Watershed Management Program (IWMP) of Department of Land Resources (DoLR) and the On Farm Water Management (OFWM) of Department of Agriculture and Cooperation (DAC).It mainly undertakes rain water conservation, construction of farm pond, water harvesting structures, small check dams and contour bunding etc. MoWR, RD&GR, undertake various measures for creation of assured irrigation source, construction of diversion canals, field channels, water diversion/lift irrigation, including development of water distribution systems
Mid-Day Meal	The Mid-Day Meal Scheme directs state governments to provide cooked mid-day meals in all government and government assisted primary schools
Paramparagat Krishi Vikas Yojana	Paramparagat Krishi Vikas Yojana addresses critical importance of soil and water for improving agricultural production. The government would support and improve the organic farming practices prevalent in India. Following cluster approach mode of farming, at least 50 farmers would form a group having 50 acres of land to implement organic farming
Rshtria Krishi Vikas Yojana (RKVY)	Launched in 2007, provides 'additional central assistance' to Central government and state schemes related to agriculture. Region-specific agriculture research and preparation of district agriculture plans, taking into account the local needs and conditions
Integrated District Approach (IDA)	It was launched in late 2004 and early 2005 in 17 districts across 14 states in the country (key strategies outlined by UNICEF India). These strategies focused on promoting community action and integrated delivery of services by establishing horizontal linkage between communities and line agencies on one hand and establishing an interface between communities and line agencies on the other, ensuring responsive, relevant and convergent delivery of services
Integrated Child Development Services	These projects strive to provide supplementary nutrition, health care and pre-school education to children below the age of six

Source: Gupta et al. (2016)

projects, wherever these are relevant and feasible. Tested tools and techniques are available for adaptation according to the specific requirements of countries, sectors and projects.

References

Anand N (2018) The mounting economic toll of Kerala's once-in-a-century floods, August 27. Retrieved from Quartz India: https://qz.com/india/1368531/kerala-floods-automobile-it-rubber-sectors-worst-hit/

Asian Development Bank (2015) Global increase in climate-related disasters. ADB Economics Working Paper Series no. 466

COP-14, U (2019) United Nations convention to combat desertification COP-14, September 1–13. Retrieved from UNCCD COP-14 New Delhi: http://unccdcop14india.gov.in/unccd

EM-DAT (2015) Global frequency of natural disasters by type (1970–2014). Center for Research on the Epidemiology of Disasters. Retrieved from http://www.emdat.be

GCRI, Battelle, & PNWD (2009) India: the impact of climate change to 2030. National Intelligence Council

GIZ (2014) State action plans on climate change: creating India's Architecture for Climate Change Action. New Delhi: Climate Change Adaptation in Rural Areas of India (CCA RAI) and Deutsche Gesellschaft für Internationale Zusammenarbeit (GIZ) GmbH

GoI (2018) National action plan on climate change 2018. PM Council on Climate Change, Government of India, New Delhi

GoI M (2010) Climate change & India: a 4x4 assessment; A sectoral & regional analysis for 2030s. MoEF&CC, New Delhi

Gupta AK, Nair S (2011) Environmental knowledge for disaster risk management – concept note. NIDM & GIZ, New Delhi

Gupta AK, Nai SS, Wajih SA, Chopde S, Gupta G, Aggarwal G (2014) Mainstreaming climate change adaptation and disaster risk reduction into district level development plans. NIDM, GEAG & ISET, New Delhi

Gupta AK, Singh S, Chopde S, Wajih S, Katyal S, Kumar A (2016) Climate resilient and disaster safe development: process framework. Institute of Social and Environment Transition, Colorado

Independent Evaluation Department (IED) (2013) The rise of natural disasters in Asia and the Pacific: learning from ADB's experience. Asian Developmental Bank, Manali

IPCC (2007a) Climate change 2007: impacts, adaptation and vulnerability

IPCC (2007b) Climate change 2007: working group II: impacts, adaptation and vulnerability, glossary. IPCC. Retrieved from https://www.ipcc.ch/publications_and_data/ar4/wg2/en/annexessglossary-a-d.html

ISDR (2009) Adaptation to Climate Change by Reducing Disaster Risks: Country Practices and Lessons. UNISDR, Geneva

MHFW (2020) National Health Mission, April 22. Retrieved from Natioanl Health Mission, Ministry of Health and Family Welfare, GoI: https://nhm.gov.in/index4.php?lang=1&level=0&linkid=445&lid=38

MNRE (2020) Wind energy, April 23. Retrieved from Ministry of New and Renewable Energy, GoI : https://mnre.gov.in/wind/current-status/

MoEF&CC (2018) National Coastal Mission, June 18. Retrieved from Society of Integrated Coastal Management, Ministry of Environment, Forest and Climate Change: http://sicom.nic.in/projects/national-coastal-mission

MoEF&CC (2020) State action plan on climate change, January 15. Retrieved from Ministry of Environment Forest & Climate Change, GoI: http://moef.gov.in/division/environment-divisions/climate-changecc-2/state-action-plan-on-climate-change/

NAPCC (2008) National action plan on climate change. Government of India. Retrieved from http://www.moef.nic.in/downloads/home/Pg01-52.pdf
NAPCC (2018) National action plan on climate change 2018. PM Council on Climate Change, Government of India, New Delhi
National Disaster Management Plan, NDMA (2016). Retrieved from http://ndma.gov.in/images/policyplan/dmplan/National%20Disaster%20Management%20Plan%20May%202016.pdf
NDMA (2009) National policy on disaster management 2009. NDMA
NDMA (2016) National disaster management plan. National Disaster Management Authority, Government of India, New Delhi
NDMA (2019) National disaster management plan 2019. NDMA, New Delhi
NICRA (2012) National Initiative on Climate Resilient Agriculture; Technology Demonstration Component Technical Programme 2011–2012. NICRA, Hydrabad
NMRE (2019) Waste to energy overview, March 31. Retrieved from Ministry of New and Renewable Energy, GoI: https://mnre.gov.in/waste-to-energy/current-status
NOAA-ESRL (2018) Earth's CO_2 Home Page. Retrieved from CO_2.earth: https://www.co2.earth/earths-co2-main-page
NRC (n.d.) Climate change evidence, impacts, and choices. The National Academics
Pahuja N, Pandey N, Mandal K, Bandyopadhyay C (2014) GHG mitigation in India, March. Retrieved from WRI: https://www.wri.org/publication/ghg-mitigation-india
Rattani V (2018) Coping with climate change: an analysis of India's National Action Plan on Climate Change
Rijiju K (2015) National policy on disaster management, December 16. Retrieved from Press Information Bureau Government Of India: http://pib.nic.in/newsite/mbErel.aspx?relid=133377
Sarkar S (2017) Climate risk perilously high in India, November 12. Retrieved from India Climate Dialogue: http://indiaclimatedialogue.net/2017/11/12/climate-risk-high-india/
TheEconomicTimes (2020) National Wind Energy Mission, February 15. Retrieved from The Economic Times: https://economictimes.indiatimes.com/topic/National-Wind-Energy-Mission
TheIndianExpress (2020). Four new missions to boost response to climate change, April 20. Retrieved from The Indian Express: https://indianexpress.com/article/india/india-others/four-new-missions-to-boost-response-to-climate-change/
UNFCCC (2007) Climate change: impacts, vulnerabilities and adaptation in developing countries. United Nations Framework Convention on Climate Change
UNFCCC (2011) Assessing the cost and benefits of adaptation options: an overview of approaches. Bonn
UNISDR (2006) Developing early warning systems, March 16. Retrieved from isdr-ppew@un.org.: https://www.unisdr.org/files/608_10340.pdf
UNISDR (2016) Press Release: the human cost of the hottest year on record - climate change and El Nino drove disasters worldwide in 2015. UNISDR
UniteForSight (2015) The sanitation crisis. Retrieved October 17, 2018, from Sanitation. http://www.uniteforsight.org/environmental-health/module7

Anil K. Gupta, Professor of Policy and Strategies, Integrated Resilience. Head of the ECDRM Division, International Cooperation, Programme Director – Centre for Excellence on Climate Resilience, Nodal Officer for Central Ministries at National Institute of Disaster Management, (Govt. of India), New Delhi. He is also a member of the expert Team for WMO Climate Statement and Core Group Member of IUCN-Council of Ecosystem Management.

Ms. Pritha Acharya has a background in environmental sciences. She has worked in the field of sustainability, climate change adaptation and biodiversity conservation while working with IUCN India. She is currently working as a research associate and manages the CAP-RES (DST) funded project.

Akhilesh Gupta is the Secretary SERB and Sr Advisor at DST coordinating two National Missions on Climate Change under National Action Plan on Climate Change. He has published over 185 research papers in various National and International Journals and conference proceedings. Dr. Gupta was a member of the National Coordination Team which drafted India's National Action Plan on Climate Change in 2008. He has been president of Indian Meteorological Society and Association of Agro meteorologists.

Chapter 2
Evolution of Disaster Risk Reduction Systems in India

Shweta Bhardwaj, Pritha Acharya, and Anil K. Gupta

Abstract India has always been vulnerable to different hazards due its diverse geographic and climatic condition. In its initial years, disaster management in India was purely driven by relief-based approach. Major activities during that period included only providing relief and rehabilitation of necessary infrastructures, focussing primarily on post disaster phases. Better understanding of disaster risk and lessons learnt from past disaster events helped India in redesigning its approach for managing disasters. It was then, when we witnessed a paradigm shift in India's disaster management, a shift from 'relief centric approach' to a more comprehensive and holistic approach for managing disasters; addressing disaster prevention, mitigation and preparedness along with disaster relief and response through dedicated and committed interventions in all the disaster phases (pre or post-disaster). This whole transition in the disaster management approach is a result of enormous initiatives taken over the years. Presently, India has a well establishment legislative, institutional and financial arrangement at national and sub-national levels (states and districts) for managing disasters efficiently. In this chapter, an attempt is made to trace this evolution of Indian disaster management systems by highlighting various national or international policies/strategies that have played a defining role in reaching where India stands today in this disaster management discourse. Along with this, efforts are made to draw an understanding of the present challenges of disaster risk reduction systems and the need for environment-based disaster management approach in the present scenario of climate change and increased disaster risk.

S. Bhardwaj (✉)
CAP-RES Project, National Institute of Disaster Management, New Delhi, India

Environment Climate and Disaster Risk Management, National Institute of Disaster Management, New Delhi, India

P. Acharya
ECDRM Division, CAP-RES DST Project, Centre for Excellence on Climate Resilience, National Institute of Disaster Management (Ministry of Home Affairs), New Delhi, India

A. K. Gupta
ECDRM Division, International Cooperation, CAP-RES DST Project, National Institute of Disaster Management (Ministry of Home Affairs), New Delhi, India

A. K. Gupta et al. (eds.), *Disaster Risk and Management Under Climate Change*, Disaster Resilience and Green Growth,
https://doi.org/10.1007/978-981-99-4105-6_2

Keywords Disaster risk reduction (DRR) · Holistic approach · Relief centric approach · Disaster vulnerability

2.1 Introduction

2.1.1 Understanding of Disaster and Vulnerability

Disasters cause enormous physical and economic losses along with severe social and environment impacts, disrupting the normalcy of a society. Disasters are outcome of complex human and environment interactions. Since ages, understanding disasters and defining them holistically have remained a challenge for all. There are plenty of definitions available, which are influenced by different school of thoughts which evolved over different phases of disaster discourse. If we look at the prehistory of disaster understanding, then disasters were perceived to be the *"act of god"*, these actions were attributed to supernatural power, implicating that nothing can be done by humans to prevent them from causing destruction (Quarantelli 2000). Over the time, the view on disasters shifted from disasters as *"act of god"* to disaster as *"act of nature"* and then to *"act of nature"* influenced by *"act of man and women"*. However, the major shift in disaster discourse came with vulnerability perspective of disasters. The idea of vulnerability came into the picture with the argument that questioned the *"naturalness"* of disasters (McEntire 2010). It was then that a new understanding of disasters came into existence which was based on social, political and economical dimensions of disasters in contrast to the traditional disaster understanding which was solely based on hazards or the agents of disaster. Vulnerability is believed to play a critical role in disaster occurrence and its consequences, and this new approach captured the idea of how a society can create conditions in which disasters can have differential impacts on different groups of people.

The evolved understanding of disasters and different disaster experiences helped in addressing the gaps in India's traditional relief centric disaster management approach. It called for a paradigm shift in our approach towards managing disasters, it was then we shifted from relief centric to preparedness centric disaster management. The new approach advocated for overall, comprehensive and holistic management of disasters through dedicated and committed interventions in all the disaster phases (pre or post-disaster). For a fact, in 1999 the Orissa super cyclone caused dead of more than 10,000 people in contrast to this during cyclone Fani in 2019, death toll remained less than 100. Undoubtedly, it very well reflects the whole transition in our planning and approach towards disaster risk management, particularly when it comes to averting human deaths but still there is a long way to go before we can manage disasters holistically and more efficiently. As on now we have been able to curb down the losses of life but over the time witnessed increase in trends of population affected, geographical area under disasters and economic damages and losses. Underlying causes of risk and vulnerability, therefore, have now attained the concern of recent researches. Dimensions of environmental damages, climate risk, livelihood and health aspects and pathways of impact to infrastructure, property and resources of the people, sought a new focus of understanding in disaster management approach.

2.2 Significance and Concept

2.2.1 United Nations-International Decade for Natural Disaster Reduction (UN-IDNDR)

The first and foremost international initiative of its kind was the United Nation-International Decade for Natural Disaster Reduction. Taking inspiration from international IDNDR and Yokohama strategy, India took number of initiatives to reduce disaster related losses and damages.

United Nations designated 1990–1999 as the International Decade for Natural Disaster Reduction (IDNDR) with aim of reducing impacts of natural disasters through international actions, especially in developing nations. IDNDR focussed on increasing capacity of nations in predicting, assessing, mitigating and preventing natural disasters through scientific and technical knowledge, sensitive to cultural and economic context of a country. The decade which started with the motivation of fostering scientific and technical abilities for disaster risk reduction later expanded to addressing social as well as economic impacts of disasters, during the decade. As the outcome of IDNDR, in the mid of the decade at the first world conference on natural disaster reduction at Yokahama in 1994, "Yokahama Strategy and Plan of Action for a Safer World" was adopted. It is the first document to provide principles and guidelines on disaster risk reduction at international level. During IDNDR, number of initiatives has been undertaken by governmental and non-governmental agencies in India. As stated by IDNDR—Indian Experiences and Initiatives (IDNDR—Indian Experiences and Initiatives 1999) some of key initiatives of the period included:

- Hazard and vulnerability mapping under the initiative taken by the Ministry of Urban Development. First edition of Vulnerability Atlas of India along with the housing vulnerability table was released in 1997.
- Various guidelines for disaster resilient construction were issued during the decade for hazards like earthquake, cyclones, landslides and floods.
- Several advancements were made in early warning, disaster forecasting and monitoring systems using science and technology.
- Establishment of institutional networks for research activities and enhancement of human resource in the field of disaster management. In 1993, a Central Sector Scheme on Natural Disaster Management Programmes (NDMP) was implemented with core objective of carrying out different activities under IDNDR which lead to establishment of National Centre of Disaster Management (NCDM) in 1955 along with various other disaster management faculties in different states across the country.

- The decade also witnessed interventions of various Voluntary and Non-governmental Organisation in different domains of disaster management. Key focus areas of their intervention remained relief and rehabilitation post disasters.

The decade 1990–2000 was an eventful decade for India, during which it witnessed number of disasters extreme including Latur earthquake (1993), Malpha landslide (1994) and Orissa Super-cyclone (1999). During the decade, India faced a total of 115 disasters in which almost 50,000 people lost their lives and 429 billion people were affected by these disasters (ADRC 2002). Despite the efforts made during the decade, Indian administration failed in managing disasters efficiently. The increased disasters events and their impacts, particularly the horrifying experiences of Latur earthquake and Orissa super-cyclone, helped government recognising the need for a comprehensive approach for managing disasters; as a result of which Government of India proposed the constitution of High Powered Committee. At the end of IDNDR, in 1999, Government of India set up a High Powered Committee (HPC) on disaster management with the initial mandate of preparing comprehensive management plans for the *'natural disasters'* which was later expanded to *'man-made disasters'* at national, state and district levels. It was the first ever attempt in the country to draw a holistic, comprehensive and systemic approach towards disaster management. The recommendation of HPC came in October 2001, soon after the 2001 Gujarat earthquake. High powered Committee recommended measures for bringing in legal, institutional and financial reforms in the managing disasters. Some of the key recommendation of HPC included establishment of legal framework through enactment of disaster management act; it emphasized on instilling the culture of prevention, strategic thinking, quick response and preparedness; establishing institutional framework by setting up National Centre of Calamity Management (NCCM) and National Institute of Disaster Management (NIDM) at the centre and disaster management departments at the states; strengthening financial arrangement by reconstitution of calamity relief fund (CRF) and setting up of a National Disaster Response (Rescue/ Relief/ Rehabilitation and Reconstruction) Fund and a National Disaster Prevention Mitigation and Preparedness Fund. Post-independence, Ministry of Agriculture was the nodal ministry for managing disasters in the country; in 2002 based on the recommendation of HPC the Ministry of Home Affairs was made the nodal ministry for overall disaster management across the country. Also, there are different nodal ministries and departments acting as lead agency for managing particular type of disaster (flood, drought, earthquake etc.), as notified by the central government.

Just at the beginning of the decade 2000–2010, India witnessed some of the worst disasters in its history, the Bhuj earthquake, 2001 and the Indian Ocean Tsunami of 2005, both of these events left the authorities to rework and redesign their traditional approach of dealing with disaster which was since its beginning was relief centric and both of these events proved instrumental in establishing/defining legislative history of disaster management in the country through enactment of National Disaster Management Act, 2005.

2.3 Strategies and Management

2.3.1 India's Response to Disaster Risk Reduction: Prior to National Disaster Management Act, 2005

Prior to National Disaster management Act, 2005, there was no legal mandate for disaster management in the country; since there is no mention of disaster management in India's constitution. However, India is a very reactive nation and time to time different disasters across the Indian history have led to establishment of measures to manage the impact disasters particularly using relief based approach. In order to assess what role National Disaster Management Act, 2005 played in evolving disaster management system in the country, it becomes necessary to first understand the scenario of disaster management before enactment of National Disaster Management Act. As mentioned earlier, there was no legal arrangement for managing disasters in the country. However, there were basic institutional and financial provisions for disaster management.

Institutionally, at the national level, the Ministry of Agriculture remained the nodal agency for managing disasters in the country through the Ministry's Natural Disaster Management (NDM) division. Apart from Agricultural Ministry some other the ministries such as the Ministry of Civil Aviation, Environment and Forest, Health and others acted as nodal agencies for some specific disasters. National Crisis Management Committee (NCMC) was the apex operational agency for ensuring implementation of relief measures effectively, headed by cabinet secretary and constituted of secretaries of different departments. As deemed necessary, NCMC has to give direction to Crisis Management Group (CMG). CMG, headed by Central Relief commissioner consisted of senior officials from various ministries, the group formed to ensure assessment of the relief measures needed during any calamity and implementing necessary measures effectively by providing necessary support to state. Also functions of CMG included yearly review of contingency plans of different departments/ministries to deal with the calamities concerning their respective sectors. Institutional mechanisms at the state level included, relief centric management of disaster taken care by revenue department which over the times passed on to the department of relief and rehabilitation and then to the department of disaster management (separate entity or sometimes within the department of revenue). At the national level, High Level Committee constituted of ministers from relevant departments (finance, agriculture and others) was responsible for taking decisions on relief assistance to be provided to state in case of calamities on recommendation of Inter-Ministerial Group (IMG). Central Crisis Group (CCG) headed by secretary from Ministry of Environment and Forest, ensured management of chemical disasters. Similar to centre, State Crisis management committee (SCMC), headed by chief secretary, remained apex entity for coordinating relief and response activities at the state. Crisis Management Group at state headed by chief secretary has been formulated to manage chemical disasters. At district level, district collector bears the sole responsibility of managing the disasters. District

Table 2.1 Institutional and financial arrangements prior to the National Disaster Management Act 2005

Arrangement	National level	State level
Institutional	• Nodal Ministries • Ministry of Agriculture—Central nodal agency for disaster management • Ministry of civil aviation—air accident • Ministry of railways—rail accident • Ministry of environment and forest-chemical disaster, forest fire Ministry of health—biological disaster • Ministry of atomic energy—nuclear disaster • National Crisis Management Committee • Crisis Management Group • High Level Committee • Central Crisis Group	• Revenue department/disaster management department/department of Relief and Rehabilitation[a] • State Crisis Management Committee (SCMC) • State Crisis Management Group
Financial	• National Calamity Contingency Fund	• Calamity Relief Fund

[a] At district level, District Coordination Committee headed by district collector

Coordination Committee responsible for ensuring effective coordination of various programmes and activities, looked at disaster management activities during any crisis event as well.

Prior to DM act, disaster management financing evolved based on the recommendations of finance commission over the years. Disaster financing was also influenced by relief centric approach, focused on providing immediate financial assistance to the affected and restoring critical infrastructures. The major funds prior to DM Act were National Calamity Relief Fund at the national level and Calamity Relief Fund at the states (Table 2.1).

2.3.2 National Disaster Management Act and National Disaster Management Policy

National Disaster Management Act, 2005, brought the paradigm shift in the India's disaster management from relief centric approach, made emphasis on adopting an integrated and holistic proactive approach addressing prevention, mitigation and preparedness along with relief and response. Past disaster experiences highlighted the need for a holistic approach for disaster management. Country needed a comprehensive legal, institutional and financial arrangements and the National Disaster Management Act played a pivotal role in establishing one such comprehensive disaster management framework in the country. Based on one of the most crucial recommendation of HPC to address the limitations and gaps of disaster management approach of that time, National Disaster Management Act was enacted in 2005, providing the much needed legal framework for disaster management in the country at state and national level. The act, despite its top-down approach proved highly useful in guiding sub-national entities i.e. states and districts in establishing their own legal, institutional and financial provisions for effective management of disasters. The act mandates the establishment of three tier institutional system for holistic management of disasters at national, state and district level. The act lays down a comprehensive institutional framework clearly defining structures, roles and responsibilities of different agencies. National Disaster Management Authority (NDMA) was designated as an apex central agency among the other three central entities at the centre—National Executive committee (NEC), National Institute of Disaster Management (NIDM) and National Disaster Relief Force (NDRF). At state level, act provides mandate for constitution of State Disaster Management Authority (SDMA) and State Executive Committee (SEC). Similarly at district level, act envisaged constitution of District Disaster Management Authority (DDMA). Apart from a comprehensive institutional framework, the act also provided financial provision for managing disasters. Act envisaged constitution of two types of funds—disaster response fund and disaster mitigation fund at national, state and district levels. Act also mandates the preparation of disaster plan at national and sub-national levels (state and district) by the respective disaster management agencies.

In 2009, National Disaster Management Policy was notified; prepared in accordance with National Disaster Management Act, 2005. The policy laid emphasis on establishing holistic and integrated approach through community based disaster management, developing capacities of the institutions as well as stakeholders, bringing in multi-sectoral synergy and cooperation among regional, national and international agencies. The DM policy envisaged promotion of culture of planning and preparedness; ensuring participation of different stakeholders for disaster mitigation and preparedness (community, local institutions—panchayati raj bodies, NGOs, CSR and others); establishment of proper institutional mechanism/chain of command; standard operating procedures and disaster management plan at all levels; development of disaster management codes/manuals, ensuring safe construction practices; building capacity of communities, institutions artisans etc. and creating knowledge network through research and documentation. The policy reinforced the existing disaster management act by laying down an integrated policy covering all aspect of disaster management, emphasizing on priority areas where interventions are needed and drawing down different mechanisms required for operationalizing these interventions in effective manner.

2.3.3 Post IDNDR: Hyogo and Sendai Frameworks: An Integrated Approach

The close look at India's national developments in the field of disaster management post IDNDR, very well reflect the synergy between these development with the international efforts made post IDNDR for mainstreaming the culture of disaster management and risk reduction. Taking into the account the international context; at the end of IDNDR, as mandated by United Nation General Assembly resolution, an International Strategy for Disaster Reduction (ISDR) was adopted and United Nation office for Disaster Risk Reduction (UNDRR) was established as the secretariat to facilitate the implementation of ISDR. UNDRR carried out the review of Yokohama Strategy. The areas of gaps and challenges as identified during the review became the key areas of intervention for upcoming framework for the decade 2005–2015. In January 2005, at the world conference on disaster reduction, Kobe, Hyogo, Japan, the *"Hyogo-Framework for Action 2005-2015: Building Resilience of Nations and communities to disasters"* was adopted. Hyogo Framework for Action (HFA) laid down five priorities for actions and guiding principles to build disaster resilience. HFA aimed at reducing impacts, losses and damage by making countries and communities resilient towards disasters. The concept of disaster risk reduction as put forward by the Hyogo Framework strongly laid emphasis on disaster preparedness and mitigation and not just the disaster relief and response like in previous decades. During the decade, national government were reported to make major progress against HFA's objectives and priorities by strengthening institutional and legislative mechanisms through adoption of framework and plan for effective management of disasters (GAR 2011). During this period, India also made a significant progress in disaster management which were in line with goals and objectives of HFA. National policy and legislative framework for disaster management i.e. National Disaster Management Act (2005), National Disaster Management Policy (2009) and National Disaster Management Plan were adopted; the three tier institutional framework proposed by national disaster management act and policy ensured participation all the stakeholders in disaster risk management (including local communities); 11th five year plan mainstreamed disaster risk reduction into national economic and developmental planning; setting up of institution like NIDM ensured capacity development, training, research, documentation and knowledge management on disaster risk management across the country; adaption of National Action Plan for Climate Change, enforcement of building codes, redesigning of financial provisions for disaster management are some of key initiatives take up during the decade 2005–2015 which drew inspiration from HFA.

At the third UN World conference at Sendai, Japan, the "*Sendai framework for Disaster risk reduction*" (SFDRR) was adopted. The successor framework of Hyogo Framework for Action, SFDRR laid down seven global targets to be achieved till 2030 which includes targets on reducing global mortality, people affected during disasters, disaster economic losses, damage to critical infrastructures and enhancing international cooperation, national/local disaster reduction strategies and access to early warning and disaster information, along with four priorities area for action. The core theme of SFDRR include building resilience, ensuring inclusion of all stakeholders (women, elderly, children, specially-abled and others) and promoting community based solutions for disaster risk reduction.

Interestingly, these international frameworks and strategies coincided with many other global developmental initiatives; the subjects addressed under these initiatives holds an important place in disaster discourse and are critical for achieving holistic/ comprehensive management of disasters. In the year 2000, at the millennium summit, 191 nations unanimously adopted *Millennium Declaration* which laid down set agendas and targets for key developmental issues such as poverty reduction, health and nutrition, education and food security for twenty century, to be achieved globally by the year 2015. Based on the framework for sustainable development as envisaged by Millennium Declaration, MDGs were replaced by *'Agenda for Sustainable Development'* in the year 2015. Agenda for sustainable development consist of a set of 17 goals and 169 targets envisioned to be achieved globally by 2030. SDGs envisage sustainable and inclusive development for all by addressing social, economic and environmental issues along with other development priority areas as focused under MDGs. SDGs have strong interconnections with disaster risk reduction, which can be established by the fact that 10 out of 17 SDGs are directly or indirectly linked to disaster risk reduction. Another such global initiative is the Kyoto protocol of 1997, first of its kind legally binding agreement for reduction and limitation of carbon emission by the ratifying countries for the commitment period of 2008 to 2020. Another important intervention in area came in 2015 at 21st Conference of Parties, held in Paris which is called the Paris agreement. It was ratified by 169 counties with the objective of combating climate change by investing in low carbon future. At its core Paris agreement aims at keeping the global temperature rise below 2 °C for the current century. Both these agreements were instrumental in bringing the global attention towards global waning which contributes to climate change, leaving us more exposed towards the extreme events. According to a report by CRED and UNISDR 2018, 91% of extreme events between the period 1998–2017 were climate related. The number, frequency and intensties of climate induced disasters is expected to increase with the emerging climate change challenges.

2.4 Conclusion and Way Forward

2.4.1 Emerging Issues: Ways Forward for Disaster Management Systems in the India

From first paradigm shift in disaster management, which was *from relief to preparedness centric*, India is currently moving towards the second shift, which widely recognises the need for environment centric approach towards reducing disaster risk. The emergence of this paradigm is drawn from evolved understanding of disasters, in which environment plays an important role; emerging environmental challenges such as changing land use land covers, climatic patterns, degrading natural resources and others are determining in increasing the exposure of people towards disasters (Gupta and Nair 2013). Presently, for holistic management of disasters, there is a need for operationalizing the second paradigm shift effectively through institutional, legislative and financial frameworks. This requires committed interventions particularly in three key focus areas: environment planning, infrastructure planning/investments and strengthening of social capital.

On environmental front, there is need to address climate change and land use challenges. As we all know, climate change is the biggest challenge that world is currently facing which can be established by the fact that during past few decades we have witnessed the rapid increase in climate induced disasters, whose impacts are exacerbated due ill planned anthropogenic activities. Climate change is influential in increasing the existing hazard risk as well as the future vulnerabilities as a result of degraded environment. Changing land use patterns and natural ecosystems plays a vital role in defining the exposure to climate changes and its consequences. Therefore, land use planning and natural ecosystem management can be used as mitigative and adaptive tools for combating climate change. Land use planning should be central at national and sub-national developmental planning as well as it must be in lines with the climate adaptive plans, ensuring to address the future climate change issues.

There is a need for a more comprehensive policy instrument for land use planning that should address land development, regulation and environment protection issues as well as should manage and acknowledge the future demands for land resources in the changing climatic scenarios. In this particular context (environmental and climatic), widely recognised nature based approach can provides a golden opportunity for addressing climate change adaption and mitigating biodiversity loss in one go. Nature based solution approach which promotes use of natural ecosystems such as mangroves, wetlands and other such systems can be helpful in reducing disaster risk. There still remains number of areas within this particular approach with enormous potential which are still underexplored or unexplored; green infrastructure, climate proofed agriculture systems are some of examples. Apart from addressing various environmental concerns for reducing disaster risk there also lies an immediate need to monitor and govern the infrastructural development. Infrastructures are critical for smooth functioning of any society and are also one of the most vulnerable assets during any disasters.

Investing in disaster resilient infrastructure sensitive to hazard profile of an area is what we need the most at present to reduce the economic and other losses during disasters. In this context, for cities in particular, this infrastructure resilience is very crucial since cities not just houses huge population but also huge economic assets as well. As predicted our cities will keep on growing spatially and would become more populated in coming times. Meeting their future demands in terms of providing living spaces, critical infrastructural services (electricity, water etc.) and waste management in sustainable way is very much needed to ensure our cities are resilient to future disaster extremes. Burgeoning population density and urbanisation pave way for huge investments in infrastructures such as bridge, dams, transportations and others, again it has to be ensured that these investments are sensitive to present and future disaster risk.

Another important area that requires more attention in India's disaster management policies and plans is the investment in social capital. Social capital, embedded in every community can play an important role in disaster risk management in each of its stages. Communities are the ones who suffer the most during any disasters and are also the first responders to these events. Due to such vital role that community plays and the changing disaster traits, wherein disaster risk is becoming so unpredictable, it become extremely necessary that our communities are well prepared and aware. The community based approach for disaster management ensures local participation and ownership. This would require interventions to increase awareness of communities about disaster risk; building and enhancing capacities of local stakeholders; strengthening local institutions and promoting indigenous knowledge or measures for effective disaster management. Not just the local communities but we need to tap into potentials of civil society organisations for managing disasters. As mentioned early, India has come a long way from where it had started in the field of Disaster Risk Management but there is still a long way to go before it can ensure holistic safeguard of its people (irrespective of their socio-economic status), infrastructures, economic activities and environment; which will require committed efforts from all the stakeholders including national and sub-national government agencies, policymakers, local communities, NGOs and others.

References

ADRC (2002) 20th century data book on Asian natural disasters. Asian Disaster Reduction Centre (ADRC)

CRED & UNISDR (2018) Economic losses, poverty and disasters: 1998–2017

GAR (2011) Global assessment report on disaster risk reduction. United Nations

Gupta AK, Nair SS (2013) Applying environmental impact assessments and strategic environmental assessments in disaster management

IDNDR (1999) Indian experiences and initiatives. Ministry of Agriculture

McEntire D (2010) Addressing vulnerability through an integrated approach. Int J Disaster Resilience Built Environ 1:50–64

Quarantelli EL (2000) Disaster planning, emergency management and civil protection: the historical development of organized efforts to plan for and to respond to disasters

Ms. Shweta Bhardwaj has done her master's in Disaster Management from TISS, Mumbai. Presently works with NIDM in CAP-RES (DST) funded project and looks after the Climate Proofing component of the project. She has worked extensively in the field of community organization and social development, environment issues, disaster risk and vulnerabilities and disaster governance.

Ms. Pritha Acharya has a background in environmental sciences. She has worked in the field of sustainability, climate change adaptation and biodiversity conservation while working with IUCN India. She is currently working as a research associate and manages the CAP-RES (DST funded project).

Anil K. Gupta, Professor of Policy and Strategies, Integrated Resilience. Head of the ECDRM Division, International Cooperation, Programme Director—Centre for Excellence on Climate Resilience, Nodal Officer for Central Ministries at National Institute of Disaster Management, (Govt. of India), New Delhi. He is also a member of the expert Team for WMO Climate Statement and Core Group Member of IUCN-Council of Ecosystem Management.

Chapter 3
Climate and Weather Forewarning Systems for Disaster Preparedness and Response

Mrutyunjay Mohapatra

Abstract India experiences several types of natural disasters including cyclones, floods, droughts, earthquake, landslides, heat wave, cold wave, thunder squalls and tornadoes. Most of these natural disasters (about 80%) are hydro-meteorological in nature. The risk management of these natural disasters may include several steps based on its severity and importance: (1) hazard analysis, (2) vulnerability analysis, (3) preparedness and planning, (4) early warning, (5) prevention and mitigation. The early warning component includes: (1) skill in monitoring and prediction of natural hazards, (2) effective warning products generation and dissemination, (3) coordination with emergency response units and (4) public awareness and perception about the credibility of the official predictions and warnings.

Though there have been significant improvement in multi-hazard monitoring and warning system in recent years due to various initiatives of Ministry of Earth Sciences (MoES)/India Meteorological Department (IMD) and policy frame work of government of India, there is still scope for improvement at state and district level in terms of (1) improving the mesoscale hazard detection and monitoring, (2) improving the spatial and temporal scale of forecasts through technological up gradation, (3) sectoral applications of early warning, (4) warning communication to last mile and disaster managers through state of art technology, (5) developing synergized standard operation procedure among the early warning agencies and user agencies and (6) upgrading and enhancing the link between early warning service provider and disaster managers. Further to stress the importance of the above, World Conference on Disaster Risk reduction (WCDRR) during 2015 at Sendai has set a target to substantially increase the availability and access to multi-hazard early warning systems and disaster risk information and assessments to the people by 2030. All these aspects have been discussed in this chapter with special emphasis on extreme weather characteristics (spatial and temporal distribution and intensity etc.), damage potential, modeling and prediction, prediction skills, information dissemination mechanisms, socio-economic impacts, achievements in recent years, existing gap

M. Mohapatra (✉)
Directorate General, India Meteorological Department, New Delhi, India

A. K. Gupta et al. (eds.), *Disaster Risk and Management Under Climate Change*, Disaster Resilience and Green Growth,
https://doi.org/10.1007/978-981-99-4105-6_3

areas and future scope in view of Sendai framework to achieve the target for India by 2030.

Keywords Climate · Early warning · Disaster risk information · Sendai framework

3.1 Introduction

It has always been challenging to predict weather, especially the extreme weather events over the South Asian region including India due to mesoscale convective processes in the tropics, varied physiography of the region and their complex non-linear interaction with large scale atmospheric circulation. The South Asian region including India is affected by different hazards including cyclone, depression, heavy rainfall, thunderstorm/squall/hailstorm, storm surge and coastal inundation etc. Most of these natural disasters (about 80%) are hydro-meteorological (water and weather related) in nature.

India is one of the few countries in the tropics that have a long record of observations that are necessary to understand tropical climate variability. These long records are essential to understand changes in climate that occur over both short and long-time scales. The accelerated global heating is disturbing the natural processes, causing frequent disasters and amplifying the risk and vulnerability associated with these events. India's diverse physical and climatic factors make it more vulnerable to climate change and related disasters (Rajeevan and Nayak 2017). The risk management of these natural disasters depends on several factors including (1) hazard analysis, (2) vulnerability analysis, (3) preparedness and planning, (4) early warning, (5) prevention and mitigation (NDMA 2016). The India Meteorological Department (IMD), Ministry of Earth Sciences (MoES) is the Nodal National Meteorological agency mandated for issuing seamless operational weather forecast and warnings for various meteorological hazards across the country. IMD also provides forecast guidance to the South Asian countries and members of WMO/ESCAP panel countries as the WMO designated Regional Specialized Meteorological Centre (RSMC). The regional guidance on severe weather events includes cyclone, heavy rain, strong wind, sea waves and storm surge (RSMC, New Delhi 2018).

The components of a forecasting and warning system include (1) detection, monitoring and warning through global, regional, national and local observations, (2) numerical weather prediction, (3) forecasts on different timescales (now-casting to several days), (4) timely dissemination of authoritative warning information, (5) risk analysis and impact assessment.

All the above aspects of early warning are analyzed and presented in the subsequent sections. The climatology and variability of weather extremes are presented in the following section.

3.2 Present Scenario

3.2.1 Climatology and Variability of Weather Extremes

India's diverse physical and climatic factors make it more vulnerable to climate change and related disasters (Rajeevan and Nayak 2017). The climatology and variability of weather and climate parameters like temperature and its extremes (heat wave and cold wave), rainfall and its extremes (heavy rainfall leading to floods and deficient rainfall leading to drought), tropical cyclones and other severe weather events like thunderstorms, heat/cold waves etc. are presented and discussed below:

3.2.2 Temperature

According to Srivastava et al. (2017), based on the analysis of surface air temperatures using gridded monthly temperature data for the period 1901–2010, the annual mean, maximum and minimum temperatures averaged over the country as a whole showed significant warming trend of 0.60, 1.0 and 0.18 °C per 100 years respectively. The rate of warming trend in the annual mean temperatures since 1980s is much sharper, 0.2 °C per decade. On the seasonal scale, the highest increasing trend is observed in the post-monsoon and winter seasons (Srivastava et al. 2017). Monsoon season shows the lowest increasing trend. The warming is higher in the northern, central and eastern/northeastern parts of the country and peninsular India experienced the least warming.

Despite solar dimming, the mean annual air temperature has been increasing (Srivastava et al. 2017) which may be due to increase in greenhouse gases.

Future regional Climate Change scenarios for the period 1950–2100 (Sanjay et al. 2017) indicates evidential temperature increases (more than 1.5 °C) across the central and northern parts of India during 2031–2060. The annual warming range over South Asia land areas is 1.8–3.0 °C by 2060.

3.2.3 Heat/Cold Waves

Heat wave is a condition of atmospheric temperature that leads to physiological stress that sometimes becomes lethal for living beings. The World Meteorological Organization (WMO) defines a heat wave as five or more consecutive days during which the daily maximum temperature exceeds the average maximum temperature by 5 °C. In India, heat wave is defined when the maximum temperature of a station reaches 40 °C or more in plains, 37 °C or more in coastal stations and at least 30 °C or more in hilly regions and the departure from normal is 5 °C (NDMA 2016). Similarly, cold wave is defined as the condition when the daily minimum temperature falls below the average minimum temperature by 5 °C as per WMO. Different countries also define cold wave differently in context of their local conditions. In India, cold wave is considered if minimum temperature of a station reaches at least 10 °C or less for plains, and the departure from the normal is 5 °C or more. The criteria are different for northern hilly regions (IMD 2012).

3.2.4 Rainfall and Floods

Flood is the overflowing of the normal confines of water bodies or the accumulation of water over areas that are not normally submerged. Over India, floods generally occur during the southwest monsoon season (June to September) associated with movement of westward moving low pressure systems such as monsoon depressions (Rao 1976). On some occasions, interaction between the monsoon flow and mid-latitude weather systems cause heavy rains and floods especially over northwest and northern parts of India (e.g. Uttarakhand floods in 2013 and Jammu and Kashmir floods in 2014). Floods also occur during post-monsoon seasons (October to December) over the south-eastern parts of the country (e.g. Chennai flood in 2015) due to low pressure systems or tropical cyclones.

The frequency of rainstorms (weather systems with potential of causing large scale floods) has shown an increasing trend of 4 rainstorms in 65 years (Guhathakurta et al. 2017) based on data of 1951–2015. A rainstorm is defined as a closed isohyetal (line of equal rainfall) pattern around its heavy rainfall centre, the grid point with maximum rainfall. The closed isohyetal pattern usually has an area of about 50,000 km^2 or more with rainfall of 25 mm/day or more than that. The above conditions should be satisfied for at least 2 consecutive days. The criterion on areal extent is considered to include large rainstorms that cause large-scale flooding.

3.2.5 Tropical Cyclones

Cyclones bring strong winds, torrential rains and storm surges along with it. The tropical warm North Indian Ocean (NIO), like the tropical North Atlantic, the South Pacific and the NW Pacific, is a breeding ground for the disastrous tropical cyclone phenomenon (Mohapatra et al. 2012), About five tropical cyclones develop over the North Indian Ocean in a year including 4 over the Bay of Bengal (BOB) and 1 over the Arabian Sea (AS) (Mohapatra et al. 2014). Out of these five, two to three tropical cyclones become severe. The tropical cyclones develop in two seasons; viz., pre-monsoon season (April–June) and post-monsoon season (October–December), with later being the primary tropical cyclone season (IMD 2013). Though there is no significant trend in this ratio over the BOB, it shows also decreasing trend for the NIO as a whole during the year. There is also significant decreasing trend in the ratio of tropical cyclone to CD during monsoon season (June–September) over the BOB, AS and NIO.

3.2.6 *Thunderstorms, Hailstorms, Squall and Lightning*

Thunderstorms are potentially risky for aviation sector. Transport, power, communication and other socio-economic sectors are also well within the dangerous fold of thunderstorms. Squall, strong updraft and down draft, turbulence and icing, cloud electrification, lightning, localized heavy rain, hailstorms etc. are the weather conditions when a thunderstorm strikes (Department of Science & Technology (DST) 2005).

Climatology of thunderstorms over Indian region studied by National Disaster Management Authority (NDMA), 2018 based on IMD data of 1980–2010 show that annual thunderstorm frequency more than 80 thunderstorm days per year occurs over north-eastern part of India and some part of Kerala. Thunderstorms' frequency and intensity is maximum in summer months (March to June).

India is a country where one of the highest frequencies of hail occurs over the year. There are about 29 hail days per year of moderate to severe intensity (DST 2005). About 25% of total occurrence in the past recorded hailstones has 3-cm or more diameter. The hailstorms are mainly observed in the winter and pre-monsoon seasons with virtually no events after the onset of the southwest monsoon (DST 2005).

Lightning is a high-energy luminous electrical discharge from a thundercloud to the ground accompanied by thunder. The study shows that the frequency of lightning and death due to lightning is increasing over the years based on the data of 1997–2011 (Singh et al. 2011). However, another study on lightning flash rate by Saha, Singh, Kamra, Galanaki, Maitra, Singh, Singh, Chakraborty and Singh, 2017 based on data of 200–2014 indicate decrease in trend over northern parts of the country and increase in trend over extreme western parts covering Gujarat and Rajasthan.

3.2.7 *Cloudbursts*

Cloud burst leads to landslides and flash floods causing loss of life and property. The cloudburst phenomenon is characterized by high intensity precipitation, usually more than 10 cm/h, within a short span of time, over a small area (Dimri et al. 2017). The impact of cloud burst depends on its location and area of occurrence. If the area involved is a small catchment characterized by steep hill slopes and high river bed gradients, the impact is more ravaging.

3.3 Significance and Concept

3.3.1 Weather Forecasting and Warning Services of IMD

The weather analysis and forecasting is the succession of following tasks:

To clearly understand the recent development and the actual condition of the atmosphere at all-time scales (weather analysis);

To obtain the pertinent information from numerical models and observational systems including satellite and radar and to assess the future evolution of the atmosphere in order to determine the most likely scenario;

To deduce the consequences of the expected synoptic situation in terms of weather elements (weather elements forecasting) and to evaluate the risk of the occurrence of hazardous phenomena (risk assessment);

To prepare the meteorology related information (weather warnings) to be directed towards the various internal or external users.

3.3.2 Monitoring Process

The entire process of extreme weather monitoring and forecasting is shown in a schematic diagram (Fig. 3.1). The observational network for monitoring consists of land-based surface and upper-air stations, observations from Doppler Weather Radars (DWRs) and data from geo-stationary and polar orbiting satellites. In addition, observations from ships and buoys are of immense importance for the analysis and forecasting of extreme weather, especially Ocean related hazards like cyclones,

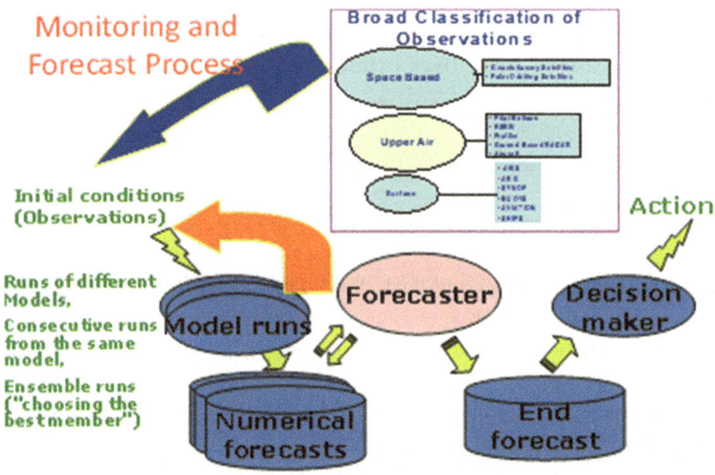

Fig. 3.1 Weather Monitoring and Forecasting process of IMD (Mohapatra et al. 2013a, b, c)

monsoonal heavy rainfall etc. It is important to correctly determine the location and intensity of the CDs over the ocean, as initial error in location and intensity can lead to increase in error in forecast of location and intensity (Mohanty et al. 2010; Osuri et al. 2012).

3.4 Strategies and Management

3.4.1 Forecasting System

A variety of observational data have been used in India till 1960s to forecast the extreme events. Satellite era, since 1960s, added another feature. There has been rapid development in objective techniques since 1970s and especially in recent years for forecasting extreme weather events.

There are three types of NWP models, viz., individual deterministic models, Multi-Model Ensemble (MME) and single model Ensemble Prediction System (EPS) for different ranges of forecast as mentioned below (Regional Specialized Meteorological Centre, RSMC), New Delhi, 2018).

3.4.2 Medium Range Forecast

Global forecasting system (GFS) global model (horizontal resolution of 12 km and forecast upto 10 days), Unified model (horizontal resolution of 12 km and forecast upto 10 days), Global Ensemble Forecasting System (GEFS) global probabilistic model (horizontal resolution of 12 km and forecast upto 10 days, Unified model ensemble prediction system (horizontal resolution of 12 km and forecast upto 10 days).

3.4.3 Short Range Forecast

There are many weather research forecast (WRF) Mesoscale model (horizontal resolution of 3 and 9 km and forecast upto 3 days. All the above models are operational at MoES (Ministry of Earth Science) using High Power Computing Systems (HPCS). The model is run twice in a day (00 UTC and 12 UTC).

The super-ensemble has also been developed based on above ensembles. In the ensemble prediction system (GEFS and UMEPS) run in MoES, various probabilistic forecast products are generated for severe weather events, probabilistic occurrence of different thresholds of rainfall (heavy rain and deficient rain), temperature (heat wave and cold wave) and wind (squally wind and gale winds) etc. at different locations/regions.

3.4.4 Decision Support System (DSS)

The analysis, prediction and decision-making process is made by blending scientifically based conceptual models, dynamical and statistical models, meteorological datasets, technology and expertise (IMD 2013). For the sole purpose of it, decision support system (DSS) is digitally used to plot and analyze weather parameters, satellite, and Radar and NWP model products etc. This is further supported by modern graphical and GIS applications for producing high quality analysis and forecast products.

3.4.5 Warning System of IMD

The goal of the warning system is to maximise actions for safety. To carry out the warning process, there is a well-defined Organization and Standard Operation Procedure (SOP) is in place (IMD 2012, 2013). The warning criteria are defined for each parameter. Linkages with Media and Disaster authorities are well defined. Scheduled time of issue and frequency of bulletins along with the content (text and graphics) are also designed as per the requirement of users. The warnings and advisories by IMD for national purpose are issued for different temporal ranges (from a few hours (nowcast) to 5 days (short to medium range forecast)) and different spatial domains (IMD 2012).

3.4.6 Cyclone Warning System

Types of bulletins and warnings issued in the interest of Mariners and coastal states in cyclone specific situations are: Bulletin for All India Radio (AIR), Warnings for Designated/Registered users, Bulletins for Press.

The cyclone warnings are issued to central and state government officials in four stages. The **First Stage** warning known as **"PRE CYCLONE WATCH"** issued at least 72 h in advance. The **Second Stage** warning known as **"CYCLONE ALERT"** is issued at least 48 h prior to onset of adverse weather condition over the coastal areas. The **Third Stage** warning known as **"CYCLONE WARNING" is** issued at least 24 h in advance of the expected commencement of adverse weather over the coastal areas. The **Fourth Stage** of warning known as **"POST LANDFALL OUTLOOK"** is issued by the concerned ACWCs/CWCs/and CWD at headquarter at least 12 h in advance of expected time of landfall.

The cyclone warning is provided on a real time basis to the Control Room of the Ministry of Home Affairs, Government of India, besides other Ministries and Departments of the Central Government, and electronic and print media and concerned state Governments.

3.4.7 Thunderstorm/Dust Storm/Lightning Warning

Severe thunder storms are localized phenomena with a life period of about a few hours. These can be detected on real time with a dense network of DWR in the country. The currently available array of numerical weather prediction system is not capable to provide accurate location specific forecast of these thunderstorms and associated weather with reasonable lead time.

In India, this nowcasting (very short range forecasting upto 3–6 h) is limited to 433 cities. At the same time the district level nowcast against the thunderstorm and associated adverse weather is issued by IMD through its state level Meteorological Centers to the disaster managers and general public. However, with the latest advances in NWP models, IMD started issuing thunderstorm, squall and lightning forecast twice a day valid for next 48 hours since April, 2018.

3.4.8 Heavy Rainfall and Flood Warning

IMD issues heavy rainfall warning at meteorological sub-divisional level through its National Weather Forecasting Centre (NWFC) located at New Delhi. The following are the terminology in use at present to indicate and forecast the rainfall intensity (24 h accumulated rainfall ending at 0830 h IST of each day) in the short and medium range weather forecasts (Table 3.1).

3.4.9 Heat and Cold Waves

The Ministry of Earth Sciences has implemented state-of-the-art coupled climate model for seasonal forecasting. The model has been used to prepare the monsoon forecasts since 2012. The model was used first time to prepare a temperature outlook for the 2016 Hot Weather Season (April–June) and also cold weather season (December–February) since 2016–17 (Benke et al. 2019). In short to medium range, heat wave and cold wave warning are issued daily valid up to 5 days. The color code system indicating severity of an expected heat wave and cold wave used by IMD is given below (Tables 3.2 and 3.3).

Table 3.1 Categorization of rainfall and snowfall

Descriptive term	Rainfall (mm)	Descriptive term	Snow depth (in cm)
Heavy rain	64.5–115.5	Heavy snowfall	64.5–115.5
Very heavy rain	115.6–204.4	Very heavy snowfall	115.6–204.4
Extremely heavy rain	>204.5	Extremely heavy snowfall	>204.5

Table 3.2 Color code for cold wave severity

Cold wave		
Green(No action)	Normal night	Minimum temperatures are near normal
Yellow Alert (Be updated)	Cold Alert	Cold wave conditions at isolated pockets persists on 2 nights
Orange Alert (Be prepared)	Severe Cold Alert for night	(i) Severe cold wave conditions persists for 2 days (ii) Through not severe, but cold wave persists for ≥4 nights
Red Alert (Take Action)	Extreme cold Alert for night	(i) Severe cold wave persists for more than 2 nights (ii)Total number of cold/severe cold wave nights > 6 nights.

Table 3.3 Color code for heat wave severity

Heat wave		
Green(No action)	Normal Day	Maximum temperatures are near normal
Yellow Alert (Be updated)	Heat Alert	Heat wave conditions at isolated pockets persists on 2 days
Orange Alert (Be prepared)	Severe Heat Alert for day	(i) Severe heat wave conditions persists for 2 days (ii) Through not severe, but heat wave persists for ≥4 days
Red Alert (Take Action)	Extreme Heat Alert for day	(i) Severe heat wave persists for more than 2 days. (ii) Total number of heat/severe heat wave days > 6 days.

3.4.10 Fog

There has been drastic improvement in the Fog monitoring and forecasting systems at Indira Gandhi Interactional Airport (IGIA) in winter of recent years. It has been possible because of successful and effective implementation of three new project, viz., (1) installation of 13 Indigenous Drishti for runway visibility measurement,

(2) Winter fog experiment (WIFEX) and (3) forecasting demonstration project (FDP) in recent years (Ghude et al. 2017).

3.5 Warning Dissemination Mechanism

The warnings are disseminated to various users through telephone, fax, e-mail, SMS, All India Radio, FM and community radios, Television and other print and electronic media. These warnings/advisories are uploaded in the website of IMD (www.imd.gov.in). Also the warning bulletins are disseminated by email and SMS to state and national disaster management authorities. In case of emergency, police wireless and telecommunication lines of railways and aviation authorities are also used. The Cyclone Advisories bulletin for WMO/ESCAP Panel countries and international airports are disseminated through global telecommunication system (GTS) and e-mail.

Recently there is initiative through NDMA for Common Alert protocol (CAP) for dissemination of thunderstorm warning. Once, implemented, it will be very useful for quick dissemination to last mile. Further it is recommended that the synergized observational and forecast and warning products in text cum visual graphics format should be generated by IMD with direct CAP-feed facility so as to reach the last mile with no loss of time. Similarly the proposed last mile connectivity through satellite based initiative if ISRO, viz., Gagan Messaging System and Navik in collaboration with MoES and state and central government will be very helpful to mitigate the disasters.

3.6 Gap Areas

Weather Prediction Models are used for generating forecasts of weather and climate parameters. However, the present day forecast models are known to have systematic errors and random biases. Inaccurate treatment of physical processes in the models (Physical parameterization schemes) is one of the largest sources of the errors in weather and climate prediction models (MoES 2015). Therefore, it is important to improve physical parameterization schemes in the models. This is an important, but difficult task. Development of new physical parameterization schemes or improvement of existing schemes will need many observational data of physical processes. These data are used for testing a physical hypothesis, validating physical processes in the models and also for tuning empirical constants used in the physical parameterization schemes. Therefore, observational campaigns, physical process studies and model development should go together.

As mentioned earlier, the weather and climate services have improved over the time due to investments in observations, modelling, and specialized trainings. At present, there is a skilful short range prediction of weather up to 3–5 days over the

Indian region. Since short range forecasts depend on initial conditions, improvement in initial conditions can make a visible improvement in prediction skill. Therefore, observational network (conventional, satellite and Radar) needs to be strengthened and the data assimilation methods to ingest the observations need to improve. More stress is needed to ensure good quality of data and their timely reception for data assimilation. The current skill level of 3–5 days can be improved to 7–8 days by employing an ensemble approach to forecasting in addition to the deterministic forecasting system at very high horizontal resolution of about 10 km.

Significant research has been carried out in the past to enhance the understanding on Indian monsoons, however, challenges and uncertainties still remain with respect to the accuracy of monsoon predictions and precipitation forecast skill improvements across different time-scales (daily, sub-seasonal, seasonal). A comprehensive understanding of the monsoon involves the study of all aspects of the physical climate system i.e., atmosphere, ocean, land, and cryosphere. The potential predictability of the seasonal Indian summer monsoon rainfall is found to be around 0.70. We need to address the issues that are required to improve the present skill of 0.55 to the potential skill.

The users also need forecasts for much smaller spatial and temporal scale, for example at district level. Even though seasonal prediction of district level rainfall may not be feasible, attempts can be made to predict seasonal rainfall over a sub-division, especially over the central plains of India. Extended range prediction system provides very useful products at time scale of 15 days. This skill needs to be improved beyond 3 pentads. At present, there are very limited applications with the extended range prediction system. The potential areas in which these forecasts can be applied are agriculture, water resources/management, power, energy and health.

On many instances of extreme rainfall having potential of causing floods, flooding events are missed causing loss to economy and life as the meteorological forecasts are not readily usable by various stake holders. Therefore, there is an urgent need to improve and customize meteorological forecasts specifically for floods. For formulating the flood forecast in the real time, the observed meteorological and flow data are used into the calibrated and validated real time flood forecasting model to forecast the flood flow and corresponding water levels for different lead periods varying from few hours to few days depending on the size of catchment and purpose of the forecast.

Accurate forecasts of the intensity of Tropical Cyclones (TCs) still remain a gap area which sets the basis for research in improving the model. There is also scope to further improve prediction skill in predicting landfall and associated rainfall, storm surge and winds. The recent research analysis suggests that models do not perform well in predicting associated heavy rainfall during the landfall of tropical cyclones.

Large research gaps include the understanding and prediction of thunderstorms and associated adverse weather. Denser Radar network not only would help detecting mesoscale convective system but also would help constraining the model parameters for better representation of convection. Assimilation of meso-network observational data would be necessary to generate mesoscale analyses/reanalyzes representing the regional heterogeneity. Predictability of convection initiation and scale interactions associated with superposition of micro-scale, meso-scale, and synoptic-scale processes needs to be studied. MoES has translated the high resolution weather forecast into lightning probabilities. This was one of the most important gap areas of forecasting. However, it needs improvement in accuracy and spatial resolution.

We need to improve in predicting prolonged heat and cold wave spells and improve forecast accuracy of flash flood events over the urban and complex topography. Another detrimental weather phenomenon during Indian winter months is the dense fog formation over the Gangetic Valley which has very high damage potential (physical as well as economical). Objective tools for short- and medium-range forecasting of timing of onset, intensity and duration of fog well in advance falls into one of the largest gap areas. A comprehensive understanding of micrometeorological and chemical processes responsible for the haze or fog formation over the Ganges Basin and fog impact on health and ecosystems is not clearly addressed yet. High-resolution mesoscale model, coupled with interactive aerosol model with chemical details of species is also the need of the hour.

Forecasting urban weather is increasingly important to manage disasters, decision making in the public sector, and for urban planning purposes, etc. Recent introduction of urban development plans such as Atal Mission for Rejuvenation and Urban Transformation (AMRUT), Smart Cities Mission, and Housing for All (Urban) by Prime Minister are modes to increase the opportunity for urban centers and especially urban areas are considered as "growth engines". India is poised to have 100 smart cities in 5 years and 2 crore houses will be constructed in next 7 years. These urban development plans also should be supported by adequate weather related support systems for information, management and mitigation of urban hazards such as heat wave, flash floods, heat islands, air pollution dispersion, etc.

3.7 Conclusion and Way Forward

The management of meteorological disasters will get improved due to advancement in forecast and warning services including accuracy, lead period and communication and dissemination. World Conference on Disaster Risk reduction (WCDRR) during 2015 at Sendai has set a target to substantially increase the availability and access to multi-hazard early warning systems and disaster risk information and assessments to the people by 2030. MoES and IMD continuously expands and strengthens activities in relation to observing strategies, forecasting techniques, disseminating methods and research relating to different aspects of extreme weather events to ensure most critical meteorological support to disaster managers and decision makers.

Demands from the public/private/government sectors are increasing for more accurate prediction of weather and climate at various temporal and spatial scales due to possible impacts of global climate variability and change. Improved and reliable forecast of weather and climate requires high resolution dynamical models. Thus, intensive monitoring of various weather systems through different platform based observing systems provide not only the necessary information about current weather systems, their effective assimilation in numerical models provide important guidance for accurate forecasts. Hence, there is scope to improve further the climate and weather services by providing high resolution customized impact based forecast and risk based warning products for different existing and emerging socio-economic sectors. There is need to:

Strengthen the observational network (conventional, satellite and Radar) on urgent basis and to ensure good quality of data and their timely reception for data assimilation into numerical prediction models.

Improve quantitative precipitation forecasts (QPF) for river basins and major cities.

To develop a meteorological support system for flood warning and forecasting and

To improve the intensity forecast of tropical cyclones (TCs).

To improve in predicting anomalous/unprecedented severe weather episodes, viz., prolonged heat and cold wave spells, thunder-storms spells, dense fog etc., and improved forecast accuracy of these episodes in particular for flash flood events over the urban and complex topography.

To further enhance the understanding on Indian monsoons and to improve the forecast skill of monsoon predictions and precipitation forecast at different time-scales (daily, sub-seasonal and seasonal).

Develop an Advanced Weather Prediction System, for block level forecasts, skill-full for next 3–5 days and develop advisories for sectors like Agriculture, Disaster Management, Water resources, Power, Tourism and Pilgrimage, Smart Cities, Renewable Energy Sector and Transport. Users, especially farmers need forecasts at least at block level for better planning of agricultural activities and renewable energy sector need forecasts at minutes, hour and day-ahead time scales.

Develop an Advanced Regional Climate Services Framework to cater to the needs of Agriculture, Water Resources, Alternate Energy Resources and Health on the basis of location specific and user specific knowledge of the climate variability and predictions of climate variables. Worldwide, there is a lot of emphasis on developing/strengthening advanced climate services for various applications.

Carry out cutting-edge research studies on climate variability including monsoons, meteorological droughts, desertification, land use changes using observations and modeling.

Develop Urban Meteorological Services with dense Meso network, observational and modeling strategy to cater to the needs of growing cities in the country.

Augment HPCS and data archival/storage facilities for MoES institutes to achieve the above objectives, particularly on modeling and observational data storage/ archival.

There is also scope for improvement of (1) sectorial applications of early warning, (2) warning communication to last mile and disaster managers through state of art technology, (3) synergized standard operation procedure among the early warning agencies and user agencies and (4) the link between early warning service provider and disaster managers.

Acknowledgement The author is thankful to the reviewers for their valuable comments for modification of the manuscript. The author also acknowledges Mrs. Monica Sharma, Meteorologist, IMD, New Delhi for her support in preparation of the manuscript.

References

Benke M, Takle J, Pai DS, Rao SA (2019) Analysis of monsoon mission coupled forecasting system (MMCFS) model simulations of sub-division scale temperatures over India for the hot weather season (April–June). J Earth Syst Sci 128:182. https://doi.org/10.1007/s12040-019-1178-6

Department of Science & Technology (2005) Science plan: severe thunderstorms – Observations & Regional Modeling (STROM) Programme; Department of Science & Technology, Government of India, New Delhi, December 2005

Dimri AP, Chevuturi A, Niyogi D, Thayyen RJ, Ray K, Tripathi SN, Pandey AK, Mohanty UC (2017) Cloudbursts in Indian Himalayas: a review. Earth Sci Rev 168:1–23

Ghude SD, Bhat GS, TharaPrabhakaran RK, Jenamani DM, Chate PD, Safai AK, Karipot MK, Prakash Pithani V, Sinha PSP, Rao SA, Dixit S, Tiwari K, Todekar S, Varpe AK, Srivastava DS, Bisht P, Ali MK, Mina U, Dharua M, Rao YJ, Padmakumari B, Hazra A, Nigam N, Shende U, Lal DM, Chandra BP, Mishra AK, Kumar A, Hakkim H, Pawar H, Acharja P, Kulkarni R, Subharthi C, Balaji B, Varghese M, Bera S, Rajeevan M (2017) Winter fog experiment over the Indo-Gangetic plains of India. Curr Sci 112:767–784

Guhathakurta P, Pai DS, Rajeevan MN (2017) Variability and trends of extreme rainfall and rainstorms. In: Rajeevan MN, Nayak S (eds) Observed climate variability and change over the Indian Region, pp 37–50

IMD (2012) Standard operation procedure: weather forecasting and warning. Cyclone Warning Division, IMD, New Delhi

IMD (2013) Cyclone warning services: standard operation procedure. Cyclone Warning Division, IMD, New Delhi

MoES (2015) Vision for 2030. MoES, Government of India, New Delhi

Mohanty UC, Osuri KK, Routray A, Mohapatra M, Pattanayak S (2010) Simulation of Bay of Bengal tropical cyclones with WRF model: impact of initial and boundary condition. Mar Geod 33:294–314

Mohapatra M, Bandyopadhyay B, Tyagi A (2012) Best track parameters of tropical cyclones over the North Indian Ocean: a review. Nat Hazards 63:1285–1317

Mohapatra M, Bandyopadhyay BK, Nayak DP (2013a) Evaluation of operational tropical cyclone intensity forecasts over North Indian Ocean issued by India meteorological department. Nat Hazards 68:433–451. https://doi.org/10.1007/s11069-013-0624-z

Mohapatra M, Nayak D, Sharma RP, Bandyopadhyay BK (2013b) Evaluation of official tropical cyclone track forecast over North Indian Ocean issued by India meteorological department. J Earth Syst Sci 122:589–601

Mohapatra M, Sikka DR, Bandyopadhyay BK, Tyagi A (2013c) Outcomes and challenges of forecast demonstration project (FDP) on landfalling cyclones over the Bay of Bengal. Mausam 64:1–12

Mohapatra M, Bandyopadhyay BK, Tyagi A (2014) Construction and quality of best tracks parameters for study of climate change impact on tropical cyclones over the North Indian Ocean during satellite era. In: Mohanty et al (eds) Monitoring and prediction of tropical cyclones in the Indian Ocean and climate change. Springer, Cham, Capital Publishing Company, New Delhi

NDMA (2016) National Disaster Management Plan (NDMP), published by National Disaster Management Authority Government of India NDMA Bhawan A-1, Safdarjung enclave New Delhi

Osuri KK, Mohanty UC, Routray A, Mohapatra M (2012) The impact of satellite-derived wind data assimilation on track, intensity and structure of tropical cyclones over the North Indian Ocean. Int J Rem Sens 33:1627–1652. https://doi.org/10.1080/01431161.2011.596849

Rajeevan M, Nayak S (2017) Observed climate variability and change over the Indian region, Published by Springer, Singapore, 305 pp

Rao YP (1976) Southwest monsoon. Met Monograph, India Meteorological Department, 1/1976, 367 pp

RSMC, New Delhi (2018) Report on cyclonic disturbances over North Indian Ocean during 2012, published by cyclone warning division. IMD, New Delhi

Saha U, DevendraaSiingh AKK, ElissavetGalanaki AM, Singh RP, Singh AK, Chakraborty S, Singh R (2017) On the association of lightning activity and projected change in climate over the Indian sub-continent. Atmos Res 183:173–190

Sanjay J, Ramarao MVS, Mujumdar M, Krishnan R (2017) In: Rajeevan MN, Nayak S (eds) Regional climate change scenarios, in observed climate variability and change over the Indian region. Springer, pp 285–304

Singh C, Mohapatra BBK, Tyagi A (2011) Thunderstorm climatology over northeast and adjoining East India. Mausam 62:163–170

Srivastava AK, Kothawale DR, Rajeevan MN (2017) Variability and long-term changes in surface air temperatures over the Indian subcontinent. In: Rajeevan MN, Nayak S (eds) Observed climate variability and change over the Indian region. Springer, pp 17–36

Dr. Mrutyunjay Mohapatra is the Director General of Meteorology, India Meteorological Department (IMD) and the Head of WMO recognised Regional Specialised Meteorological Centre for Tropical Cyclones, New Delhi. He has 25 years of experience in weather forecasting and made significant contribution in modernization and improvement of early warning services of IMD. Dr. Mohapatra is known as Cyclone Man of India.

Chapter 4
Community Based Issues and Opportunities in Climate Change Adaptation and Disaster Risk Reduction

Hari Krishna Nibanupudi

Abstract Intergovernmental Panel on Climate Change (IPCC) has given enough evidences to justify the fact that current warming trend is the result of human activities. Climate Change has and will continue to have a profound impact on ecosystems, agricultural production and human well-being; in several ways that impose new threats and challenges to the survival and sustainable development of human beings. Climate Change is not only triggering an increase in frequency and intensity of cyclones/hurricanes, floods, associated land erosion and landslide hazard in some regions while it also causes prolonged periods of heat waves, water scarcity and droughts in other regions where there are reports of drastic depletion of underground water table, drying up of natural water springs and decreasing trends of streams discharge. Mitigation and Adaptation are the two most important approaches to counter Climate Change impacts. While efforts focus on carbon sequestration and greenhouse gas reduction, adaptation measures lend strength to resilience and coping capacities at several levels and scales. Powerful countries took decades to acknowledge the reality of Climate Change which is result of their relentless economic and industrial pursuits. After failing to deliver on the Kyoto Protocol, they finally came together again to deliberate and finally agree on suitable mitigation actions at 2015 climate conference in Paris. On the contrary, the simple farmers, fishers, coastal communities, hill tribes, nomads in the high altitudes in all parts of the world have felt and talked about the impacts of Climate Change, much before climate scientists and the IPCC alarmed the world about it. These local communities in remote corners of the world are at the front line to receive the impact of the Climate Change. They have been dealing with it on their own with their resilience, adaptation and coping strategies. In this background, this chapter discusses Community-based Adaptation (CBA) practices and suggests that the wisdom of simple communities should guide policy measures for boosting adaptation and resilient capacities.

H. K. Nibanupudi (✉)
International Expert on Climate Change Adaptation, Disaster Risk Reduction and Sustainable Development, Hyderabad, India
e-mail: hari.nibanupud@un.org

A. K. Gupta et al. (eds.), *Disaster Risk and Management Under Climate Change*,
Disaster Resilience and Green Growth,
https://doi.org/10.1007/978-981-99-4105-6_4

Keywords Community based adaptation · Climate change adaptation · Faith · Gender · Ecosystem · Disaster preparedness · Sustainable development · Kyoto protocol · IPCC

4.1 Introduction

Adaptation is defined as preparing for the impacts of observed Climate Change, which include actions to both reduce adverse effects of Climate Change and to exploit opportunities that might provide socioeconomic benefits (IPCC 2007). Adaptation processes are analysed based on autonomous, planned or natural. Autonomous adaptation is defined as actions taken by private parties, including communities on their own in response to Climate Change and planned adaptation actions are taken by both private parties and the Governance systems together in terms of policies and programs while natural adaptation is seen as actions that appear within the ecosystem because of Climate Change (Chambwera et al. 2014).

A wide variety of perspectives are discussed in defining community based adaptation. In a nutshell, community based adaptation can be understood as an approach that acknowledges peoples capacity in understanding crisis and development challenges, their resilience in dealing with them and their ability to over come them with suitable and proportionate external support and assistance, that is focused on participatory processes. This chapter will discuss a variety of ways, the communities demonstrate adaptive capacities, the Issues and challenges they place, the role of the external actors, including the Governments in creating enabling environment and boosting adaptive capacities with appropriate policies and actions.

4.2 Present Scenario

The IPCCs new report on global warming underlines that we are already seeing the consequences of 1 °C of global warming through more extreme weather, rising sea levels and diminishing Arctic sea ice, among other changes (IPCC 2018). The rising climate concerns of, communities, in the last few decades, in all parts of the world, corroborate these findings who have been experiencing the rapid changes in average temperature, unforeseen changes in the seasons, increasing frequency of extreme weather events and other Climate Change impacts. As scientists have long predicted, increasing global surface temperatures lead to more droughts and increased intensity of storms. Further, the excessive evaporation of water in the atmosphere adding fuel for more intense wind speeds and tropical storms (USGS).

Over the last twenty years, the majority (90%) of disasters have been brought by floods, storms, heat waves and other weather-related events taking 606,000 lives, with 4.1 billion people injured, left homeless or in need of emergency assistance (UNISDR 2015). South Asia is considered to be the most disaster-prone region, where, natural disasters have caused thousands of deaths, affecting over a billion people. In 2015, South Asia alone witnessed 52 disasters and 14,647 deaths—which is 64% of the global fatalities (UNESCAP 2015). India alone witnessed 10 extreme floods between 1998 and 2017, causing economic damage of approximately US$45 billion, killing over 27,000 people and affecting more than 370 million people. IN the same period, droughts impacted over 680 million people in the country (Devika 2019).

The hardest hit by these extremes are those who have least or absolutely no role in the global climate mess that has been created by the rich countries and communities in their reckless pursuit of unsustainable life styles. Different impacts of the climate have been dealt with by communities in different parts of the world with their unique ways of adaptation and inherent strength of resilience. United Nations Framework Convention on Climate Change (UNFCC) defines adaptation as the process, practices, and adjustments in social, ecological and economic domains to moderate the impacts of Climate Change (UNFCC, undated). Adaptation appears to be the only option for communities especially in the developing countries since the impacts of already damaged climate are going to be long lasting, even if further damages to the climate is minimized. At the same time, the adaptive options for the communities in the developing world are made tougher due to opportunistic trade and economic policies of the west and native country compounded by chronic socio-economic inequalities in the developing countries.

The recent spate of global commitments for climate adaptation has indeed been a sigh of relief. However, the pace of Climate Change is much faster than the speed of implementing the commitments, which is making adaptation more difficult and expensive. The simple communities world over, whose livelihoods are rooted in nature, soil and environment have been responding to these climate challenges in their own way with flood defenses, early warning innovations, crop diversifications, sustainable fishing, etc., without waiting for policy or financial incentives from national governments and international actors. Gupta (2007) explains on how "communities" living close to nature invariably evolve a language to understand and interpret the variations and discontinuities in nature. Tribal societies whether in Latin America or Asia or Africa have had a tremendously rich knowledge, based upon local resources and nature (Gupta 2007) and understanding climate-environment and life systems linkages.

For instance, the communities in the Koshi river basin in Bihar state of India, in the wake of inadequate policy and infrastructure support have evolved their indigenous strategies to cope with the seasonal and intense flooding. Before the onset of flooding, the communities erect heightened floor made of bamboo (machan) to live there during prolonged days when flood water stays in the villages during the period of flooding when they move to the machans, each family stock and carry food (roasted maize floor for longevity), water, and ration for 15 days, along with first aid kits prepared according to the household needs. Further, pastoral communities in many parts of the world have conserved indigenous livestock breeds for thousands of years through very careful evaluation and crafting of socio-cultural and economic institutions (Gupta 2007). Such socio-cultural-technological and institutional innovations evolved by communities for over centuries helped them withstand harshest weather and climate events, diseases, wars and other calamities.

However, such local climatic knowledge and adaptation capacities of communities are under tremendous stress for the last decade due to fast-paced worsening of climate coupled with reigning scientific and political leadership that for decades condemned anything indigenous to be obscure and unscientific and several government regulations that served, snatched the management of natural resources from the hands of local communities. In this backdrop, this chapter discusses a few options of community adaptation and highlights the issues of adaptation, adaptation practices, and challenges because of the rapid pace of Climate Change in the last decade.

4.3 Significance and Concept

4.3.1 Community-Based Adaptation

As discussed previously in the chapter, communities have evolved a wide variety of adaptation strategies to deal with the times of crisis caused by droughts, other disasters and shocks; with a sense of dignity, self, and community respect. For instance, following classification of community's portfolio adjustments to a variety of risks (Gupta 1988) shows that communities make best possible autonomous adaptation efforts to deal with the crisis before depending on planned and externally driven adaptation solutions (Table 4.1):

It's clear from the above analysis, externally driven adaptation approaches must not overlook community priorities, dignity, and capacities and should focus more on enhancing and enabling adaptation capacities than providing programmatically controlled short-term dole outs. As UNFCCC (United Nations Framework Convention on Climate Change) suggests, enhancing adaptation capacities depend not only on Government policies and agency programs but also on effective management of

Table 4.1 Communities Portfolio Adjustment to adapt to crisis situations (Gupta 1988)

Adaptation strategies	Adaptation actions
Intra-Household adaptation strategies	Asset disposal, migration, reduction in consumption, reallocation of resources and
Inter-Household adaptation strategies	Labour, credit, land related bi-lateral contracts, exchange of favours and their impacts
Intra-Community adaptation strategies	Reliance on Common Property Resources, group ploughing, sowing, group trading, group resources of food, fuel
Cultural adaptation strategies	Drawing strategies and options from myths, folklore, sanctions on private profits during stress periods, use of traditional systems and calendars to anticipate and cope with risks, etc.
Planned-policy adaptation strategies	Public interventions such as drought or flood relief from public institutions, seed supply, disaster recovery, infrastructure development, community capacity building, etc.

community knowledge and resources. Further, parties to the UNFCCC and Paris agreement recognize that adaptation practices should be guided by the best available science combined with traditional, local and indigenous knowledge systems, ecosystem approaches and gender perspectives, for integrating adaptation into relevant socioeconomic and environmental policies and actions (UNFCCC 2018). It is therefore important to make adaptation interventions, recognize community interdependency with the ecosystem, their traditional knowledge, cultural strengths, while prioritizing on institutional technological and education type of interventions inclusive of the ecosystem, traditional knowledge. Further, in the context of rapid Climate Change events, the role of local knowledge, in mitigating the adverse impacts of Climate Change will become even more important.

The subsequent sections will discuss the five approaches of community-based adaptation for a resilient and sustainable response to Climate Change:

4.3.2 Ecosystem Links to Community Adaptation

It is increasingly evident that development interacts with nature and therefore with natural hazards. When development pursuits ignore this reality, they contribute to turning natural hazards into disasters and destroying lives, livelihoods, environment, infrastructure, enhancing hazard risks and socio-economic vulnerabilities. The environment and ecosystem services highly affect and determine human well-being. A robust and healthy ecosystem serves as an absorber of natural shocks by mitigating the intensity of natural hazards and reduces people's exposure to hazard risks. The environment can prevent or accelerate disasters depending on how the development policies and practices treat the environment.

The World Health Organization estimates that 24% of the global human disease burden is caused by the environmental factors whilst the Millennium Ecosystem Assessment clearly shows the close linkages between ecosystem services like provisioning (e.g., food, water and medicines), regulating (e.g., climate regulation and flood control), and cultural services (e.g., recreational activities and spiritual benefits) and human well-being (MEA 2005). In other words, human well-being is intimately linked to the state of an ecosystem that sustains life.

Community well-being and adaptive capacities are also directly linked to the climate, changes in climate with impacts on the environment. The way humans interact with the ecosystem for survival, progress and recreation has continuously changed over time. However, pace and magnitudes of human interaction with the ecosystem have increased over the past century and have even more rapidly intensified in the last five decades. These changes have caused a profound impact on the climate which in turn affected ecosystems and changed ecosystem services ultimately challenging the adaptive capacities of the communities.

For instance, research into the negative impacts of Climate Change on land and environment and further impact on the well-being of Chepang communities in Nepal has shown that diminishing of wild food that constitutes a vital source of nutrition for

Chepang communities, due to environmental degradation resulted in widespread malnutrition. Further, environmental degradation has also driven Chepang communities to previously unexplored urban territories exposing them to air and water contamination and resulting in a variety of water and airborne diseases, including diarrhea. Similarly, research conducted in Tibet has shown that degradation of grasslands has not only created health hazards to the local communities but also to the ecosystem due to abruptly changed lifestyle of local communities (Nibanupudi H., Climate Risks and Eco system health based adaptation in the Hindu Kush Himalaya Mountain Region, 4th International Disaster and Risk Conference, 2012).

The environment that is sensitive to Climate Change can reduce or aggravate disasters, depending on how development policies and practices treat the climate. Therefore, the outcome report of Rio+20 called for DRR to continue to be addressed in the context of sustainable development and placed within the post-2015 development agenda. It also called for increased coordination among national, regional and international levels for a robust response to environmental emergencies and improved forecasting and early warning systems and their integration into development policy.

4.3.3 Faith Is Part of Community Adaptive Capacity

Faulty interpretation of modern science in the education system made us believe that faith and traditions are regressive. True, faith and traditions can obstruct scientific thinking, progress, and the pace of development to an extent. But it's wrong to believe that all that humans want are material progress and logical thinking in every sphere of life. Faith continues to be the biggest healer and harbinger of hope for human beings whose life is riddled with so many questions and challenges that science can't answer. There are a few traditions linked to faith in different parts of the world that teach us the value of sustainability and tell us the way out from the climate crisis. These traditions show the power of faith and spirituality in connecting communities across religions in building and enhancing adaptation to climate challenges. Many traditional cultural practices are born as an adaptive mechanism to local natural variances and hazards. For instance, communities in Indonesia continue cultural practices for thousands of years that are rooted in the context of the local ecosystem. These practices are born out of a scientific understanding of local ecosystem and aimed at protecting the ecosystem, adapting to local environmental challenges, deriving livelihood benefits from the ecosystem and being protected by ecosystem from climatic shocks.

Further, spiritual teaching and cultural practices that are rooted in faith also play an important role in promoting livelihoods and lifestyles that are in harmony with nature and contribute to ecosystem resilience to extreme natural events. Therefore, faith, spirituality, and cultural heritage are important social capital for sustainable livelihoods, CCA and disaster resilience. In many parts of the world, ecosystems are recognized as sacred places, for example, rivers, forests, and mountains. These places are historical, ecologically and culturally interconnected with local people living and depending on those environments. Respect and faith in these sacred places are derived from the understanding and relationship of humans with nature. Therefore, usually many places of worship are built in the lap of nature. Consequently, investments in eco-system-based development will provide multiple benefits for increasing resilience to climate shocks, enhancing adaptation, livelihoods, human well-being, and ecosystem health.

Ancient Indian traditional wisdom signifies tree as an important resource for flood risk reduction. This tradition tells people to worship trees as it provides them with fruits, cover them from heat, dust, and rains. Even when floods uproot it, it still helps humans and animals to survive by helping them to float. In Indonesia, *Beringin* tree is considered sacred. People believe that gods live on it. Ecologically this tree grows in arid and drought-prone areas and is vital to the local ecosystem. Local traditions forbid cutting of this tree and people believe that God will curse if anybody harms this tree. Similarly, Bodhi tree has a great ecological significance in low and mid hills areas. This tree helps hold the soil, prevents landslides and increases water table. Buddhism forbids cutting this tree as it considered sacred for its association with Lord Buddha. Buddhism, as practiced in Sri Lanka, says—Before cutting any tree for livelihood needs, the cutter should plant ten trees first. Otherwise, Godwill curse. However, the misguided and arrogant belief that humans are the masters of all natural resources on the earth is a major reason for the way we have been abusing the nature all these years and have invited a catastrophe upon ourselves.

Faith leads a community to see the light amidst darkness, to see hope when there are calamity and destruction all around. Such a hope triggered by the faith also forms a vital element in communities' adaptive capital and strengthens the resiliency of the people which allows them to bounce back with a positive outlook of life. Further, in the case of any disaster events, much before professional humanitarian organizations step in, the affected communities are supported with preliminary relief support and much needed psycho-social support by the local faith institutions such as the leadership of Churches, Temples, and Mosques with their resources and infrastructure. Communities trust them, feel comfortable with them and most importantly find much needed connection with them, as these institutions have been very much part of their adaptive capital.

The faith workers, who may be called priests, religious teachers, etc., have much-needed motivation to associate with communities, strengthen their adaptive capacities as they see it as a divine duty. Most importantly the priests who are the representatives of God can play and indeed are playing an important role in many countries as powerful motivators for climate action and adaptation. In this backdrop, it is no surprise that Pope Francis in his 2015 encyclical Laudato Si′ underlines the importance of social and cultural adaptive capacities in protecting world climate. Pope Francis says that the sustainability of physical, moral, spiritual and social ecology hinges on culture and moral conditions created by it (SAPIENS, 201). The ill-informed education systems and standardized tools of development wrongly orient development practitioners to condemn faith (that is eco-friendly) as an obstacle to human progress. These standardized approaches drive the development practitioners to "listen to what they already assumed" and ignore how faith is a lot more powerful in the lives of communities than the financial resources that agencies mobilize for them.

4.3.4 Women's Adaptive Capacity Is a Key to Deal with Climate Crisis

It is wrong to believe that women are born vulnerable. Women are great survivors, adapters and resilient protectors of families against natural shocks as well as societal cruelty. Women's connect and knowledge of nature and environment make them front-line soldiers to deal with the impacts of Climate Change. Despite the negligence and marginalization, women continue to be the agents of adaptation to environmental change. Women play a key role in the management of water, watersheds, forests, and other common pool resources (Meinzen-Dick et al. 2011). Women and other marginalized groups hold indigenous knowledge of low-impact, low-cost methods and coping strategies that can prove appropriate in strengthening capacity for resilient farming practices in response to Climate Change. Different other studies (Mitchell et al. 2007; Panda 2007; Van Koppen and Hussain 2007) have also shown the women's negotiation skill in the public, private, and community spheres of social life and their coping strategies for dealing with downward mobility, and how they accumulate new roles and balance these additional roles through a process of negotiation.

Women are considered the managers of natural resources with the knowledge and skills that are critical for sustaining the environment. Though they are often excluded and under-represented in decision-making institutions and policy processes regarding Climate Change, women are active agents who have developed locally adapted, appropriate and sustainable coping strategies and responses within the scope of limited access to resources and disadvantageous gender power relations (Nellemann et al. 2011).

Women and girls have always maintained a complex balance of multiple roles, both traditional and non-traditional. Meinzen-Dick and Zwarteveen states the involvement of women in water user organizations in South Asia, enhanced the effectiveness of water resource management. A study from Bangladesh revealed, instead of growing paddy which requires more water, women grew off-season vegetables and fruits in a home garden that fetch income (Mitchell et al. 2007).

With their valuable knowledge about ecosystem, ability to harness the environment sustainably, technical knowledge and skills in agriculture, livestock caring and management of water for domestic as well as for agricultural uses and managing all these alongside domestic chores and dealing with day-to-day shocks in all these domains, women have been the core capital of adaptation for human societies for many hundreds of years.

There are several studies that documented the cases of women, who have shown enormous adaptive capacities despite the disproportionate amount of challenges they face after disasters. In the absence of an overarching international legal framework that protects the rights and dignity of disaster-affected populations, the fate of millions of climate and disaster-affected women around the world depends on the discretion of local authorities, the capacities of local agencies, and existing short-sighted management policies and practices. Common models of climate and disaster intervention have yet to adequately recognize women-specific needs and their adaptive capacities in the aftermath of climate extremes, including sexual abuse, trafficking, unavailability of compensation for home-based livelihood loss, and maternal and reproductive health care requirements.

Gender research studies, conducted in Tamil Nadu, India, in the aftermath of Indian Ocean Tsunami in 2004, reveal several cases, where men spent all the compensation money they received on personal pleasures, such as drinking, gambling, etc. while women used every cent received on livelihood and food security needs of the families. To meet the basic needs of the families many single and widowed women doubled their labor work, reared kitchen gardens, grew and sold vegetables, etc. (Pincha 2008). Similarly, when Koshi River floods in 2008, caused sedimentation of thousands of hectares of croplands in Nepal, men migrated to cities and some to other countries to find work. The women who were left behind decided to reclaim the lost soils from the flood sedimentation. Without the help of engineering or any other support from the Government, hundreds of women toiled hard to reclaim at least a portion of fertile soil so that they can restart the agriculture to meet family's food security needs (Nibanupudi H., Climate Risks and Eco system health based adaptation in the Hindu Kush Himalaya Mountain Region, 4th International Disaster and Risk Conference, 2012).

However, common models of post-disaster intervention have yet to adequately recognize women-specific needs and vulnerabilities in the aftermath of the disaster which includes a greater risk for abuse and trafficking, lack of compensation for the loss of home-based livelihoods, and maternal and reproductive health care requirements that are often ignored. They have also failed to capitalize on women's unique capacities for risk reduction and disaster recovery.

Further, several dynamics make adaptation more difficult for some women due to lack of access to formal education, economic poverty, discrimination in food distribution, food insecurity, limited access to resources, exclusion from policy and decision-making institutions and processes and other forms of social marginalization.

Although their contributions are often overlooked, and they experience disproportionate challenges after climate and natural disasters, women continue to play a key role in preparing families and communities deal with climate extremes and recover with dignity. Therefore, it's vital to recognize and respect women's adaptation capacities and tap into women's knowledge of environmental resources, enhance their livelihoods skills and empower them to have a greater space in the policy and polity to reverse vulnerability and strengthen adaptive capacities. Community-based adaptation and disaster risk reduction initiatives should be used as opportunities to increase women's access to and control over resources at all levels. Access to finance and asset creation for women is vital to breaking the vicious circle of gender-based discrimination of women in all sphere of life.

4.4 Strategies and Management

4.4.1 Community-Based Disaster Preparedness (CBDP), a Planned and Institutionalized Adaptation Approach

For Disaster Management what matters most is- time gap between the disaster event and the response that follows. Most of the loss of lives and properties occur during these crucial periods. Hence, leaving the responsibility of saving lives to the government agencies is not a wise and right way out for disaster management. The response to a disaster should start when the disaster strikes and if organized local group immediately strikes the action, the loss of lives and assets can be minimized. Without empowering local communities for the impacts of changing climate, disaster preparedness will always be less efficient and invite more loss of lives and capital. The risky habitations, poor housing and sanitation conditions etc. are the few reasons behind the vulnerability of local communities. Sustainability of disaster preparedness lies in addressing these vulnerable conditions and changing their context while organizing them to deal with disasters. The CBDP approach works for communities where it is needed, where there is a favorable environment, communities have some basic capacities and are willing to adopt the approach. These conditions must be complimented with a planned approach to skill building, community organization, stakeholder's participation and policy and institutional support (Fig. 4.1).

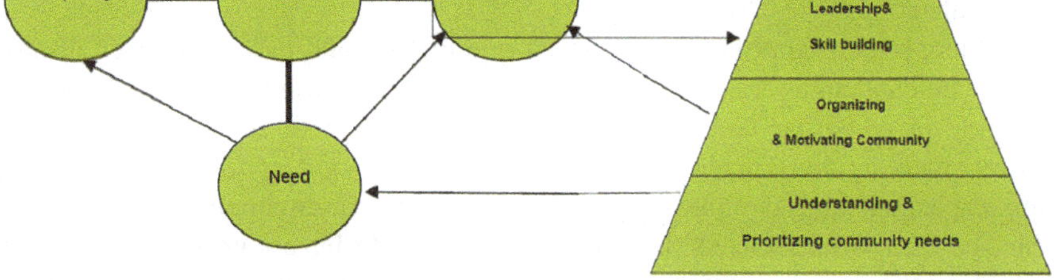

Fig. 4.1 Conditions for the successful adaptation of CBDP approaches (Nibanupudi 2004 (a))

Table 4.2 Successful CBDP adaptation approaches (Nibanupudi 2004)

Initiator	Focus	Strategies	Adaptation Activities
Oxfam in South India Since 1998	Address factors of Cyclone and Flood Disaster Risks	Community empowerment Multi-stakeholder involvement Vertical and Horizontal expansion of disaster preparedness	Contingency planning Task force groups formation. Livelihood promotion Disaster risk sharing Health, hygiene promotion, housing, and drinking water
UNICEF in West Bengal Since 2000	Address survival needs of communities in the first few weeks after massive flooding. Sustaining these initiatives with or without localized needs-based interventions	Create equal ownership of Government and Communities. Use Women's groups as an entry point. Integrate with polio eradication programs.	Contingency Planning. Task force groups formation. Family survival, child survival, and pregnant woman survival kits. Promotion of low-cost lifesaving kits, *Machans and elevated tube wells*
UNDP in India Since 2002	Local Level Risk Management	Capacity building to Institutionalize Disaster Risk Management with the involvement of local Government units	Multi-hazard preparedness and mitigation plans for Disaster Risk Management at all levels of Governance from state to village
CUBA Government Since 50 Years	Culture of Preparedness	A development model that reduces disaster risk	Appropriate legal framework. National and local level contingency plans. Creation of social and human capital Disaster preparedness in schools and colleges

Typically, a CBDP oriented adaptation program needs a survey of the villages, as the initial step. This survey should be based on the past experiences of the community when it faced disasters. This sharing is further ascertained in multiple community meetings followed by articulating the community's strengths and weaknesses concerning a disaster situation. Within this process, the community identifies threats and needs during a disaster and develops adaptation plans and delegate responsibilities among multiple groups. They approach the government and private institutions for necessary resources and other forms of support to make adaptation plan achievable. Further, CBDP oriented adaptation approaches are bottom up by nature but must have gender-equal leadership and participation, wider stakeholders support policy and institutional backing and should be pursued as integral to the local development agenda. There are several successful CBDP approaches in India and other countries (Table 4.2) all of which were built by meeting the conditions analyzed in Fig. 4.1.

As depicted in Table 4.2, CBDP oriented adaptation approaches can't be pursued in isolation either by the government, NGOs or by the communities. Community-Based Disaster Preparedness oriented adaptation approach requires collaborative action and commitment by all important stakeholders- governments, NGOs, communities, media, and the academic community. As amply demonstrated by Cuban experience, disaster preparedness helps reduce risks comprehensively only when it is adapted into daily lives consciously and consistently. The above approaches further highlight the importance of adaptation approaches like CBDP to be integral to our development planning. Disaster preparedness should become a culture and a conscious practice. It is necessary that potential hazard risks are determined, and risk management approaches are considered while designing development plans i.e. poverty reduction strategies/socio-economic development plans. Communities provide much-needed commitment and creativity in developing strategies for integrating adaptation needs into specific development plans and practices.

4.5 Conclusion

Communities at the front line of Climate Change demand nothing, for; they can't rely on uncertain global support system or ineffective governance in their countries. They continue to evolve their own adaptation mechanism to deal with emerging challenges. As discussed in the chapter, communities adaptive capacity is evolved through generations of communal living, collective spirit, bonding with eco system, joint planning and resource management, faith and acceptance of challenges of life and most importantly the critical leadership roles women play in holding the families and communities together and providing invisible leadership in creating resilient communities.

However, just because, they are adaptive and resilient to deal with crisis silently and peacefully, communities should not be left to fend for themselves. Conference of Parties (COPs) seventh meetings from 2001 in Marrakesh to the COP-21 meeting in Paris have acknowledged adaptation needs of communities in the developing countries and have created finance, technology and capacity to support community adaptation. These initiatives include the National Adaptation Plan, Local Adaptation Plan, and Least Development Countries Fund with technical support arrangement called Least Developed Countries Group, Nairobi Work Program, Green Climate Funds, etc.

However, whatever has been committed or made available for climate action is far less than what is needed and what has been agreed. Developed countries are far from fulfilling the pledge of $100 billion they made at the Paris convention in 2015. Although the Paris agreement has gained support from almost all the countries in the world, the most powerful and most polluting nation, the USA was able to withdraw from it because of the absence of legal binding.

4.6 Way Forward

It is high time that Climate Change Adaptation and mitigation actors, humanitarian and DRR institutions push for an international climate legal framework on the lines of human rights law that has the teeth to push for the rights and dignity of the climate affected people. The available funding for Climate Change Adaptation should focus on boosting adaptation capacities of communities and also reducing their exposure to the ever-increasing risk of climate extremes.

Greater emphasis should be on improving women's livelihoods, and strengthen adaptation by ensuring women's access, control and ownership of resources (such as land, livestock, property and income opportunities) and access to development resources such as credit, information, training and outreach, and culturally appropriate and labor-saving technology. Further, there should be designated focal institutions at all levels of the government to streamline and coordinate community-based adaptation planning coming from multiple agencies to optimize on the investments and harness collective knowledge, technology and resources.

References

Chambwera M, Heal G, Dubeux C, Hallegatte S, Leclerc L, Markandya A, Neumann J (2014) Climate change 2014: impacts, adaptation, and vulnerability. Part A: global and sectoral aspects. Contribution of working group II to the fifth assessment. IPCC, New York

Devika S (2019) From droughts to floods: the cost of climate change to India continues to mount. Acclimatise News, 03 September, 2019. https://www.acclimatise.uk.com/2019/09/03/from-droughts-to-floods-the-cost-of-climate-change-for-india-continues-to-mount/

Gupta AK (1988) Survival under stress: socio ecological perspective on farmers' innovation and risk adjustments, W.P. No. 738, 1988. International Congress on Plant Physiology revised version published in Capitalism, Nature and Socialism, pp 79–96

Gupta AK (2007) Indigenous knowledge and innovations for managing resources. Institutions and Technologies Sustainably: A Case of Agriculture, Medicinal Plants and Biotechnology. Indian Institute of Management Ahmedabad, Research and Publication Department, IIMA Working Papers

IPCC (2007) Climate change 2007: working group II: impacts, adaptation and vulnerability. Glossary. IPCC. Retrieved from https://www.ipcc.ch/publications_and_data/ar4/wg2/en/annexessglossary-a-d.html

IPCC (2018) Summary for policymakers. In: Masson-Delmotte V, Zhai P, Pörtner H-O, Roberts D, Skea J, Shukla PR, Pirani A, Moufouma-Okia W, Péan C, Pidcock R, Connors S, Matthews JBR, Chen Y, Zhou X, Gomis MI, Lonnoy E, Maycock T, Tignor M, Waterfield T (eds) Global Warming of 1.5°C. An IPCC Special Report on the impacts of global warming of 1.5°C above pre-industrial levels and related global greenhouse gas emission pathways, in the context of strengthening the global response to the threat of climate change, sustainable development, and efforts to eradicate poverty. Cambridge University Press, Cambridge, pp 3–24. https://doi.org/10.1017/9781009157940.001

MEA (2005) Millennium Ecosystem Assessment, 2005. In: Ecosystems and human well-being: synthesis. World Resource Institute, Washington, DC

Meinzen-Dick RAQ, Behrman J, Biermayr-Jenzano P, Wilde V, Noordeloos M, Beintema N (2011) Engendering agricultural research, development, and extension. International Food Policy Research Institute, Washington, DC

Mitchell T, Tanner T, Lussier K (2007) We know what we need: South Asian women speak out on Climate Change adaptation. IDS, Sussex

Nellemann C, Verma R, Hislop L (2011) Women at the frontline of climate change: gender risks and hopes. A rapid response assessment. United Nations Environment Programme, GRID-Arendal

Nibanupudi H (2004) Culture of disaster preparedness. A monograph. Oxfam India Trust, Hyderabad

Nibanupudi H (2012) Climate risks and eco system health based adaptation in the Hindu Kush Himalaya Mountain region, 4th international disaster and risk conference. IDRC, Davos

Panda S (2007) Mainstreaming gender in water management: a critical view. Gender Technol Dev 11(3):321–338

Pincha C (2008) Indian Ocean Tsunami through the gender lens: insights from Tamil Nadu, India. I. Oxfam America and NANBAN Trust

UNESCAP (2015) Disasters in Asia and the Pacific: 2015 year in review. United Nations Social Council for Asia and Pacific, Bangkok. Retrieved from https://www.unescap.org/sites/default/files/2015_Year%20in%20Review_final_PDF_0.pdf

UNFCCC (2018) What do adaptation to climate change and climate resilience mean?

UNISDR (2015) *The human cost of weather-related disasters 1995–2015.* United Nations International Strategy for Disaster Risk Reduction. Geneva & Center for Research on the Epidemiology of Disasters (CRED), Brussels

USGS (n.d.) How can climate change affect natural disasters. The United States Geological Survey website. https://www.usgs.gov/faqs/what-are-long-term-effects-climate-change-1?qt-news_science_products=0#qt-news_science_products

Van Koppen B, Hussain I (2007) Gender and irrigation: overview of issues and options. Irrig Drain 56:289–298

Hari Krishna Nibanupudi is a Senior Adviser-Resilience, Climate Change, Anticipatory Action and HDP Nexus, World Food Programme. He is a specialist with over two decades of experience of working with communities, developing programs, advocating policies, building capacities and fostering regional cooperation. He helmed senior managerial and technical portfolios in South Asia, South East Asia, Middle East, and Africa with several International organizations such as Oxfam, World Bank Institute, ICIMOD, and United Nations Development Program (UNDP).

Chapter 5
Flood Management: Present Practices and Future Revisions Under Climate Change

Archana Sarkar

Abstract Flood disasters are the largest cause of economic losses in India in the recent years. Climate Change will influence stream flow patterns mainly through changes in the volume and timing of precipitation as well as snow melt from glaciers, and changes in the type of precipitation, snow or rain. The present design criteria/ policies of Flood Management in India do not account for the impact/scenarios of Climate Change on flood discharge in rivers and streams. The present chapter presents a brief review of the riverine flood problems of the country including the general practice of Flood Management, government initiatives and policies on flood and case studies attempting to study the impact of Climate Change on floods and Flood Management.

Keywords Flood management · Climate change · National water policy · Design flood · Return period

5.1 Introduction

Many of the observed changes to the global climate system since 1950s are unprecedented. These observed changes indicate warming of atmosphere and ocean, diminishing amounts of snow and ice, rising of the sea level, increase in the number of extreme precipitation events in various regions, decrease in the number of cold days and increase in the number of warm days and nights on the global scale. In respect of extreme precipitation events, there are likely more land regions where the number of heavy precipitation events has increased than where it has decreased. The increasing trend in extreme precipitation, which usually incurs extreme discharges in some watersheds, implies greater risks of flooding at a regional scale. Meanwhile, impacts of the climate-related extremes, such as extremely heavy precipitation, heat

A. Sarkar (✉)
National Institute of Hydrology, Roorkee, Min of Jal Shakti, Department of Water Resources, RD&GR, Govt of India, New Delhi, India
e-mail: archana.nihr@gov.in

A. K. Gupta et al. (eds.), *Disaster Risk and Management Under Climate Change*, Disaster Resilience and Green Growth,
https://doi.org/10.1007/978-981-99-4105-6_5

waves, droughts, floods, cyclones, thunderstorms and wildfires, may endanger some ecosystems and many human communities (IPCC 2014).

Floods include river floods, flash floods, urban floods and sewer floods, and can be caused by intense/heavy and/or long-lasting precipitation, snowmelt, dam break, or reduced conveyance due to ice jams or landslides. The severity of floods is dependent on various factors such as timing, volume and intensity of precipitation; antecedent river flow conditions; antecedent soil moisture and soil conditions in the drainage basin of rivers; antecedent snow and ice in snow-fed river basins; presence of manmade structures such as dams/dikes/reservoirs across the rivers; and extent of urbanisation within the drainage basin of the river. Encroaching of flood plains increases the damage potential of floods and there is a lack in the flood response plans.

5.2 Present Scenario

5.2.1 Global Trends

The damaging effects of floods are complex. The direct impact of floods can be seen in the form of damages to major infrastructure disrupting normal life in the form of damage to supply systems of electricity and water, disruption of roads, railway lines and airports affecting the transport system. The sewage disposal system also gets disrupted leading to health hazards. As such, the economic losses due to floods are much bigger than the figures calculated based on the direct damages to buildings and other infrastructures (Hallegatte 2015). Indirect economic losses in terms of damage to agriculture, horticulture and livestock etc. are typically spread well beyond the flooded area and may last much longer than the flood itself. A major flood disaster may affect the local as well as regional economy which in turn may affect the national economy seriously. Figure 5.1 shows the top 10 countries/territories of the globe in terms of absolute losses (billion US$) during 1998–2017.

5.2.2 Flood Issues and Existing Flood Management Mechanisms in India

In India also, many parts of the country face recurrent floods which cause damage to public property as well as loss of lives. As reported by Marchand et al. (2015), about five million people are exposed to riverine floods every year in India. However, the economic impact of floods is much larger due to its impact on the economic activities which in turn affect the national economic growth. It can be noticed from Fig. 3.1 that for India in absolute monetary terms, floods are responsible for majority of the losses amounting to US$79.5 billion, reflecting high asset values as well as frequent

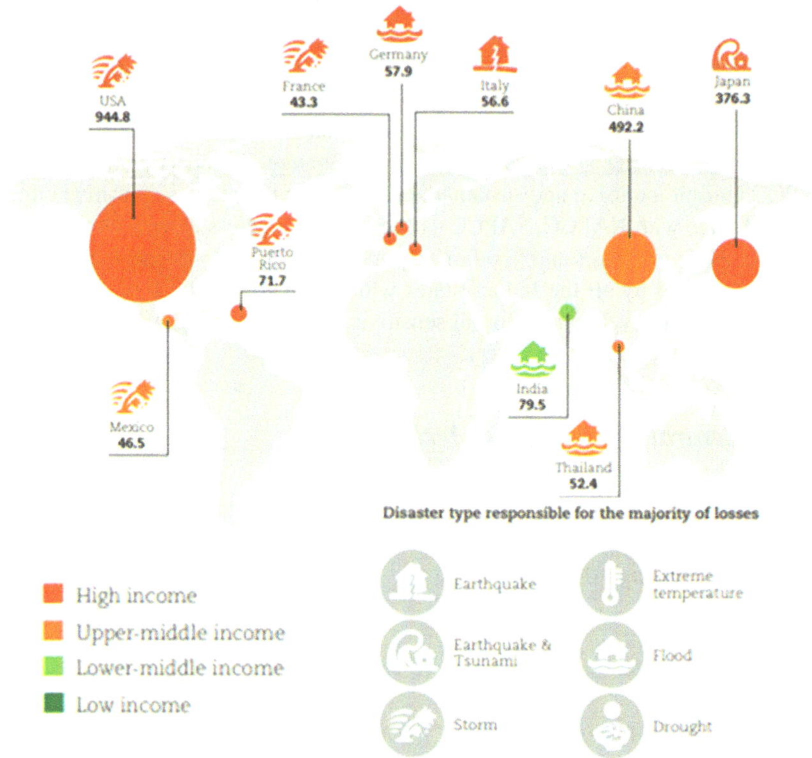

Fig. 5.1 Top 10 countries/territories in terms of absolute losses (billion US$) 1998–2017 (CRED and UNISDR 2018)

events. Besides, Indian continent has peculiar climatic conditions since it has floods in some parts whereas drought in other parts. India has a two tier Flood Management system exists briefly described in the following sections.

5.2.3 State Level Mechanism

State level mechanism includes the State Water Resources Departments, State Technical Advisory Committee and State Flood Control Board. However, in some of the States in India, the Public Works Departments and the Irrigation Departments take care of the flood matters. As per the Indian constitutional arrangements, Flood management is a State subject, therefore planning, funding and execution of various schemes for flood control are taken up by the State Governments themselves as per their own priorities out of their State plan funds which are made available to them through Planning Commission. Central Government provides an assistance which is technical, catalytical and promotional in nature (Marchand et al. 2015).

The State Specific Action Plan on Climate Change (SAPCC) is a recent initiative of the Government of India under the National Water Mission (NWM) which is one of the eight missions of the National Action Plan on Climate Change (NAPCC) under Prime Minister's Council of Climate Change established in 2008. The NAPCC aims at inclusive and sustainable development strategy to sensitive climate change. In line with NAPCC, SAPCC can help States address climate change issues including water resources and flood management. There is a common framework of SAPCC followed by all the Indian States which would be helpful in harmonizing national and state level actions for all sensitive sectors including water.

5.2.4 Central Government Mechanism

In order to enable the State Governments to take up the flood problems in a comprehensive manner, the Union Government has set up various organizations and expert committees described below:

5.2.4.1 Central Water Commission (CWC)

The Central Water Commission (CWC) was set up by the Government of India in 1945 with a mandate of furthering and promoting measures of conservation and utilization of water resources as well as flood control throughout the country in the areas of river conservation, Flood Management, irrigation and hydropower generation, and beneficial. CWC is also entrusted with the work of flood forecasting and warning in India. Presently, CWC has a network of 226 flood forecasting stations comprising of 166 Level Forecasting Stations for towns/important villages and 60 Inflow Forecasting Stations for Dams/Reservoirs (Central Water Commission 2018).

5.2.4.2 Brahmaputra Board

Under Brahmaputra Board Act, 1980 (46 of 1980), the Government of India set up Brahmaputra Board under the then Ministry of Irrigation (now Ministry of Water Resources, River Development and Ganga Rejuvenation). The function area of Brahmaputra Board includes all NE States in Brahmaputra and Barak Basin. Flood Management master plans for Brahmaputra and Barak Rivers have been prepared by the Brahmaputra Board. Survey and investigations for of master plan preparation addressing the problems of flood, erosion and drainage congestion including DPRs for multipurpose projects have also been taken up by the Brahmaputra Board.

5.2.4.3 Ganga Flood Control Commission

The Government of India set up the Ganga Flood Control Commission (GFCC) in 1972. GFCC has prepared Flood Management plans of the 23 sub-basins of Ganga Basin apart from implementing them in phased manner including updating of 22 plans once and 5 plans twice. GFCC carried out studies for assessment of the adequacy of existing waterways under road and rail bridges for 21 river system in Ganga Basin and apprised about 300 flood management schemes since XI Plan. GFCC has also monitored several flood management schemes implemented in the Ganga basin with central funding (GFCC website: http://gfcc.bih.nic.in/achievements.htm). However, impact of climate change on flood management has not been considered in any of the schemes/studies of GFCC so far.

5.2.4.4 Farakka Barrage Project Authority

Farakka Barrage is located in Murshidabad and Malda districts of West Bengal at about 300 km North of Kolkata. It is one of the largest barrage of its kind in the country having a Feeder Canal for a flow of 40,000 cusec (1135 cumec) whose bed width is wider than that of Suez Canal. The anti-erosion as well as the riverbank protection works by the project authority do not consider the impact of climate change in their design criteria.

5.2.4.5 National Disaster Management Authority (NDMA)

Under the Chairmanship of Hon'ble Prime Minister of India, the Government of India has set up a National Disaster Management Authority (NDMA) in 2005 for prevention and mitigation effects of disasters (including flood disasters) as well as for undertaking a holistic, coordinated and prompt response to any disaster situation. In January 2008, the NDMA has issued Flood Management guidelines and specified the roles of various Central and State agencies for preparation of flood mitigation plans including taking up of relief measures during flood disasters.

5.2.4.6 National Disaster Response Force (NDRF)

The National Disaster Response Force (NDRF) was constituted in 2006 under the Disaster Management Act for the purpose of specialized response to natural and man-made disasters. The NDRF has a unique distinction of being the only dedicated disaster response force in the world. Presently, NDRF consist of ten battalions.

5.3 Government's Initiatives and Policies on Floods

The Government of India took several initiatives after the unprecedented floods of 1954 and constituted a number of Committees to study the problem of floods in the country. Some of the important expert committees are listed below:

- Policy Statement—1954
- High Level Committee on Floods—1957 and Policy Statement of 1958
- National Flood Commission (Rashtriya Barh Ayog R.B.A)—1980
- Expert Committee to Review the Implementation of the Recommendations of National Flood Commission-2003 (R Rangachari Committee)
- National Water Policy (1987)
- National Water Policy (2002)
- National Water Policy (2012)

Detailed account of the recommendations of above committees can be found in various literatures.

5.4 Need of Revised Flood Management Practices

Flood damages in the future will be governed greatly by the decisions of land use plans, patterns of settlement, the accuracy of flood forecasting, the quality of flood warning and response systems, and the economic value of infrastructure located in vulnerable areas (Mileti 1999; Pielke and Downton 2000; Changnon 2005), as well as on climatic changes (Schiermeier 2006). Kleinen and Petschel-Held (2007) reported through their study of the impact of global warming on flood hazard vulnerability of various river basins of the world that the river basins likely to be affected by increased flood hazard by the 2080s house about 20% of the world's population.

In spite of all the measures undertaken in India, flood damages, losses and deaths continue to be an annually recurrent phenomenon and seem to be increasing. There are other contributing factors for the flood damages like societal changes, including demographic and economic growth and climate change. Moreover, Climate Change could become a key driver of change in population exposure to river floods in India in the coming decades. Many authors have carried out future climate studies for South Asia which are based on climate model simulation (Lal et al. 2000; May 2004; Meehl and Arblaster 2003; Rupakumar et al. 2006; Trenberth 2011). These studies point towards the intensification of monsoon precipitation, due to global warming over a broad region covering South Asia. A recent World Bank Report (2013) predicts that India will be threatened by a more erratic monsoon season, extreme floods, rising sea levels and very high temperatures due to Climate Change. Over past several years, many expert committees constituted by the Government have undertaken the flood studies and come up with Flood Management measures and plans. However, the trend of flood disasters and damages has been increasing over the past five decades despite the various measures undertaken posing challenge to the Government as well as to the people (Ahmad 2018). The approaches to Flood Management presently exercised in India also need to give a re-look to have an integrated strategy for policy and management related to floods with reference to Climate Change.

5.5 Causes and Drivers

5.5.1 Indian River Systems and Associated Flood Problems

In the context of flood problems, the Indian Rivers have been classified into four broad regions (Mohapatra and Singh 2003): (1) Brahmaputra Region; (2) Ganga Region; (3) North West Region; and (4) Central India and Deccan region. Figure 5.2 shows the flood prone area in various river basins of the country.

5.5.2 Brahmaputra River Region

The Brahmaputra River region comprises of the Brahmaputra and Barak River system. Seven States of the country namely, Assam, Meghalaya, Arunachal Pradesh, Mizoram, Manipur, northern parts of West Bengal, Tripura and Nagaland fall into this region. The catchments of these rivers receive very heavy rainfall ranging from 110 cm to 635 cm at various stations within the catchment (India-WRIS wiki 2015) a year which occurs mostly during the months of May/June to September resulting in very frequent and severe floods. The major flood problems in this region are the flooding rivers by overflowing their banks due to drainage congestion and tendency of some of the rivers to change their courses. River bank erosion in the Brahmaputra in recent years has assumed serious proportions.

5.5.3 Ganga River Region

The Ganga River region comprises of the Ganga River system. Ten States of the country namely, Uttarakhand, Himachal Pradesh, Uttar Pradesh, parts of Haryana, Rajasthan, parts of Madhya Pradesh, Delhi, Jharkand, Bihar, South and Central parts of West Bengal fall into this region. Major tributaries of the Ganga River region are the Ghaghra, the Gandak, the Kosi, the Yamuna, the Sone, and the Mahananda. The normal annual rainfall variation in this region is from 60 cm to 190 cm of which more than 80% falls during the south-west monsoon (India-WRIS wiki 2015). Similar to rainfall, the direction of increase of flood problems is from south to

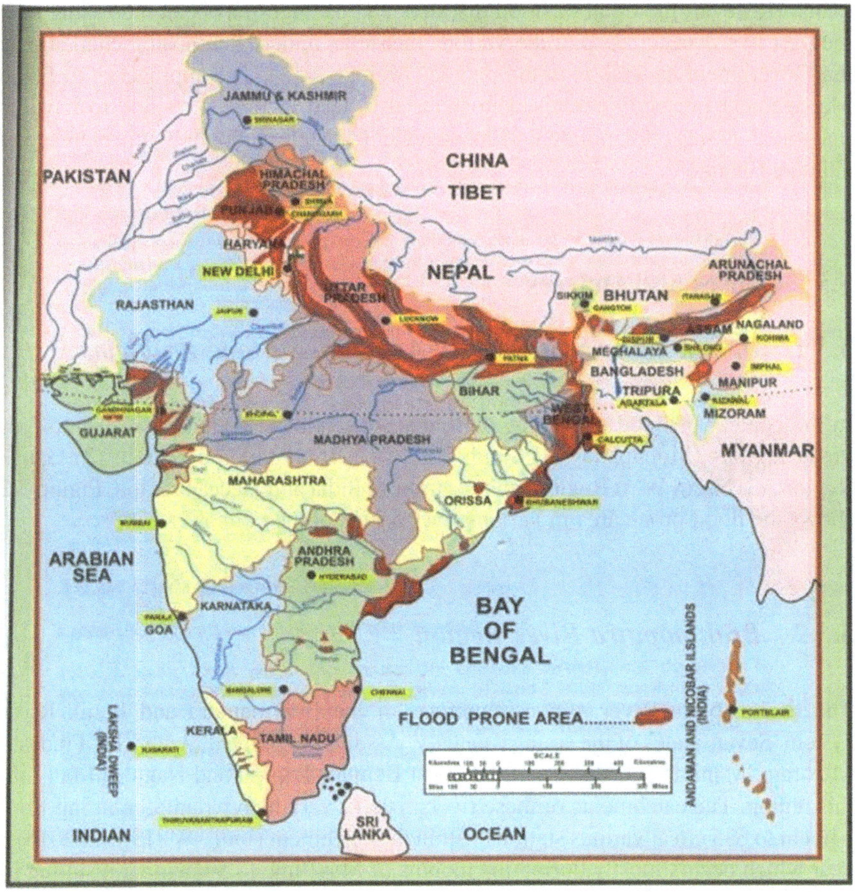

Fig. 5.2 Flood Prone Area (depicted by RED color). Source: India-WRIS wiki (2015)

north and from west to east. The problem of drainage congestion is prevalent in the North Western as well as some eastern parts of the region. On the other hand, the States located in the downstream of the Ganga region face serious problems of flooding and erosion.

5.5.4 North West River Region

The North West River Region comprises of the rivers Sutlej, the Beas, the Ravi, the Chenab and the Jhelum, as well as the tributaries of Indus. Five States of the country namely, Jammu and Kashmir, Punjab, parts of Himachal Pradesh, parts of Haryana and parts of Rajasthan fall into this region. All the rivers of this region originate in the Himalayas and carry heavy discharge during the monsoon period and also huge amount of silt. These rivers frequently change their courses and leave behind tracts of sandy waste. The flood problems in the North Western River region are less compared to the regions of Compared to the Ganga and the Brahmaputra River. Vast areas of this region face a major problem of inundation and water logging caused by inadequate surface drainage. There have been incidents of flash floods due to cloud bursts as well as heavy floods due to continued heavy rains in the J&K State of this region in recent past.

5.5.5 Central India and Deccan Region

The Central India and Deccan Region comprises mainly of the rivers namely, the Tapi, the Narmada, the Mahanadi, the Krishna, the Cauvery and the Godavari. The river courses of these rivers are mostly stable and well defined. Except for the delta area, natural banks of all the rivers have adequate to carry the flood discharge. However, the important rivers on the east coast have embankments in their lower reaches which largely eliminate the problems of floods. However, the recent floods in Kerala in August 2018 are an exception. As per the analysis carried out by the Central Water Commission (CWC 2018), the August 2018 flood in Kerala was due to severe storm occurrences during 8–9, August 2018 and 15–17, August 2018. It has been further clarified in the above report that the flood situation due to the 1-day rainfall of above 200 mm in most of the area and the continuation of such severe rainfall for another 3–4 days could not have been mitigated even with the 75%-filled reservoir conditions. Further investigations on Climate Change attribution have been taken up by many researchers.

5.6 Flood Management Measures

Various measures of reducing the flood losses and protecting of the flood plains have been adopted in India. The Flood Management measures can be broadly classified in two categories: (a) Engineering/Structural Measures; and (b) Administrative/Non-Structural Measures.

5.6.1 Engineering/Structural Measures

The engineering measures in the form of physical structures attempt to modify/control the floods which bring relief to the flood prone areas by reducing flood flows. Various engineering measures like reservoirs, detention basins, embankments, drainage improvement, channel improvement, diversion of floods interlinking of rivers etc. are briefly described below:

5.6.1.1 Reservoirs

Reservoirs are storage structures which help in controlling the river floods to a certain extent by moderating the intensity and timing of the incoming flood wave. Typically, the waters of high discharges in rivers are stored in the reservoirs and released after the cessation of the critical high flow condition thereby preparing them to receive the waters of subsequent flood waves. Flood protection efficiency of reservoirs depends on their available water storage capacity at the time when there is a flood wave and also on the distance/proximity of the reservoir from downstream infrastructure/population likely to be affected. Reservoir operation is based on a regulation schedule carefully planned to take into account the safety of the dam and related structures as well as the safe carrying capacity of the lower reaches of the river in their present condition.

Reservoirs with specifically earmarked flood space are more effective in Flood Management compared to the incidental flood moderation available for any type of storage on a river, as in the case of Damodar Valley Corporation (DVC) dams across the Damodar and its tributaries. Therefore, in light of the increased flood events due to heavier precipitation in recent past, for new dams, a separate flood cushion space should be provided as a policy. For existing dams, it is very important to review the rule curve/operation schedule prescribed for filling during monsoon period to ensure space for flood moderation but which can be filled for conservation later on when the condition of high flows is over. It has been reported by Tehri Hydro Development Corporation Limited (THDC) that the Tehri Dam reservoir (Fig. 5.3) was useful in absorbing part (Bhagirathi River flood) of the Uttarakhand flood of June 2013 and saved the downstream areas by moderating the flood discharge.

Fig. 5.3 Part of Tehri Dam Reservoir as seen from the Chamba-Chinyalisaur Road in Uttarakhand Himalayas, India

However, it is very important to have arrangements for an accurate inflow forecast for enhancing the reservoir efficiency and improving upon the schedules of reservoir operation for the provision of any specific/incidental flood moderation effects.

5.6.1.2 Detention Basins

The natural lakes and swamps/depressions can be used for flood moderation in the form of detention basins by constructing embankments all around besides increasing their storage capacity and making provisions for installation of suitable regulating structures for release of stored water as and when required. Detention basins are cheaper compared to other structures as the land under the low depressions/marshes may hardly require compensation and rehabilitation measures. A good example is from the Rajasthan State, the Ghaggar detention basin. Another example is from the Srinagar city of J&K State, where the depressions on the left bank of Jhelum River available in the upstream of the city have been utilized. Other examples of natural basins providing flood moderation are the hoars/beels of Barak River basin, Kotla lakes in Haryana and Mokama group of Tals in Bihar. With increasing frequency of heavy rainfall events and subsequent flooding, it is important for the State Governments to identify possible good sites for detention basins and have actionable programmes for making use of such identified sites.

5.6.1.3 Embankments

Flood damage can also be reduced by confining the flood flows of rivers within the river banks thereby preventing overbank flows by constructing embankments. Embankments have been the most popular structures for flood protection as they can be constructed quickly incurring comparatively low budget especially in the lower reaches of many large rivers. There are places where embankments are the only feasible option of flood protection by checking inundation. The design and construction of embankments is limited to provide flood protection against floods of certain intensity and frequency or limited to the maximum recorded historical flood at the time of design when detailed hydrological data of long periods are not available. In the case of the meandering rivers, hydraulic/physical modelling is very important for proper planning, design and implementation of embankments. In many of the flood prone States, strengthening and raising of existing embankments have also been taken. In order to adequately redesign the embankments, the flood frequency approach should be adopted taking into account the historical data of floods, which is now available.

Another serious problem which incurred huge expenditure (Marchand et al. 2015) in the past is the river bank erosion along the natural banks of rivers as well as the constructed embankments especially the embankment systems in U.P, Bihar, Punjab, Assam and West Bengal facing serious erosion problems. Such river

a. Details of Proposed Land Spur

b. Construction of Land Spur

c. Porcupines to Protect River Bank Erosion

Fig. 5.4 Engineering Protection Measures in Brahmaputra River in Guimara Futuri area, near Guwahati in Assam, India. (**a**) Details of Proposed Land Spur. (**b**) Construction of Land Spur. (**c**) Porcupines to Protect River Bank Erosion

embankments need to be protected by suitable anti-erosion measures. On some embankment systems, like the Piprasi-Pipraghat embankment in Bihar on River Gandak and the Kosi embankment also in Bihar on Kosi River, the river attacks are so relentless that the protection measures cannot be taken up in the normal maintenance work and require special anti-erosion programmes. Similar is the case with Brahmaputra River near Guwahati (Fig. 5.4).

While many of the embankment systems on various rivers have provided flood protection in a sustained manner, there are cases in some rivers, where the embankments actually had negative effects and exasperated the flood situation by drainage congestion arising from the rise in river bed levels, decrease in the carrying capacity of river channels and alterations in the gradient/level of the outfall points. Therefore, flood control by embankments may be undertaken only after carrying out detailed studies regarding their favorable and adverse effects including flood protection under Climate Change scenario of heavier precipitation vis-à-vis heavier flood discharges in the rivers.

5.6.1.4 Channelization of Rivers

For tackling the extensive meandering problem of rivers, some of the affected States are considering channelization of rivers, at least in identified reaches, which will train these river stretches into their original courses and also activate the navigational channels. However, the channelization of rivers must take into consideration the natural waterway of the River where it has a right to pass the floodwater and sediment load. During the XI Plan, major works of river channelization were taken up in the States of Himachal Pradesh and Punjab.

5.6.1.5 Channel Improvement

The method of improving the channel by improving the hydraulic conditions of the river channels by de-silting, dredging, lining etc., to enable the river to carry its discharges at lower levels or within its banks. These methods of channel improvement provide limited benefits but involve high cost along with other problems and therefore, could not be adopted extensively. Dredging in certain stretches of a river could be economically justifiable when taken up as an anti-erosion measure or channel improvement for navigation or clearing river mouth/narrow constrictions. In early 70s, dredging operations of River Brahmaputra were taken up on experimental basis but discontinued due to excessive cost and limited benefits.

5.6.1.6 Drainage Improvement

Surface water drainage congestion produces similar damages as produced by floods and sometimes it is difficult to distinguish between the two. When natural or artificial drainage channels are inadequate to carry the storm water discharge, then surface water drainage congestion occurs. By providing new channels or improving the existing drainage system, drainage can be improved and this technique is recommended as an integral part of the Flood Management programme in the country (India-WRIS wiki 2015). The natural systems like the 'dhars' or 'old channels' which are efficient in drainage of surface flows and the spillage generated by local rains, must be preserved. The blocking of these natural drainage channels as well as natural depressions in the process of land reclamation for development must be strictly discouraged. Also, in the areas affected by drainage congestion, a review of the adequacy of existing sluices and drainage channels must be undertaken and any inadequacies must be addressed. It is also required to re-visit the design criteria under of climate change and re-design of the network to accommodate the higher intensity floods.

5.6.1.7 Diversion of Flood Waters

By physical diversion of the flood waters, a part of the flood discharge is taken either to another basin or to the downstream of the problem area in the same basin. The flood discharge could also be taken to a depression where it could be stored and released subsequently. This type of measure is useful for management of unusual urban floods as done in the Srinagar City through nearby flood spill channel. Another application of this measure is in the lower reaches of a river in close proximity to Sea as done in the Krishna Godavari drainage scheme. Some more examples at various stages of planning and execution are the Damodar River in the lower reaches in West Bengal, the supplementary drain in Delhi, the Kolleru lake diversion into the sea in Andhra Pradesh, the Thottapally Spillway diversion in Kerala, the Hulwaa drain in

Uttar Pradesh, the Kama-Pahari drain in Rajasthan (India-WRIS wiki 2015). However, the flood diversion plan needs to consider the impact of climate change with studies using revised design flood and assessment of flood inundation area downstream.

5.6.1.8 Watershed Management

Watershed management can be undertaken by construction of check-dams, detention basins, and diversion channels, etc. for conservation and development of the soil and vegetative l covers within the watershed. In case of watershed management of upper catchment, land treatment through grass land development practices and afforestation need to be supplemented by structural works (physical structures) for reducing the velocity of water flow and arresting the sediment.

5.6.1.9 Interlinking of Rivers

The Inter-Linking of Rivers (ILR) is a proposed large-scale engineering project of the Government of India with the objective of effective water resources management in the country through linking the rivers by a network of canals and reservoirs thereby addressing the persistent problems of floods in some parts and water shortages in other parts. Under the National Perspective Plan (NPP) prepared by Ministry of Water Resources, River Development and Ganga Rejuvenation, Government of India, 14 links under Himalayan Rivers Component and 16 links under Peninsular Rivers Component have been identified for inter basin transfer of water based on field surveys and investigation and detailed studies. As the ILR is still in the DPR stage, it will be prudent to consider the impact of climate change on riverine discharge and any probable increase in the flood scenarios could be addressed by transferring such waters to deficit river basins.

5.6.2 Administrative/Non-structural Measures

The administrative/non-structural measures attempt to mitigate the flood damages by facilitating in-time evacuation of the people and shifting of their movable property to safer location through flood forecasting and flood warning. Mitigation of flood damages can also be achieved by discouraging creation of valuable assets/settlement of the people in the areas subject to frequent flooding i.e. enforcing flood plain zoning regulation. Other methods of non-structural Flood Management include flood modelling, disaster preparedness and response planning, flood proofing, flood insurance etc. The non-structural Flood Management methods are briefly described below:

5.6.2.1 Flood Plain Zoning

The concept of flood-plain zoning is based on flood plain management by considering the basic fact that the River's flood plain is essentially its domain and any intervention/development activity into must recognize the 'right of way' of the River. The objective of flood-plain zoning is demarcation of zones/areas within the flood plain likely to be affected by floods of different magnitudes or frequencies and probability levels, and specify the types of permissible infrastructure developments in each zone, so that damage can be minimised in the event of a flood, if not avoided.

Unfortunately, the flood plain zoning approach has been accepted in principle by all, very little consideration has been given to its actual practice, which is leading to an increase in flood damages. A model draft bill for flood plain zoning legislation was circulated by the union government in 1975 to all the states. However, there has been passive resistance on the part of the states to follow up the various aspects of flood plain management including possible legislation.

5.6.2.2 Flood Modelling

Flood modelling involves hydrologic, geomorphologic, and environmental analysis/modelling. Various hydrologic/hydraulic models for flood simulation have been developed and undergoing constant improvements. An understanding of the behaviour of the floods is required to evaluate flood characteristics and understand the underlying hydrological processes. Therefore, it is required have adequate knowledge on the flood such as occurrence temporally and spatially in order to mitigate the damages caused. Flood inundation modelling provides an ability to map the actual extent of inundation, timing and propagation as well as its intensity along with predicting or forecasting flooding extent. These models have also been integrated with latest technologies such as remote sensing and geographic information system (GIS) for visualisation and time series analysis.

5.6.2.3 Design Flood Estimation

Design flood estimation is required for the design of various flood protection engineering structures such as storm sewers, spillways, diversion works, bridges, culverts and other flood control works at a certain location in order to estimate the size and cost of those structures. From economic considerations, it is not practical to provide for the safety of various structures against maximum possible flood within the basin/catchment. Small structures such as culverts and storm drainages can be designed for less severe floods as the consequences of higher than the design flood may not be very serious. Failure of small structures can cause temporary inconvenience like traffic disruption but very rarely severe damage to property and/or loss of

Table 5.1 Criteria for classification of dams based on size and hydraulic head and Design flood for safety of dam

Classification	Gross storage (million m3)	Hydraulic head (m)	Inflow design flood for safety of Dam
Small	Between 0.5 and 10	Between 7.5 and 12	100 year flood
Intermediate	Between 10 and 60	Between 12 and 30	Standard project flood (SPF)
Large	Greater than 60	Greater than 30	Probable maximum flood (PMF)

life. On the other hand, storage structures such as dams need greater attention to the magnitude of floods used in the design. The failure of storage structures causes huge damage to property and loss of life in the downstream region. Therefore, the design criterion for choosing the flood magnitude is based on the type and importance of the structure as well as economic development of the surrounding area. For optimal design of water resources projects, applications of advance deterministic and probabilistic modelling based approaches for design flood estimation is essential. Most commonly used methods of flood estimation are: Rational method; Empirical methods; Flood frequency methods; Unit hydrograph techniques; and Watershed models.

The hydrologic design criteria for fixing spillway capacity are mentioned in IS 11223-1985, "Guidelines for fixing Spillway Capacity". The criteria for classification of dams is based on size of the dam and the hydraulic head (MWL—average flood level on downstream). The classification for the dam is greater of the two indicated by the two parameters as given in Table 5.1. The inflow design flood for safety of the dam would be as given in Table 5.1.

Design flood estimation using various flood return periods should be re-looked after considering precipitation scenarios based on various climate impact models while addressing the uncertainties at various levels of modelling. Therefore, design of all new water resources structures should consider the revised design criteria based on Climate Change scenarios. Design of existing structures need to be checked for their flood protection capacities by considering the Climate Change scenarios and the structures may be re-designed, if required.

5.7 Case Studies: Impact of Climate Change on Flood/Flood Management

Many studies have been taken up across the globe, which attempts to study the impact of Climate Change on floods as well as Flood Management. It was done focussing on different aspects of floods like, assessment of flood discharge in rivers under various scenarios of precipitation, temperature etc. and the associated flood damage, modelling of the changes in the recurrence intervals of floods and associated damages, assessment of the impact of Climate Change on high flows at basin scale using various RCPs. Some of the available work is reported in the following sections including highlights of the results of a study on the Ganges basin in which the author was also involved.

Schreider et al. (2000) estimated the impact of Climate Change on flood damages in Australia from modelled changes in the recurrence interval of present-day 20- or 100-year floods, and estimates of the damages of present-day floods as determined from stage-discharge relations [between gauge height (stage) and volume of water per unit of time (discharge)], and detailed property data. The authors using this methodology projected the average annual direct flood damage for the three Australian drainage basins to increase by a factor of 4–10 under conditions of doubled atmospheric CO_2 concentrations.

In a study of the 16 worldwide basins, Milly et al. (2002) illustrated that, for 15 out of the 16 large basins considered, the control 100-year peak volumes (at the monthly time-scale) are projected to be exceeded more frequently as a result of CO_2 quadrupling. The authors found that in some of the study areas, the present 100-year flood (in the control run), is projected to occur much more frequently, even every 2 to 5 years, although with a large uncertainty.

Choi and Fisher (2003) studied some selected regions in the USA in order to estimate the expected change in flood damages under two climate-change scenarios in which there was an increase in mean annual precipitation by 13.5% and 21.5%, respectively, with either no change or a proportional increase in the standard deviation of annual precipitation. The estimates given by the authors suggested a relation between flood losses and exposure because of an 82% explanatory power of population and wealth which increased to 89% by adding precipitation.

A scenario study of the damage due to river and coastal flooding in England and Wales in the 2080s combined four emissions scenarios with four scenarios of socio-economic change in an SRES-like framework. Damages by flood were predicted to hike with on-going inefficient Flood Management policies, practices. For a 2 °C temperature increase in a B1-type world, by the 2080s annual damage was estimated to be £5 billion as compared to present £1 billion, while with approximately the same Climate Change, estimated damage was only £1.5 billion in a B2-type world. On the other hand, with the same temperature increase of 2 °C in a A1-type world, the predicted annual damage amounted to £15 billion by the 2050s and £21 billion by the 2080s (Hall et al. 2005; Evans et al. 2004).

An assessment undertaken by Indian Scientists of the Indian Institute for Technology (IIT) Delhi, India as part of the Indian National Communication (NATCOM) on Climate Change project under the Ministry of Environment and Forests (MoEF), India, using the HadRM2 daily weather data to determine the spatial-temporal availability of water in the river systems in India has indicated that the severity of floods under the projected Climate Change is likely to intensify (Gosain et al. 2006). The study shows that western India may experience very high river discharges more frequently than it does at present. Apart from changes in snowmelt and precipitation in the Himalaya, changes in the pattern, intensity and frequency of rainfall will have significant impacts on floods. Climatic changes could result in more frequent high-intensity rainfall events.

Sarkar (2015) analysed the probable impact of Climate Change using hypothetical climate scenarios to understand the behavior of total stream flow as well as snowmelt runoff under the changed climatic conditions for the trans-boundary Subansiri sub-basin of Brahmaputra river basin. It has been found from studies that total stream flow changed for all the scenarios of temperature and precipitation. The observed maximum % increase in mean annual stream flow was about 6% for (T + 2 °C and P + 10%) scenario and the minimum % decrease in mean annual stream flow observed was about 11% for (T + 1 °C and P-10%) scenario. Results of seasonal analysis of stream flow under the scenario of warmer climate and enhanced precipitation (T + 2 °C, P + 10%) indicated marginal increase in annual water availability with a reduction in water availability during pre-monsoon, winter and post-monsoon seasons, but substantial increase in water availability during monsoon, thereby imposing an increased risk of floods in the already flood prone basin.

Marchand et al. (2015) assessed the possible impacts of climate variability and Climate Change on the flood hazards in two selected basins of India i.e., Burhi-Gandak basin of Bihar and Brahmani-Baitarani basin of Odisha using three GCM model outputs under RCP 6.5 scenario for two time periods viz., mid (2040s representing climatology over 2030–2059) and long-term (2080s representing climatology over 2060–2099) with 1961–1990 as the baseline period and APHRODITE as the reference data for bias correction. The results show that for the Burhi-Gandak, increase of 5% in flooding extent could be expected in 2040 and even 22% in 2080 for annual floods (occurring once every 25 years). As far as Brahmani-Baitarani is concerned, 25% and 29% increase in flooded surface, is predicted at once every 25 years. Increased errant rainfall and projected sea level rise is the main causes speculated behind these conditions.

A recent collaborative study (completed in 2017) named Water Rain-Him by SMHI and SEI from Sweden and NIH Roorkee and IIT Delhi from India focused on assessment of the impacts on the water fluxes due to change in climate in the Indian-Himalayan basins, i.e. Ganges up to Farakka with an aim to use the impact modelling results to develop holistic and robust adaptation strategies to drive planning for proper and efficient water resources management (Bueker et al. 2017).

The hydrological model HYPE (Lindström et al. 2010) has been used to simulate daily stream flow. An ensemble of 20 climate future projections consisting of different combinations of General Circulation Models (GCMs), and representative concentration pathways (RCPs) to represent different emission scenarios has been used. In total, five GCMs from the CMIP5 archive and three different RCPs have been used. The details of the GCM models are given in the following Table 5.2. RCPs are numbered after their increased radiative forcing until year 2100 (+2.6, +4.5, and +8.5 W/m^2, respectively). The models provide daily values for the period 01/01/1950 to 31/12/2099, and are available from ISI-MIP.

The climate projections (mean precipitation and temperature) were bias-adjusted to the WATCH Forcing Data for the period 1971–2001 at 0.5° resolution (Hempel et al. 2013). The WATCH data are derived based on a combination of the ERA-40

Table 5.2 Earth system models (RCP2.6, RCP4.5 and RCP8.5)

Model name	Institution
HadGEM2-ES	Met Office Hadley Centre Earth System Modeling group, England
IPSL-5 CM5A-LR	Institut Pierre-Simon Laplace, France
MIROC-ESM-CHEM	Japan Agency for Marine-Earth Science and Technology, Atmosphere and Ocean Research Institute, Japan
GFDL-ESM2M	Geophysical Fluid Dynamics Laboratory of the National Oceanic and Atmospheric Administration, USA
NorESM1-M	Norwegian Climate Centre, Norway

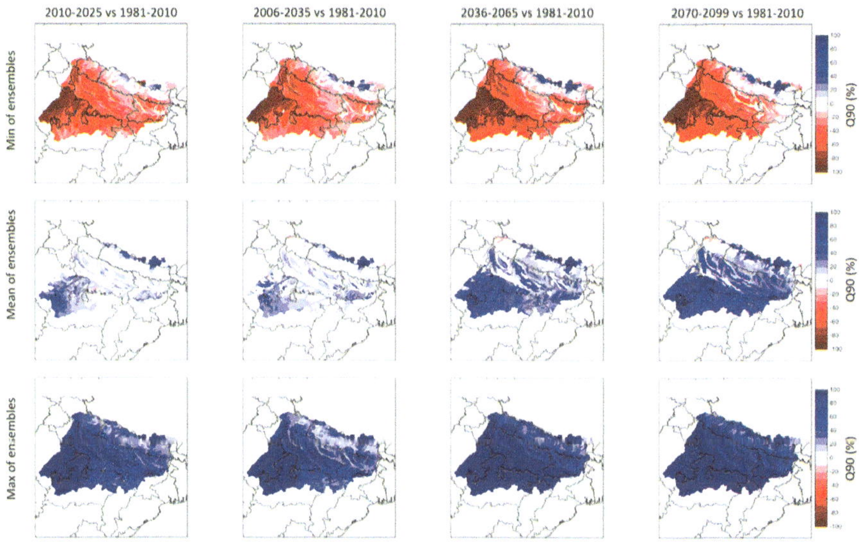

Fig. 5.5 Changes in High Flows (90th Percentile) (Source: Presentation by the project leader of Water RAIN-Him in the International workshop on changes in water resources and adaptation options in the Indian-Himalayan basins, NIH, Roorkee, India, March 9–10, 2)

(the 40-year reanalysis of the European Centre for Medium Range Weather Forecasts (ECMWF) and the Climate Research Unit TS2.1 dataset (CRU) (Weedon et al. 2011). This investigation is based on five time periods: Reference period (1981–2010): Very Early Century (2010–2025): Early Century period (2006–2035); Mid Century period (2036–2065); End Century period (2070–2099).

Analysis of the basin runoff showed high spatial variability of change in runoff (from 10% up to +40%) in the Ganges basin. This is consistent at the end of century over the entire basin. Analysis also showed that the high flows/floods (90th percentile of daily flow) can increase up to 40%. The results of changes in high flows over the entire Ganges basin within India for different time periods considering mean, minimum and maximum of GCM model ensembles are shown in Fig. 5.5.

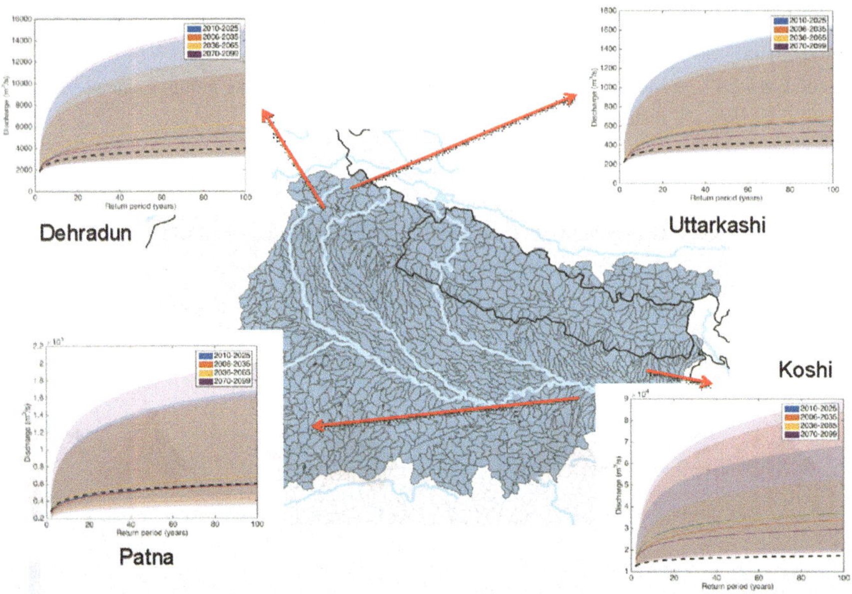

Fig. 5.6 Changes in Discharge return period (Source: Presentation by the project leader of WaterRAIN-Him in the International workshop on changes in water resources and adaptation options in the Indian-Himalayan basins, NIH, Roorkee, India, March 9–10, 2017)

A regional analysis focussing on four regions, namely Dehradun, Uttarkashi, Patna, Koshi (Fig. 5.6) has also been carried out for assessing changes in precipitation, discharge etc.

The regional analysis showed that the pattern of annual cycle for precipitation, temperature and discharge will remain the same for Dehradun, Uttarkashi, Patna and Koshi. There is a posibility of a shift in the onset of precipitation for Dehradun and Uttarkashi. The analysis further showed that an overall positive change in high flows is observed for all time horizons and all four regions. Trends in change is subject to the climate model used, however the magnitude of change does not seem dependent on the emission scenario. It is also evident from the analysis (Fig 5.6) that the discharge return periods (X-axis) will be altered, in which more extreme are expected to occur more frequently, and hence affecting the planning of infrastructures for water resources management.

5.8 Conclusions

Floods are recurrent phenomena in India. The main cause of flooding is the inadequate capacity within the banks of the rivers to contain the high flows brought down from the upper catchments due to heavy rainfall. With increasing population and developmental activities, there has been a tendency of floodplains encroachment over the years resulting in more serious damages. Areas which were not traditionally prone to floods are also experiencing severe inundation because of the varying rainfall distribution.

Flood Management can be achieved through structural as well as non-structural techniques. In India, the main thrust of the flood protection programme has been through physical (engineering/structural) measures mainly in the form of embankments which provide quick and reasonable protection to vulnerable areas with locally available material and labour. The multi-purpose dams also provide flood protection with one specific objective of flood control. Some reservoirs in India have been constructed with a specifically allocated flood cushion (Damodar system in Jharkhand and the Hirakud and Rengali dam in Orissa). However, many other large storage dams without any specific flood cushion have also helped in flood moderation to some extent. Besides, there are other structural measures of Flood Management. Among the non-structural measures for Flood Management which rely on the modification of susceptibility to flood damage, flood forecasting and warning is gaining attention of government/planners as it enables forewarning as to when the river is going to use its floodplain, to what extent and for how long. Other non-structural measures for Flood Management like flood plain zoning, flood modelling etc. have been addressed by researchers and policy makes but they are not in use in India as of now.

5.9 Way Forward

A pragmatic and cost effective approach in Flood Management is required for building of structural and non-structural measures. Such approaches have also been highlighted in the policy documents of the country at various platforms.

It has been shown by many researchers through their studies that Climate Change is expected to cause heavier rainfall events and can lead to significant increases in flood extent and flood hazards. Therefore, it will be highly prudent to re-look into the Flood Management aspects with a reference to Climate Change. Design of all new water resources structures should consider the revised design criteria based on Climate Change scenarios. Design of existing structures need to be checked for their flood protection capacities by considering the Climate Change scenarios and the structures may be re-designed, if required. It is also required to consider the Climate Change impacts on flood hazards using downscaling methods for river basins prone to flooding. Such climate impact analysis should be incorporated in the policy documents for planning and operation of water resources to achieve the desired outcomes in terms of environmentally sustainable Flood Management measures that are cost effective. Needless to say, flood management is a continuous process and requires further improvements especially in the light of climate change.

References

Ahmad MK (2018) Flood mitigation in developing countries: a case study of India. Asian Rev Soc Sci (ARSS) 7(1):91–95

Bueker P, Winther H, Ran Y, Sarkar A, Pechlivanidis I, Gosain AK, Rao S (2017) Report of the international workshop on changes in water resources and adaptation options in the Indian-Himalayan basins, NIH, Roorkee, March 9–10, 2017 (Unpublished)

Central Water Commission (2018) Flood forecast. http://www.india-water.gov.in/ffs/index.html. Acccssed 31 Aug 2018

Changnon SA (2005) Economic impacts of climate conditions in the United States: past, present, and future – an editorial essay. Clim Chang 68:1–9

Choi O, Fisher A (2003) The impacts of socioeconomic development and climate change on severe weather catastrophe losses: mid-Atlantic region MAR and the US. Clim Chang 58:149–170

Cred & UNISDR (2018) Economic losses, poverty and disasters 1998–2017. www.unisdr.org/files/61119_credeconomiclosses.pdf, pp 29

CWC (2018) Study report: Kerala floods of August 2018. http://cwc.gov.in/main/downloads/KeralaFloodReport/Rev-1.pdf, pp 46

Evans E, Ashley R, Hall J, Penning-Rowsell E, Saul A, Sayers PC, Thorne C, Watkinson A (2004) Future flooding: scientific summary. Volume 1: Future Risks and Their Drivers. Foresight, Office of Science and Technology, London. http://www.foresight.gov.uk/Previous_Projects/Flood_and_Coastal_Defence/Reports_and_Publications/Volume1/Contents.htm

Gosain AK, Sandhya R et al (2006) Climate change impact assessment on hydrology of Indian River Basins. Curr Sci 90(3):346–353

Hall JW, Sayers PB, Dawson RJ (2005) National-scale assessment of current and future flood risk in England and Wales. Nat Hazards 36:147–164

Hallegatte S (2015) The indirect cost of natural disasters and an economic definition of macroeconomic resilience. Policy Research Working Paper 7357, World Bank Policy Research Working Paper, 38 p

Hempel S, Frieler K, Warszawski L, Schewe J, Piontek F (2013) A trend-preserving bias correction – the ISI-MIP approach. Earth Syst Dynam 4:219–236. https://doi.org/10.5194/esd-4-219-2013

India-WRIS wiki (2015) Water Resources Information System of India. http://www.india-wris.nrsc.gov.in/wrpinfo (downloaded on August 21, 2018)

IPCC (2014) Climate change 2014. Synthesis report. Contribution of working groups I, II and III to the fifth assessment report of the intergovernmental panel on climate change [Core writing team, R.K. Pachauri and L.A. Meyer (eds.)]. IPCC, Geneva, 151 pp

Kleinen T, Petschel-Held G (2007) Integrated assessment of changes in flooding probabilities due to climate change. Clim Chang 81:283–312

Lal M, Meehl GA, Arblaster JM (2000) Simulation of Indian summer monsoon rainfall and its intra-seasonal variability in the NCAR climate system model. Reg Environ Chang 1(3–4):163–179

Lindström G, Pers C, Rosberg J, Strömqvist J, Arheimer B (2010) Development and testing of the HYPE (hydrological predictions for the environment) water quality model for different spatial scales. Hydrol Res 41:295–319

Marchand M, Sethurathinam S, Dahm R, Lal M, Ekande TG, Kumar Manhacheri D, Upadhyay V, Muralikrishna M, Singh U, Nansey J, Dhar S (2015) Operational research to support mainstreaming of integrated flood management under climate change (Main report), Deltares in association with RMSI and JPS, pp 97

May W (2004) Simulation of the variability and extremes of daily rainfall during the Indian summer monsoon for present and future times in a global time-slice experiment. Clim Dyn 22(2–3):183–204

Meehl GA, Arblaster JM (2003) Mechanisms for projected future changes in south Asian monsoon precipitation. Clim Dyn 21(7–8):659–675

Mileti DS (1999) Disasters by design: a reassessment of natural hazards in the United States. Joseph Henry Press, Washington, District of Columbia, 351 pp
Milly PCD, Wetherald RT, Dunne KA, Delworth TL (2002) Increasing risk of great floods in a changing climate. Nature 415:514–517
Mohapatra PK, Singh RD (2003) Nat Hazards 28:131. https://doi.org/10.1023/A:1021178000374
Pielke RA Jr, Downton MW (2000) Precipitation and damaging floods: trends in the United States, 1932–97. J Clim 13:3625–3637
Rupakumar K, Krishna Kumar SA, Patwardhan SK, Mishra PK, Revadekar JV, Kamala K, Pant GB (2006) High-resolution climate change scenarios for India for the 21st century. Curr Sci 90:334–345
Sarkar A (2015) Impact of climate change on the runoff regime of an Eastern Himalayan river basin. Global NEST J 17(2):323–333
Schiermeier Q (2006) Insurers' disaster files suggest climate is culprit. Nature 441:674–675
Schreider SY, Smith DI, Jakeman AJ (2000) Climate change impacts on urban flooding. Clim Chang 47:91–115. https://doi.org/10.1023/A:1005621523177
Trenberth K (2011) Changes in precipitation with climate change. Clim Res 47(1–2):123–138
Weedon GP, Gomes S, Viterbo P, Shuttleworth WJ, Blyth E, Österle H, Adam JC, Bellouin N, Boucher O, Best M (2011) Creation of the WATCH forcing data and its use to assess global and regional reference crop evaporation over land during the twentieth century. J Hydrometeorol 12(5):823–848
World Bank (2013) Turn down the heat: climate extremes, regional impacts, and the case for resilience. Washington, DC

Dr. Archana Sarkar is a scientist with the National Institute of Hydrology, Roorkee with research experience of more than twenty five years. A Civil Engineer by education and PhD from IIT Roorkee, She has also worked at Imperial College, London as a Commonwealth Professional Fellow. Her area of specialization is Multi-basin multi-model Hydrological Modelling; Remote sensing and GIS applications; and Climate Change studies.

Chapter 6
Drought Disaster: Issues, Challenges and Risk Mitigation Strategies

Sreeja S. Nair, Anil K. Gupta, and M. S. Nathawat

Abstract There is no universal definition for drought and it is not possible to identify the beginning and end of a drought season, unlike earthquake or flood. Drought is the most complex of all the natural disasters Drought and associated famines were reported in history as a chronic problem to mankind and still remains an unconquered issue especially in developing and lesser developed regions of the world. Droughts are classified into four types based on the impacts as meteorological, hydrological, agricultural and socio economic drought. This paper provides a brief overview of the global and national drought scenario, causes and impacts of drought and mitigation and management strategies with focus on proactive approaches. A case study on drought vulnerability and mitigation strategy analysis for the Bundelkhand region has been added at the end of the paper. This case study is an unique attempt of using scientific data and methods coupled with social science methods to carry out drought vulnerability analysis and effectives of mitigation strategies.

Keywords Drought · Famine · Shortage of soil moisture · Agriculture · Forestry · Loss of livelihood

S. S. Nair (✉)
National Disaster Management Authority, New Delhi, India

A. K. Gupta
ECDRM Division, International Cooperation, CAP-RES DST Project, National Institute of Disaster Management (Ministry of Home Affairs), New Delhi, India

M. S. Nathawat
Institute of Science, IGNOU, Delhi, India

A. K. Gupta et al. (eds.), *Disaster Risk and Management Under Climate Change*, Disaster Resilience and Green Growth,
https://doi.org/10.1007/978-981-99-4105-6_6

6.1 Introduction

Drought is generally defined as a deficiency of precipitation over an extended period of time (Eriyagama et al. 2009). Drought ranks first among natural disasters in terms of the extent of the effect on population, livelihoods, environment, society, and economy (Hewitt 1997). Drought ranks first among natural disasters in terms of extent of effect on population, livelihoods, environment, society and economy (Hewitt 1997). Climate change affects results not only in change in rainfall patterns and availability of water, but also the settings of land, ecology, forests and thereby aggravating the vulnerability of the natural and anthropogenic systems.

Droughts differ from other natural hazards such as cyclones, floods, earthquakes, volcanic eruptions, and tsunamis in several ways. The key differences are listed in the Box 6.1.

Box 6.1 Difference Between Drought and Other Natural Hazards

- Slow-onset, creeping phenomenon that makes it difficult to determine the onset and end of the event;
- Duration may range from months to years and the core area of epicentre will change over time, reinforcing the need for continuous monitoring of climate and water supply indicators;
- No single indicator or index can identify precisely the onset and severity of the event and its potential impacts-multiple indicators are more effective;
- Spatial extent is usually much greater than for other natural hazards, making assessment and response actions difficult, since impacts are spread over larger geographical areas;
- Impacts are generally non-structural and difficult to quantify;
- Impacts are cumulative and the effects magnify when events continue from one season or year to the next.

(Ministry of Agriculture and National Institute of Disaster Management, 2009)

6.1.1 Definition and Typology of Drought

Drought is a prolonged period of low precipitation in a region which results in acute scarcity of surface and ground water, significant shortage of soil moisture leading to agricultural drought i.e. reduction in crop yield. Drought impacts agriculture, forestry and ecosystems adversely and leads to land degradation, crop damage, and reduced yield, loss of livestock or reduced productivity with dire consequences on people's livelihoods. Developing countries are more vulnerable to the impact of drought as compared to the developed countries. Drought can lead to famine, a situation in which there is not enough food for a great number of people causing illness or death. Drought in long term leads to the loss of productivity of land in the form of due to land degradation. Climate change known to aggravate rainfall patterns, temperature regimes besides altering land-ecology regimes and interface with society has crucial influences on drought disaster management transitioning towards risk mitigation.

Drought has been classified into four types based on the cause and impact on the natural and human systems. Although a fourfold classification has been done, it is not possible to draw lines between the different types since the one may lead to other (NOAA, n.d).

- Meteorological drought is defined by the deficiency of precipitation from expected or normal levels over an extended period of time. Meteorological drought usually leads to other kinds of drought.
- Hydrological Drought is defined as deficiencies in surface and subsurface water supplies leading to a lack of water for meeting normal and specific water demands. The conditions of hydrological drought arise, even in times of average (or above average) precipitation, when increased usage of water diminishes the reserves.
- Agricultural Drought arises when soil moisture and rainfall are inadequate during the crop growing season to support healthy crop growth to maturity; this situation causes extreme crop stress and wilting. The conditions of agricultural drought can arise, even in the times of average precipitation, owing to soil conditions or agricultural practices and technologies.
- Socioeconomic drought is the impact of other drought conditions (meteorological, agricultural, or hydrological drought). Socioeconomic drought occurs as the result of the reduced supply of goods and ecosystem services. This hampered availability of natural resources and associated goods and services due to the deficit in water supply. This is eventually the result of low precipitation that is cause by hydrological and agricultural droughts.

6.2 Present Scenario

6.2.1 Global Scenario

As per the EM-DAT international disaster database of Centre for Research on Epidemiology of Disasters (CRED) drought and famines accounted for 86.9% of total deaths caused by hydro-meteorological disasters during the period of 1990–1999 globally. Drought contributes to 26% of the total population affected due to weather related disasters (Centre for Research on The Epidemiology of Disasters 2016). Drought affected 50.5 Million people in 2015 well above the 10-year average of 35.4 million people (UNISDR & CRED, 2015). This was the hottest year on record with temperature rise due to global warming and EL NINO weather phenomenon, 32 major droughts were recorded during 2015 (UNISDR

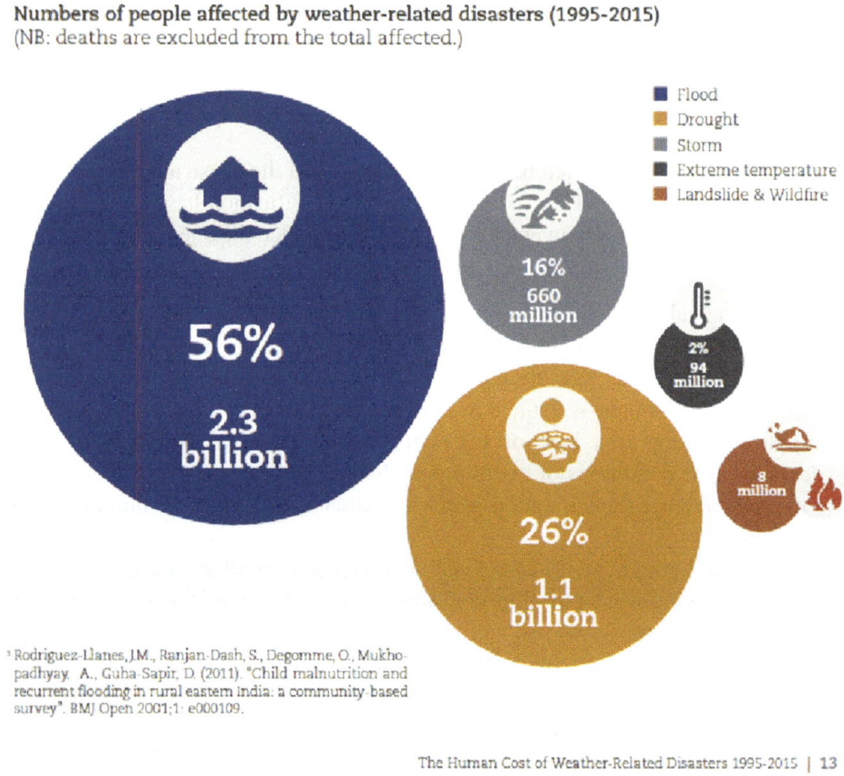

Fig. 6.1 Number of people affected due to weather related disasters (1995–2015). Source: (CBD, kein Datum)

2016). Most drought-related mortality in recent years has occurred in countries experiencing extreme poverty and civil unrest.

As per EM-DAT data, Africa is the worst hit by droughts more than any other continent, that says 136 drought events took place between 1995 and 2015 (which alone is 41% share of the global drought event). Out of these 136 droughts, 77 droughts alone were observed in East Africa. Droughts are associated with hunger, poverty, perpetuation of under-development, widespread agricultural failures, and loss of livestock, water shortages and outbreaks of epidemic diseases. Droughts can last for few days to few years causing extensive and long-term economic impacts. Succeeding failures of seasonal rainfalls in East Africa in the year 2005, led to food insecurity for at least 11 million people. EM-DAT recorded more than one billion affected population by droughts from the period 1995–2015; which is more than a quarter share of all people who are affected by all types of weather-related disasters worldwide (Fig. 6.1). Yet drought globally accounted for less than 5% of all natural hazards. While EM-DAT data also show that just 4% of weather-related disaster deaths were due to drought (Fig. 6.2), this figure is rather

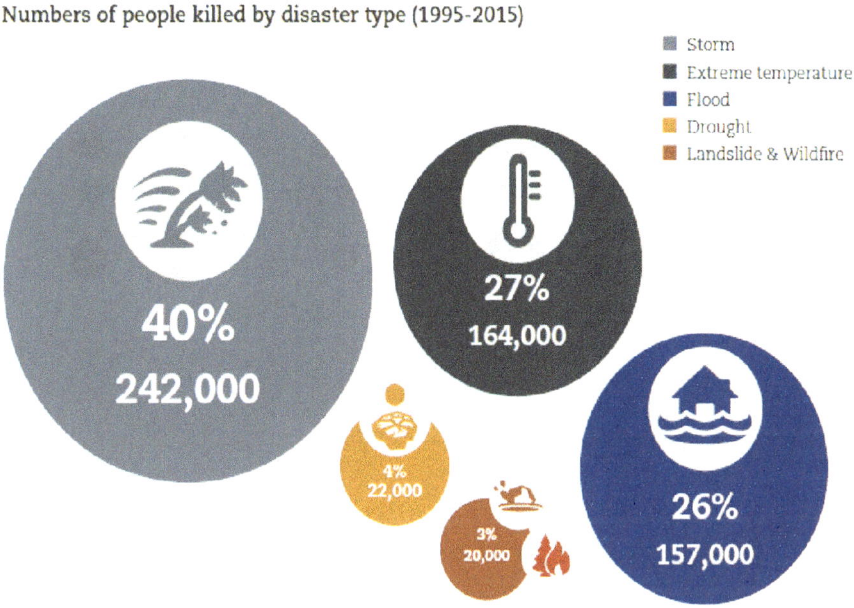

Fig. 6.2 Number of people killed by weather related disasters (1995–2015) (CRED 2019)

undermining the mortality due to drought as it excludes indirect deaths from malnutrition, disease, displacement and related humanitarian crisis.

Natural Disasters 2018 (Fig. 6.2) shows that the mortality (direct) due to drought and famine is reduced to zero. However, drought is continuing as a major cause for human suffering especially in African and Asian regions. Drought affected 10.8 million in the year 2018 alone and 73.9 million people during the period 2008–2017 (CRED 2018) (Figs. 6.3 and 6.4).

A global assessment of the disaster hotspots has been carried out by World Bank in the year 2005. This assessment shows that almost 38% of the world's land area has some level of drought exposure. The WASP index assesses the precipitation deficit/surplus over a three-month running average for the 21-year period from 1980–2000 (Dilley et al. 2005) (Fig. 6.5).

6.2.2 *National Scenario*

India has a long history of droughts and famines. At present, around 68% of India's net sown area is prone to drought in varying degrees, occurs mainly due to the failure of South-West monsoon (from June to September when Kharif crops are grown). However, many parts of the country also face northeast monsoon failure and suffer a second cycle (Rabi crops) of drought risk. Droughts are almost an annual feature affecting one or the other part of the country. As per the Irrigation Commission

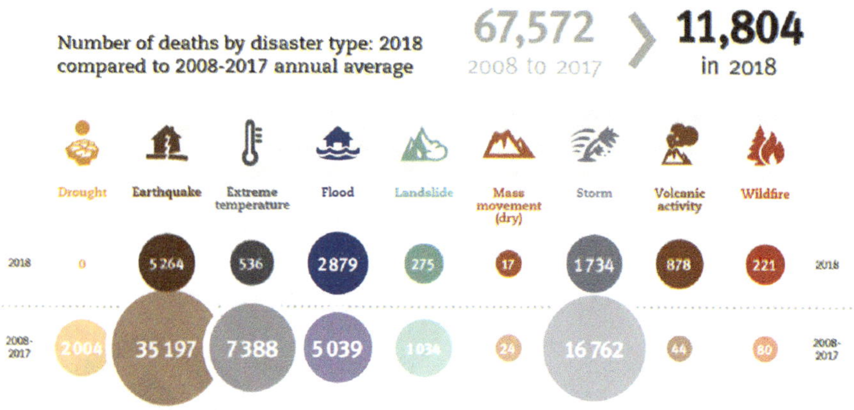

Fig. 6.3 Number of people killed by natural disasters 2018 compared to 2008–2017 period (CRED 2019)

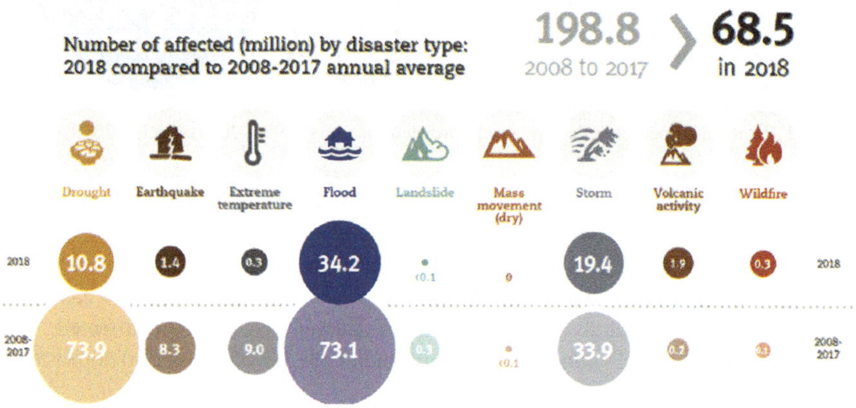

Fig. 6.4 Number of people affected in million by natural disasters 2018 compared to 2008–2017 period (CRED 2019)

Report, 102 districts across 13 states of India are chronically drought prone (Ministry of Agriculture 2009).

In the National Drought Manual released in the year 2015 it is reported that during 1871–2015, there were 25 major drought years viz. 1873, 1877, 1899, 1901, 1904, 1905, 1911, 1918, 1920, 1941, 1951, 1965, 1966, 1968, 1972, 1974, 1979, 1982, 1985, 1986, 1987, 2002, 2009, 2014 and 2015. Drought years are defined as years with All India Summer Monsoon Rainfall (AISMR) less than one standard deviation below the mean (i.e. anomaly below −10%).

Nearly 68% of cropped area in India is vulnerable to drought, 33% of which receives less than 750 mm of mean annual rainfall. These are classified as "chronically drought-prone" 35% cropped area of the country receives mean annual rainfall of 750–1125 mm which is considered as "drought-prone". The drought-prone areas

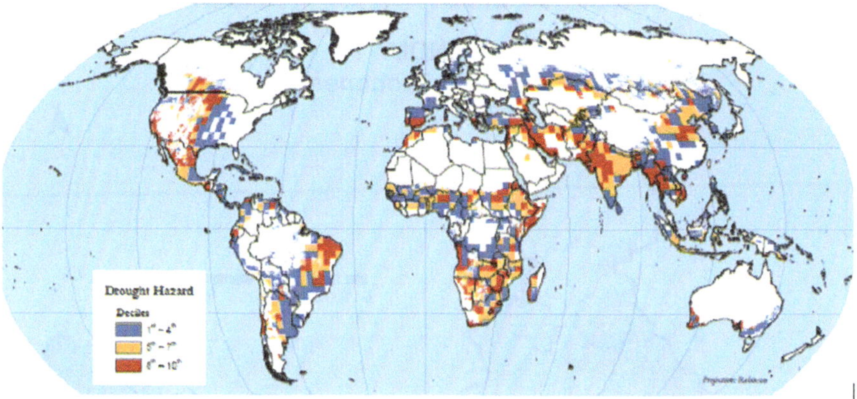

Fig. 6.5 Drought hazard distribution map (Dilley et al. 2005)

of the country are confined primarily to the arid, semi-arid, and sub-humid regions of peninsular and western India (Ministry of Agriculture 2016) (Fig. 6.6 and Table 6.1).

6.2.3 Causes of Drought

The following are the main causes of drought in India.

6.2.3.1 Deficiency of Rainfall (Hazard)

Deficiency in precipitation in the form of rain or snowfall lead to decrease in surface and ground water levels and resulting decrease in soil moisture levels. This leads to insufficient availability of water for the crops, especially during the critical stages of plant growth and resultant agricultural drought.

6.2.3.2 Increase Temperature

Increase in temperature can lead to increase in evapotranspiration and net loss of water content in plants/crops leading to physiological damage to crops and increased demand of water.

6.2.3.3 Changes in Weather and Air Circulation

El Nino impacts the weather patterns and monsoon rainfall.

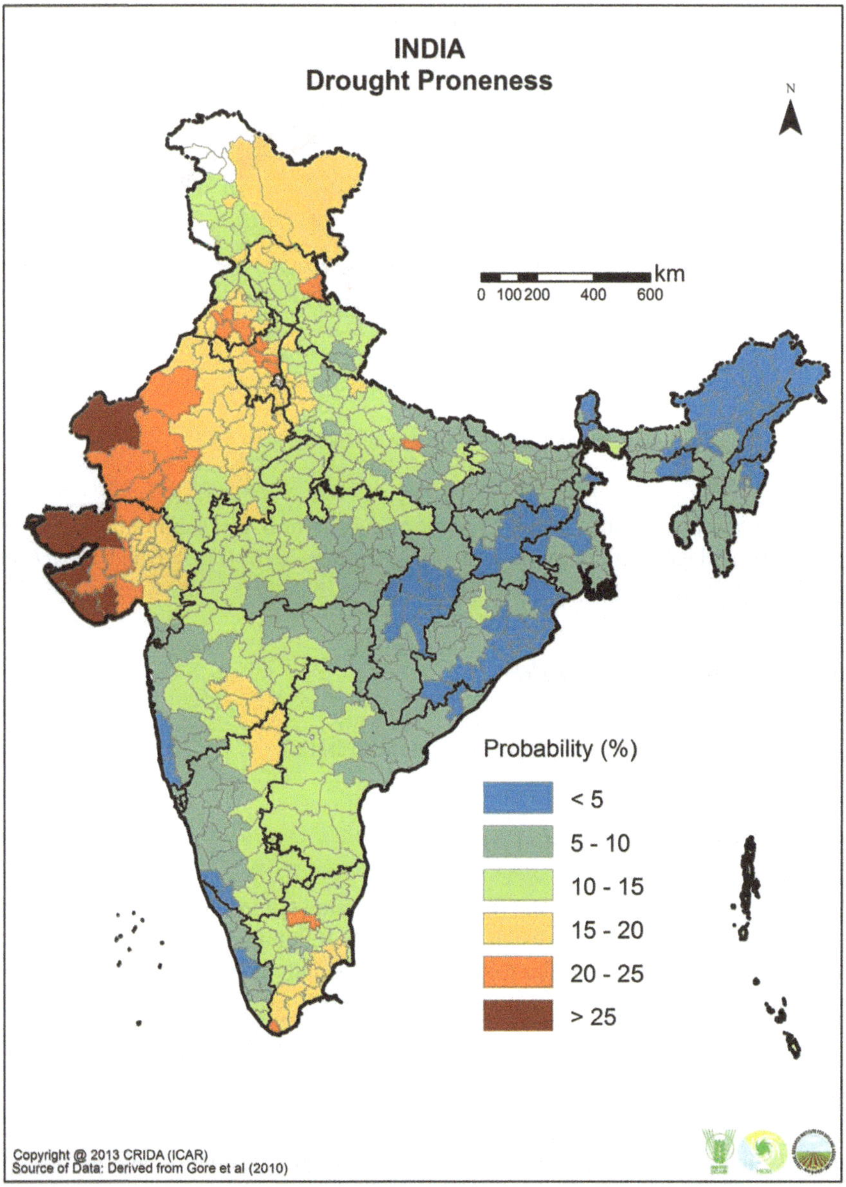

Fig. 6.6 District Level agricultural drought frequency (2000–15). Source: Farmer's portal

6.2.3.4 Unsustainable Agricultural and Land Use Practices

Land use practices, regions growing demand for water, excess withdrawal of ground water for irrigation.

Table 6.1 Probability of occurrence of drought in different states and regions of India (Ministry of Agriculture 2016)

Regions	Frequency of deficient rainfall (75% of normal or Less
Assam	Very rare, once on 15 years
West Bengal, Madhya Pradesh, Konkan, Bihar and Orissa	Once in 5 years
South Interior, Karnataka, Eastern Uttar Pradesh and Vidarbha	Once in 4 years
Gujrat, East Rajasthan, Western Uttar Pradesh	Once in 3 years
Tamil Nadu, Jammu and Kashmir, Telangana	Once in 2.5 years
West Rajasthan	Once in 2 years

6.2.3.5 Timing of Water

Unsynchronized availability of water and agricultural seasons leads to using too much water when one does not need it and too little when one does not need it.

6.3 Vulnerability Factors and Impacts

6.3.1 Vulnerability Factors

Though deficient rainfall is the primary causative factor for drought, the occurrence, spread and intensity of drought is determined by multiple factors including susceptibilities introduced by Climate Change, hydrological and soil profiles, availability of soil moisture, choice of crops and agricultural practices, availability of fodder, socio-economic profile and so on. India is one of the countries which will be worst affected climate change. It has one of the highest population densities in the world with large numbers of poor people who depending on the natural resource bases livelihoods like agriculture, animal husbandry, dairying and so on. Although irrigated land is increased substantially dependency of Indian agriculture on monsoon is still evident. By 2020, pressure on India's water, air, soil, and forests is expected to become the highest in the world (World-Bank, Climate Change, 2009). One of the most significant ways that climate change will impact the lives of people in India will be through its water resources. While water sustains life, it all too often wreaks havoc through devastating floods and droughts. All-India Summer Monsoon (June–September) Rainfall Anomalies during 1871–2015 clearly shows indicate that since 1950s, not only has the incidence of droughts increased, but rainfall in the excess of 10% has also decreased markedly (Mathew).

6.3.1.1 Physical Vulnerability

Physical Vulnerability is the condition attributed due to poor infrastructure facilities. In the context of drought, lack of irrigation infrastructure, water supply systems including storage and distribution, non-availability of electricity for pumping water as well as limited access to roads can increase the vulnerabilities.

6.3.1.2 Socio-Economic Vulnerability

Socio economic vulnerability refers to social, economic and demographic factors that affect the resilience of communities. It is inclusive of factors like low land-holding, poor family with low buying capacity during the period of drought, limited livelihood diversification and dependency on agriculture, lack of education, gender issues, lack of social cohesiveness and access to markets.

6.3.1.3 Environmental Vulnerability

Environmental vulnerability is the result of depletion of natural resources and ecosystem services as well as due to environmental degradation. Examples include—loss of wetlands, depletion of surface and ground water resources, deforestation, soil erosion, land degradation, etc.

6.3.1.4 Institutional Vulnerability

Institutional vulnerability refers to institutional and systemic weaknesses with in the community that makes them more susceptible to the adverse impact of hazards.

6.3.2 Impacts of Drought in India

Drought has widespread impacts that span across many sectors of the economy. Drought affects the overall economy of the country at different levels. The impacts can be direct as well as indirect. Direct effects of drought are mainly decline in agricultural productivity or yield and loss of food security, particularly among poor and vulnerable sections; depleted water table; decline in reservoir levels; higher livestock and wildlife mortality; cattle and animal migration; damage to ecosystem from indiscriminate exploitation; increased fire hazards etc. Indirect impacts of drought can be manifested in the form of reduction in incomes of farmers and other agricultural and natural resources related sectors, increased prices for food and fodder, reduction in purchasing capacity and slump in consumption, default on agricultural loans, distress sale of agricultural land and livestock, rural unrest, shrinkage in avenues for agricultural employment etc. These harmful impacts of drought do have huge cascading effects on the economy and the social system of a country.

Broadly, the impacts of drought can be classified as social, economic and environmental:

1. Economic impacts refer to the production losses in agriculture and related sectors, especially animal husbandry, dairy, poultry, horticulture, and fisheries. Livelihood and well-being of the entire section of the society is affected due to the slowdown of economy. Most of the farmers, farm labourers, artisans depending upon small rural businesses and rural agriculture etc. are adversely affected. Industries also suffer because of their dependency on the primary sector for raw materials suffer on account of reduced supplies and hardening prices. Drought can lead to slow economic growth by reducing the profits, drying up of income, revenue options, reducing employment opportunities through disruption caused to supply chain managements, slowing down flow of credit and tax collections, depressing industrial and consumer demand, increased dependence on imports, and lowering of overall market sentiments.
2. Environmental impacts can be gauged in terms of frequent situations of low levels in ground water and surface reservoirs, lakes and ponds, reduced flows in springs, streams and rivers, loss of forest cover, migration of wildlife and stress on biodiversity. Drying up of stream flow and loss of wetlands may affect salinity level. Rapid and increased groundwater depletion rates without sufficient recharge damages aquifers and adversely affects the quality of water (e.g., salt concentration, acidity, dissolved oxygen, turbidity). This leads to a permanent loss of biological productivity of soils. Long periods of dry spell lead to forest fire. Another environmental impact of drought is desertification and land degradation.
3. Social impacts are manifested in the widespread disruption of rural society on the account of-the massive migration of the rural population from drought affected areas, hike in school dropout rates, greater indebtedness, decline of land resources, and loss of livestock, malnutrition, starvation, and loss of social status among the most vulnerable sections. Drought and resultant migration can lead to the increased pressure on the existing resources. This situation can lead to the lack of social cohesiveness and eroding of social capital. Typical examples are civil unrest and conflicts as evident in many parts of Horn of Africa due to prolonged periods of scarcity and famine can lead to distress migration.

6.4 Mitigation and Management

The need for shifting the public policy from drought relief to drought mitigation measures has been felt considering the increase in the frequency of droughts in different parts of the country. These measures are important for adapting to Climate

Change, restoring ecological balance, and bringing development benefits to the people. Various structural and non-structural mitigation strategies are adopted at various levels by central government, state governments, NGOS and community to reduce the impact of drought. Various mitigation, preparedness and relief related interventions can be adopted during and before drought to reduce the magnitude of risk to human life, property, and productive capacity.

6.4.1 Policy Framework

6.4.1.1 International Framework

United Nations Convention to Combat Desertification (UNCCD)—1994

The United Nations Convention to Combat Desertification (UNCCD) links environment and development to sustainable land management. It specifically addresses the arid, semi-arid and dry sub-humid areas, known as the drylands, where some of the most vulnerable ecosystems and peoples can be found. The new UNCCD 201 8–2030 strategic framework is a global commitment to achieve Land Degradation Neutrality (LDN). The UNCCD focuses on avoiding, minimizing, and reversing desertification/land degradation and mitigating the effects of drought at all levels to achieve a land degradation-neutral world consistent with the 2030 Agenda for Sustainable Development (UNCCD 1994).

United Nations Framework Convention on Climate Change –1992

The United Nations Framework Convention on Climate Change (UNFCCC) is an international environmental treaty adopted on 9 May 1992 and opened for signature at the Earth Summit in Rio de Janeiro from 3 to 14 June 1992. It then entered into force on 21 March 1994, after a sufficient number of countries had ratified it. The UNFCCC objective is to "stabilize greenhouse gas concentrations in the atmosphere at a level that would prevent dangerous anthropogenic interference with the climate system". Preventing "dangerous" human interference with the climate system is the ultimate aim of the UNFCCC. The ultimate objective of the Convention is to stabilize greenhouse gas concentrations "at a level that would prevent dangerous anthropogenic (human induced) interference with the climate system." It states that "such a level should be achieved within a time-frame sufficient to allow ecosystems to adapt naturally to climate change, to ensure that food production is not threatened, and to enable economic development to proceed in a sustainable manner". In 2015, 197 parties to the convention came together for the UN Climate Change Conference in Paris 30 November–12 December and adopted by consensus the Paris Agreement, aimed at limiting global warming to less than two degrees Celsius, and pursue efforts to limit the rise to 1.5 °C. The Paris Agreement entered into force on 4 November 2016 (UNFCCC 2018).

Convention on Biodiversity

The Convention on Biological Diversity (CBD) was opened for signature at the Earth Summit in Rio de Janeiro on 5 June 1992 and entered into force on 29 December 1993. The objectives of this Convention, to be pursued in accordance with its relevant provisions, are the conservation of biological diversity, the sustainable use of its components and the fair and equitable sharing of the benefits arising out of the utilization of genetic resources, including by appropriate access to genetic resources and by appropriate transfer of relevant technologies, taking into account all rights over those resources and to technologies, and by appropriate funding (CBD, n. d.).

6.4.1.2 National Policies

National Disaster Management Guidelines for Drought

National Disaster Management Guidelines for drought was released in the year 2010 (National Disaster Management Authority 2010). Participatory approach involving all the stakeholders to take forward the task of reducing drought risk is the highlight of the document. This document includes geographical and temporal distribution, typology, causes, impacts of drought and measures to tackle drought situation in the country. The guidelines also capture a broader picture, covering policy issues, institutional aspects and financial arrangements for drought management in general—Emphasizes on emerging concerns as well including Climate Change, need to strengthen the structural and non-structural measures for drought management, as proposed in the Crisis Management Plan and Drought Manual 2009. Section 4.2 of the guideline is on Climate Change and drought and Sect. 4.3 is on mitigation actions (NDMA 2010).

National Forest Policy, 1988

Several provisions related reducing the disaster risks is inbuilt in the National Forest Policy. The policy highlights the importance of retaining environmental stability through preserving and restoring ecological balance. It is evident that ecological balance is disturbed by serious depletion of the forests of the country. Control soil erosion and denudation in the catchment areas of rivers, lakes, reservoirs in the "interest of soil and water conservation, for mitigating floods and droughts and for the retardation of siltation of reservoirs are key recommendations in the policy". This policy by conserving the forests indirectly and directly helps in reducing the risks of floods, landslides and enhanced the livelihood options of the people depending on forest. From the basic objectives of the policy (Sect. 2) provides for reducing DRR by forest conservation (MoEF&CC 1988).

National Water Policy, 2012

The objective of the National Water Policy is to have an accountable and accurate understanding of the existing situation, to propose a framework for working out a system of laws and institutions and to prepare plans of action with a single national perspective. Nation Water Policy emphasizes that large parts of India have already become water stressed. Speedy growth in demand of water because of exploding population growth, urbanization etc. is posing serious challenges to water security. Section 10.2 of the policy highlights that "land, soil, energy and water management with scientific inputs from local, research and scientific institutions should be used to evolve different agricultural strategies and improve soil and water productivity to manage droughts". It is also proposed to integrate agriculture systems and non-agricultural developments for diversifying of livelihood and reducing poverty (MoWR&GR 2012).

National Agricultural Policy 2000

India's agriculture sector accounts for 18% of GDP, and employs around 60% of the workforce. The key target of the policy is to achieve annual growth rate of 4% in the agricultural sector during 2000–2020 period. This policy also addresses the issues pertaining to protection and management of natural resources and ecosystems in rural areas for increasing resilience to Climatic Change and resultant hydro-meteorological disasters. The new policy also emphasized on sustainable agriculture, food and nutritional security, technology generation and transfer, risk management and organizational framework. Agro-ecosystems include mainly the man-made ecological production systems—farms, plantations, ponds, etc. also the natural systems used for bio-productivity purposes, covering the purposes of food, dairy, fisheries, and other livestock, etc. (MoA 2000).

National Environment Policy 2006

The National Environmental Policy, 2006 calls for an integrated approach for protection and improvement of environment and ecosystems by adopting various approached viz. management of coastal zones, wetlands and river systems; conservation and development of mountain ecosystems; land use planning; watershed management and reducing the impacts of natural hazards like, flood, landslides, storm surges and climate change. Besides EIA, notifications also have provisions which mandate the carrying out of Hazard Mapping, Vulnerability and Risk

Assessment Report as a part of environment management plan for all the projects. On-farm water management i.e. a systems approach towards controlling water on a farm in a manner that provides for the beneficial management of water for the irrigation and drainage needs, consists of components such as irrigation, drainage, water sources and sinks etc. must be selected and operated in accordance with the needs (MoEF 2006).

National Action Plan for Climate Change Adaptation

The Prime Minister's Council on Climate Change is in charge of the overall implementation of the plan. The plan document elaborates on a unique approach to reduce the stress of climate change and uses the poverty-growth linkage to make its point. Published in 2008 by the then-Prime Minister's Council on Climate Change (Government of India), the National Action Plan on Climate Change (NAPCC) aims at creating awareness among the representatives of the public, different agencies of the government, scientists, industry and the community as a whole, on the threat posed by climate change and the steps proposed at the level of India to counter these changes. There are eight govt missions under National Action Plan on Climate Change (NAPCC) (1) National Solar Mission, (2) National Mission for Enhanced Energy Efficiency (3) National Mission on Sustainable Habitat (4) National Water Mission (5) National Mission for Sustaining the Himalayan Ecosystem (6) National Mission for a Green India (7) National Mission for Sustainable Agriculture and National Mission on Strategic Knowledge for Climate Change.

6.4.2 Structural and Non-structural Mitigation Measures

6.4.2.1 Water Conservation and Judicious Use of Surface and Ground Water

These mitigation measures are aimed at integrating soil, water, and forest resources via soil conservation, watershed development, and forestry programs. In drought prone areas rainwater is the major source of surface and ground water recharge. Due to non-judicious use of ground water without recharging the groundwater and aquifers all across the country conservation of water resources has become ever so alarmingly important.

6.4.2.2 Water Conservation, Storage Structures and Management

Water harvesting and conservation is very effective mitigation strategy, as during the period of water resource depletion, the drought affected area still has significant potential for harvesting and conserving water if an integrated water resources management approach is adopted, and proper policies and investment actions are implemented using recent technologies. The scarcity of water resource during the drought period needs to be managed optimally. Water harvesting structures and storage capacities needs to be promoted where ever possible to address climatic risks to water regimes.

6.4.2.3 Cloud Seeding in Drought Prone Regions of India

Monsoon rainfall in the interior part of peninsular India is lower as compared to all India mean monsoon rainfall. Most of these regions are in rain shadow and agriculture is rainfed type. In some of the states like Karnataka and Andhra Pradesh, cloud seeding was done in the past. To replicate this technique in other areas of the country, it is necessary to carry out more studies and experiments to assess the aerosol characteristics, suitability of nuclides for cloud seeding and alternative types of cloud seeding. It is also recommended to formulate a cloud seeding policy at National level and State level for creating required environment to regulate this activity.

6.4.2.4 Micro Irrigation Systems

Irrigation projects of major and medium scale have contributed to the development of water resources. Though this conventional method of water conveyance has proven to be highly inefficient, that has not only led to wastage of water, but also other problems like water logging, salinization, and soil degradation rendering productive agricultural lands unproductive. Uses of irrigation technologies like drip and sprinkler irrigation are found effective in the use of surface as well as ground water resources efficiently. Such micro-irrigation techniques not only assure efficient use of water but also help in reducing evapo-transpiration losses.

6.4.2.5 Post-harvest Management

One of the key factors, that multiply the impacts of drought, is the absence of or inadequate facilities for proper pre and post-harvest crop management. It is necessary to create a chain of cold storages for post-harvest management with other practices like pre cooling, cold storages and refrigerated transport. Pre-harvest losses due to diseases and pests need to be reduced through better management practices.

6.4.2.6 Drought Forecasting and Early Warning Systems

Drought forecasting and Early warning systems needs to be strengthened and make more focused on reaching out the information to end users. Risk assessments shall be integrated in the EWS where ever available. Emphasis shall be given on timely forecasting and development of advisories for concerned departments. Space based inputs shall be used in forecasting and efforts needs to be made to integrate the ground-based information with the space-based information for comprehensive reporting. Additional vegetation related parameters for estimation of agro-meteorological parameters such as rainfall, soil moisture and evapo-transpiration are also required of lately. This is important considering the low density of the network of ground-based observations available in the country for monitoring the parameters. Warning systems should be made efficient by assuring multi-stake holder participation and user needs assessment. Last mile connectivity needs more emphasis and the information should be reached timely to the farmers. Wherever possible the traditional knowledge and scientific knowledge need to be integrated. EWS should also have mechanisms to evaluate the effectiveness of the system for improving incrementally.

6.4.2.7 Crop Insurance

Crop Insurance is one the most effective risk transfer mechanism to reduce the negative impact of drought. Government of India has taken several initiatives towards increasing its coverage and reach, although they are not yet successful and benefits to farmers are marginal. Some of the key initiatives of government are given below (Box 6.2). Promoting private sector insurance companies and micro insurance schemes needs to be implemented to make the crop insurance more effective. The scope of the insurance policies needs to be widened enough to include pre-sowing and post-harvest losses as well. Possibility of use of satellite based information will be explored to facilitate settlement of claims.

Box 6.2 Key initiatives by Government related to Crop Insurance

- All-India Comprehensive Crop Insurance Scheme (CCIS) for major crops was introduced in 1985.
- National Agricultural Insurance Scheme (NAIS) in 1999 subsequently replaced the CCIS.
- The Government of India (GoI) also set up the Agriculture Insurance Company of India (AIC) in 2003 to serve the needs of farmers better and to move towards a sustainable actuarial regime.
- On 18 February 2016, the Pradhan Mantri Fasal Bima Yojana (Prime Minister's Crop Insurance Scheme) was launched by Prime Minister of India. Uniform premium of only 2 % to be paid by farmers for Kharif crops, and 1.5 % for Rabi crops. The premium for annual commercial and horticultural crops will be 5%. The scheme provided income protection to the farmers by insuring production and market risks.

6.5 Drought Vulnerability and Mitigation Analysis Framework: Bundelkhand Case Study

A Bundelkhand region lies in the heart of India and is one of the poorest regions in the country. Bundelkhand administrative region span across 13 districts, seven from Uttar Pradesh (Jhansi, Jalaun, Lalitpur, Hamirpur, Mahoba, Banda and Chitrakoot) and six from Madhya Pradesh (Datia, Tikamgarh, Chattarpur, Damoh, Sagar and Panna). Total population is 18.3 million, and 79.1% of the total population live in rural areas (Census 2011). Eighty prevent of the total population lives in rural villages. Urban population is highest (42%) in Jhansi and lowest (10%) in Chitrakoot). Livelihood of people in Bundelkhand region is dependent heavily on ecosystem services and, therefore, most severely affected by conditions and factors limiting natural resource availability and access. Agriculture is the main livelihood. 60% are working in the agricultural sector as cultivators and agricultural labourers. About 45% of the area is irrigated but not adequate. Despite the region being highly vulnerable systematic study integrating scientific and socio-economic is carried out by only few organizations. Most of these studies are focusing on socio economic, physical or environmental component alone.

NIDM with the support of ICSSR carried out a study at three spatial levels i.e. macro level for the entire Bundelkhand Region, Meso level for two districts in the region, and micro level for four villages; two each in Lalitpur and Datia (Gupta et al. 2014). During the macro level analysis spatial and temporal analysis for Meteorological, Hydrological and Agricultural drought has been carried out for all the 13 districts of Bundelkhand region at district level. Macro level analysis for the entire region has been carried out with the objective of understanding the spatio-temporal inter-relationship with different type of droughts (Singh et al. 2013; Nair et al. 2013b).

In the meso-level study, block level analysis has been carried out for hydrological and agricultural drought for two selected districts, Lalitpur and Datia one each from MP and UP Bundelkhand. For analysing meteorological drought block level rainfall data was not available. TRMM data also was not found useful in deriving percent by normal since departure from normal at block level could not be computed in the absence of data (Fig. 6.7).

During the micro-level study a detailed household survey has been carried out in four villages, two each in Datia and Lalitpur districts. The present study is a combination of scientific and social science methods. Both quantitative and qualitative analysis has been carried out.

Macro level i.e. district level analysis of Meteorological, Hydrological and Agricultural drought for the entire Bundelkhand region has been carried out based on the drought indices derived from meteorological, hydrological and vegetation data (Nair et al. 2013a). Macro-level analysis depicts that meteorological drought may not necessarily lead to hydrological or agricultural drought. Drought declaration and occurrence of years of meteorological drought do not always coincide which is evident from the example of Lalitpur district. There was no major drought in the

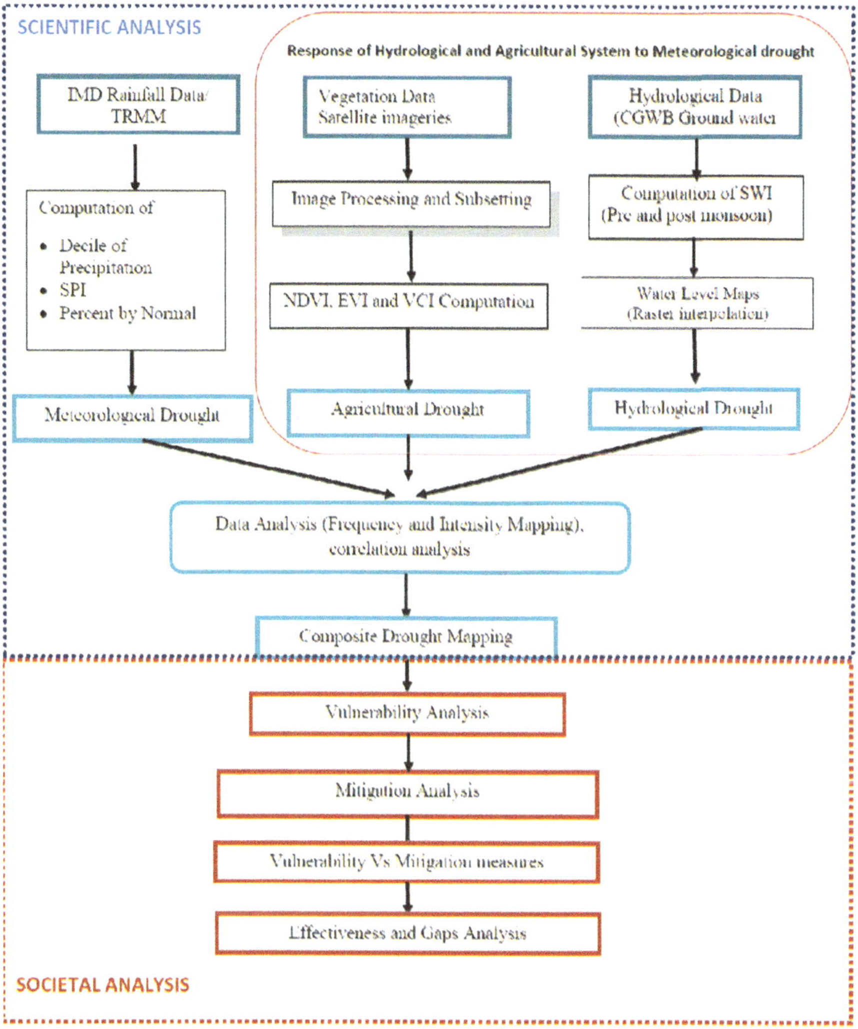

Fig. 6.7 Scientific and Societal analysis steps (Nair 2018)

region after 2009 and hence the data up to 2009 was analysed at regional level for identifying two districts for meso level analysis.

Drought related risk assessment during the macro level analysis was focused primarily on the impact of hazard i.e. i.e. the low rainfall on hydrological i.e. ground water and agricultural systems only. Standard Water Level Index derived from ground water level data is used for two reasons viz. (1) Reservoir level depends not only on the rainfall in the district but more on the catchment (2) practical difficulties in calculating the reservoir water level or surface water level at district or block level. 'Composite Drought Risk Map' for the Bundelkhand region has been developed by integrating hazard and vulnerability maps (Fig. 6.8).

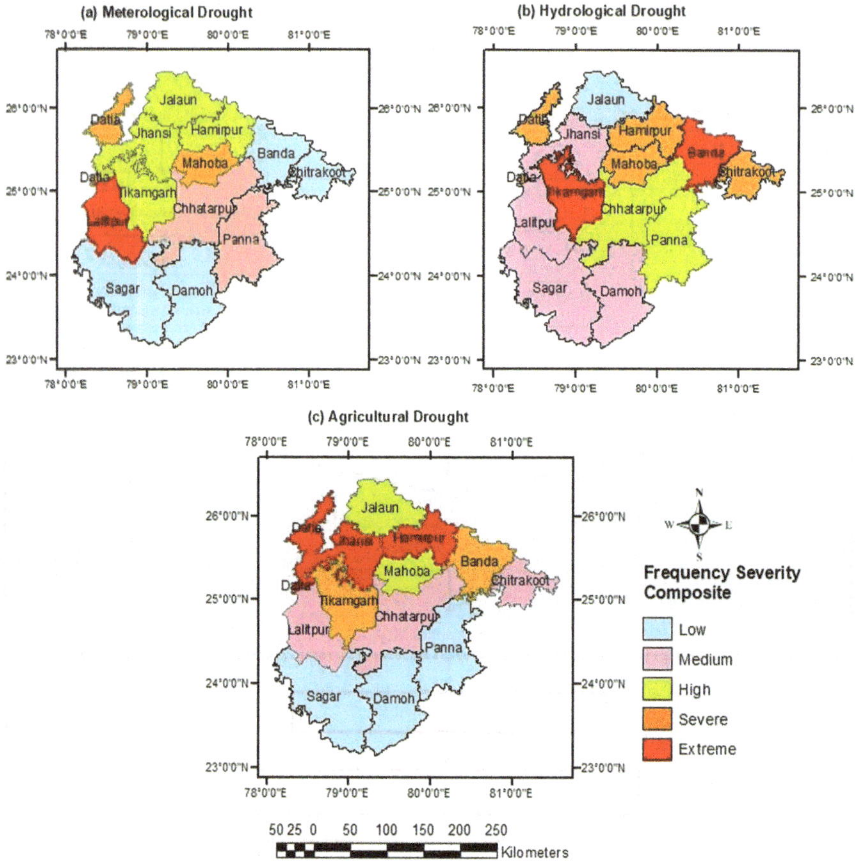

Fig. 6.8 Maps of meteorological, agricultural and hydrological drought (frequency and intensity composite)

The risk map has been prepared combining the effects of all the three types of droughts i.e. meteorological, hydrological, and agricultural for the entire region at district level (Gupta et al. 2014).

A methodological framework for carrying out the vulnerability assessment at village level has been developed. The new framework has been devised after extensive review of literature on vulnerability assessment, pilot visits and interaction with experts. This framework is intended to capture 4 components under vulnerability and 22 indicators under the 4 components. The indicators are specific to drought vulnerability in a predominantly agrarian society. The methodological framework and approach was field tested in 3 villages of Lalitpur district and 6 villages of Datia to make sure that the approach is contextualized and suitable for the drought vulnerability analysis. Comparative analysis of villages in Lalitpur and Datia district reveals that villages in Datia are relatively more vulnerable than

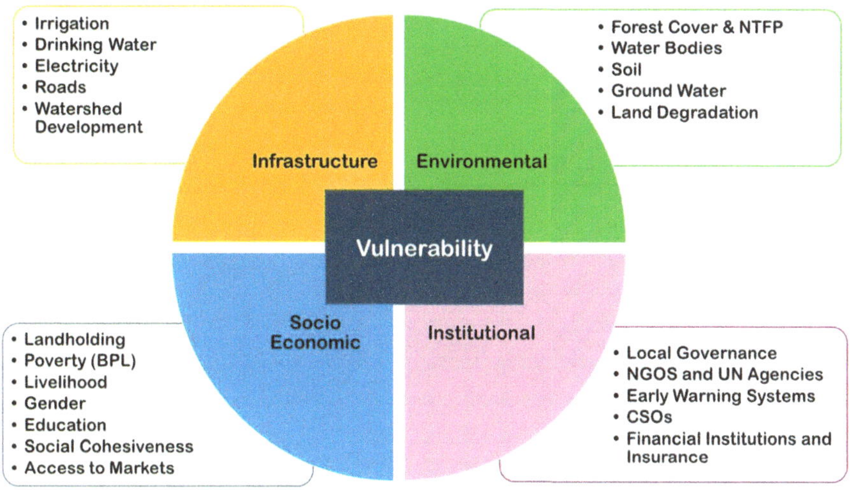

Fig. 6.9 Vulnerability analysis framework for Bundelkhand Region

villages of Lalitpur district. Mitigation analysis has been carried out in all the four villages to understand the progress and effectiveness of interventions and how far the measures are addressing the existing vulnerability scenario. Gap analysis has been carried out following the vulnerability and mitigation analysis for all the 4 components and 22 indicators (Fig. 6.9).

It is interesting to observe that vulnerability picture is changing drastically while scaling down to block, villages, and household levels. At district level the scientific study shows that Lalitpur is less vulnerable. However, village level analysis and household analysis shows that Masora Kalan in Lalitpur is the most vulnerable, and interventions are not appropriate and hence not addressing the socio-economic and institutional vulnerability context. The study has helped in understanding the spatio-temporal context of drought vulnerability at different levels in the region. The study is a unique attempt to combine the scientific analysis with socio-economic analysis (Nair 2018).

6.6 Mainstreaming Drought Risk Reduction by Convergence

Drought risk can be reduced by integrating the elements of risk reduction on sector specific programmes. This will help in mitigating the impacts of drought in the pre-disaster phase, alternative employment and livelihood options during drought period and also for early recovery, and enhancing resilience. DRR can be mainstreamed through programmes like Backward Region Grant Fund (BRGF), Bundelkhand Special Package, Integrated Watershed Management Programme

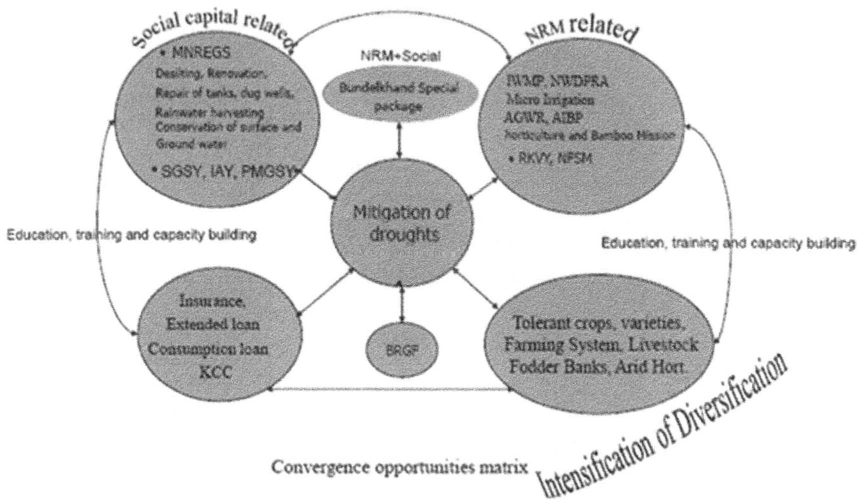

Fig. 6.10 Mainstreaming drought risk reduction through convergence (Samra 2008)

(IWMP), Rashtriya Krishi Sinchayi Yojna (RKSY) etc. Many of these schemes and programmes offered diversification of livelihood options from the core and helped people move to non-crop occupations. Alternative employment opportunities particularly during drought hit periods in such areas could come as a drought relief under the programmes and schemes such as Swarnajayanti Gram Swarozgar Yojana (SGSY).

Interventions aimed for promoting agriculture productivity were, for example, Village Level Pulses Seed Development Programme, Mechanization programme. Projects, viz. Swajal and Swajaldhara, Total Sanitation Campaign, Accelerated Rural Water Supply Programme (ARWSP), Indira Awas Yojana (IAY), Urban Infrastructure.

Development Scheme for Small and Medium Towns (UIDSSMT), Pradhan Mantri Gram Sadak Yojana (PMGSY), National Food Security Mission (NFSM), Mahatma Gandhi Rural Employment Guarantee Scheme (MNREGA) contributed both directly and indirectly in reducing the risk. Certain programs like Rashtriya Krishi Vikas Yojana (RKVY) and Sarva Shiksha Abhiyan (SSA), etc. focused on research extension and capacity building. There are certain social welfare programs like Integrated Child Development Services (ICDS), Mid-day Meal Scheme, which help in assuring food security during drought period. An indicative not holistic framework for possible mainstreaming drought risk reduction through convergence is given in Fig. 6.10.

6.7 Conclusion

Drought has always been chronic issue worldwide, including in India due to the dependency on agriculture particularly rain-fed for livelihood. Although substantial progress has been made in the irrigation sector, several areas of the world agriculture is still untouched. A global assessment of the disaster hotspots has been carried out by World Bank in the year 2005. Findings show that about 38% of the world's land area has some level of drought exposure.

India has a long history of drought and famine with 68% of the net sown area of the country prone to drought. As per the irrigation commission report and the recent National Drought Manual Published in the year 2016, 105 districts of India are drought prone. Apart from this there are chronically drought prone areas like Bundelkhand region of central India Marathwada region of Maharashtra, KBK region of Odisha and part of Telangana state. The main cause of drought in India is deficiency of rainfall. Other causes are increase in temperature, changes in weather and air circulation, El Nino impact, the weather patterns, inconsistent monsoon rainfall, unsustainable agricultural, land use practices, and timing of water availability. The impacts are felt by the society and economy much beyond the areas actually experiencing the meteorological drought because agricultural production and water resources are integral to our ability to produce goods and services. Drought affects the overall economy of the country at macro and micro economic levels, both directly and indirectly. Direct impacts are visible in falling agricultural production and heightened food insecurity among poor and vulnerable sections; depleted water levels; higher livestock and wildlife mortality; cattle and animal migration; damage to ecosystem from indiscriminate exploitation; increased fire hazards etc.

There has been a large gap between the impact of the drought and the response of delivering relief, helping recovery and building resilience. Functioning of public distribution system, mid-day meals, Integrated Child Development Services, the Mahatma Gandhi National Rural Employment Guarantee Act and other social security schemes are critical in times of emergency. Implementation of these plans needs to be executed with full efficiency. In rural areas, water bodies, the common properties and other natural resources are depleting fast and the traditional knowledge and practices of local communities in protecting these sources are being neglected. Focus is on large scale engineering investment projects (e.g. Kaleshwaram lift irrigation in Telangana); big dams are prioritized over minor irrigation and alternative energy options. There is a need to give more focus on risk reduction centric approach for mitigating drought at various levels. Risk Assessments for drought is yet be carried out in many states of India although drought prone districts are identified. Segregating and analysing drought based on the causes and the typology is needed since different type of droughts needs different interventions. For example, in case of Met-drought irrigation focused interventions are important. How-ever in the case of hydrological drought priority should be given to watershed development, soil and water conservation etc.

Various structural and non-structural mitigation strategies are adopted at various levels by central government, state governments, NGOs and community to reduce the impact of drought. Mitigation actions, programs, and policies are implemented during and before drought to reduce the magnitude of risk to human life, property, and productive capacity. Structural mitigation measures include construction of dams, reservoirs, canals (large scale, medium scale) and micro irrigation projects, water supply systems, infrastructure of rainwater harvesting, de-silting and rejuvenating of water bodies, power supply for irrigation projects, digging of wells and so on. Besides the structural interventions various non-structural measures including afforestation, soil and water conservation, formulation of evidence based policies, plans, programs and implantation strategies can also reduce the drought related sufferings. Other than policies, technological solutions like risk assessment, forecasting and warning systems, communication systems, cloud seeding as well as drought resilient crop varieties can reduce the negative impact of drought. There is an urgent need to adopt a convergence approach for mainstreaming DRR in drought risk reduction through policies, programs, projects and schemes at all levels.

6.8 Way Forward

Large number of initiatives were implemented in India for reducing impacts of drought. However there been a large gap between the impact of the drought and the response of delivering relief, helping recovery and building resilience. Integrated approach of carrying out interventions to reduce the impact of drought in changing climatic conditions is the need of the time. The following are the key recommendations.

(i) **Research**

Despite the enormous economic, environmental, and societal impacts of droughts, drought related research is yet to achieve momentum in India. Due to the slow on set nature and creeping phenomena it is difficult to assess the impacts. Climatic changes in likely to have serious consequences and can lead to extreme weather events, increase in evapotranspiration, erratic rain fall and increase in frequency and intensity of droughts. So there is a need to carry out holistic drought assessment factoring climatic changes.

(ii) **Documenting practices**

To increase resilience to droughts and water scarcity crises, the present situation calls for greater and more effective coordination and knowledge sharing between central, state- and district-level governments, inter-sectoral expertise (climate, agriculture, water), development sector players, communities and all other stakeholders. There is a need to document best practices and use of indigenous knowledge in water shed management, forest management, community based disaster risk management and other ecosystem centric approaches and how it helps in reducing drought risks.

(iii) **Developing and implementing EWS**

People centric End-to-End EWS for drought is yet to become a reality in our country. There is a huge disconnect between forecasting and early warning systems as well as dissemination mechanisms. Forecasting and warning systems are focussed on the hazard maps and hardly consider the vulnerability and risk factor. Further the user end is neither having the capacity to make use of the warning message nor having enough capacity to respond. So there is need to focus on EWS with enough linkages between the four elements.

(iv) **Irrigation Systems**

There are various types of systems of irrigation practices in different parts of the counry. Irrigation is been carried out through reservoirs, canals, wells, tanks, perennial canal, multi-purpose river valley projects. Micro-level irrigation systems particularly at farm level needs more attention to avoid water loss and conjunctive use of water.

(v) **Resilient Crops**

Drought resilient crop varieties needs to promoted and market for those products to be improved. Drought tolerant crops including sorghum, millets, pigeon pea and groundnut are smart crops because of their high nutritional value, high resilience under extreme weather and the potential to improve incomes of smallholder farmers living in very dry areas.

(vi) **Policies**

Strategic Environment Impact Assessments of policies related to watershed interventions, crop pricing, providing free electricity etc. to be carried out in drought prone areas. For example free electricity for farmers can lead to over utilisation of ground water for irrigation and can lead to lowering to water table. Similarly assured price of some crop variety including paddy can have negative consequences.

References

Centre for Research on the Epidemiology of Disasters (2016) The human cost of natural disasters 2015: a global perspective. CRED, Brussels

Centre for Research on the Epidemiology of Disasters (2019) Natural disaster 2018. CRED, Brussels. Accessed from https://emdat.be/sites/default/files/adsr_2018.pdf

Dilley M, Chen R, Deichmann U, Lerner-Lam A, Arthur L, Arnold M, Yetman G (2005) Natural disaster hotspots: a global risk analysis. The World Bank, Washington D.C.. Disaster risk management series

Eriyagama N, Smakhtin V, Gamage N (2009) Mapping drought patterns and impacts: a global perspective. International Water Management Institute, Colombo

Gupta AK, Nair S, Ghosh O, Singh A, Dey S (2014) Bundelkhand drought : a restrospective analysis and way ahead. National Institute of Disaster Management, New Delhi

Hewitt K (1997) Regions of risk: a geographical introduction to disasters. Longman, Harlow

Ministry of Agriculture (2009) Manual for drought management. Department of Agriculture and Cooperation, Ministry of Agriculture, New Delhi

Ministry of Agriculture (2016) Manual for drought management. Ministry of Agriculture & Farmers Welfare, Government of India, New Delhi

MoEF&CC (1988) National Forest Policy. Ministry of Environment and Forests, Government of India

MoEF&CC (2006) The National Environment Policy. Ministry of Environment and Forests and Climate Change, Government of India

MoWR&GR (2012) National Water Policy. Ministry of Water Resources, RD & GR, Government of India

Nair SS (2018) Climate change drought risk and mitigation strategies, study of Bundelkhand region. India School of Science, IGNOU, New Delhi, unpublished doctoral thesis

Nair SS, Nathawat M, Gupta AK (2013a) Indices for drought hazard mapping monitoring and risk assessment: analysis of existing tools techniques and approaches. Disaster Dev:81–95

Nair S, Singh A, Gupta A (2013b) Drought risk and vulnerability analysis for Bundelkhand region of India. Int Geoinform Res Dev J:1–19

National Disaster Management Authority (2010) National disaster management guidelines: management of drought. National Disaster Management Authority, New Delhi

NOAA (n.d.) Definition of drought. https://www.ncdc.noaa.gov/monitoring-references/dyk/drought-definition

Samra J (2008) Report on drought mitigation strategy for Bundelkhand region of Uttar Pradesh and Madhya Pradesh. Inter Ministerial Team, New Delhi

Singh A, Nair S, Gupta A (2013) Comprehensive drought hazard analysis using geospatial tools: a study of Bundelkhand region, India. Disaster Management and Risk Reduction – Role of Environmental Knowledge

UNCCD (1994) United Nations convention to combat desertification (UNCCD). Secretariat, UNCCD, Bonn. https://www.unccd.int/convention/about-convention

UNISDR (2016) Press release: the human cost of the hottest year on record - climate change and El Nino drove disasters worldwide in 2015. UNISDR

UNISDR, CRED (2015) The human cost of weather-related disasters 1995–2015. United Nations International Strategy for Disaster Risk Reduction, Geneva & Center for Research on the Epidemiology of Disasters (CRED), Brussels

Sreeja S. Nair is presently working as Fellow with the Earth Sciences and Climate Change Division of The Energy and Resources Institute (TERI). She has more than 20 years of experience in the area of Disaster Reduction and Climate Change adaptation and work with National, International and Regional Organisations including National Institute of Disaster Management, UNDP, UNDRR, UN-Habitat, ADPC, RIMES, Aga Khan Agency of Habitat and so on.

Anil K. Gupta , Professor of Policy and Strategies, Integrated Resilience. Head of the ECDRM Division, International Cooperation, Programme Director—Centre for Excellence on Climate Resilience, Nodal Officer for Central Ministries at National Institute of Disaster Management, (Govt. of India), New Delhi. He is also a member of the expert Team for WMO Climate Statement and Core Group Member of IUCN-Council of Ecosystem Management.

Professor M. S. Nathawat is currently working as Director, School of Science, in Indira Gandhi National Open University, (IGNOU), New Delhi, India. His research interest area are—Environment Geomorphology, Desertification, Glaciers and Climate Change. He worked as professor and headed the Department of Remote Sensing and Geo-informatics at Birla Institute of Technology, Mesra, Ranchi in Jharkhand, India for more than a decade.

Chapter 7
Cyclone Disaster Mitigation and Management in India: An Overview

Pradeep K. Goyal

Abstract The Indian subcontinent is one of most affected cyclone prone region in the world. Indian coast is about 7156 long km out of which 5400 km along the mainland are highly vulnerable to cyclone and related hydro-meteorological hazards. The coastal states namely, Odhisha, Andhra Pradesh, West Bengal, and Tamil Nadu are more prone to cyclone related hazards. Tropical cyclones are associated with strong wind, heavy rainfall, flood and storm surge. They can cause devastating damage to structures and huge loss of lives. Nearly one third of the population living in the coastal region of India is at a high risk due to Cyclone related hazards. The increase in the intensity and frequency of cyclones and rise in sea level due to climate change have led to significant increase in the vulnerability of coastal region.

Learning from the past experience of super cyclone Odhisha in 1999, it has been observed that the loss of lives due to cyclone has been reduced significantly by improving initial warning systems and information dissemination, preparedness and undertaking several risk mitigations measures. But the damages to infrastructures and economic losses were very high due to past cyclones. Therefore, there is a need for integrating disaster risk reduction measures across all sectors of development. This paper presents a brief overview of cyclone disaster mitigation and management in India.

Keywords Natural hazards · Cyclones · Vulnerability · Damage · Wind · India

7.1 Introduction

Tropical Cyclones are among the most damaging natural hazards causing significant loss of life and property when they hit the coastal region worldwide. It has been observed that around 80% of natural disaster economic losses are caused due to impact of cyclones (Tamura 2009; Tamura and Cao 2012).

P. K. Goyal (✉)
Department of Civil Engineering, Delhi Technological University Delhi, New Delhi, India
e-mail: pkgoyal@dtu.ac.in

A. K. Gupta et al. (eds.), *Disaster Risk and Management Under Climate Change*, Disaster Resilience and Green Growth,
https://doi.org/10.1007/978-981-99-4105-6_7

India is highly prone to natural hazards such as cyclones, earthquakes, floods, drought and landslides leading to huge losses of life and property. Tropical cyclones are the most disastrous hazard in the coastal region. The east coast of India is more prone to cyclone as compared to the west coast. The ratio is approximately 4:1. It has been observed that about five to six cyclones occur in a year in the Indian coastal region, out of which two or three may be severe leading to large number of deaths, loss of property and severe damage to infrastructure.

Nearly one third of the India's population lives in coastal region and is highly vulnerable to Cyclone related hazards (NDMA 2016). Almost 250 million populaces reside within 50 km of Indian coastline (ICZMP 2010) while over 40% of our total population resides within 100 km of our coastline.

Vulnerability of the coastal population has also increased due to climate change resulting in the rise of sea level and on account of increase in the frequency and intensity of cyclones.

Developing countries are more vulnerable to cyclones than the developed countries. Hence, the problem of cyclone disaster mitigation is of national importance. As such it is not possible to prevent or control the natural hazards but the impact of hazard can be minimized by implementing mitigation measures such as, efficient early warning system, construction of safe, shelters for cyclone and houses resistant to cyclones, culverts, road links and bridges as well as the strict implementation of Coastal Zone Regulations.

7.1.1 Tropical Cyclones

Tropical cyclones, generally known as cyclones are tropical weather systems (World Meteorological Organisation (WMO) 2009) in which winds equal or exceed the minimum of 34 knot (62 kmph). In Northern Hemisphere, wind blows in counter clockwise direction whereas clockwise in the Southern Hemisphere due to the Coriolis force. The formation of a tropical cyclone depends on several conditions which include:

- A warm sea surface temperature (above 26.5 °C)
- High relative humidity in the atmosphere.
- Atmospheric instability
- Lower vertical wind shear
- Coriolis force
- An initial low-pressure disturbance

A tropical cyclone is a large funnel-shaped storm with a wide top of the order of 1000 km in diameter and narrow bottom of order of 300–500 km in diameter. The height of the storm is order is of 10–15 km. The diameter of a cyclone encompassing the region of relatively strong wind is order of 500 km. The centre part of a cyclone is called the eye. The eye normally has an average radius of about 20–30 km. The pressure is the lowest and temperature is the highest in this region. The eye is surrounded by a ring of very strong wind extending on an average up to 30–50 km beyond the centre, known as eye wall. The strongest deep convection and often the most powerful winds can be found in the eyewall. Outer part of a cyclone is known as rain bands. These bands may be hundreds of kilometres long and a few kilometres wide. The intensity of a cyclone can be derived either by the low sea level pressure at the tropical cyclone pressure centre or the maximum sustained winds in the eyewall.

The cyclones that develop between Tropics of Cancer and Capricorn are knowns as Tropical Cyclones. The tropics are warm all year, averaging 25–28 °C. Cyclone is not formed at the equator because of zero Coriolis force which exists at the equator however, Coriolis force increases with latitude so cyclones are reported as we move away from equator. Coriolis force at 5° latitude is significant enough to create a storm. That is the reason most of the tropical cyclones are formed within the belt of 5° to 30° N and 5° to 30° S.

Hurricanes, typhoons, cyclones, and so on are different names for the same type of severe storms occurring in different geographical regions. Tropical Cyclones are known as hurricanes in the Atlantic Ocean, typhoons in the Pacific, Ocean and Willy-Willies in Australia seas and Cyclones in the North Indian Ocean. There are three hazards associated with a cyclone, which cause tremendous amount of destruction in coastal region. These are strong winds, storm surge and heavy rainfall causing flood.

7.2 Present Scenario

7.2.1 Global Trends

Cyclones were second only to earthquakes in terms of loss of lives, killing 233,000 people during 1998–2007. Storms were found to be the most damaging type of disaster, with estimated losses of US$1300 billion over the past 20 years, which is two times the losses for either flooding or earthquakes. In terms of occurrences, 91% climate-related disasters were observed during 1998 and 2017. Within this total, floods (43.4%) and storm (28.2%) were the two most frequently occurring type of disasters as shown in Fig. 7.1 (CRED and UNISDR 2018).

During 2018, 19.4 million people were affected due to storm, out of which 17.1 million people (88% of the global total) were located in Asia. Storms have been the costliest type of disaster in terms of economic losses. The economic losses were estimated to be $14 billion and $16 billion in Hurricanes Florence and Michael respectively in the United States. During this period, extensive damage from multiple storms was observed in Asia, China, India, Japan, Philippine (CRED and USAID 2018). During 2019, several notable and costly tropical cyclones formed within different parts of the world. Typhoon Faxai (Japan), Hurricane Dorian (Bahamas), Typhoon Lekima (China), Cyclone Fani (India), and Tropical Storm Imelda (United States) were highly damaging disaster in terms of economic losses. It

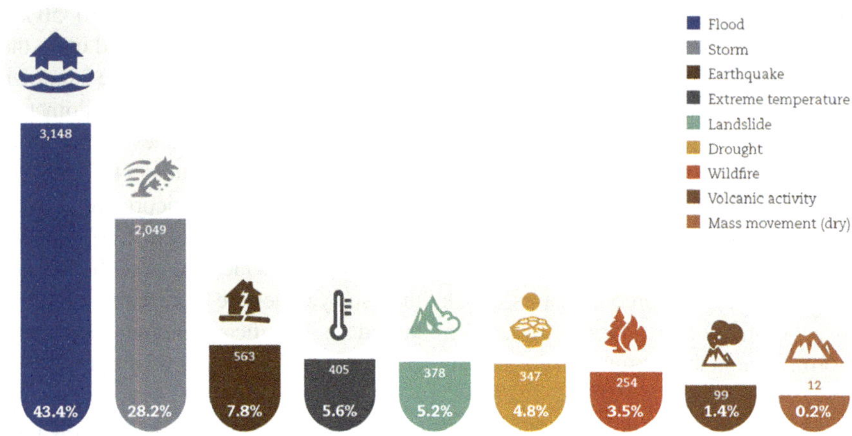

Fig. 7.1 Frequency of Disaster from1998–2017 (CRED and UNISDR 2018)

has been estimated that the economic loss was nearly USD58 billion in these six storms (Weather, Climate and Catastrophe Insight, Annual Report 2019).

7.2.2 National Level

India is one of the most cyclone prone region of the World causing huge loss of lives and properties. The Orissa super-cyclone of 1999 was the worst cyclone of the country which killed over 10,000 people, and caused widespread damage to property and economic losses. During the period of 1980 to1990, the average annual loss was around Rs. 200 crores (44 million US$). The estimated loss was around Rs. 8000 crores between 1996 to 1999 (Goyal et al. 2012).

It has been observed that 308 cyclones crossed the east coast of Indian region, out of which 103 were severe during the period 1891–2006. Within this period, 48 cyclones have crossed the western coast, it was reported that among these almost 24 cyclones were categorized as being of severe intensity (NDMA 2016).

"Phailin" a very severe category of cyclone hit Odisha's coast in 2013. It resulted in extensive damage to properties and destroyed the power as well as communication infrastructure within the State's coastal districts. According to Rapid Damage and Needs Assessment Report (2013), the monetary loss on account of Cyclone Phailin was the order of about Rs 89,020 million (this amount was almost equivalent to 1450 million USD).

Over 13.2 million residents were affected within 171 blocks located in 18 districts over the state. Approximate 30 people lost their lives during Phailin. It was observed that the death toll reduced significantly because of timely evacuation of a large populations during Phailin in 2013 and Hudhud in 2014, however, there were significant physical and financial losses.

Cyclone Fani, hit the coastal region in 2019, the estimated total damage amounted to Rs 16,465 crores (almost equivalent to 2352 million USD) it killed about 64 people and damaged more than 3,60,000 houses in the state's coastal districts (Damage loss and need assessment Report 2019).

The year 2019 was recorded one of the most active cyclone seasons in the North Indian Ocean (NIO). During 2019, 8 cyclonic storms formed over the Indian North oceans. Out of which 5 cyclones formed over Arabian Sea against the normal of 1 per year. The frequency of cyclones was abnormally high for the Arabian Sea due to increase in temperature and other reasons.

The Super Cyclonic Storm "AMPHAN" was the first Cyclonic Storm formed over the BoB, after the Odisha super cyclone of 1999, that cause considerable damage in eastern part of the country, especially in West Bengal in May 2020. After Amphan, another severe cyclone storm, Nisarga formed over the Arabian sea, was the strongest cyclone to strike the Maharashtra coast in the month of June, 2020 since 1891.

7.3 Classification of Tropical Cyclones

The India Meteorological Department (IMD), is the nodal agency which is responsible for monitoring and forecasting of cyclones in India. Cyclones are categorized on the basis of wind speed in their circulation. However, categorization of cyclones differs in every region. In United States, Hurricane are categorized into five different classes on the basis of their wind speeds as captured over the Saffir–Simpson Hurricane Wind Scale.

The categorization of cyclones is based on maximum sustained winds associated at the surface level, it is presented within Table 7.1. These systems with low-pressure and variable intensity are usually referred to as cyclonic disturbances (Mohapatra 2015).

A table of the damage caused by tropical cyclones in the coastal region, increased storm intensity is associated with greater damage are available in Standard Operation Manual of Cyclone Warning in India published by, IMD, Delhi in 2013.The

Table 7.1 Indian system for categorization of cyclones

S. no	Classes	Highest sustained wind speed
1.	Low pressure area	Lower than 17 knots (<31 kmph)
2.	A depression	17–27 knots (31–49 kmph)
3.	Deeper depression	28–33 knots (50–61 kmph)
4.	Cyclone storm	34–47 knots (62–88 kmph)
5.	Severe cyclone storm	48–63 knots (89–118 kmph)
6.	Very severe intensity cyclone storm	64–119 knots (119–121 kmph)
7.	Super cyclone storm	120 knots and greater than 120 knots (≥222 kmph)

Source: IMD (2013)

information is based upon the prevailing scientific knowledge of tropical cyclones and expert opinions. From this table, it is possible to predict the potential damage at a particular intensity of storm, accordingly the measures can be undertaken to minimise the impact of cyclones.

7.3.1 Classification of Cyclone Prone Areas of India

The extent of damage caused by a tropical cyclone in any coastal area is determined by storm surge and wind. Among these two factors strong winds are responsible for causing the major damage. Hence it becomes significant to classify regions on the basis of cyclone-wind risk vulnerability for undertaking mitigation measures. This provides assistance in identifying locations in which lifeline systems are prone, selection of pockets for the provision of important community services, preparing insurance policies and consider those areas which require strict provisions for code.

The macro-level wind speed zones have been given for the country IS 875 (Part 3):2015. India has been divided into six zones, namely, 55, 50, 47, 44, 39 and 33 m/s on the basis of basic wind speed (V_b). From wind hazard zonation of point of view, these zones can be categorised as shown in Table 7.2.

Wind Hazard Maps have been published in Vulnerability Atlas of India which divides the area into six zones as given in Table 7.2. The Wind Hazard Map of India is given in Fig. 7.2. The cyclone affected coastal region of India come under 50 and 55 m/s zones.

Wind Hazard maps for both India and state/UT levels are available in Vulnerability Atlas of India (third edition) published by *BMTPC* in 2019. The district-wise housing vulnerability risk Tables based on wall types and roof types are also available as per 2011 Census Housing data. It is a useful tool for urban managers, town planners, Government agencies for taking necessary steps related to disaster management. It is useful in policy planning, preventive actions for housing and related infrastructures, warning system installation, land use planning, etc. thereby aiding in better cyclone management.

Table 7.2 Wind hazard zonation of country

Basic wind speed (V_b)	Risk zone
55 m/s	Very High Damage Risk Zone—A
50 m/s	Very High Damage Risk Zone—B
47 m/s	High Damage Risk Zone
44 m/s	Moderate Damage Risk Zone—A
39 m/s	Moderate Damage Risk Zone—B
33 m/s	Low Damage Risk Zone

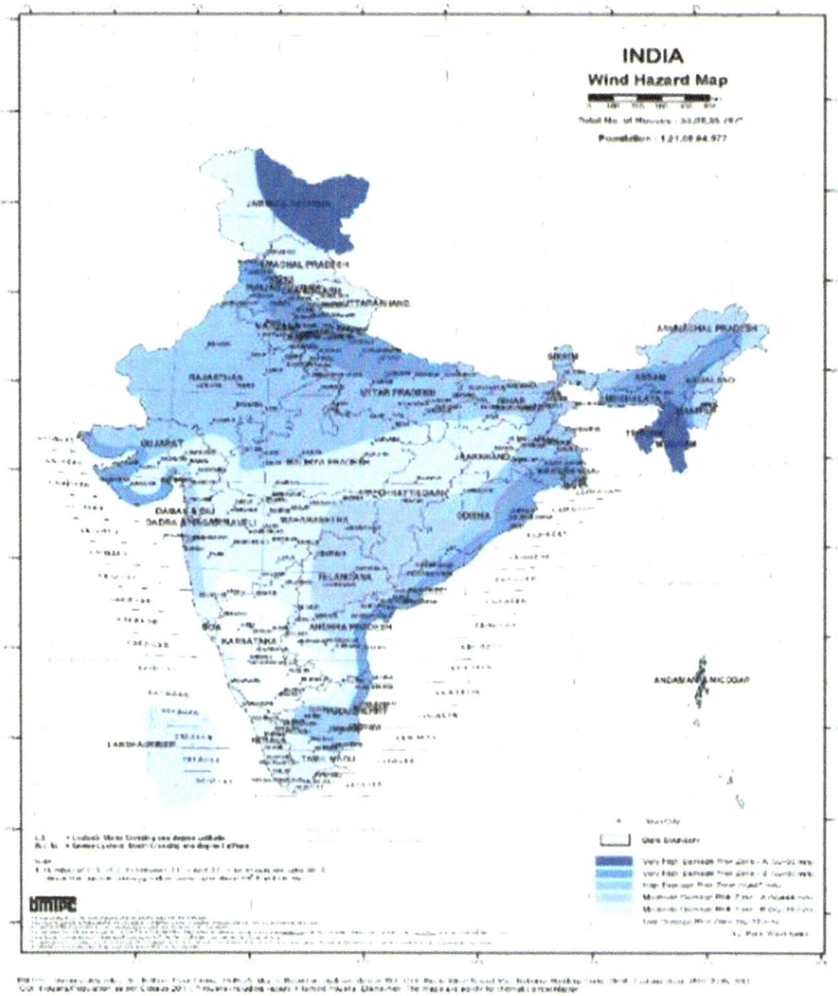

Fig. 7.2 Wind Hazard Map of India (Source: Vulnerabilty Atlas of India 2019)

7.3.2 *Cyclone Occurrence Map of India*

Cyclone occurrence map for the Indian coastal region has been developed by Building Materials and Technology Promotion Council (BMTPC). The map has been prepared based on maximum 3-min average Maximum Sustained Wind of cyclone storms that affected coastal districts of India during 1891–2008. The Vulnerability Atlas of India, 2019 (third edition) has published the cyclone occurrence map of India as shown in Fig. 7.3.

The classification of cyclone hazard prone districts by adopting hazard criteria in India was conducted in the year 2015 (Mohapatra et al. 2015). The districts are

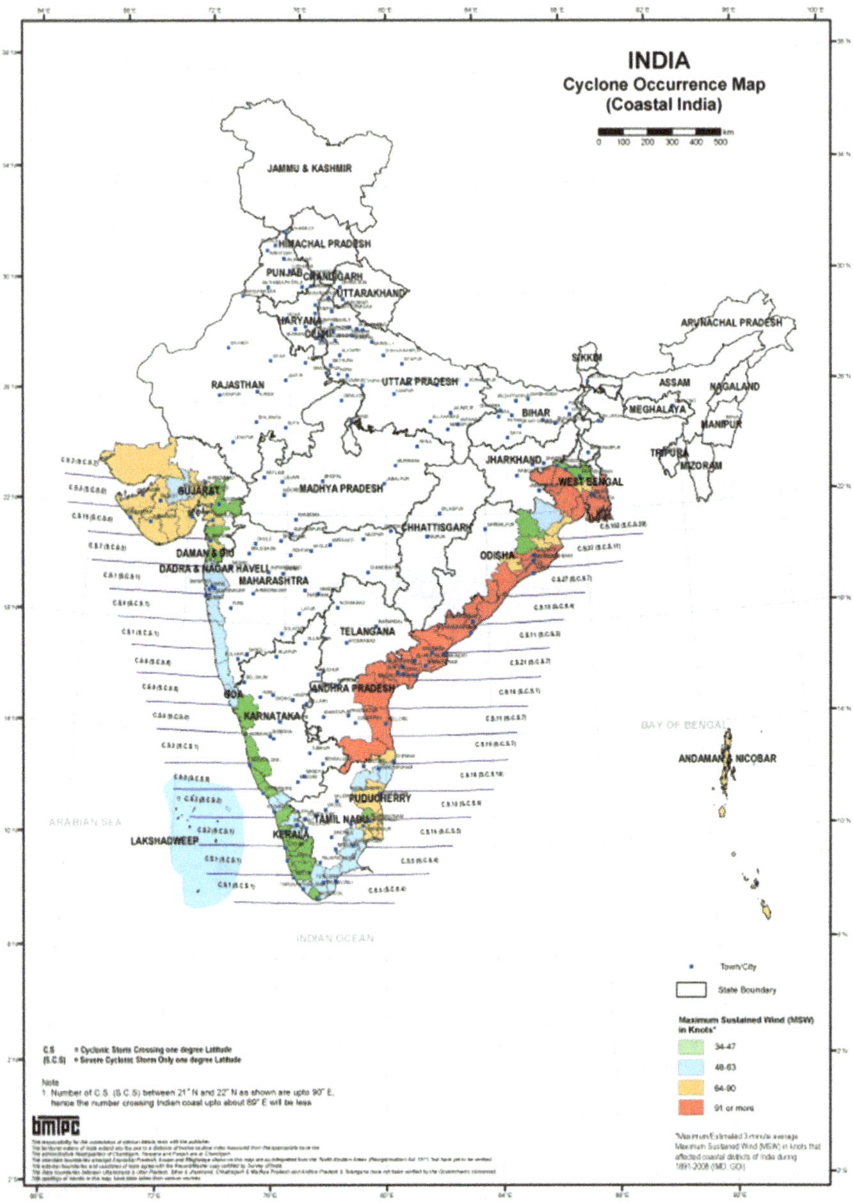

BMTPC . Vulnerability Atlas- 3rd Edition, Peer Group, MoHUA, Map is Based on digitised data of SOI, GOI. Maximum Sustained Wind (MSW) Data from IMD, GOI. Disclaimer: The maps are solely for thematic presentation

Fig. 7.3 Cyclone occurrence map (Source: Vulnerability Atlas of India 2019)

categorised as P4 (Low), P3 (Moderate), P2 (High) and P1 (Very high) based on the degree of cyclone hazard proneness. The degree of proneness was determined by undertaking the analysis of the cyclone frequency along with the intensity of the land falling cyclones, in addition other hazards such as wind and storm surge. This classification of districts may be considered for coastal zone management and planning. Goyal and Datta (2012) presented a method for cyclone microzonation of Indian coastal areas with assistance from cyclone historical records. This procedure is very useful for those regions where measured parameters such as wind speeds are not available particularly within the coastal areas of the developing as well as the lesser developed nations.

7.4 Damage Assessment of Buildings

All buildings have been classified into three main categories in India as non-engineered (NE), semi-engineered (SE) engineered (E). Buildings, which are constructed using the conventional building materials and local construction practices are classified as non-engineered building. Buildings constructed from structural materials without proper design are classified as semi-engineered buildings. Buildings, which have been designed and constructed as per Indian standards are termed as Engineered buildings (IS 15498: 2004).

In the coastal region of India, non-engineered and semi-engineered houses are the most common residential dwelling and are a major source of casualties due to cyclones.

Study conducted by Shanmugasundaram et al. (2000) shows that failure of non-engineered houses ranged from thatch cover's complete blowing off to total lifting of the roof and the bamboo structure, several times the mud walls survive while other times there is a total breakdown of the mud walls. With regard to the semi-engineered dwellings the extent of damage ranges from breakage of the tiles and AC sheets to a total lifting-off of the roof's structure resulting inconsiderable damage to the walls. Hence, assessment of vulnerability or damage of different forms of dwellings in a coastal region is important in relation to cyclonic disaster mitigation. The damage assessment methods are classified as qualitative and quantitative methods. Qualitative damage is determined using damage scale according to Indian standard procedure (IS 15499: 2004). Guidelines for survey of housing for vulnerability assessment in cyclone prone areas is given in IS 15499:2004 (Reaffirmed 2015). Damage can be categorised as None, Marginal, Medium, Heavy and Total based on failure of houses.

It has been observed that damage to houses is mainly responsible for loss of life in the coastal region. Therefore, there should be greater emphasis on the safety of houses to minimise the damage to property and loss of life. Bureau of Indian Standards IS 15498:2004 (Reaffirmed 2015) has published guidelines for construction of houses in cyclone prone region.

There are two quantitative methods for vulnerability and damage assessment for a group of houses (1) Direct method and (2) Component-based method. In the direct method, the model is developed for the entire house. In the component-based method, the vulnerability of roof, secondary roof members, wall, etc. are computed

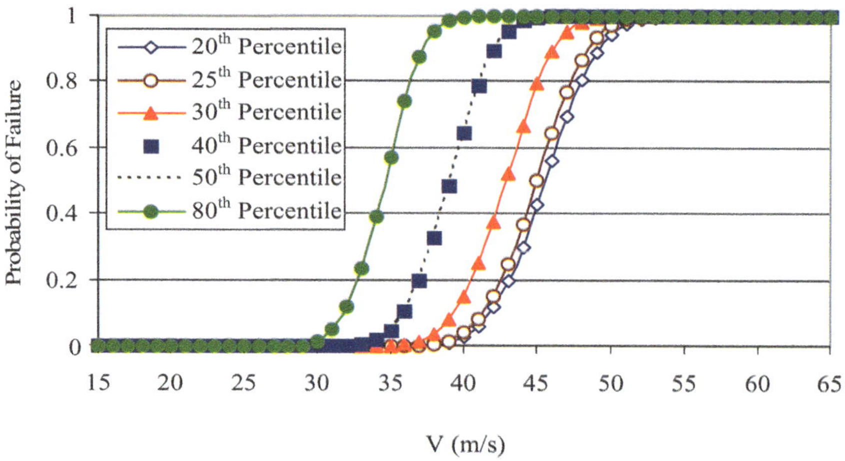

Fig. 7.4 Fragility curves for damage state III for NE:SE = 3:1 (Source: Goyal et al. 2012)

for each component. The vulnerability model of the structure is developed by combining components vulnerabilities of structure and their interactions.

A procedure has been presented by Goyal et al. (2012) for the estimation of the susceptibility of a rural house cluster to cyclones. It involved undertaking a fragility analysis analogous to varying damage states such as damage state I, state II and state III respectively of the rural houses within a cyclone susceptible region, lacking systematic data for carrying out such a statistical analysis. This method is of use in getting an initial assessment of the probability of collapse of a rural house cluster for varying states of damage and accordingly, in optimally allocating the resources for the mitigation of cyclone. Fragility curves depicting the overall failure probability for percentiles ranging from 20th–80th percentile corresponding to damage state III for NE to SE ration of 3:1 are shown in Fig. 7.4.

A vulnerability assessment of rural houses using component-based method has been presented for the coastal region of India. Vulnerability curve for ratio of non-engineered and semi-engineered houses (n_s) equal to 3:1 is shown in Fig. 7.5 (Goyal and Datta 2014). A study was also conducted by Goyal and Datta for analysing the effect of wind directionality on vulnerability of rural houses (Goyal and Datta 2013).

7.5 Cyclone Risk: Mitigation Measures

There are structural as well as non-structural ways to reduce the impact of cyclones. Mitigation means measures taken prior to the impact of a disaster to minimise damage and injury from the storm. The structural measures imply any physical

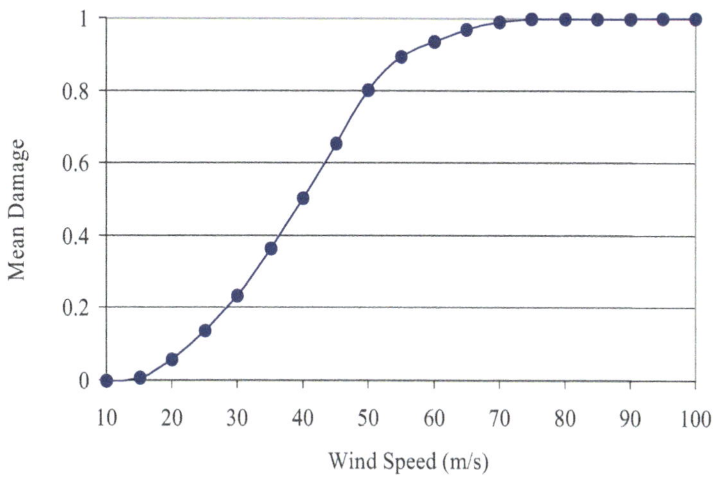

Fig. 7.5 Combined vulnerability curve for non-engineered and semi-engineered houses (Source: Goyal and Datta 2014)

form of construction to minimise the possible impacts of hazards to achieve hazard-resistance and protective the structures and infrastructure.

Structural measures consist of:

- cyclone shelter construction
- culverts, road links and bridges construction;
- drains, canals and water tanks construction,
- embankments for saline water;
- towers for communication and networks for power transmission etc.

Non-structural methods are the means to reduce cyclonic risk with the help of public awareness, training, policies, regulations, education, and vulnerability assessment of coastal regions. These measures consist of

- Early warning systems,
- Systems for communication as well as dissemination
- Coastal zones management
- Creating awareness and capacity building
- Disaster risk management

During cyclones, it is significant to ensure the provision of safe cyclone shelters for the protection of human life. It has been observed that loss of life is large due to the lack of cyclone shelters in coastal region. Many buildings, schools, places of worship, community halls, etc. have been utilized for providing shelter to population that is evacuated from the cyclone affected areas. It was found that several previously constructed cyclone shelters did not have pucca all-weather roads for access that need to be provided to affected people for shifting during emergency evacuation.

There is a need to review and improve the quality of existing road network, culvert and bridges. The coastal villages should be connected with the help of all-weather approach roads for proper coordination of relief and rescue operations and effective evacuation. The conditions of roads, bridges, and culvert should be review by the concerned agencies for smooth functioning of operations related to disaster management during cyclones. Roads, bridges, culvert should be constructed for the last mile connectivity.

The network of canals and drains plays a very important role in reducing the cyclone impact through the receipt, accommodation and return back of surge waters towards the sea. Coastal canals may also be used as an alternate to road during a cyclone or flood. Further it was reported that even the surface water tanks support in lowering the effect of cyclones as well as storm surges. Construction and renovating the existing saline embankments are required to protect people, live stocks and agricultural fields from saline water inundation/storm surge. Based on failure of communication and transmission towers in past cyclone, it has been recommended that towers should be constructed based on the 100-year return period of cyclone's wind velocity.

On the basis of past experiences, it has been observed that early warning system along with community-based management have played an important role in minimising the loss of lives, and reduce economic losses.

7.5.1 Early Warning System

Early warning systems form an important part of disaster management. Past experiences have highlighted their role in managing disasters particularly with in the South Asian region. Increased lead interval along with improved accuracy in forecasting of cyclone genesis, tracking of its intensity and adverse weather would be useful in improving disaster management. A crucial short-term cyclone mitigation measure is provision of timely warning to the population as well as the officials in the area. These predictions include cyclone's arrival time, intensity and the location. It is of significance to regularly upgrade the constituents of this early warning system on the basis of latest technology for better response. Such advances have greatly enhanced the capability of EWS and as per estimates this has been instrumental in reducing the human lives loss to even below 100.

Several significant steps have been taken to modernize and upgrade the cyclone monitoring and early warning systems in India by IMD. The warning in graphics bulletins has been introduced in addition to text bulletins. However, even at present few scenarios present difficulties in forecasting of cyclones such as rapidly becoming severe or weak cyclones close to the land.

In India the early warning systems for cyclone have gradually evolved over a time period. The storm warning system was first established for the Bay of Bengal storms at Calcutta in 1865 and for the Arabian Sea storms in Mumbai by 1882. After the establishment of India Meteorological Department (IMD) in 1875, cyclone warning started to be issued to various stakeholders (Sikka 2019). Now a days the warnings are being issued by Cyclonic warning organisation of IMD. In India, the Cyclone Warning Organization of IMD has a 3-tier system to cater to the needs of the coastal states. Cyclone Warning Division (CWD) set up at IMD Head Quarters Delhi is also functioning as Regional Specialised Meteorological Centre—Tropical Cyclones (RSMC—Tropical Cyclones), New Delhi to liaising with Central Government organisations and other government agencies. Cyclone Warning Division supervise the cyclone warning operations and provide the cyclone advisories to WMO/ESCAP Panel countries. There are three Area Cyclone Warning Centres (ACWCs) at Chennai, Mumbai and Kolkata and four Cyclone Warning Centres (CWC) at Visakhapatnam, Ahmedabad, Bhubaneswar, and Thiruvanathapuram.

The observational network for monitoring tropical cyclones should determine the correct location and intensity of cyclones. Presently, observations are collected from Surface Observatories, Upper Air Observatories, Radars and satellites monitoring and forecasting of tropical cyclones.

According to report of RMSC (2019), at presentism has 560 surface observatories. The data obtained from these stations are used on real time basis for operational forecasting. In, Upper Air Observatories, there are 62 Pilot Balloon Observatories, 43 Radiosonde/Radio wind observatories. All the 43 stations are latest of the art-GPS based observatories. The data collected all over the country are used on real time basis for tropical cyclone forecasting. There are 24 Doppler weather radars (DWRs) in weather radar network managed by IMD. IMD also utilizes the DWR installed by ISRO at Thiruvananthapuram, Cherrapunji and Sriharikota. Radars of IMD are being used for detection of rainfall, hail storm, dust storms, thunder storms and tracking of cyclonic storms. Within a phased manner IMD has planned the induction of almost 55 DWRs within the network so that the whole country including the coasts is covered under DWR. Now a days, two Indian satellites namely INSAT-3D and INSAT-3DR are used to receive and processing meteorological data. INSAT-3D and INSAT-3DR have an advanced imager with six imagery channels. The products derived from the satellite are useful for monitoring of tropical cyclones. IMD and ISRO have jointly developed a tool RAPID (Real-time Analysis of Product and Information Dissemination) for visualization, which can monitor and analyze satellite imageries and other products of INSAT 3D and 3D(R).

Four stages warning system has been introduced by IMD for tropical cyclones to issue cyclone warnings to state government and disaster managers from 1999.

- Pre-Cyclone Watch is the first stage of warning system, which is issued at least 72 h in advance. It is issued when a cyclonic disturbance in the north Indian ocean (NIO) is likely to intensify into a tropical cyclones and coastal region is likely to affect in future.
- Cyclone Alert is the second stage of warning system issued at least 48 h in advance of the expected commencement of the bad weather over the coastal areas (WMO and ESCAP 2009). It provides information regarding the storm's location, its intensity, the probable direction of movement, severity and coastal districts to be impacted by storm severity. System issues waning to fishermen,

media and the district officers. Colour code 'yellow' is used for the second stage of warning., 'cyclone Alert' and it is given by ACWC (area cyclone warning centre), CWC (cyclone warning centre) and the CWD (cyclone warning division)
- Cyclone Warning is the third stage of warning system issued at least 24 h before the commencement of the bad weather over the coastal areas at 3-h intervals (WMO and ESCAP 2009). This information also contains the expected damages and action suggested by giving the advice to media, fisherman, general public and agencies involved in disaster management. These warnings are issued by the ACWCs/CWCs/and CWD. Colour code 'orange' is used for 'Cyclone Warning'.
- Post landfall outlook is the fourth stage of warning system issued at least 12 h before cyclone landfall. At this stage, Information about time and place of landfall is more accurate and specific. The possible impact due to cyclone in interior areas are warned at this stage. Colour code' red' is used for forth stage of warning system.

Subsequently once the cyclone weakens into a depression, a 'de-warning' message is disseminated. Warnings are shared with public via fax, telephone, SMS, email, press release, GTS (Global Telecom System), WIS (WMO Information System), radio, community radio, TV, IVRS (Interactive Voice Response system) and other print and electronic media. The weather information and forecast of important Indian cities may also be accessed by dialling this Toll-free number 1800 1801717. A central IVRS caters to major cities weather updates. Over the years several initiatives have been undertaken by IMD to improve the weather forecasting and the warning systems on the basis of recent tools and techniques. From 2009, IMD has launched SMS service for weather and alerts via AMSS at New Delhi. IMD has also made use of Digital India platform to utilize the Mobile Seva provided by the Department of Electronics and Information Technology (DeitY). For the warnings based on SMS and weather-related information to varied users the assistance has been taken from the Ministry of Communication and Information Technology. On 25th Dec, 2014 cyclone-based SMS alert service for registered users from the public was launched. The GMDSS (Global Maritime Distress and Safety System) message also is highlighted on the RSMC website (URL: www.rsmcnewdelhi.imd.gov.in) and is also transmitted via the GTS. The WIS Portal–GISC based at New Delhi is an additional system for the dissemination of cyclone warning. Warning messages may also be accessed by users from the website URL: http://www.wis.imd.gov.in. Since 30th March, 2016 IMD has begun issuance of the NAVTEX bulletins for the coastal regions on both the east and west Indian coasts. It supports the functioning of fisherman and lightships. Along with these networks, for rapid dissemination of warnings to public and officials against upcoming cyclones, special receivers have been installed by IMD in the susceptible coastal areas through the utilization of broadcast capacity of the INSAT satellite. This direct broadcast service of cyclone warning is issued in the local regional languages.

7.5.2 Community-Based Cyclonic Risk Management

Community based disaster risk management is important aspect to achieve sustainability for managing disaster risk. It is not possible to manage and handle disaster without active participation of vulnerable community. It is decentralized bottom-up approach involve vulnerable people in implementation of disaster management and planning. The local community is the first to respond the disaster can play an important role to deal with the impacts of disasters. Communities should be supported to understand disaster risk. The emphasis on community-based disaster management is given in the National Disaster Management Plan (NDMA 2016, 2019). Capacity of community should be enhanced through awareness, sensitization, mock drills and developing skills. During cyclone Fani, volunteers acted immediacy and supported in evacuation of people from the vulnerable area. They also supported in clearing roads, providing medical aid, food distribution and relief materials to the cyclonic victims. Therefore, the community-based approach of disaster management is strongly relevant in DRR and the climate change mitigation. It is based on managing cyclone risks, provision of early warning and preparing the community to act against a cyclone.

7.5.3 Relevant IS Codes

There are several IS codes which are referred for the planning of houses, survey of houses and construction of various type of structures, taking into account wind loads in cyclone prone region. There are also Indian Road Congress (IRC) specifications used for maintenance and guidelines for roads/culverts/bridges. These are:

- IS: 15498-2004. "Design and Construction of Buildings and Structures in Cyclone Prone Regions".
- IS: 15499-2004. "Guidelines for Survey of Housing and Building Typology in Cyclone Prone Areas for Assessment of Vulnerability of Regions and Post Cyclone Damage Estimation".
- IS 456: 2000(R2016). "Plain and Reinforced Concrete—Code of Practice".
- IS 800: 2007. "General construction in steel—Code of practice".
- IS 875(Part-3):2015. "Design Loads (other than Earthquake) for Buildings and Structures -Code of practice".
- IS 8237: 1985. "Code of Practice for Protection of Slope for Reservoir Embankment (reaffirmed 2017)".
- IS 12094: 2000. "Guidelines for Planning and Design of River Embankments (Levees)".
- IS 12169: 1987. "Criteria for design of small embankment dams".
- IS 10635: 1993. "Freeboard requirements in embankment dams—Guidelines (reaffirmed 2003)".

- IS 11532: 1995(R2001). "Construction and maintenance of river embankments (levees)—Guidelines".
- IRC: SP 18:1978. "Manual for highway inspection".
- IRC: SP35:1990. "Guidelines for inspection and maintenance of bridges".
- *IRC*: *SP33:1989*. "Guidelines on Supplemental Measures for. Design, Detailing and Durability of Important. Bridge Structures".
- SP 7: 2016. "National Building Code of India 2016".

7.5.4 National Cyclone Risk Mitigation Project (NCRMP)

The first substantial work in the area of disaster management in India was done through the constitution of the High-Powered Committee (HPC) in August, 1999, just a few months before the Orissa super cyclone in November, 1999. In the immediate aftermath of the Orissa's Super Cyclone in 1999, the Orissa State Government set up the Orissa State's Disaster Mitigation Authority or the OSDMA. The aim of the authority was to focus on the state residents to become more disaster resistant by systematic planning of disaster management.

Subsequent to the December 2004 tsunami, DM Act 2005 was made by the parliament to set up an institutional mechanism for proper management of disasters in the country.

The institutional framework was created at National level, State Level and District level under DM Act 2005. At the national level, National Disaster Management Authority (NDMA), at state level, State Disaster Management Authority (SDMA) and at district levels. District Disaster Management Authority (DDMA) has been were created.

After the formation of NDMA, the management of National Cyclone Risk Mitigation Project" (NCRMP) was transferred to NDMA by MHRD in September 2006. The Government of India has initiated the NCRMP to address cyclone risk in the coastal states. The overall objective of the project is to strengthen the structural and non-structural cyclone mitigation efforts to minimise the losses due to impact of cyclones in the coastal states of India. These States/UTs were classified into two categories, namely category I and category II, based on their vulnerability to such risks. In category I, the higher vulnerable coastal states of Andhra Pradesh, Gujarat, Orissa, Tamil Nadu and West Bengal were selected. In categories II, the coastal States/UTs of Maharashtra, Goa, Karnataka, Kerala, Aman& Diu, Puducherry, Lakshadweep and Andaman and Nicobar Islands were identified based on lower vulnerability to cyclones.

National Cyclone Risk Mitigation Project Phase I covers states Andhra Pradesh and Odisha, has been stared form Jan. 2011. NCRMP Phase—II covers States of Gujarat, Maharashtra, Goa, Karnataka, Kerala, and West Bengal.

The phase I of National Cyclone Risk Mitigation Project (known as NCRMP-I), is a scheme sponsored by the Central government and is being funded by the World Bank in form of Adaptable Programme Grant. This scheme is being implemented in

Fig. 7.6 Approach Road in West Godavari District Bridge (Source: NDMA Annual report, 2018–19)

the states of Odisha and Andhra Pradesh from January, 2011. NDMA is the Nodal Agency at the National Level for monitoring the implementation of the project while the participating states include Odisha and Andhra Pradesh.

The major objectives of this project are the up gradation of cyclone early warning and dissemination systems (EWDS), building of the cyclone risk mitigation infrastructure for reducing the vulnerabilities of coastal communities by constructing multi-purpose cyclone shelters, bridges, roads and saline embankments and undertaking capacity building for disaster risk management.

As per the Annual report (2018–19) of NDMA various measures in form of construction of535 cyclone shelters, 32 bridges, 1086 km of approach roads and 88.12 km of saline embankment have been finished under NCRMP Phase-I. 275 Early warning dissemination systems, 476 Alert sirens, digital mobile radio and 34 satellite terminals have been procured and installed. Under this project, three studies on the assessment of coastal hazards, risk and vulnerability, the preparation of modules for long term training as well as capacity building; and finally Post Disaster Needs Assessment (PDNA) have been conducted. Some photographs of assets created in Andhra Pradesh and Odisha under NCRMP Phase-I are given in Figs. 7.6 and 7.7.

National Cyclone Risk Mitigation Project Phase-II has been approved by Government of India in July, 2015 for a 5-year duration till March, 2020. NCRMP-II includes the Indian States of Gujarat, Goa, Kerala, Karnataka, Maharashtra and West Bengal.

For providing assistance for Implementation of Early Warning Dissemination System (EWDS), TCIL has been engaged by the states of Maharashtra, Goa, Kerala and Karnataka in the form of their partner. PWC and KPMG have been engaged by West Bengal and Gujrat as their knowledge partner (NDMA Annual report,

Fig. 7.7 Multi-purpose Cyclone Shelters in Puri District (Source: NDMA Annual report, 2018–19)

Fig. 7.8 Multipurpose cyclone shelter in West Bengal

2018–19). Photograph of multipurpose cyclone shelter constructed under NCRMP-II IN Gujrat and West Bengal are shown in Figs. 7.8 and 7.9.

Fig. 7.9 Multipurpose cyclone shelter in Gujrat

7.5.5 *Role of Shelterbelts; Mangroves and Buffer Zone: Coral Reef in Cyclone Disaster Mitigation*

For protection from the wrath of cyclone nature has ensured biological means in the form of mangroves and shelterbelts. It has been recognised that mangroves and shelterbelt are able to reduce the energy of storm surges and tsunami waves. They reduce the impact of tsunami waves, storm surge, and high winds resulting in fewer loss of lives and damage to property from storms and cyclones. Therefore, mangroves and shelterbelt are important measures to mitigate the impact of cyclones.

Das and Vincent (2009) have conducted a study to identify the role of mangroves in cyclone risk mitigation in 1999 Orissa Super Cyclonic Storm. It had been found that the villages, which were in the shadow of mangrove forest experienced significantly fewer death as compared to the villages with narrower or no mangrove cover.

As per report on Cyclone Hudhud, complied by NDMA 2015, the mangroves played a significant role in protecting the coastal areas from storm surge. It has also been recommended that there is acute need to maintain more shelterbelts and mangrove in coastal region.

Coral reefs act as a buffer; protect our coastal areas from the impacts of waves, floods and storms. With the buffering action of coral reefs, the shorelines are prevented against adverse impacts of waves, floods and the storms which minimize the loss to lives and damage to property. In the absence or damage caused to such a natural barrier, the damage caused by storms to the coastal communities may increase manifold.

The wellbeing of coral reefs is dependent on the implementation of sustainable developmental practices for the coast which guard these susceptible coral ecosystems along with the organisms which reside in these areas. Climate change presents a significant threat to coral reefs. It is evident from existing scientific evidence that anthropogenic activities have resulted in green house gas emissions which have led to the warming of Earth's atmosphere and the oceans. Climate change results in warming of oceans causing direct heat stress to coral reefs. Even indirectly it alters patterns of oceanic circulation, boosts the power of cyclones which leads to extensive destruction of coral reefs along with the associated organisms. Though the cyclone frequency is not likely to increase its intensity has been predicted to rise on a global scale which would result in occurrences of damaging cyclones with severe damage caused to coral reef ecosystems and the services which are provided by these ecosystems to populations across the globe.

The Ministry of Environment and Forest and Climate Change, Govt. of India had taken up activity under the Integrated Afforestation and Eco-development Project to form a shelterbelt along the coastline (GIDM 2020).

7.5.6 Coastal Regulation Zone

The transition area between marine and territorial zones is called as coastal zone. Owing to the continuous exploitation of the coastal regions the mangroves, fish breeding and coral reefs have been adversely impacted. This has resulted in reduction of the livelihood of 200 million villagers and other communities residing along the country's 7516 km long coastline. As such it is not possible to prevent or control natural hazards, but the losses of life and damage to property can be considerably minimised by coastal zone management. Such measures include properly protecting and restoring natural bio-shields, proper planning for finding the affected communities and building infrastructure within safer areas in coastal region, etc., Hence, it has been considered as essential to initiate a plan of action for coastal area for the purpose of overcoming the environmental problems.

The Coastal Regulation Zone (CRZ) was notified by the Ministry of Environment and Forest notification, Govt. of India in 1991 under the Environmental Protection Act, 1986 to protect and conserve the coastal environment in coastal areas of India. Coastal area has been categorised as CRZ-1, CRZ-2, CRZ-3, CRZ-4 under this notification. The Ministry of Environment, Forests, and Climate Change (MoEF&CC) was replaced the Coastal Regulation Zone (CRZ) Notification, 1991 with two Notifications namely;

1. Coastal Regulation Zone (CRZ) Notification, 2011 applicable for Mainland India to ensure livelihood security to fishing and other local community.

2. Islands Protection Zone Notification, 2011 applicable for A&N islands and Lakshadweep

A Coastal Regulation Zone Notification has been issued in 2018 on the basis of inputs received from several coastal states, the Union Territories as well as the other stake holders. The notification has been issued with a perspective to conserve as well as to protect the distinctive environment existing within the coastal and the marine areas. In addition, the focus is on provision of livelihood to coastal communities and for the promotion of sustainable development considering the risks of natural disasters and that of sea level rise on account of global warming.

The land area from High Tide Line (HTL) to 500 m on the landward side along the sea front has been declared as Coastal regulation zone as per CRZ notification 2018.

CRZ shall be applicable to the land area between HTL to 50 m or width of the creek, whichever is less on the landward side. The land area between the HTL and the Low Tide Line (LTL) is termed as "intertidal zone".

7.5.7 Classifications of Coastal Zones

The coastal regulation zone under CRZ notification (2018) for the purpose of conserving and protecting the coastal areas are classified as under:

- **CRZ-I** These are environmentally most critical areas. The CRZ-I is further categorised into CRZ-I A consisting of the ecologically sensitive areas (ESAs) and geomorphological features (like, Mangroves, Corals and coral reels, sand dunes, etc.) and CRZ-I B covering the area between Low Tide Line (LTL) and High Tide Line (HTL), i.e., intertidal zone.
- **CRZ-II:** Areas which are developed up to or close to the shoreline.
- **CRZ-III:** Areas that are relatively undisturbed (viz. rural areas, etc.) and do not fall under CRZ-II. The CRZ-III is further classified into CRZ-IIIA and CRZ-IIIB.

CRZ-III A–Rural areas which are densely populated with a population density of more than 2161 per km^2 according to 2011 Census. CRZ-III A will have a No Development Zone (NDZ) of 50 meters from the High Tide Line on the landward side as against 200 m from the High Tide Line (HTL) mentioned in the CRZ notification, 2011. CRZ-III B includes rural areas with a population density of less than 2161 per km^2 square according to 2011 Census.

- **CRZ- IV:** These areas are the water area. The CRZ-IV is further categorised into CRZ-IVA and CRZ-IVB. CRZ- IVA includes the water area and the sea bed area between the Low Tide Line up to 12 nautical miles on the seaward side. Areas which cover the water area and the bed area between LTL at the bank of the tidal influenced water body to the LTL on the opposite side of the bank are categorised as CRZ-IVB.

The CRZ Notification, 2018 will lead to the economic growth by enhancing the activities in the coastal regions following the conservation principles of coastal regions. It will boost tourism and environment and add value to the economy of India.

Sundarban region of West Bengal and other islands are categorised as Critical Vulnerable Coastal areas. These areas are managed with the participation of coastal communities. Nature based solution like, Mangrove plantations is one of important mitigation measure for cyclones.

The National Coastal Zone Management Authority (NCZMA) at national level and CZMAs at State level have been constituted for effectively monitoring and enforcing the notification.

7.5.8 *Integrated Coastal Zone Management Project (ICZMP)*

The Government of India with the assistance of World Bank has initiated the Integrated Coastal Zone Management Project to implement the various provisions of the CRZ Notification and IPZ Notification, 2011. It includes formulation of plan for Integrated Coastal Zone Management of State. The ICZMP Phase-I is being implemented by MoEF&CC in Gujarat, Odisha and West Bengal to address most of the issues relevant to impact of Climate Change in coastal areas.

7.6 Case Study: Cyclone Fani

A comprehensive report on the loss and damage along with the need assessment for Cyclone Fani which occurred in Odisha in 2019 was prepared by collaborative efforts of Odisha state Government, OSDMA, the World Bank, the agencies of United Nations along with the Asian Development Bank. Based on this report, a case study of cyclone Fani presented.

On 3rd May 2019 morning, extremely severe cyclonic storm, the Fani, a rare form of summer cyclone, reached the coast of Odisha quite close to Puri. Wind speeds of 175–180 km per h (kmph) reaching to over 205 kmph were reported during the landfall of cyclone at Satpada as per IMD. The Fani was the strongest tropical cyclone to strike Odisha, an eastern state of India since 1999 Odisha Super Cyclone. Approximately 10,000 people were killed during super cyclone of 1999, while 64 people lost their lives during the Fani. As compared to super cyclone 1999, human causalities were very low in the Fani.

7.6.1 Disaster: Effects and Impact

About 16.5 million people (about 36% of total population of Odisha) in over 18,388 villages in 14 of the 30 districts in the state were affected. The most impacted districts from Fani included Khurda, Puri, Cuttack, Kendrapara and Jagatsinghpur. During cyclone Fani the Power, telecommunication infrastructure, and road services were severely affected. About 3.62 lakh houses were damaged. Approximately 19,734 Ha of agriculture area was affected. The worst impacted sectors from Fani included power, telecom, housing, and the livelihoods in the form of agriculture, fisheries and livestock. According to the report the total damages were of the order of Rs. 16,465 crore almost 2352 million USD, while the recovery needs were estimated to be INR 29,315 crores.

7.6.2 Preparedness

Upon receiving the early warnings from IMD, Government of Odisha evacuated almost 1.55 million residents within 9177 shelters. With the help of 23 special trains as well as 18 buses over 25,000 tourists were successfully evacuated. Two days before the landfall all sorts of fishing operations were suspended. For the rescue and relief activities, 20 units of the Odisha Disaster Rapid Action Force (ODRAF), 335 units of Fire Service and 25 units of the National Disaster Response Force (NDRF) were employed into service. For 24 × 7 the block and district level emergency centres were put into action. Almost 4.5 lakh polythene sheets (temporary shelter materials) were positioned at district and sub district level. In susceptible areas the Govt's disaster response forces were brought in service. Food packets were made ready for Air force helicopters to drop to people.

Volunteers, officials, workers and other civil society organisation, NDRF, ODRAF, other agencies worked together round the clock to evacuate 1.55 million people in 24 h. It was the largest evacuation program ever done in the World.

7.6.3 Relief and Rescue Operations

A massive rescue and response operation were started immediately after the landfall. Sixty teams from NDRF, 18 units of ODRAF, 585 fire team and nearly 45,000 volunteers carried out relief operations. State Government was also supported by Eastern Naval Command of the Indian Navy undertook the rescue as well as the relief operations. Almost 10,000 packets of food were dropped from Airforce helicopters. With the help of local people, more than 6000 were opened. Major roads within Puri and Bhubaneshwar were cleared and power supply was restored within 48 h.

7.6.4 Reasons for Successful Management of FANI

Improvements in early warning systems, forecasting capabilities, preparedness and disaster management significantly helped in mitigating the impacts of cyclone Fani.

7.6.4.1 Early Warning System (EWS)

The India meteorological department has taken several imitative to modernize and upgrade the early warning systems for cyclones and built an effective service to predict accurate timing of cyclone formation in the North Indian Ocean. Prior to formation of low-pressure area, the cyclone was monitored and predicted from 18th April onwards by IMD. Information about the formation of low-pressure area over EIO and adjoining south BoB was given on 25th April, 2019. An extremely severe cyclonic storm (ESCS) with wind speed of 175–185 kmph gusting to 205 kmph would cross around Puri on 3rd May, was continuously predicted on 30th April, 1st and 2nd May 2019. This EWS enabled the state to get prepared and steps were taken in advance to minimise the loss of lives. After issuing of warning, people followed the instruction given by the authorities. A massive evacuation drive was undertaken prior to landfall in the state of Odisha.

7.6.4.2 Cyclone Shelters

Before Odisha super cyclone 1999, state didn't have a proper plan for disaster management. Many multi-purpose cyclone shelters, and other infrastructure have been built under National Cyclone Risk Mitigation Project/other scheme. These were utilized in managing the cyclone Fani in 2019.

7.6.4.3 Proper Communication Plan

Warning and advisory through social media (face book and tweeter) was given regularly (every 6 h) during the life period of the system. Roughly 2.6 million text messages were delivered to locals people before cyclone Fani hit. Press and electronic media were given daily updates by officials to update people of the approaching cyclone. People were advised not to get panicked.

Huge campaigns for awareness were organised for informing regarding the basics of cyclone by the government. This helped in the record evacuation of over 1.55 million residents within 9177 shelters.

7.6.4.4 Community Participation and Proper Co-ordination of Various Agencies

Government agencies, volunteers, local panchayats and self-help groups (SHGs) worked together in preparedness to response, evacuation, distribution of relief, and shelter management.

Community Participation is an important aspect within any disaster management strategy.

Senior Government officers were deployed at the district level to coordinate the activities of various agencies for proper management. Forces of Disaster response (ODRAF and NDRF) were pre-positioned within the districts in coastal areas. For ensuring availability of food to the community during the period of cyclone, Food supplies were pre-positioned in cyclone shelters and food packets were prepared for air force helicopters to drop to people.

7.7 Conclusion and Way Forward

India's coastal regions are highly vulnerable to the impact of cyclones and other related hazards. It is important to note that over 40% of our total population resides within 100 km of our coastline. Vulnerability of the coastal population has also increased due to climate change resulting in the rise of sea level and on account of increase in the frequency and intensity of cyclones.

In order to minimize the loss of lives and damage to houses and structures, various Structural and non-structural measures are important aspects. In the aftermath of Odisha's super Cyclone, 1999, several mitigation measures have been undertaken. As per recent trends of impact of cyclones, the number of lives lost has declined, while the goal of zero mortality is close to realization, it needs to be highlighted that the economic losses and damage to infrastructure have been on a rise. Frequency and intensity of cyclones has also increased due to climate change. Mangroves and shelterbelt are powerful tools to prevent the impact of cyclones. The coastal belt plantation drive should be promoted with the involvement of civil society and community-based organisations. The focus should be to involve community by massive awareness campaigns, training of fisherman, farmers and capacity building to prepare a resilient community. Clearly the goal of minimizing the loss of life is achieved, but we are far away from achieving the minimizing the damage to houses and infrastructures. There is urgent need to reduce the economic losses due to impact of cyclones. This can be achieved by adopting Resilient Housing, Resilient Livelihood and Resilient Infrastructure. Houses, schools, hospitals, cyclone shelter should be cyclone resistant and safe structures. Existing guidelines should be used for construction of these houses. Quality of housing can be improved through better building standards and regulations.

References

CRZ Notification (2018) Ministry of Environment, Forest and Climate Change Notification, 2018
Cred & UNISDR (2018) Economic losses, poverty and disasters 1998–2017
Damage loss and need assessment report (2019) Cyclone Fani, Odisha May, 2019. https://www.recoveryplatform.org/assets/publication/PDNA2019/Web-version%20Cyclone%20Fani%202019%20Odisha%20DLNA%202019.07.16%20(Re-revised%20-%20Final).pdf
Das S, Vincent JR (2009) Mangroves protected villages and reduced death toll during Indian super cyclone. Proc Natl Acad Sci 106:7357–7360
GIDM (2020) Training module on cyclone risk management
Goyal PK, Datta TK (2013) Effect of wind directionality on the vulnerability of rural houses due to cyclones. Nat Hazard Rev Am soc Civil Eng 14(4):258–267
Goyal PK, Datta TK (2014) Probabilistic damage estimate of rural houses due to Cyclonic wind. Second International Conference on Vulnerability and Risk Analysis and Management (ICVRAM) and the sixth International Symposium on Uncertainty Modeling and Analysis (ISUMA), 13–16 July 2014, University of Liverpool, UK, ASCE, pp 1475–1483
Goyal PK, Datta TK, Vijay VK (2012) Vulnerability assessment of rural houses to cyclonic wind. Int J Disaster Resilience Built Environment Emerald 3(1):20–41
http://www.rsmcnewdelhi.imd.gov.in/images/pdf/sop.pdf
http://www.sicommoef.in/Data/Sites/1/skins/gec-sicom/images/docs/pad.pdf
https://www.wmo.int/pages/prog/www/tcp/documents/PTC36_FinalDraftReport.pdf
ICZMP (2010) Document of world bank—project appraisal document for an integrated coastal zone management project report no: 54612-IN
IMD (2013) Cyclone Warning in India: Standard Operation Procedure
IS:15498 (2004) Guidelines for improving the cyclonic resistance of low-rise houses and other buildings/structures. BIS, New Delhi (Reaffirmed 2015)
IS:15499 (2004) Guidelines for survey of housing and building typology in cyclone prone areas for assessment of vulnerability of regions and post-cyclone damage estimation, BIS, New Delhi (Reaffirmed 2015)
Mohapatra M (2015) Cyclone hazard proneness of districts of India. J Earth Syst Sci 124:515–526. https://doi.org/10.1007/s12040-015-0556-y
NDMA (2015) Cyclone Hudhud: strategies and lessons for preparing better & strengthening risk resilience in coastal regions of India. NDMA, New Delhi
NDMA (2016) National disaster management plan. NDMA, New Delhi
NDMA (2019) National disaster management plan. NDMA, New Delhi
Rapid Damage and Needs Assessment Report (2013) India Cyclone Phailin in Odisha. http://documents1.worldbank.org/curated/en/168471468257979992/pdf/838860WP0P14880Box0382116B00PUBLIC0.pdf
RMSC (2019) Report on cyclonic disturbances over north Indian ocean during 2018. RSMC-Tropical Cyclones, New Delhi
Shanmugasundaram J et al (2000) Cyclone damage to building and structures – a case study. J Wind Eng Ind Aerodyn 84:369–380
Sikka DR (2019) Major advances in understanding and prediction of tropical cyclones over North Indian Ocean: a perspective. Mausam 57(1):165–196
Tamura Y (2009) Wind-induced damage to buildings and disaster risk reduction. 7th Asia-Pacific conference on wind engineering, APCWE-VII Taipei, Taiwan
Tamura Y, Cao S (2012) International group for wind-related disaster risk reduction (IG-WRDRR). J Wind Eng Ind Aerodyn 3(11):104–106

Vulnerability Atlas of India (2019) Earthquake, wind, flood, landslide, thunderstorm maps and damage risk to housing, BMPTC, Ministry of Urban Affairs and Employment, Government of India
WMO (2009) Tropical Cyclone operational plan for the Bay of Bengal and the Arabian Sea; Tropical Cyclone Programme Report no. TCP-21, Edition 2009, WMO/TD No 84
WMO and ESCAP (2009) WMO/ESCAP panel on tropical cyclones

Dr. Goyal is an associate professor at Delhi Technological University Delhi. He has extensive expertise and experience in the field of cyclone disaster mitigation, microzonation, hazard, vulnerability analysis and risk assessment.

Chapter 8
Heat Wave Disaster Risk Management Action Planning: Experience and Lessons

Sushma Guleria and Anil K. Gupta

Abstract Heat waves are now recognized as a major cause of extreme weather-related deaths. With the current concern for global warming and Climate Change altering weather patterns, adverse Climate Change impacts will increase in frequency, severity, duration or areal extent in the future. With most weather stations still struggling to provide an adequate definition bracket for heat waves, it is difficult to assess either the past changes or probable impacts and consequences for the future. Rises in death tolls due to erratic weather conditions have been attributed to heat waves over the past few years. Heat waves have been reported as a disastrous phenomenon linked to their origin in hydro-meteorological events. Despite the increase in severity and duration of heat waves combined with other environmental interactions, countries have failed to develop proactive associated risk perceptions; preparedness and lethargic emphasis on mitigation measures have made the problem worse. Many studies have evidenced that health risks associated with exposure to excess heat increase significantly. However, heat-wave related fatalities can be prevented by establishing precautionary measures, which can be simple as planting more trees. Therefore, the purpose of a Heat Wave Action Plan (HAP) is to mitigate the adverse health impacts of excessive heat by raising public awareness and sensitizing the vulnerable groups and also to trigger preventive actions and build a strong social care system.

Keywords Extreme-weather · Climate change · Heat-action plan · India

S. Guleria (✉)
Environment, Climate and Disaster Risk Management, National Institute of Disaster Management, New Delhi, India

A. K. Gupta
ECDRM Division, International Cooperation, CAP-RES DST Project, National Institute of Disaster Management (Ministry of Home Affairs), New Delhi, India

A. K. Gupta et al. (eds.), *Disaster Risk and Management Under Climate Change*,
Disaster Resilience and Green Growth,
https://doi.org/10.1007/978-981-99-4105-6_8

8.1 Introduction

Heats waves is known as the major cause of weather-related fatalities in the United States, England, Europe and Asian Nations as well over recent years. Detailed analysis of severe events and their impacts are conspicuous. However, relatively not much is known about the climatic behaviour of heat waves. Changes in lifestyle, ignorance of risk perception and inadequate attention towards preparedness of local governments, people and institutions, including abeyance from traditional cultural practices in the backdrop of Climate Change impacts, have multiplied and complexified the risk of heat impacts on people and other living beings.

While heat waves do occur naturally as part of the earth's climatic systems, anthropogenic greenhouse gas (GHG) emissions have contributed to them occurring more frequently and severely. These greenhouse gases settle in the stratosphere and trap the heat within the atmosphere, which occurs over a period of time. Hence, the increased frequency and intensity of heat waves is considered a long-term effect of anthropogenic GHG emissions. As a result, the increased GHG levels in the stratosphere are causing global warming and changing climatic patterns, which include the shifting of seasons, an increase in temperature globally, sea level rise, and changing agricultural patterns. Further, rapid depletion of forest natural resources at a rampant pace has added to the unprecedented changes in the global climate resulting in serious implications on the very question of human survival. In turn, these climatic changes that cause natural disasters contribute to other side effects such as population migration and major public health hazards.

8.2 Present Scenario

Globally, 2015 was the hottest year on record, making it the fourth time this century that a new high temperature record was set (National Oceanic and Atmospheric Administration, United States of America, 2016: www.noaa.gov). The proportion of the earth's surface area experiencing extremely hot summers increased from 1% over the 1951–80 period to 10% over the 1981–2010 period. Record-breaking heat events were witnessed across the globe in Australia, Argentina, China, Central Asia, and large parts of Europe, Mexico, Japan, South Korea and the USA.

Additionally, India is feeling the impact of climate change in terms of increased instances of heat waves that are more intense in nature with each passing year. These intensified heat waves have a devastating impact on human health, thereby increasing the number of heat wave casualties. In 2015, the heat wave in India is known for more than 2300 deaths, making it the fifth worst globally in terms of number fatalities as per government records. Most of the deaths were reported in the States of Andhra Pradesh, Telangana, Punjab, Odisha and Bihar. In 2016, the month of April recorded highest average global temperature causing a rise in temperature creating conditions favourable for the development of severe heat wave during this month alone contributing too many heat related deaths.

Heat waves occur mostly over areas with dry and warmer air and have the maximum insulation during the summer season. Therefore, they tend to develop over northwest parts of India and then spread gradually eastwards and southwards. Bay Islands, Lakshadweep, Tamil Nadu, Kerala, Coastal and South Karnataka generally do not get affected by heat waves due to the occurrence of maritime air over these regions, but, at times, heat waves may also develop over any region in situ under the favourable conditions.

8.3 Significance and Concept

Heat wave as a hazard is a physical phenomenon of high heat conditions characterized as a complex of hydro-climatic risks coupled with social, occupational and public health threats. A heat wave is a period of abnormally high temperatures, more than the normal maximum summer temperature or prolonged period of excessive heat that typically occurs between March and June, and in some rare cases even extends until July. The extreme temperatures and resultant atmospheric conditions adversely affect people living in these regions as they cause physiological stress, sometimes leading to death. However, as such there is no universal definition for heat wave.

While definitions for what constitutes a "heat wave" vary across and even within countries, these are generally measured relative to the usual weather in the area and relative to normal temperatures for the particular season. The World Meteorological Organization (WMO) defines heat wave as **'when the daily maximum temperature of more than five consecutive days exceeds the average maximum temperature by 50 °C'.**

The Indian Meteorological Department (IMD) (www.imd.gov.in) has the following criteria for defining Heat Waves:

- Heat wave need not be considered till maximum temperature of a station reaches at least 40 °C for plains and at-least 30 °C for hilly regions
- When normal maximum temperature of a station is less than or equal to 40 °C Heat Wave Departure from normal is 5 °C to 6 °C and Severe Heat Wave Departure from normal is 7 °C or more
- When normal maximum temperature of a station is more than 40oC Heat Wave Departure from normal is 4 °C to 5 °C and Severe Heat Wave Departure from normal is 6 °C or more
- When actual maximum temperature remains 45 °C or more irrespective of normal maximum temperature, heat waves should be declared.

Such extreme heat events are becoming increasingly more common. Generally, a linear relationship can be drawn between temperature and health impacts that are observed with each degree of increase in temperature. In a moderate heat wave condition, it is mainly the high-risk vulnerable groups who are affected. However, during an extreme heat wave scenario, even the fit and healthy people can be affected.

Over the years, excess summer deaths show regional variations, which relate largely to differences in temperature levels across regions. An inability to adapt to heat waves, especially during the first few days, lead to excessive deaths and illnesses. Therefore, there is a need to be an emphasis on adequate preparedness for the first hot period of any season and the very beginning of a heat wave, as well as for prolonged and sustained periods of unusually hot weather.

The Inter-governmental Panel on Climate Change (IPCC) has played a key role in connecting Climate Change induced impacts and the causes of natural disasters. IPCC's fourth Assessment Report (2007) proved to be instrumental in bringing recognition globally for converging Climate Change Adaptation and resilience building into Disaster Risk Management (IPCC 2007). With evident climatic variations often threatening human survival, advocates for promoting disaster risk reduction actions have called for the inclusion of climate resilience and sustainable development components. Disaster risk management, Climate Change Adaptation, and sustainable development have interlinked goals, as embedded in the three major agreements India is signatory to—Sendai Framework for Disaster Risk Reduction (SFDRR), Paris Climate Agreement and Sustainable Development Goals (SDGs).

In particular, the development arena encompassing developing nations and especially the under-served low-income groups are threatened by their exposure to increased frequency and intensity of hydro-meteorological hazards such as floods, prolonged dry spells, heat waves, intense wind speeds and storm surges. The alteration of ecosystem services and accelerating natural resource degradation imposes negative impacts on people's livelihoods and hence, enhancing their vulnerabilities.

This vulnerability is compounded by socio-economic conditions, which exacerbate the impact of disasters. A serious notice, review and insight into the implications of climate change started with many South Asian Nations being engulfed by devastating urban floods in cities like Mumbai, Dhaka, Islamabad, Bangkok, Surat, Uttarakhand, Kashmir, and Chennai. Increasing frequency and intensity of cyclonic disasters off late such as Phailin and Hudhud affecting Indian coastal and sub-coastal states, other nations and islands in the region, intense heat wave in Indian states of Maharashtra, Telangana, Odisha and Andhra Pradesh, wide spread coverage of drought regions year by year are on an exponential rise. Countries of Asia Pacific region are not uniform in terms of their development status and few are yet under development and subsequently, suffer intense loss of natural and socio-economic resources. Never was this challenge for enhancing resilience with explicit focus on people, their health and livelihoods been more compelling with the substantial increase in natural hazards and their vagaries.

8.4 Impacts

Extreme events like heat waves have the greatest impacts on sectors that are closely linked with or dependent on the climate, for example water, agriculture and food security, forestry, health and tourism. This challenge is particularly more critical for developing countries like India, where majority of economy thrives on agrarian income generation options. Given India's topography and environment, the increasing risk of glacial lake melting, more severe heat waves, and devastating

Table 8.1 Classification of Heat Index (HI) and health impacts

Heat Index	Category	Health Effects
27–32 °C	Warm	Caution- fatigue possible with prolonged exposure and/or physical activity. Continuing activity could result in heat cramps
32–41 °C	Hot	Extreme caution- heat cramps and heat exhaustion possible with prolonged exposure and/or physical activity
41–54 °C	Very hot	Danger- heat cramps or heat exhaustion likely and heatstroke possible with prolonged exposure and/or physical activity
> 54 °C	Extremely hot	Extreme danger- heatstroke highly likely with continued exposure

precipitation spells, desertification and drought, windstorms and disease epidemics have become of paramount importance. For India, the 2010 heat wave spate can be considered a wakeup call that all stakeholder actions, preparedness, adaptation including community outreach become indispensable imperative to save lives especially towards harsh heat wave spells. Again, the year 2014 surpassed 2010 as the warmest year for average global temperature since the1880s. Studies based on increase in mean temperatures suggest that the rise in minimum temperatures will have detrimental impact by 2030s. As global temperatures rise, the warmer atmosphere will have more capacity to hold water, which can produce more intense rainfall events as well as longer dry periods or decreased rainfall spells between these events.

These severe heat waves have a significant impact on human livelihoods. The majority of the work place occupations in developing nations is tremendously influenced by the outdoor temperatures and will prove to have aggravated heat stress impacts in addition to physical and social vulnerabilities. This adverse climate change impacts will affect a large work force population, with many occupations requiring direct sunlight exposure. Even a moderate increase in ambient temperature can engulf large labor population into the realm of heat stress related health risks.

Four main medical disorders resulting from excessive exposure to heat waves are heat syncope, heat cramps, heat exhaustion and heat stroke, as can be seen in detail at Table 8.1. The signs and symptoms are as follows:

Heat rash: Also known as sunburn phenomenon with pinkish skin resulting in burning of skin and pain sensation.

Heat syncope: Giddiness, vertigo headache and sudden onset drowsiness/unconsciousness.

Heat Cramps: Edema (swelling) and Syncope (Fainting) generally accompanied by fever below 39 °C i.e.102 °F.

Heat Exhaustion: Fatigue, weakness, dizziness, headache, nausea, vomiting, muscle cramps and sweating.

Heat Stoke: Body temperatures of 40 °C or more along with delirium, seizures or coma. This is a potentially fatal condition.

The increase in heat waves has also caused a sharp rise in the number of cases of acute gastroenteritis and food poisoning due to spoilage of food and reduction of its shelf life. Other medical issues caused by heat waves include the cases of anxiety, palpitations, nervousness and behavioural changes linked to extreme temperature rise. More vulnerable groups such as toddlers, pregnant women, and elderly, those with critical medical care need, having occupational profile of as agricultural labourers, coastal community dwellers, and people living Below Poverty Level (BPL) category with mostly outdoor occupations feel the negative impacts most significantly.

Common heat health impacts commonly treated during heat waves include dehydration, cramps, exhaustion and heatstroke. The European heat wave of 2003 led to more than 70,000 excess deaths in France, Germany, Italy, Spain and other countries and the Russian heat wave in 2010 resulted in over 55,000 excess deaths (EM-DAT, 2015: International Disaster database at www.emdat.be). Significant deaths have also been reported in countries such as Australia, Japan, Mexico and the USA in recent years. Increasing heat-related fatalities have also been recorded in India over the last two decades. Since 1992, nearly 23,000 people have died due to heat waves.

Periods of severe heat waves are also linked to sparser pre-monsoon season rains in many areas, thus with reduced moisture content. As a result, the drier conditions leave large portions of areas more arid than usual, which contribute significantly to heat wave periods. This weather pattern, coupled with the El Niño effect, which often increases temperature in the Asian region, combined to create the record-high temperatures. Further, as per the National Oceanic and Atmospheric Administration record (US-NOAA: www.noaa.gov), the year 2016 was recorded as the hottest year globally—higher than 2015, the previous record-holder. The other nine warmest years on record are 2009, 2010, 2003, 2002, 2014, 1998, 2006 and 2007.

Besides health, extreme dry spells have been found to profoundly affect crop production both in terms of quantity and quality. Primarily, crop loss is encountered due to flower drop and higher mortality in new plantations. In particular, Kharif crops are more impacted than Rabi crops owing to variability in rainfall associated with heat wave periods as any extreme change in temperature will affect crop productivity. Serious challenges arise when extreme heat events linger for prolonged periods, especially, for states where majority work in unorganized and informal sectors whose livelihoods depend on daily earnings. So, when heat waves are prolonged, these daily wage earners face difficult choices. They are either forced to stay indoors and compromise their source of income or run the risk of succumbing to heat wave related illness upon continuing to work. Therefore, this situation necessitates the need to revisit and strengthen the existing heat wave preparedness and response plans to make it more effective.

Heat health impacts in India are serious. Estimates suggest that there have been over 22,000 heat-related fatalities in India since 1992. In 2015, the country witnessed the fifth deadliest heat wave in history (EM-DAT, 2015: International Disaster database at www.emdat.be). In addition to the vulnerable populations

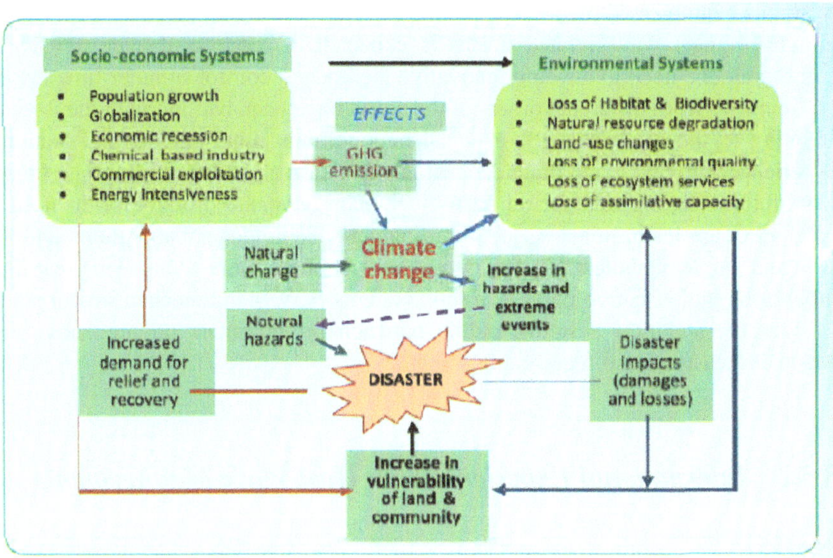

Fig. 8.1 Causative interactions of socio-economic, climate change, and disasters (Gupta 2015)

identified above, the poor may be differentially impacted on the account of gaps in health services, housing and basic amenities.

Loss of nearly 5000 lives in the recent past has been attributed to heat waves since the last 4 years. Approximately 1600 people lost their lives due to irregular weather conditions, out of which 557 casualties were the result of heat wave. Extreme heat can lead to dangerous, even deadly, health consequences, including heat stress and heat stroke. Until recently, communities and government were oblivion to this fact and had turned 'moon-eyed' about this emerging threat— it is another grim reminder about the vagaries of nature and our callous outlook towards this vicious web of disasters, climate change, socio-economic and environmental systems as can be seen in Fig. 8.1.

8.5 Strategies and Management

8.5.1 Management of Heat Waves

While heat waves have occurred in the past, in recent years significant initiatives of planned approach to counter heat waves and mitigate the associated risks have also been taken at various levels—cities, districts, states and national levels as a core concern. Governments and communities need to effectively prepare for developing Heat (risk) Action Plans for countering heat wave through coordinated adaptive and preparedness plans on pilot basis. Then, governments should replicate the best workable examples with adequate monitoring mechanisms in place as well. Rising number of heat wave related deaths is expected to serve as danger signal and corroborate the need to develop innovative methods to control summer-time losses.

Countries from different corners of the world have been battling with the issue of adapting to the rapid changes in weather conditions. International and National experiences provide many practical examples to learn and adopt appropriate framework for developing Heat Action Plan. Such plans advocate strongly for the need to have adequate institutional set up coordinating stakeholder collaborations and the stringent implementation of actions required amongst them along with real-time monitoring and effective Early Warning Systems (EWS) and mechanisms in place. Besides the technical systems, public outreach programmers on preventive and adaptive measures also require due attention.

8.5.2 International Experience with Heat Wave Management

Climate Change will cause more frequent and severe heat waves and in turn, serious impacts on human health, livelihoods, and the environment—so, countries around the world are taking action to implement heat wave management planning. For example, Canada and about twenty European Nations have initiated Heat Wave Management Planning at the federal, provincial and/or local levels. Canada has invested in heat health science; heat monitoring and health surveillance; vulnerability assessments; best practices for developing Heat Alert and Response Systems (HARS); and, communicating with the public to enable behaviour change. Under the HARS, sub-national governments are encouraged to develop programs emphasizing community mobilization and engagement (via a coordinating agency); alert protocols; community response planning through broad stakeholder engagement; communication planning; and periodic system evaluation.

On the other hand, Australia does not have a National Heat Wave Management Plan, as of now. In Australia, provincial and local governments are being encouraged and pushed to develop a national heat wave management plan through federal guidance. Key focal point for it will be: role-clarity among stakeholders; heat health alert systems; public communication; and, improved heat health intelligence.

European nations have broadly followed WHO's Guidance on Homeless House Assistance Programs (HHAPs). The Heat Wave Plan for England, for example, prioritizes: long-term multi-agency strategic planning to combat climate change; alert systems; institutional preparedness; public communication; partnership with service providers; community engagement; and, monitoring and evaluation.

In August 2014, New York City officials from the Mayor's office, the Department of Health and Mental Hygiene, the Office of Emergency Management, along with 30 additional partners conducted a heat wave table top simulation exercise. The simulation discussed preparedness efforts, communication availability to the public and health providers, surveillance of heat-health planning, and development of temperature thresholds and identified gaps in the emergency response plan. Another example of a State Government taking action to prevent heat-wave related fatalities is California's Climate Action Team Working group, which includes two sub-sections addressing heat waves and public health. The work group's recommendations included identification of a lead agency which would discuss and implement various strategies for tackling heat wave health impacts. They also recommended multi-disciplinary efforts and partnerships between various constituents of the health sector including social service agencies, business, labor, utilities and representatives of vulnerable populations. Importantly, the work group acknowledged the need for further research on heat wave impacts on health, economic, ecological, and social determinants and on evaluation of warning systems.

8.6 Case Study: India's Experience on Heat Wave Management

In India, the first systematic attempt at heat wave management planning has happened with Ahmedabad Heat Action Plan (AHAP), launched in 2013 and subsequently being updated each year thereafter. This plan was developed by Ahmedabad Municipal Corporation (AMC) in partnership with domestic and international experts focusing on four key strategies;

- Building public awareness on risk of heat waves through mass outreach program in local language (Gujrati)
- Execution of response system to prevent heat-related death and illness at the onset
- Starting an early warning system to alert citizens on predicted extreme temperature and
- Capacity building of city officials and healthcare professionals

Additionally, the heat wave management planning also led to identification of adaptive measures such as mapping high-risk areas and cooling spaces during extreme heat days and coordinating utility services such as water and electricity to support life. The Ahmedabad HAP worked well and was eventually replicated by the States of Maharashtra, Andhra Pradesh, Telangana and Odisha which are affected the most. Other notable initiatives include the launch of several studies related to impacts of heat stress on health, especially with respect to direction by the Department of Science and Technology in 2011. Similarly, the India Meteorological Department (IMD) has remarkably supported more than 100 heat affected cities and town by providing frequent forecast on heat conditions as they do for rainfall and intense winds not only comprising of early warnings on heat waves but issuing specific advisories for authorities for rigid implementations.

Further, customized heat alert plans for various heat-affected States, along with public awareness through government mechanisms, are a few steps that are required in addendum throughout the nation. With this backdrop, India has an indispensable need for a National Heat Action Plan, which should not only reach affected cities and states also. A National Heat Action Plan should also drive a national agenda to embed adaptation planning for rising temperature in our design for land-use space, utilities, infrastructure and industries planning as seen in Fig. 8.2.

8.7 Heat Action Planning Framework

Heat wave management issues are being addressed at the national, state and sub-state levels; yet, there is a need for comprehensive and streamlined planning to prepare for future heat events in any country. Then, this action will facilitate preparedness and responses to extreme heat events. To ensure frameworks are effective, a HAP calls for individuals, families and all related stakeholders in a state to tackle possible heat impacts. Such HAP frameworks can promote the desired management planning for all vulnerable locations and facilitate prevention and preparedness to avoid detrimental health impacts. Such plans commonly emphasize the need for:

- Institutional role clarity, coordination and collaboration;
- Special focus and emphasis in planning for the vulnerable groups;
- Stakeholder communication on preventive, preparatory and adaptive behaviours;
- Real-time monitoring and Early Warning Systems (EWSs);
- Short and medium-term strategies to lower indoor heat exposure; and
- Long-term preventative measures related to urban land-use planning, energy and transportation.

Based on the Heat Wave Action Plan prepared individually by the states of Gujarat (Ahmadabad), Andhra Pradesh (Telangana) and Odisha, the following can be used as a template for a Heat Wave Plan in terms of planning and execution.

Crucial nodal departments such as IMD, Information & Public Relations, Medical, Labour, Animal Husbandry, Transport, Education, IT, NGO'S can be integrated as key stakeholders in implementing HAPs. Institutional set up for heat action plan and its efficient management can be seen in Fig. 8.3.

Based on the observation of various States, an ideal HAP can be executed in the following phases:

Phase I-Documenting existing initiatives and practices undertaken against protection from heat waves

Phase II-Assessment of the magnitude of the heat wave related morbidity and mortality

Phase III-Determination of threshold values (thermo-hygrometric index) which causes heat wave related adverse health outcomes and facilitate IMD in developing an Early Warning System (EWS) based on the threshold values.

Phase IV-Mapping community vulnerability to assess and identify the most vulnerable sections towards impact of heat waves and recommend contextualized resilience building models.

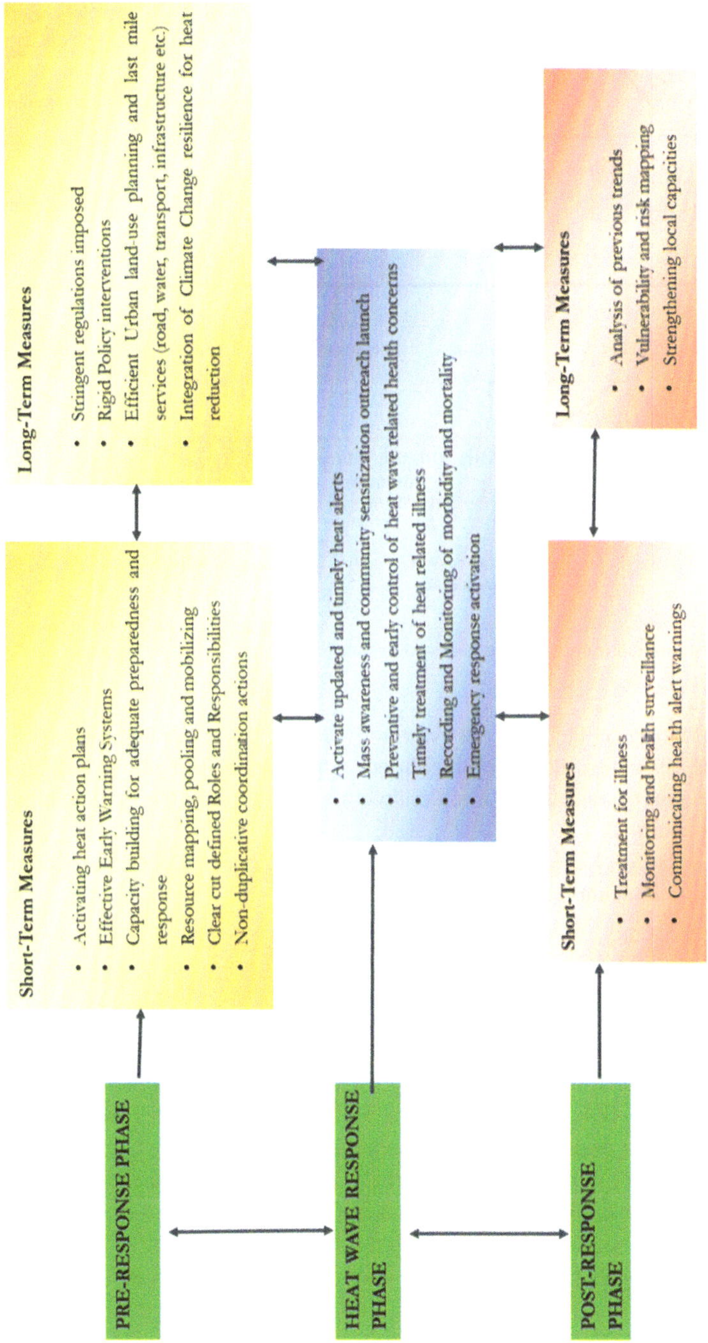

Fig. 8.2 Framework for heat wave risk management

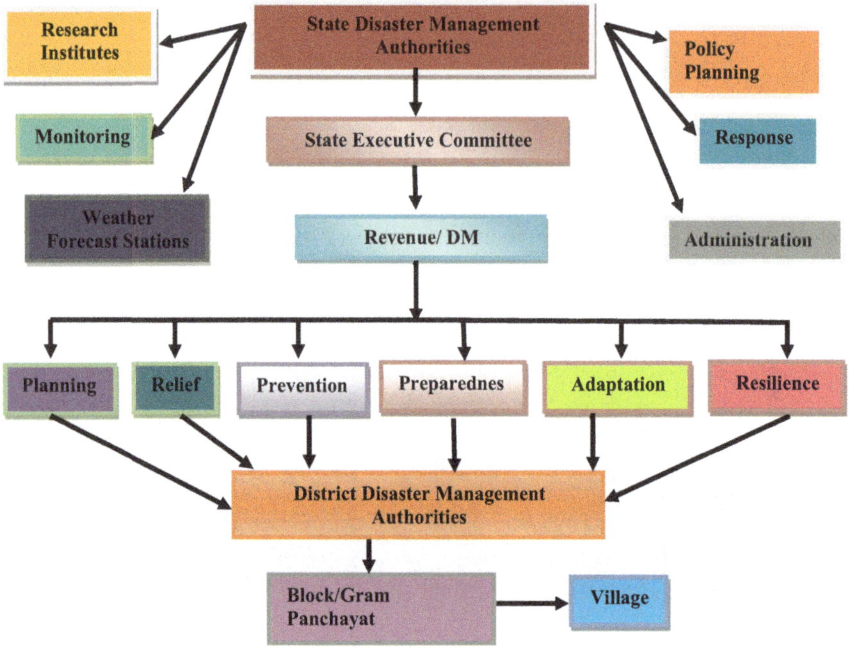

Fig. 8.3 Institutional framework for heat wave management

Phase V-Improvise the plan on regular basis by incorporating improved early warning systems, alternate livelihood opportunities for those affected and involvement of various stakeholders/sectors to support in the implementation of Heat action plan.

Phase VI-Monitor, Review, Update and Evaluate the Heat Action Plan time-to-time and its implementation (Table 8.2)

8.8 Way Forward

It is more evident from Climate Change Risk Assessment studies that the future summer months will have higher temperature, which corroborates the fact that heat-induced health impacts will significantly rise by 2020s. This information also indicates that, *very likely* heat wave spells will increase in frequency, duration and intensity. Although, substantive progress in Climate Change awareness in various sectors and the associated weather variations, the increase of heat waves necessitates adaptation to Climate Change across all communities for their betterment. Thus, the challenge remains for long-term planning and preparedness along with convergence of multi-agency partners. Such preparations can be broadly divided into short- and

Table 8.2 Implementation of HAP by various departments

Department/ Organization	Immediate (before and during Summer)	Short term (Six months of summer)	Long term (throughout the year)
SDMAs' and DDMAs'			
IMD and local weather stations			
State& District EOC			
Road and transport			
Water resources			
Environment and Forest			
Health and family welfare			
Women and child development			
Animal husbandry			
Fisheries			
Tourism			
Fire			
Housing (rural/ urban)			
Private partners			
Tribal affairs			
Panchayat raj			
Education			
Public information			
Labor			
Energy			
Steel and industries			

medium-term strategies to mitigate heat impacts and long-term adaptive strategies eventually to build resilience and reduce vulnerability.

8.8.1 *Short Term (0–5 Years)*

Embedding preventive measures of extreme weather planning into the social care structures is imperative to prevent short-term heat wave fatalities and related health concerns, especially since the Sendai Framework emphasizes the same. Multi-agency local resilience building forums can have a critical role in the preparations and response to heat wave management and ensure maximum adaptation to reduce harm from heat waves.

Joint Strategic Needs Assessments (JSNAs) committees can be formed to identify the challenges posed by extreme weather related deaths locally, and Joint Health and Well-being Strategies (JHWSs) can be used to design customized state-specific actions to reduce them. Both these processes can be prioritized across the local social health care system. Departments of Public Health should proactively be prepared to take an active role in setting a local agenda. These departments must work in partnership with local authorities and primary health centres to identify vulnerable populations and geographical areas to target long-term planning and interventions during a heat wave spell. High temperatures during a heat wave may require affected wards to move patients to cooler areas; extra beds must be made available in hospitals if, need arises.

8.8.2 Medium Term (10–30 Years)

Heat wave management programs should focus on building design of hospitals and other healthcare establishments to aid passive cooling where possible, and target vulnerable areas (patients, medications, IT) with air-conditioning. Review external hospital surroundings for ways to aid cooling—for example, consider constructing underground car parks and maximize green space and trees surrounding hospital premises. Such works can be monitored in partnerships with local authorities to identify and focus on vulnerable urban areas and populations—for example, urban areas which may be affected more by high temperatures etc. Monitoring the implications of, new diseases arising due to rise in temperatures during summer months. Cool roofing system for houses, schools, community halls, etc. should also be focused.

8.8.3 Long Term (30+ Years)

While planning for new hospitals and health care facilities may ensure maximum green space and water conservation, surrounding such new construction with passive cooling mechanisms, such as more trees. Also, any new constructions should absolutely avoid building on flood plains because of the dangers of changing rainfall patterns and the unpredictability of flash floods. More resources must be focused on the research and development of temperature-resistant drugs and laboratory materials to ensure their longevity. In addition, a summary of implications of more severe and frequent heat waves and their impacts on other sectors can be documented in the HAPs.

8.9 Implementable Actions for Future

8.9.1 Urban Heat Island (UHI)

Urban areas and mega cities tend to be hotter than their rural counterparts, creating UHI effects. The reasons for such an effect include: increased absorption and reflection of the sun on concrete and hard surfaces as against green or brown spaces; reduced airflow and cooling from breezes in built-up areas due to thick constructions; and increased heat produce from energy consumption in residences, industry, businesses and vehicles. Such effects call for taking up local actions towards mitigation measures to tackle hot weathers which may include incorporation of modifications in surface properties and integration of green infrastructure, for example 'cool roofs', 'green roofs' and 'cool pavements' etc.

8.9.2 Creating Cool Environments with Green Spaces

Tree cover can make urban micro-climates cooler by creating canopy shades and allows for cooler air to circulate at ground levels. Large-scale vegetation plantations, especially trees with white or pale leaves, have the potential to reflect heat upwards. Thereby, planting more trees will increase the cooling effects and possible for creation of green spaces to enhance evaporation. Additionally, more green spaces and plantations facilitate drainage thereby reducing the risk of flooding. Studies suggest that air-conditioning demand can be reduced by up to 30% with having thick tree covers. Water features such as lakes, ponds and fountains also help to cool the surroundings. Urban Local Bodies and other land-use practitioners can also be sensitized about the benefits of green infrastructure.

8.9.3 Insulation

Promoting programs that focus on improving insulation in buildings and homes is particularly effective for preventing overheating, particularly when combined with measures to keep surroundings cool. For instance, setting shading or using reflective external wall surfaces can cool the surrounding areas as well as the dwelling. Interestingly, few such measures are low cost—for instance, usage of light shaded curtains with white reflective linings are not expensive and are very effective for cooling the inside of a building. External wall insulations have proved to be better than internal wall insulation as it prevents heat from getting in and trapped, so, the way insulations are installed, and the way occupants use heavily insulated homes, can help protect against overheating. This type of program may be particularly important for instance with elderly occupants and other vulnerable groups at home during the daytime.

8.10 Conclusion

To be able to address the key concerns of vulnerability reduction of heat wave-related detrimental impacts, emphasis must be laid on increasing the social and financial status of communities, especially those located in regions of heightened risks from such hazards. Globally, greater attention and more pre-emptive intervention are necessary for planning on the part of Governments and private stakeholders. The suggested adaptation options may be considered in building a heat resilient nation. Lessons must be learnt from past experiences and the post-heat wave periods and viewed as an opportunity to document and implement good practices. These lessons provide crucial information for improving Heat Wave Plans and integrating flexible mitigation measures. Implementing lessons learned instead of rushing to rebuild that would increase the vulnerability of local communities to such future events, would be the prudent course of action to ensure that India's government make the best choices for their people in the face of Climate Change.

References

Ahmedabad–Gujarat Municipal Corporations' Heat Action Plan (n.d.). Retrieved from https://www.nrdc.org/sites/default/files/ahmedabad-heat-action-plan-2017.pdf

Challenges (2015) Concept paper–retrofitting precincts for heat wave resilience: challenges and barriers in Australian Context by Gertrud Hatvani-Kovacs and John Boland

Every year could bring a heat wave if Climate Change continues (2016) Reuters. Retrieved from http://www.reuters.com/article/us-climatechange-heatwave-emissions-dUSKCN0VX26N

Gupta AK (2015) Yojana-A development monthly. Special Issue. Retrieved from http://yojana.gov.in/Yojana%20Decembe%202015%20FINAL%20PDF.pdf

Heat Health Action Plans: World Health Organization (2008). Retrieved from http://www.euro.who.int/__data/assets/pdf_file/0006/95919/E91347.pdf

Heat wave (2016) National disaster management authority. Retrieved from http://www.ndma.gov.in/en/media-public-awareness/disaster/natural-disaster/heat-wave.html

Heat Wave Plan for England (2013) Making the case: the impact of heat on health–now and in the future. Public Health England. Local Government Association and Met Office. 2013

Heat waves and Health: Guidance on Warning-System Development, WMO-No. 1142 (2015) World meteorological organization and world health organization. G.R. McGregor, lead editor P. Bessemoulin, K. Ebi and B. Menne. Retrieved from: http://www.who.int/globalchange/publications/WMO_WHO_Heat_Health_Guidance_2015.pdf

Hoverter SP (2012) Adapting to urban heat: a tool kit for local governments. Georgetown climate center: a leading resource for State and Federal Policy.

IPCC (2007) Climate change 2007: working group II: impacts, adaptation and vulnerability, glossary. IPCC. Von https://www.ipcc.ch/publications_and_data/ar4/wg2/en/annexessglossary-a-d.html abgerufen

Robinson PJ (2000) On the definition of a heat wave. J Appl Meteorol 40:762–765

Dr. Sushma Guleria is working as an Assistant Professor with the Environmental Disaster Risk Management Division National Institute of Disaster Management (Govt. of India), New Delhi. She has more than 21 publications to her credit and in her current profile, she looks after the Centre for Climate Resilience and Environment and Centre for Water and Land Disaster Risk Reduction (DRR). She is the recipient of the World Bank's Young Researcher's Grant Award for Disaster Risk Reduction for the year 2005-06 from India.

Gupta Anil K , Professor of Policy & Strategies, Integrated Resilience. Head of the ECDRM Division, International Cooperation, Programme Director - Centre for Excellence on Climate Resilience, Nodal Officer for Central Ministries at National Institute of Disaster Management, (Govt. of India), New Delhi. He is also a member of the expert Team for WMO Climate Statement and Core Group Member of IUCN-Council of Ecosystem Management.

Chapter 9
Impact of Climate Change on Forest Fires in India and Climate Adaptive Management Strategies

Sweta Baidya Das and Vijay Kumar Dhawan

Abstract Forest fire is a growing problem worldwide; especially the tropical forests are in constant danger of wildfire typically during the respective summer season. The extent of fire and the amount of loss are increasing day by day. India is also not an exception of this situation. Though the reasons of Indian forest fires are mostly attributed to the anthropogenic causes, globally researches have shown that along with anthropogenic activities accelerated climate change has almost equal contribution to these wildfires. Not only in India, worldwide now it is accepted that with increase in climate change scenario, more devastating wildfires are going to take place in near future. Therefore, now it is high time to rethink about the forest fire management strategies and climate adaptive forest fire management strategies are the new solution for the future resilience building towards wildfires.

Keywords Climate change · Forest fire · Climate adaptation · Strategy planning · Management · Resilience building

9.1 Introduction

9.1.1 Forest Cover, Types of Forest Fire and its Impacts

The extent of forest within any specific region or country is considered as the forest cover. Normally vegetation that extents more than a hectare and has greater than 10% canopy density including orchards, bamboo, palms, trees, grasses etc. within community, private, government or institutional lands are considered as forest cover. Forest Survey of India has classified Indian forest cover into five categories, such as (a) Very Dense Forest, (b) Moderately Dense Forest, (c) Open Forest, (d) Scrub and (e) Non-forest. The estimation of forest cover throughout the country is done on the

S. B. Das (✉)
National Disaster Management Authority, New Delhi, India

V. K. Dhawan
Forestry Expert, Forest Research Institute, Dehradun, India

A. K. Gupta et al. (eds.), *Disaster Risk and Management Under Climate Change*, Disaster Resilience and Green Growth,
https://doi.org/10.1007/978-981-99-4105-6_9

basis of data acquired by Indian Space Research Organization (ISRO) through indigenous LISS III sensor of IRS Resoursesat series of satellites. Later, these data is corrected through series of processing and validated through ground truth observations. The successive assessment of forest cover on 2017 and 2019 has revealed an overall gain of 3976 sq. km of forest. States and Union Territories like Karnataka (1.025 sq. km), Andhra Pradesh (990 sq. km), Kerala (823 sq. km) and Jammu & Kashmir (371 sq. km) has gained forest cover due to better conservation, protection, afforestation activities etc. On the other hand some of the North Eastern Hill states have shown loss in forest cover, such as Manipur (499 sq. km), Arunachal Pradesh (276 sq. km) and Mizoram (180 sq. km) mostly due to practice of shifting cultivation, natural calamities like storm, flood etc., felling of trees, forest fire and anthropogenic developmental pressure (State of Forest Report 2019).

Generally three types of forest fire has been identified (a) Ground fire or peat fire, (b) Surface fire on forest surface due to dry fuels and (c) Crown fire on trunks and crown of tree (Földi and Kuti 2016). But Indian forest fires are categorized into two broad types, (a) surface fire and (b) crown fire. Ground fire or peat fire occurs at alpine forests especially in Taiga and Tundra forests. The surface fire spreads along the ground but the crown fire burns the trees from bottom to top and therefore, crown fire is the most dangerous and causes maximum destruction (https://himachal.nic.in/).

9.1.2 Global Scenario

Global forest fire analysis (Giglio et al. 2013) has shown that over all global burnt area has decreased 2% per year within 2003–2012. Regional data provides different scenarios between 1996–2012 over the continents, like, Australia and New Zealand has shown 5% per year reduction in burnt area, although the year 2012 showed the largest area burnt. But on the other hand in Southeast Asia, Middle East and Boreal North America, the burnt area has increased by 3–4% and for temperate North America the increase is 0.1% per year (Giglio et al. 2013). But over all study of USA shows that, within 1991–2015, the burned area by forest fire has increased by 5% per year, though the number of fire has decreased (NIFC 2016) (Figs. 9.1, 9.2, and 9.3; Tables 9.1 and 9.2).

9.1.3 Asian Scenario

9.1.3.1 Indonesian Forest Fire

In 2015 summer (between June to October) the severe forest fire event engulfed around 2.6 million hectares of forest of Indonesia, which cost Indonesia USD 16.1 billion, which is close to 1.9% of the country's GDP. The amount of loss was two

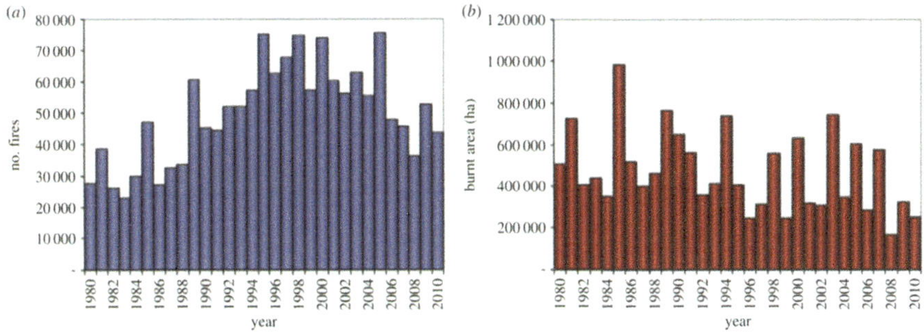

Fig. 9.1 Number of forest fires (**a**) and corresponding area burnt (**b**) in the European Mediterranean region over the period of 31 years (2010–1980). Source: San-Miguel-Ayanz et al. 2013

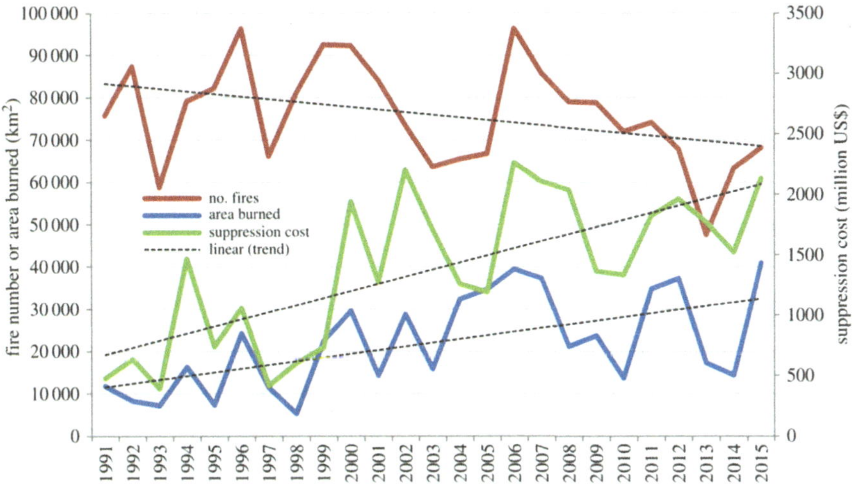

Fig. 9.2 Area burned, number of fires and suppression costs (inflation adjusted to 2016 equivalent) for the USA with linear trend lines (1991–2015). Data: National Interagency Fire Center, NIFC 2016

times more than the reconstruction cost after Aceh tsunami and 1.5 times more than the profit made by country's palm oil production in 2014 (Dogra et al. 2018). The extent of fire is comparable to the size of the state of Tamil Nadu. Usually in Indonesia fire is used as cost effective tool for field preparation for palm oil cultivation but without proper control burning measures, which caused havoc in 2015. A research by Centre for International Forestry Research (CIFOR) has shown that 85% profit of this type of land preparation goes to local elites and plantation developer (Fig. 9.4).

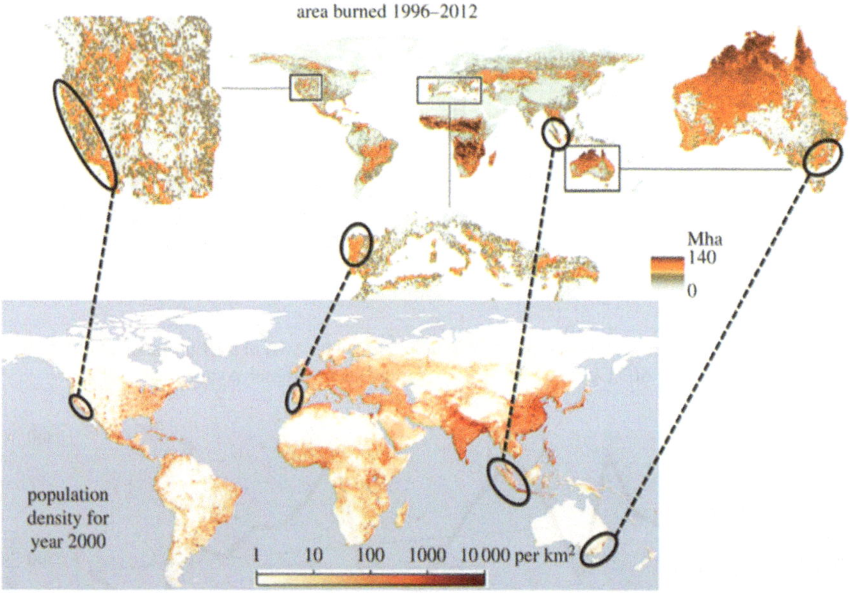

Fig. 9.3 Global area burned between 1996 and 2012 and global population densities are marked here. High proportion of burned area coincides with high density of population. Source: Doerr and Santin 2016

Table 9.1 Global comparison of human and economic losses caused by wildfire, earthquake and floods within 1901 to 2014. (Source: Guha-Sapir et al. 2015)

	Wildfires	Earthquakes	Floods
No. of Events	387	1291	4481
People killed	3753	2574627	6947908
People injured	6812	2614875	1329923
People affected (million)	6	190	3604
Risk of Death (%)[a]	0.06	1.4	0.02
Total direct damage (million US$)	54828	774771	681427
Cost per event (million US$)	142	600	152
Cost per person affected (US$)	9138	4078	189

[a]No. of fatalities per no. of people affected (%)

9.1.4 *Impact of Climate Change on Forest Fire and Vice Versa*

The entire world is facing the effect of climate change, therefore, forests around the world will not be an exception and going to face the heat due to rapid climate change (Mukhopadhyay 2009). It is well accepted that forest fire activity increases during the dry season (Liu et al. 2010; Abatzoglou and Kolden 2013; Riley et al. 2013; Aragao et al. 2018). Climate change has significant influence on the frequency,

Table 9.2 Continent wise economic and human loss caused by forest fire between 1984 and 2013. (Source: Guha-Sapir et al. 2015)

	No. of events	People killed	Total people affected	Death rate/ event	Economic costs (million US$)
Africa	25	272	21672	11	440
America	118	234	1229 175	9	25 229
Asia	50	748	3 188 257	30	11892
Europe	89	462	1295 562	18	12619
Oceania	21	224	74 320	9	2121
Total	303	1940	5 808 986	78	52 301

Category of damage/loss	US$ millions
Agriculture	4,839
Estate crops	3,112
Food crops	1,727
Environment	4,253
Biodiversity loss	287
Carbon emissions	3,966
Forestry	3,931
Manufacturing and mining	610
Trade	1,333
Transportation	372
Tourism	399
Health	151
Education	39
Firefighting costs	197
Total	**16,124**

Fig. 9.4 Damages caused by the forest fire of Indonesia in 2015. (Source: Excerpted and adapted from World Bank (2016). Dogra et al. 2018)

intensity, number and duration of the forest fire (Flannigan et al. 2008). Statistical analysis of long term forest fire events and climatic data sets from Himachal Pradesh and Uttarakhand has shown that weather severity like drought situations are one of the major contributors of the forest fires (Ahmad and Goparaju 2018). Climate Change has a bearing on forest fire. Forest fire is a hydro-meteorological related disaster. Climate change and forest fire impacts each other. Climate change leads to lesser precipitation and high temperature which ultimately triggers forest fire. On the other hand, if forest fire takes place, the black carbon generated by the fire covers the ice covers and helps in fast melting and ultimately gives rise to warming and lesser deposition of ice caps which will reduce the water flow of glacier fed rivers which will increase local dryness. The smoke and flying ash generated by forest fire also reduces the insolation for few days and photosynthesis activity by trees is also disrupted. Forest fire triggers positive feedback for even drier conditions. The layer of ash created by the fire burning is water repellent which means for next 1 or 2 years that ash covered soil cannot absorb any rainwater as rainwater does not infiltrate through this ash layer. Consequently it increases surface runoff which leads to debris flow or mud flow during heavy rains (Wohlgemuth et al. 2012). The rain water infiltration rate is also reduced due to canopy burning during burning of crown which increased splash erosion and caused flash flood. Fire also damages the micro organisms and macro and micro and macro fauna living over the soil. The micro and macro fauna help in decomposition of soil and increase soil aeration that leads to increase soil organic matter, soil nutrients and increase productivity of the soil. Thus forest fire impacts on the natural succession of the forest species. Forest fire retrograde the succession stages and continues forest fire may lead to change of original forest type. Therefore, forest fire ultimately complexes the situation. Forest fire can change the amount of insolation and the smoke and ash affects the health of people and health of ecosystem. Sometimes, forest fire also causes the extinction of endangered species of flora and fauna which is an irreversible loss. Climate change also leads to change in composition of species as well as promotes spread of Alien Invasive Species (AIS) like *Lantana, Ageratum, Eupatorium, Parthenium, Chenopodium* etc. These invasive species spread at very faster rates after forest fire.

Tropical forests around the globe are most important as it gets maximum affected and many of these serve large number of population for their livelihood. These are the most vulnerable ecosystem towards the long term changes in temperature and rainfall and resultant climate change (CIFOR 2007). If due to climate change these forests face longer dry spells in combination with other pressures like industrial forestry etc. could bring in the possibility of forest fire (Mukhopadhyay 2009).

9.1.5 Impact of Climate Change on Himalayan Forests

All countries across the globe by now are able to feel the heat of the climate change especially the tropical countries like India (Bahuguna 2018). The analytical results have brought out the fact that, climate change indeed has impacts on forests and vegetation. The impacts are even severe in last 30 years (1991 to 2020). Meteorological data have shown an average 0.7 to 1^{0}c temperature rises in Indian forests. Moreover the total number rainy days and intensity of rainfall has also decreased throughout Indian forest areas by 5 days and 241 mm respectively. On the other hand, excess rainfall within a short period of time and prolonged monsoon break is enhancing the drought situation throughout the country (Zhisheng et al. 2015). This has given rise to the compositional change of the forests' vegetation types; distribution and species density, particularly the Alpine flora of the higher Himalaya are among the most affected hot spots of Bio-diversity. As a whole the Himalayan forests are turning xeric due to massive change in local hydrological set up due to climate change and resulting vegetation type change (Bahuguna 2018).

9.1.6 Impact of Climate Change on Western Ghats' Forests

Studies from Western Ghats of Southern Peninsular India, have shown strong impact of climate change and drought on wildfires. The modelling results of the studies carried out over Western Ghats have proved droughts caused by water deficits potentially increases the vulnerability of forests towards fire both at a landscape and regional scale. In this region, climate variability strongly influences the forest fire activity at 3 to 6 month timescale (Kodandapani and Parks 2019). Impact of continuous forest fires changed succession pattern of Shola Forests of Nilgiri hills. Some pockets Shola forests retrograded into grasslands.

9.2 Impacts of Forest Fire in India

Some studies have shown that forest fire in India has grown by 1.5 times in six years time (Paliath, 2018) Forest fire plays few important roles in maintaining the forest ecosystem like low intensity fire removes low-growing underbrush cleans the forest floor of debris and opens it up to sunlight which allows spp. sprout after fire and helps pines shed their seeds from cone etc. Forest fire also kills diseases and insects that prey on trees and nourishes the soil, allows established trees to grow stronger and healthier and clears heavy brush, leaving room for new grasses, herbs and shrubs that provide food and habitat for many wildlife species.

But even then, forest fires are the most significant cause for damages of country's forest resources which leads to invaluable loss of forests and ecosystem services. Not only the direct impacts, forest fire has many other indirect impacts on the ecosystem, like landslide (especially in the hilly regions), soil erosion, loss of regeneration, habitat degradation, loss of forest produces etc. (State of Forest Report 2017 & 2019). Most of the Indian forest fire is caused by intentional and unintentional anthropogenic activities like shifting cultivation, forest floor clearance for NTFP collection, promoting grass growth for grazing animals, careless activities like throwing of burning match sticks or unattended cooking fire by temporary road workers, criminal activities by poachers to forcefully bring out animals from their safe habitats etc. The interspersed habitation of various population sizes and fragmentation of the forests intensifies the probability of fire incidences. Forest Survey of India (2019) reported that 54.40% of Indian forest faces occasional fire incidences, 7.49% moderately frequent fires and 2.40% high level of fire incidences. But good thing is that 35.71% of Indian forest are still well maintained and experience comparatively insignificant fire incidence. The forests of Deccan Plateau and North East India are the worst affected from fire. The world wide climate change and resultant drought situation aggravates the possibility of the fire incidences and vulnerability of the forests (National Action Plan on Forest Fire). Within the 13 year period from 2004 to 2017 a total number of 2,77,758 fire points throughout the country has shown repeated fire detection through MODIS data (Fig. 9.5).

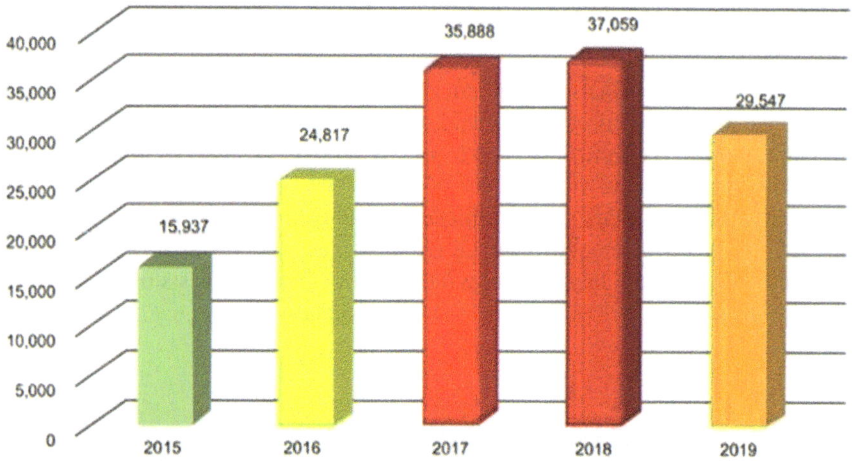

Fig. 9.5 Forest fire detections based on MODIS in last five years (Source: State of Forest Report 2019)

9.2.1 *Damage Assessment*

Every year due to forest fire India faces loses worth almost INR 550 crore (Hindustan times, 2018). Forest fire devastated more than 4500 hectares of forest (which is 1.3 times the size of the entire Shimla city) in Northern Himachal Pradesh since seventh April, 2016. In Uttarakhand also around 4000 hectares of forest were engulfed by fire in 2016 itself. During the period of 1990 to 2001, the worst forest fire happened in Uttarakhand in 1995 which burnt 3,75,000 ha of forest. This was followed by Ganga-Yamuna watershed area where in 1999, 80,000 hectares of forest was devastated. In the year of 2010, 19,109 hectares of forest in Himachal Pradesh and in 2008 and 2010, 10,000 hectares of land in Maharashtra were burnt by wildfires. As per the Government data submitted to Parliament on July 20, 2018, forest fire of India has increased by 1.5 times over last 6 years to 35,888 fire incidences (Indiaspend 2018). In 2017–18 and 2018–19, INR 35 crore and INR 38 crore were released by Government for forest fire prevention management (Pandey 2019). In spite of that, 30,000 forest fires were reported in 2019. In February 2018, an estimated 4800 hectares of forest were burnt at Bandipur Tiger Reserve in Karnataka (Srivastava 2020). Between November 2018 and February, 2019, the number of forest fire increased from 4225 to 14,107 according to the real time forest alert system of Forest Survey of India (FSI). Between first January, 2019 and 26th February, 2019, 209 among the 558 forest fires took place in Southern Indian states—Andhra Pradesh, Karnataka, Kerala, Tamil Nadu & Telengana (Pandey 2019). As on May 2020, 46 forest fires were reported from Uttarakhand, where 71 hectares of land have already been destroyed by fire (India.com 2020) (Table 9.3).

Table 9.3 Reported monetary loss caused by forest fire in selected states from 2007 to 2016

	2007	2008	2009	2010	2011	2012	2013	2014	2015	2016
Total Losses, INR lakh (US$ 1,000)										
Chhattisgarh	0	0	0	0	0	0	0	0	0	0
	0	0	0	0	0	0	0	0	0	0
Himachal Pradesh				255.23	97.69	43.08	276.83	52.31	113.27	134.78
				-558.17	-209.33	-80.62	-472.42	-85.71	-176.56	-200.58
Kerala						1.89	1.04	2.46	1.11	0.25
						-3.53	(1.77)	-4.04	(1.73)	-0.38
Uuarakhand	3.84	2.68	4.79	1.90	0.3	42.89	4.39	23.58	7.94	46.5
	-9.29	-6.16	-9.9	-4.16	-0.64	-80.27	-7.5	-38.63	-12.38	-69.2
Losses per hectare, INR (US$)										
Chattisgarh	0	0	0	0	0	0	0	0	0	0
	0	0	0	0	0	0	0	0	0	0
Himachal Pradesh				1,027.10	1,246.47	2,450.24	1,332.56	1,615.75	1,683.74	2,343.97
				-22.46	-26.71	-45.85	-22.74	-26.47	-26.25	-34.88
Kerala						33.42	43.91	93.54	65.54	12.87
						-0.63	-0.75	(1.53)	(1.02)	-0.19
Uttarakhand	240.79	113.04	116.42	118	128.25	1,518.91	1,144.09	2,534.27	1,132.19	1,048.82
	-5.82	-2.6	-2.41	-2.58	-2.75	-28.42	-19.52	-41.53	-17.65	-15.61

Note: Numbers are in nominal terms, not adjusted for inflation; assuming official exchange rate for INR to US$. Source: Dogra et al. 2018

9.2.2 *Impact on Livelihood*

Around 68% of Indian population lives in rural areas. A bulk share of these rural populations is directly or indirectly dependent on forests for their livelihood like bamboo, fodder, fuel wood, small timber and NTFPs. As per census 2011 among the 6,50,000 villages of the country, 1,70,000 villages are located in the proximity of forest areas. These villages are known as Forest Fringe Villages (FVVs) (FSI 1999). Forest is an integrated part of the socioeconomic and cultural practices of these villages.

Forest fire not only stops the collection of livelihood from forest, but poses various other risks directly or indirectly. For example, when fire has caught the forest on the steep slopes, it leads to rockfall even if it is still burning. If the top soil is burnt by the fire, the destabilized stones start rolling down slopes (Wohlgemuth et al. 2012), also later on mud-flow and debris flows are triggered due to slope failure during the heavy rain. These poses life and property risk to the people staying in FVVs of hilly region. On the other hand the smoke generated from fire and the particulate ash poses respiratory health risks to the local residents.

The forest fire also increases the possibility of man animal conflict as fire incidences may drive the animals towards the nearby locality to save their lives from fire and hunt for food and water to drink. This eventually increases the possibility of man animal conflict leading to the loss of animal and human life (Tables 9.4 and 9.5).

Table 9.4 Poverty rates and fire density in rural forested districts, grouped by quartile

	1 (Lowest poverty rate)	2	3	4 (Highest poverty Rate)
Poverty head count Ratio	4.5	14.1	27.2	48.2
mean, 2011				
(% population below national poveny line)				
Poverty head count ratio,	0.0-8.8	8.9-20.4	20.5-35.2	35.3-78.7
range, 2011				
(% population below national poverty line)				
Average forest cover, 2000	60.2	55.9	51.1	53.5
(% total district area with >				
10% tree canopy cover)				
Average forest fire density,	I.I	2	2.7	3.8
2003-2016				
(annual fire detections per I 00 kmt treed area)				

Note: The poverty head count ratio is of 2011 and average forest fire density is from 2003–2016. Average forest cover and fire density are weighted by the forested area per district; sample is limited to rural districts where population density is <1000 per km2 and having at least 10% forest cover in 2000 (554 of 638 districts with data); Forest area is defined as an area having minimum 10% tree canopy cover. Source: Dogra et al. 2018.

Table 9.5 State/UT wise annual forest resource collection by the people living in FVVs (Source: Table reproduced from State of Forest Report 2019)

S. No.	State/UTI	Fuelwood ('000 tonnes)	Fodder ('000 tonnes	Sm11U Timber (cum)	Bamboo ('000 tonnes)
1.	Andhra Pradesh	2,789	25,043	81,808	14.74
2	Arunachal Pradesh	44	528	1,314	0.4
3	Assam	1,411	11,712	32,972	14.44
4	Bihar	821	4,338	13,766	11.34
5	Chhattisgarh	3,608	82,771	852,164	392.49
6	Goa	30	35	2,699	0.31
7	Gujarat	4,983	119,054	1,192,475	291.75
8	Haryana	500	6,840	16,471	0.04
9	Himachal Pradesh	593	3,256	11,264	0.59
10	Jammu & Kashmir	1,299	14,018	19,763	0.09
11	Jharkhand	7,372	55,482	183,240	50.54
12	Karnataka	6,323	21,501	41,098	0.4
13	Kerala	3,390	3,472	100,259	0.85
14	Madhya Pradesh	7,663	222,720	1,473,754	630.66
15	Maharashtra	9,539	157,136	862,138	128.67
16	Manipur	39	262	8,618	2.92
17	Meghalaya	93	220	5,821	0.9
18	Mizoram	18	23	849	0.42
19	Nagaland	278	488	12,225	1.09
20	Odisha	9,186	56,035	376,521	110.79
21	Punjab	456	4,269	18,758	0.09
22	Rajasthan	8,560	112,708	82,433	3.7
23	Sikkim	82	440	1,320	0.07
24	Tamilnadu	1,752	20,123	102,566	2.35
25	Telangana	1,969	15,958	1,541	6.86
26	Tripura	700	1,588	8,468	3.5
27	Uttar Pradesh	5,141	59,335	159,587	109.51
28	Uttarakhand	4,076	32,119	38,801	2.43
29	West Bengal	2,519	21,209	134,946	45.47
30	A & N Islands	22	83	2,506	3.74
31	Dadra & Nagar Haveli	33	274	8,057	3.1
	Total	**85,290**	**1,053,039**	**5,848,204**	**1,834.25**

9.2.3 Fire Environment

This diagram nicely describes the factors influencing the forest fires. The long term dry spell resulting from decreased rainfall and humidity and increased temperature increases the probability of forest fire. Direction and speed of winds decides the extents of the fire. Many researchers have used El Nino Southern Oscillation (ENSO) to be the indicator of the severity of the fire season, as El Nino corresponds to the comparatively drier monsoon (lesser rainfall). But in Indian Scenario the correlation between the ENSO and fire events are not straightforward. Also in recent decades the correlation between the ENSO and monsoon rain has weakened (An et al. 2015) (Fig. 9.6).

Fuel is the second important factor which includes the dry leaves, dead trees and biomass over soil and some invasive species which attracts fire. The total amount of such materials along with the dryness will decide the intensity of the fire.

Topography is the third important factor for forest fire. Topography influences the fire prevention and suppression and could raise the possibility of spreading especially in steep hilly terrain whereas spread is much lesser over gentle terrain. Wind on rugged terrain can change the direction quickly to engulf more areas. Remote

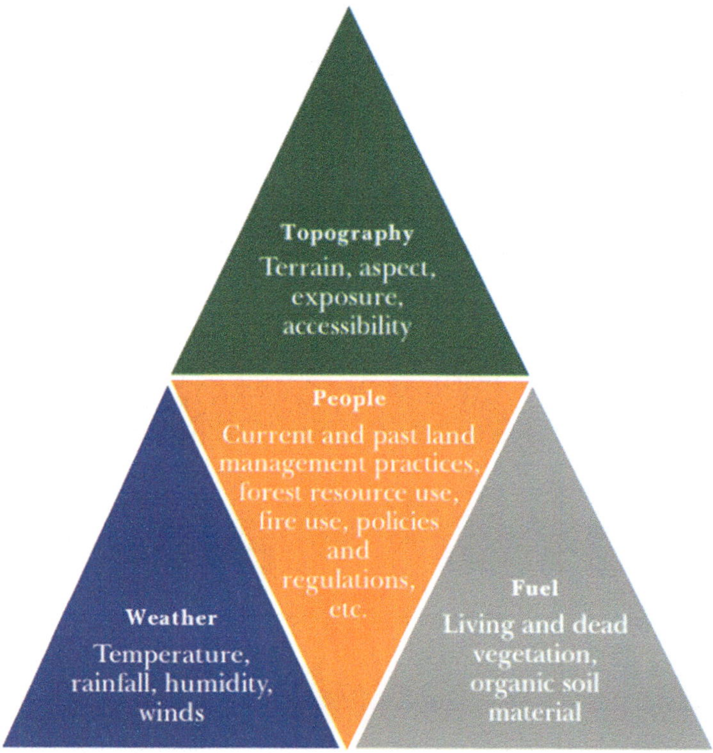

Fig. 9.6 Causes of forest fire. Source: Dogra et al. 2018

location and rugged terrain works as the barrier for the fire crew to reach timely and save the forest.

Worldwide researches have shown though there could be many other natural reasons which could trigger the forest fires but, in maximum cases, forest fire is caused by human either knowingly or unknowingly due to lack of knowledge, carelessness or for personal benefit. Sometimes the faulty conservation or adaptation methods could also lead to the fire incidences (Smith 2017) (Tables 9.6 and 9.7).

9.3 Current Plans, Policies and Practices for Forest Fire Management

The objective of fire management is to minimize the occurrence of forest fire incidences by early warning, awareness generation, enabling and empowering the communities living in the fringes of the forests and provide incentives to make them work in tandem with the forest departments to reduce the vulnerability of forest and humans. The management also aims towards increasing the capability of community as well as other personnel and fire fighters to speed up recovery after the fire events. A National Action Plan on Forest Fire (NAPFF) has been prepared by Forest Protection Division of Ministry of Environment, Forests and Climate Change, Government of India. The NAPFF has provided guidelines to assist the administrators, policy makers, frontline staff, forest user agencies, forest fringe communities, tourists, and the entire civilian societies for a holistic management of forest fires (NAPFF) (Table 9.8).

9.3.1 Fire Risk Management

Fire risk zonation mapping is done to identify the priority areas for management intervention like zone mapping, monitoring, resource allocation and other fire risk controlling measures. There are various risk factors, assessed to identify and quantify the forest fire risks like, fuel load (the types of vegetation), proximity to habitation, property and other assets including infrastructure, agricultural lands, railways roads etc., topography (slope, aspect, elevation, terrain ruggedness etc.), dependence of habitats for livelihood, historic fire occurrence pattern, area of special ecological or cultural importance etc.

Table 9.6 Causes of forest fires on the basis of survey by State Forest Department Officers

	Second most important cause of fire						Most important cause of fire					
Category	Cause	Assam	Chhattisgarh	Himachal Pradesh	Jharkhand	Kerala	Madhya Pradesh	Meghalaya	Odisha	Telangana	Tripura	Uttarakhand
	Unknown		3					38				
Natural	Natural, not specified	38		5	8	22		2		4	10	6
	Lightning			1						2	5	
	Other natural			1								
Acci dent	Accident, not specified				6	36		52	5	2		5
	Electric power			7			6				2	5
	Works											1 0
	Self-ignition											1
Negligence	Negligence, not specified		3	15	10	64		17			7	35
	Negligence, use of fire	38	32	100	42	100		69	22	36		100
	Negligence, glowing objects	100	18	51	46	19	8	21	1	19	24	33
Deliberate/ incendiary	Deliberate, not specified				8			21			21	13
	Responsible (arson)	58	12	16	19	92	11	2	12	30	10	24
	Not responsible (e.g., fires set by minors)			6								
Resource collection	NTFP collection		100	8	100	72	100		100	100	7	3
	Grazing or fodder		8	73	8	25	22	38	1 1	64	1 2	70

Wildlife	Burning to deter wildlife			7	10	39		1 4	3			17
	Hunting		1 2	5	4		19	1 9	1 4		1 0	
	Traditional Practice, not Specified					6	3					1
	Shifting cultivation	85	5					100	45	9	100	6
	No. of responses	**6**	**7**	**14**	**10**	**10**	**3**	**11**	**9**	**9**	**9**	**16**

Note: Numbers provided in the boxes are index of importance (1–100). Source: Dogra et al. 2018

Table 9.7 Region wise fire detection (Observed) & stage of drought (Measured by Keetch-Byram drought index (KBDI) from 1st February to 31st May 2012–2016

		Share of forest fire detections by region (% of districts and days for which a forest fire was detected when KBDI was in the					
KBDI	Drought Stage	Central	North	Northeast	South	West	w. Himalaya
0-99	0 (Saturated Soils)	0	0	1.55	0.02	0.14	6.52
100-199	1	0.01	0	2.06	0.02	0	13.38
200-299	2	0	0.19	3.44	0.35	0	5.13
300-399	3	0.02	1.08	3.74	2.01	0	5.73
400-499	4	0.1 1	4.26	5.93	5.88	0	9.79
500-599	5	0.6	9.6	13.62	13.64	3.24	15.91
600-699	6	7.47	19.71	24.31	24.02	14.19	26.23
700-800	7 (severe drought)	91.79	65.16	45.34	54.05	82.43	17.31

Note: regional averages are constructed by weighting districts by size of forest area

9.3.2 *Prevention of Forest Fires*

9.3.2.1 Awareness and Sensitization

In India most of the forest fires are human induced and caused by the livelihood activity of the forest fringe community. Therefore, it is always a best practice to create awareness among them for prevention and prepare them for the worst situations. As a mass communication strategy publicity and sensitization programmes are organized for local cattle herders, farmers, students, women, NTFP collectors, Infrastructure professionals, urban dwellers including tourists pilgrims etc. and other related stake holders through campaigns, radio, television, social media, mobile, portals and by organizing various festivals.

9.3.2.2 Empowering Communities for Practicing their Responsibilities

Not only awareness programmes are enough for the prevention of forest fires enforcement of existing laws are also important. Through this enforcement the participation of local communities in prevention and management forest fire is ensured. Similar responsibilities are given to the shifting cultivators, Van Panchayats, JFMCs etc. by delegating authorities and funds. Provisions for providing incentives are also there to enhance the participation of local communities (Table 9.9).

9.3.2.3 Capacity Building of the Communities

Training and capacity building programmes are organized frequently to enhance the knowledge and capacity of the EDC, JFMC and Van Panchayat members.

Table 9.8 Agencies involved in forest fire policy management in India

Central Government Entities	
MoEFCC	• Overall policy guidance and standard setting for FFPM • Administers centrally-sponsored schemes and provides funding to states
FSI (under MoEFCC)	• Issues pre-warning alerts for high fire danger to state forest departments nationwide • Nationwide monitoring and alerts for active fires, provided to state forest departments and the public • Nationwide estimation of burnt forest area
ICFRE (under MoEFCC)	• Apex research organization for forestry in India • Research institutes under ICFRE include FRI, which has developed training modules for the SFDs and firefighting equipment kits
DFE (under MoEFCC)	Coordinates training for frontline staff across the country, including forest rangers and state forest service officers
NRSC	Provides near-real time satellite data to FSI for fire monitoring
NDMA	Policies and planning for disaster management across the country So fa1 has played minor role in FFPM Deployed NDRF during 2016 forest fires in Uttarakhand Organized mock drill for forest fire response in April 2017
Military, Paramilitary, and Home Guards	Local units may be called by the SFD to assist in response to large forest fires from time to time
State Government Entities	
State forest department (SFD)	• Primary agency responsible for implementing FFPM • Approves forest working plans for forest divisions within the state, laying out required forest fire prevention activities • Issues state-specific instructions, standard operating procedures, and manuals for field staff • Monitors and collects field-reported data on fire occurrence, burnt area, damages, and forest offences in forest divisions across the state
SDMA	• Policies and planning for disaster management at the state level • Approves district-level disaster management plans • So far, has played minor role in FFPM
District Government Entities	
District Magistrate	• Coordinates among different departments like revenue, health, fire brigade in the event of a large fire • Approves the district-level fire management plan
Community/ village-level institutions	
Joint forest management committee:1:1	• Primary institution for community-based forest management in India and entry point for SFD engagement with communities on FFPM • Responsible for developing forest micro-plans for J FMC areas, with technical support from the SFDs • Carries out FFPM activities in coordination with the SFD and may organize labor from the local community for clearing fire lines, conducting controlled area burning, seasonal firewatchers, etc.

(continued)

Table 9.8 (continued)

Central Government Entities	
Other Community Institutions	• A diverse variety of other community-level institutions have evolved in the different states for community-based forest management, such as the Van Panchayats of Uttarakhand and Village Fire Protection Committees in parts of the Northeast

Note: *DFE* Directorate of Forest Education; *FRI* Forest Research Institute; *FSI* Forest Survey of India; *ICFRE* Indian Council on Forestry Research and Education; *MoEFCC* Ministry of Environment, Forest and Climate Change; *NDMA* National Disaster Management Authority; *NRSC* National Remote Sensing Centre, ISRO; *SDMA* State Disaster Management Authority

Firefighting equipment like leaf litter blower, protecting clothing and other tools are distributed to them. There members are also engaged to participate in the mock drills on forest fire before the fire season every year to share knowledge about the location of available water sources nearby and other infrastructure availability (like effective use of fire lanes etc.) to fight fire (Figs. 9.7 and 9.8).

9.3.2.4 Resilience Building

1. Water and Moisture Conservation

Dry deciduous forests with predominance of grasses are most vulnerable to fire. Therefore strategies are implemented to retain the moisture longer after rainy season. Long term plans are implemented to improve the water bodies of the forest area and along with this eco-rehabilitation and catchment area treatment plans are also part of the scope.

2. Forest Floor Biomass Management

The litters like dry leaves and woods on the forest floor are potential fuel for fire. Therefore, this forest floor biomass has to be managed on regular basis for gainful utilization of the biomass and to reduce the fire risks. To promote biomass collection, royalty and transit fee is waived and micro entrepreneurship through Self-help groups is to be promoted. Policy implementation has to be done to promote biomass based electric generation through off-grid micro power plants in remote locations. Sustainable removal of dead bamboos during mass flowering is another option to conserve the ecosystem from fire. In Himalayas chirpine needles are very vulnerable to fire. Forest department of Uttarakhand is using the needles in making check dams and creating micro power stations for villages with the help of Electricity department. Villagers are preparing briquettes. Scientists of FRI are carrying research works on preparation of fibers and methanol from the needles. The needles are collected by villagers and forest department fixed the rate of pine needles.

Table 9.9 Characteristics of community held forests in Meghalaya, Mizoram & Nagaland

Cluster	Forest type	Local Name	Size (ha)	Management institution	Degree of Protection	Access to forest resources	Shifting cultivation	Collection of Timbers	Collection of NTFPs	Collection of fuelwood	Hunting	Grazing
1	Raid forest	A1	35-50	Group ofVillages council	Low	All	Allowed	Allowed	Allowed	Allowed	Prohibited	Allowed
2	Village forest	A2	20-27	Villagecouncil	Low	All	Allowed	Allowed	Allowed	Allowed	Prohibited	Allowed
3	Restricted forest	A3	4-10	Villagecouncil	High	Prior Permission	Prohibited	Prohibited	Allowed	Prohibited	Prohibited	Prohibited
4	Sacred forest	A4	1-100	Villagecouncil	Very high	None	Prohibited	Prohibited	Prohibited	Prohibited	Prohibited	Prohibited
5	Clan forest	A5	5-20	Clan council	Very low	Clan Members	Allowed	Allowed	Allowed	Allowed	Allowed	Allowed
6	Cemetary forest	A6	1-30	Church	High	All	Prohibited	Prohibited	Allowed	Prohibited	Allowed	Prohibited
7	Regeneration forest	A7	3-5	Villagecouncil	Very high	None	Prohibited	Prohibited	Prohibited	Prohibited	Prohibited	Prohibited
8	Bamboo reserve	A8	10-15	Villagecouncil	Low	All	Allowed	Allowed	Allowed	Allowed	Prohibited	Allowed
9	Recreation forest	A9	10	Villagecouncil/ YMA	High	None	Prohibited	Prohibited	Prohibited	Prohibited	Prohibited	Prohibited
10	Reserve forest	A10	5-10	Villagecouncil/ YMA	High	Prior permission	Allowed	Allowed	Allowed	Allowed	Prohibited	Prohibited
11	Medicinal plantation	A11	50	YMA/YLA	Very high	Prior permission	Prohibited	Prohibited	Prohibited	Prohibited	Prohibited	Prohibited

Note: *YMA* Young Mizo Association; *YLA* Young Lai Association; YMA and YLA are indigenous civil society organizations. Source: Tiwari et al. (2013)

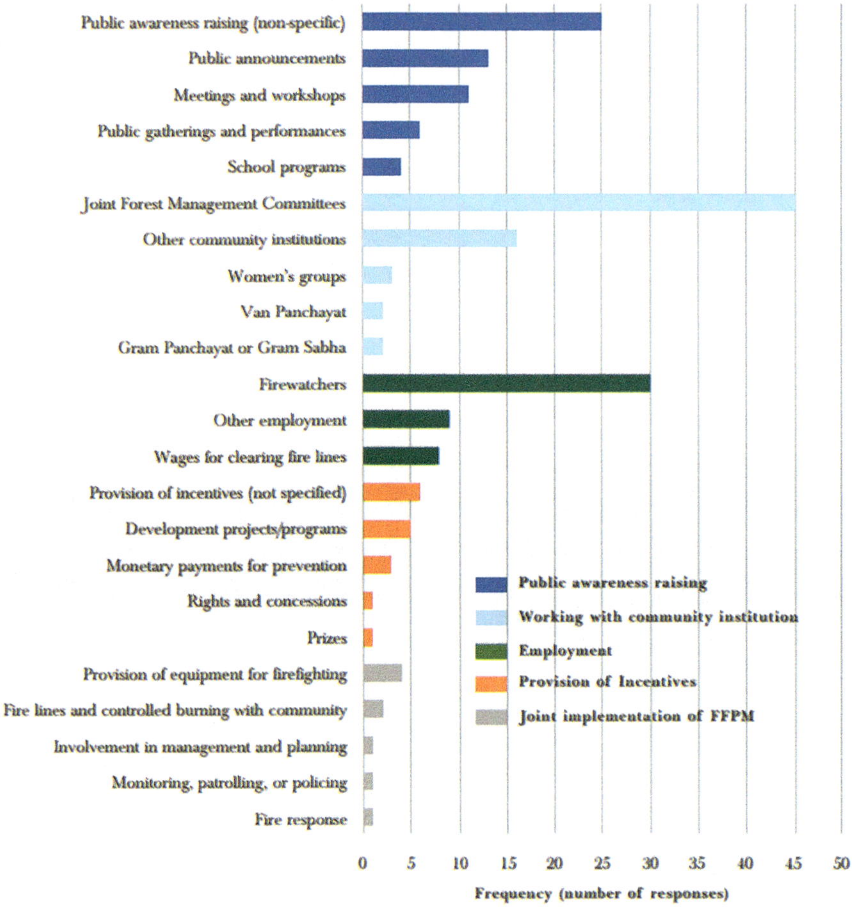

Fig. 9.7 Methods of community engagement used by state forest departments. Number of responding officers are 94. Officers may have indicated more than one methods of engagement. Source: World Bank survey of state forest department officers, April–August 2017

3. Weed Management

The weeds like *Lantana, Eupatorium, Chinopodium* etc. are covering a vast forest areas of the country starting from south to north and east to west. The weeds not only reduce native biodiversity but re-grow and spread after forest fire at a very faster rate and in dry season prone to fire also. To mitigate this, weeding is required for removal with proper techniques (manual, biological, chemical) and followed by land treatments. The treated land has to be regenerated with native fire resistant species for covering the area.

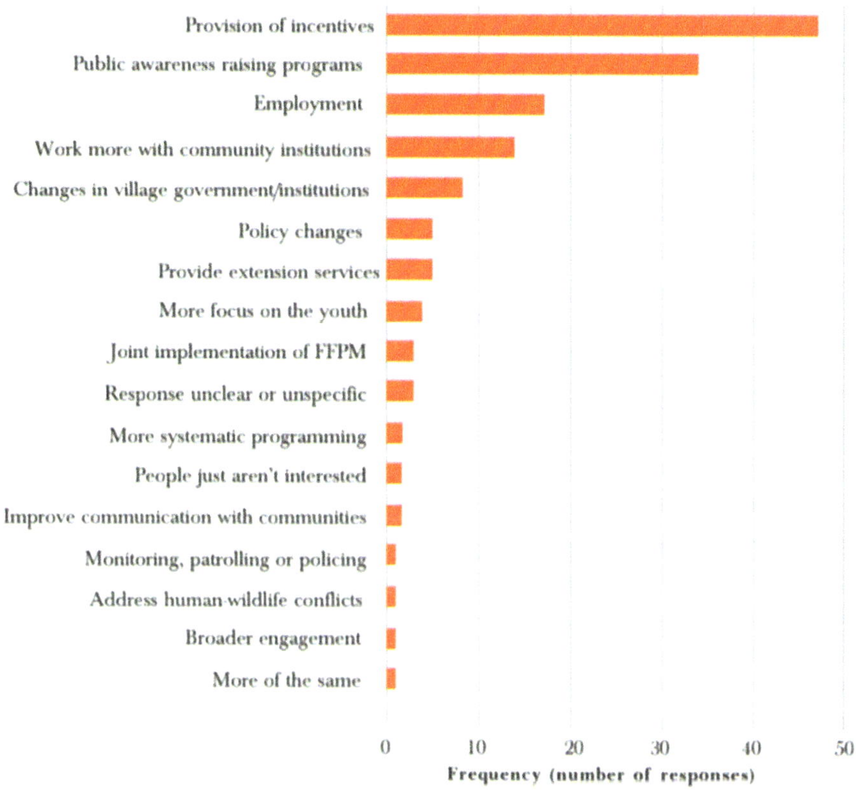

Fig. 9.8 Ways to improve local community engagement. Number of responding officers are 91; Officers may have told more than one way for better engagement. Source: World Bank survey of state forest department officers, April–August 2017

4. Forest Landscape Restoration and Agroforestry

Forest Landscape Restoration (FLR) is a planned series of action towards restoring the ecological functionality and increase human well being in the fire affected forests (WWF 2020). In most of the cases there are opportunities for restoring the pristine biodiversity. Fringed people or tribal may be encouraged to practice agroforestry in fire prone vulnerable areas of the forests so as to protect the forests from fire. They may be allowed to raise cash crops like tea, zinger, turmeric, coffee etc. along with forests on some kinds of leased agreement.

9.3.2.5 Preparedness to Fight the Forest Fire

5. Forest Fire Detection and Alert

Forest Survey of India has materialized the active satellite based *forest fire alert system* all over the country. The entire system of identifying the forest fire and alerting concerned people involves digitization of forest boundaries, promoting adoption of improved Forest Fire Alert System, improving ground based fire detection, increased engagement of local communities, dedicated toll free telephone line and wireless network for smooth communication, monitoring and evaluation of the existing tools and instruments (Table 9.10).

6. Saving Critical Resources and Assets

Along with establishing new assets, it is important to save and manage already existing resources and assets for better preparedness and future planning. To fulfill this purpose the IT cells of the forest departments needs to prepare inventory map and gather information regarding the already existing firefighting and overall management resources including control rooms, ground crew stations, fire lines, nearby roads, railways and telecommunication networks, nearby water bodies, natural fire breaks etc.

7. Forest Fire Lines

Fire lines are the first safeguards to pause or stop the fire situation; therefore it is most important to maintain the status, functionality and adequacy of the fire lines. State Forest Departments has to gather updated information regarding the already existing fire lines like roads, rail lines, rivers or water bodies, transmission lines etc. to decide on the need of the new fire lines or maintaining the old one depending on the forest type, old fire data, new growth of habitats etc. The fire lines should be cleared from any type of fuel (including trees) before fire season (Table 9.11).

8. Control Burning

In some of the forests, control burning is necessary to prevent spread of fire. For the consistent performance and effective control burning state specific guidelines and prescriptions for control burning have to be prepared. Timely release of funds, prioritization of the areas for control burning, monitoring the burning activity is among the other important factors to be taken care of.

9.3.2.6 Controlling and Management of Fire Situations

9. Training

Immediate response to fire can change the entire scenario of the grave results of the forest fires and can save enormous resources, ecosystem, life of animals and people and many more. To respond immediately we need to train field staff, fire watchers and other community fire fighters, field officers, seasonal fire watchers and other relevant stake holders and equip them with proper tools. They should be trained about the most effective fire suppression techniques and place and problem specific solutions and their respective responsibilities. After proper consultation with the responsible authorities like Directorate of Forest Education (DFE) of the MoEFCC,

Table 9.10 Field reporting on satellite based fire alert in eight states

	Andhra Pradesh	Chattisgarh	Himachal Pradesh	Kerala	Madhya Pradesh	Meghalaya	Telangana	Uttarakhand
Field reporting rate for satellite-based lire alerts (%)	89.4	3	100		70	60	9	
Rate or false alerts for Satellite based detections (%)	0.5	1	70	75.3	10	17	13	
Fires reported by field staff but not detected by satellite (%)		0	30	1.6	0	20		20
Source	1	2, 3	4	5	6	7	8	9

Notes and sources: [1] Data are for 2017, Andhra Pradesh Forest Department Geomatics Information System, http://www.fgis.ap.gov.in/AP/FIRE/FIRE17/Fire17.htm (accessed 30 Sept 2017); [2] Feedback reported for 979 of 33,179 fire locations between February and June 2017, Chhattisgarh Forest Department, http://www.fmisonline.org/fire.aspx (accessed 30 September 2017); [3] Rate of false alerts as per Chhattisgarh Forest Department data sheet, sent to World Bank study team August 2017; [4] Himachal Pradesh Forest Department data sheet, sent to World Bank study team August 2017; [5] Kerala Forest Department data sheet, sent to World Bank study team August 2017; [6] Interviews with Dr. Atul Kumar Srivastava, Add'l PCCF, and Mr. Anurag Srivastava, Add'l PCCF (IT), Bhopal, Madhya Pradesh, 28 January 2017; Meghalaya Forest Department data sheet, sent to World Bank study team September 2017; [8] Average of reporting rate and false alert rate for 2016 and 2017, data from Telangana Forest Department Geomatics Information System, http://www.tgfgis.com/ (accessed 30 September 2017); [9] Uttarakhand Forest Department data sheet, sent to World Bank study team August 2017. Table source: Dogra et al. 2018

Table 9.11 Details of forest fire line provided by state forest departments

Length of fire lines (km)	Chhattisgarh	Himachal Pradesh	Kerala	Telangana	Tripura	Uttarakhand	Meghalaya
Fire lines on SFD lands (per working plans)	91,001	2,750	No info	3,866	No info	16,443	Not prescribed
Fire lines mapped and digitized on a GJS layer	0	0	0	3,866	0	0	0
Fire lines maintained annually	91,001	1,000	15,000	3,866	No info	No info	274
Fire lines on non-SFO lands	No info	No info	0	No info	No info	No info	10

Note: *SFD* State Forest Department. Source: Dogra et al. 2018

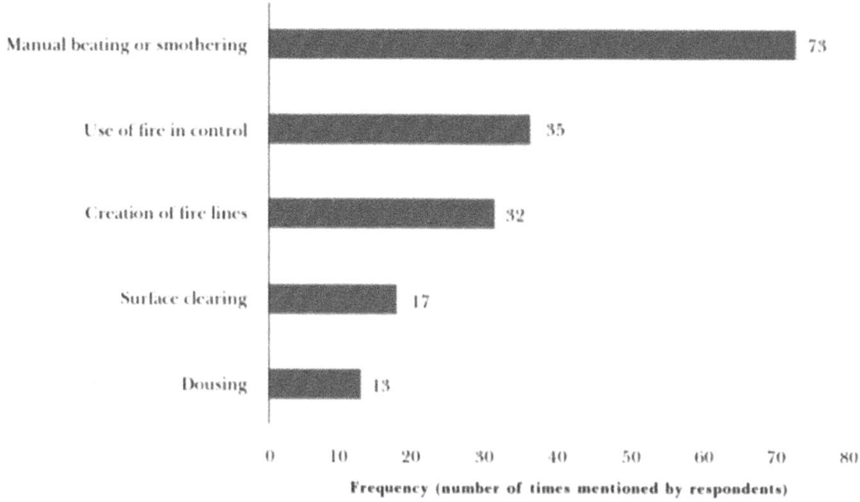

Fig. 9.9 Techniques used to suppress forest fires by fire fighters and local people. Number of responding offices: 85; Officers may have indicated more than one suppression techniques. Source: Dogra et al. 2018

Police, Fire Department, National Disaster Management Authority (NDMA) NDRF etc. The state forest departments may prepare the standard training curriculum for training. Mock drills should be organized involving the District administration, Police, Fire Department, NDRF, SDRF and State Disaster Management Authorities.

10. Equipping the Firefighters

The State Forest Department has to consult with Indian Council of Forestry, Research & Education (ICFRE), Dehradun, to identify and provide the suitable firefighting equipments and tools to all the above mentioned fire fighters as per the local need along with some common equipment like leaf litter blowers, protective clothing etc. (Figs. 9.9 and 9.10; Table 9.12).

11. Infrastructure Development

An effective communication system has to be established in all fire prone forests. Drones could be used to identify the exact location, direction and extent of the fire. State Forest Department should develop and enforce a protocol for accessing its vehicles for smooth movement of the fire fighters to the fire spot. Proper maintenance of the forest roads are of utmost importance in this regard. Prior arrangements of trained firefighters and community volunteers have to be ensured.

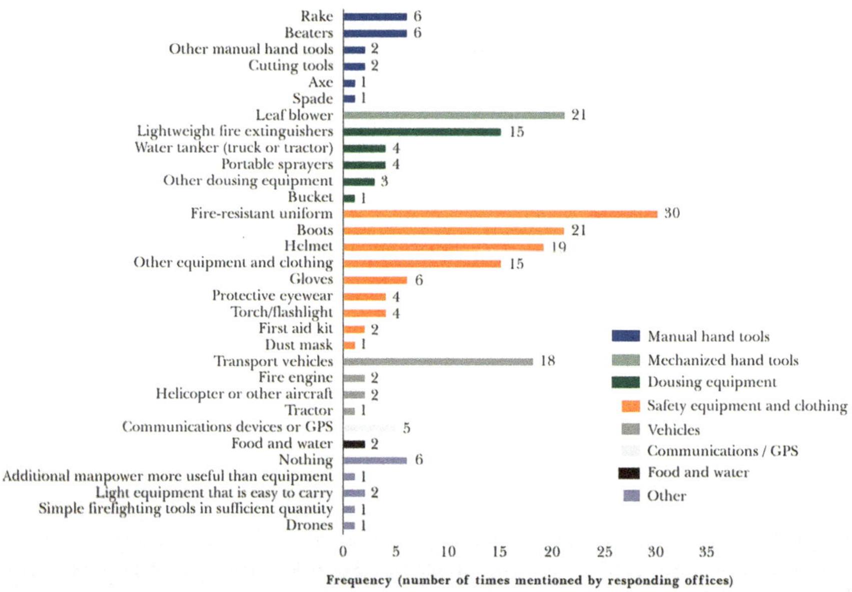

Fig. 9.10 Additional equipment needs mentioned by Forest Officers. Number of responding officers: 74. Officers may have mentioned more than one type of equipment needs. Source: Dogra et al. 2018

9.3.3 Post Fire Management

12. Loss Assessment

FSI develops National level yearly data base on burnt areas. Standardized protocols and procedures have to be developed to facilitate the reporting of the extent of affected area and amount of resource loss. Standardized methodologies for assessing loss have to be developed. Adequate training programmes are also needed to organize for capacity building of the forest officials regarding the assessment of the loss due to forest fire.

13. Identify the Cause of Fire

It is an important task for the State Forest Department to investigate the real reason behind the fire and take immediate and proper measures including legal actions as per need (Table 9.13).

14. Mobilization of Funds

State Forest Department (SFD) should make arrangements for financial resources at district, range and local level before the fire season. SFD should get access of the funds under climate change provided by global bilateral and multilateral as

Table 9.12 Forest Fire hand tools developed by FRI being used by Forest Departments, Van Panchayats and JFMCs

Hand Tool Kit for controlling forest fires	
1. Fire rakes (Kanghi): This tool is designed for raking and cutting of small bushes for construction and maintenance of forest fire lines. There are three kinds of fire rakes made of stainless steel sheets are exhibited below as photographs with specifications:-	
A. Arrow type rake: It is used to remove twigs and small branches, thickness of angle iron 10 mm, weight 700 gm, number of rods per rake 12, length of rake including angle iron 12 cm	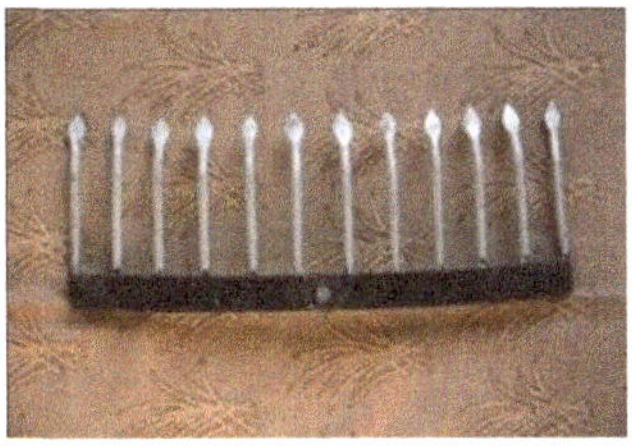
B. Nail type rake: It is used for raking light litter, thickness of angle iron10 mm Weight 700gm, number of rods 12, Length of rake including angle iron 12 cm	
C. Peg tooth type rake: It is used for raking thick litter. Gauge 3 mm Weight 800 gm Peg tooth angle 32–35^0	
2. Fire broom (Jhaapa): It is specially designed to beat the surface fire and control the ground fire. The handiest tool to beat the fire is made from spring steel wire reinforced with steel cup connection to club with adjustable rod. The purpose of developing this tool is to avoid loss of green branches (jhaapa) by the fire fighters. Steel wire dia 2 mm, weight 800 gm, length of broom 45–50 cm, breadth of broom 35 cm	
3. Fire beater (fire pressed): It is developed to put off fire hidden under lumps. It will be of much use in post fire operations. It is effective and widely used in beating the fire in grass and hard inflammable material. It is made from steel or iron. Length 24 cm, breadth 24 cm, thickness of steel 3 mm, weight 900 gm	

(continued)

Table 9.12 (continued)

Hand Tool Kit for controlling forest fires	
4. Torch: It is a portable torch which can be used as head lamp by crew members during forest fire operations. It can also be mounted on the wrist of crew members. It is light weight and uses two small rechargeable batteries with LED lamp.	
5. Adjustable rod: This multiple use rod can be fitted with all types of above shown tools. The rod is very light in weight and is adjustable to various lengths. It is made up of high class conduit pipe with wooden grip. Adjustable length is of about 5–9 feet. Diameter of rod 2.5 cm, length of rod 1.5 m, weight 500 gm	
6. Water bottle: A light weight plastic bottle is essential element in the tool kit. A bottle that can carry about 2 lt. Of water is very useful for crew members during fire fighting operation.	
7. Pathal (modified sickle): It is designed to cut the small branches and twigs for clearing way for crew members. Length of iron blade 17.50 cm, length of wooden handle 33.50 cm, length of iron handle support 26 cm, weight 350 gm	
8. Forest fire control tool kit bag: This canvas bag is designed to keep various tools in such a way that the carrier does not feel discomfort. The inside of the bag is designed to evenly spread out of the weight of the tools. Length of bag 2 ft., breadth of bag 1 ft., material used Melli fabric.	

Table 9.13 Source of forest fire ignition in Kerala from 2011 to 2016. (Source: Dogra et al. 2018)

Cause	2011	2012	2013	2014	2015	2016
Accidental	37	391	112	103	83	96
Incendiary	1	2	1	4	2	4
Deliberate	0	2	3	0	1	1
Lightning	0	2	16	0	3	2
MFP collection	1	14	3	2	5	1
Natural	9	117	71	30	53	121
Not ascertained	49	320	91	79	30	67
Power line	0	2		6	4	2
Settlements	1	2	3	3	1	
Travelers/tr uckers	8	66	42	I I	14	27
Fringe dwellers	1	4	0	0	2	I
Forest offenders	13	95	92	98	74	162
Graziers	0	0	0	0	2	1
Unknown	340	0	69	189	65	0

increasing forest fire is one of the results of climate change. SFD should also get access of the funds for Information and Communication Technology and MNREGA & other community Development and welfare programmes. Crowd funding, funds from environmental management plans and other foreign investments are sources of economic resources (Table 9.14).

15. Mitigation Methods and Tactics (for Dousing Fire)

The suppression of fire is dependent upon the location, type of vegetation and local weather conditions. A few important fire suppression methods include (a) control line method, (b) area method, (c) counter firing, (d) water, (e) chemicals (f) air-borne tankers etc. Various traditional methods are also being employed to control the forest fires. To cite a few, in Uttarakhand, officials of forest departments with the help of local communities make use of Jhapa (broom made of local bushes) for fire beating which is found to be very effective in difficult hilly terrains. All the *Panchayats/Van Panchayats*, Mahila Mangal Dals, Self Help Groups and Youth Mangal Dals, Vanagni-Prahari and Gram-Prahari are mobilized to control the forest fires in their respective areas (Fig. 9.11).

16. Restoration of Affected Areas

After assessing the reason and extent of loss, case specific restoration plans has to be prepared and practice silviculture for mass afforestation to restore the previous ecosystem. While doing this proper soil moisture conservation measures has to be adopted. Indigenous vegetative barriers should also be identified and planted along with other vegetation. Coordination with other ministries and departments would facilitate the entire forest management system (NAPFF) (Fig. 9.12; Tables 9.15 and 9.16).

Table 9.14 Allocated funds under central govt. sponsored schemes and other programmes, 2011–2016

	2011-12		2012-13		2013-14		2014-15		2015-16	
Funding/expenditure Category	INR crore	USD Million	INR Crore	USD Million	INR Crore	USD Million	INR Crore	USD Million	INR Crore	USD Million
NAP funds released GI M funds released	303	57	193	33	258	42	244	38	94	14
GIM funds released					13	2	o*	o*	70	10
IFMS funds released	63	12	41	7	59	10	56	9	52	8
Ad-hoc CAM PA funds released	942	176	1,029	176	1,085	178	1,980	309	1,402	209
Total MoEFCC bud get/ expenditures	1,982	371	1,753	299	1,890	310	1,514	236	1,521	226
Total central government expenditures	1,286,997	240,843	1,393,577	237,821	1,541,466	252,577	1,670,220	260,354	1,761,812	262, 193

Note: GIM funds before 2015–16 were allocated for preparatory activities prior to formal approval of the mission; *NAP* National Afforestation Programme, *GIM* National Mission for Green India; *IFMS* Intensification of Forest Management Scheme; *CAMPA* Compensatory MoEFCC: Ministry of Environment, Forest and Climate Change; *INR* converted to US$ at official exchange rate for period average; INR and US$ are in nominal amounts not adjusted for inflation

Fig. 9.11 Of Jhapa for putting off forest fires

9.4 Case Studies

9.4.1 Uttarakhand Forest Fire 2016

In 2016 Uttarakhand was hit by the worst forest fire, when around 4000 hectares of forest cover across 13 districts were completely burnt. This incident devastated the biodiversity and forest ecosystem of that area. More over 9 people were killed and 17 were injured by the fire. 6000 forest officers along with NDRF team were deployed to contain the fire (Upadhyay 2016a, b). The fire incident in 2016 was much more severe than the fire in 2015; only in Nainital and Udham Singh Nagar respectively 1 and 6 fire points were identified. But in 2016, 13 districts were engulfed by fire and total 1270 fire points were identified among which, 32% were in Pauri Garhwal and around 22% were in Nainital itself (Dalei 2016).

Lessons Learned

As per experts scanty rainfall during the winter months and long dry spells of summer months and low vapour pressure are the principle driver of the fire (Dalei 2016). It is also said that the El-Nino condition prevailing on 2015–16 has influenced this fire event by decreasing the precipitation amount and increasing the temperature (Aggarwal and Mehta 2016). Therefore, climate change could be an important factor which is going to bring the similar situation in future as well (Dalei 2016). But on the top of those anthropogenic activities like extraction of timber, collection of honey sal

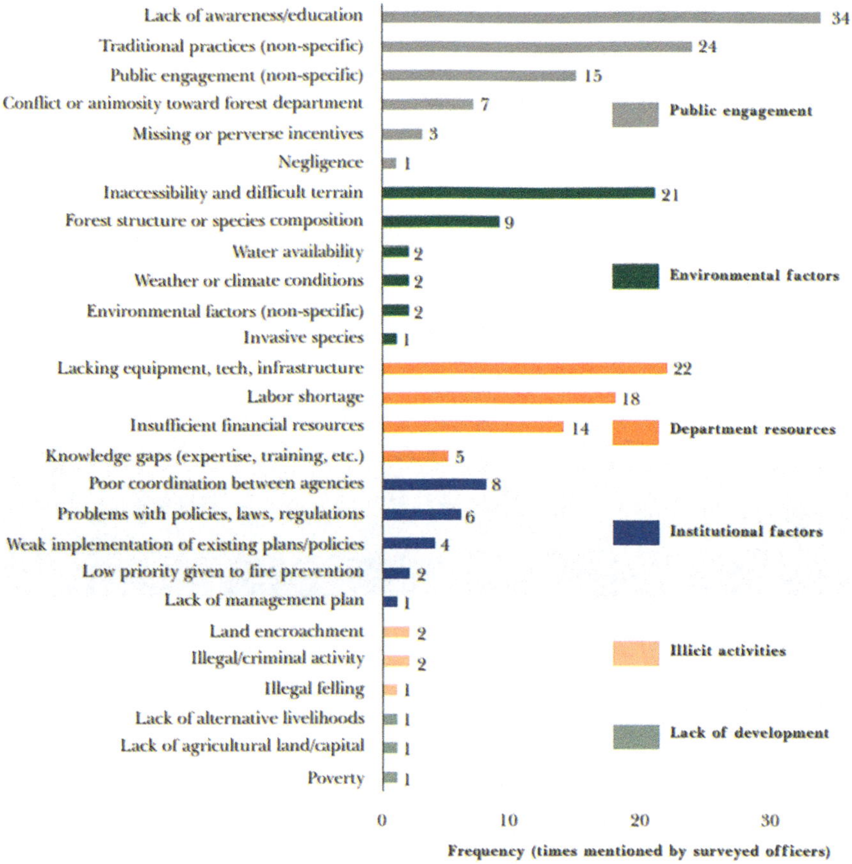

Fig. 9.12 Challenges faced by responding officers towards effective forest fire prevention. Responding officers = 96; each responding officer may mention more than one challenge. Source: Dogra et al. 2018

seeds, hinting wild animals, encroaching forest lands and accidental fires are also additional reasons (Dalei 2016).

9.4.2 Australian Bush Fire

Although in June 2019, Queensland Fire and Emergency Service warned and gave a forecast of early advent of bushfire, but in spite of that as of March 2010 around 18.6 million hectares of (186,000 km$^{2)}$ of forest land (Burton 2020) has been burnt. It also destroyed 5900 buildings which include 2779 homes and at least 34 people were killed (Green 2020; ABC News 2020; SBS News 2020; Henriques-Gomes 2020; The Sydney Morning Herald 2020). Almost one billion animals were killed and

Table 9.15 Ranking of challenges faced by responding officers towards forest fire prevention

Category of challenge	Uttarakhand	Tripura	Telangana	Odisha	Meghalaya	Madhya Pradesh	Kerala	Jharkhand	Himachal Pradesh	Chhattisgarh	Assam
Public engagement	1	1	1	1	1	1	2	2	1	1	1
Environmental factors	3	3	2	3	3	3	1	3	3	4	4
Lack of department resources	2	2	2	2	1	1	3	1	2	2	2
Institutional issues	4	4	2	3	4		4		4	2	
Illegal Activity		4	2								2
Economic development		4		3							

Note: 1st = Most mentioned. Source: Dogra et al. 2018

Table 9.16 Recommendation and Priorities for Forest Fire Management

Recommendation	Lead Implementer	Priorities and Timing
FFPM guidelines to cover: • revised working plan code • development of standard Operating procedures (SOPs) by the SFDs (see below) • fire lines, siting and maintenance controlled burning • Silvicultural practices (prevention and post-fire restoration or rehabilitation) • common classification scheme for the causes of forest fires • standard protocols for post-fire reporting, the investigation of fire causes, and standard methods for assessment of damages • incentivizing accurate reporting by field staff on fires occurrence, burnt area, and damages	MoEFCC (in consultation with relevant stakeholders)	MoEFCC to begin drafting these immediately, and to finalize them in consultation with relevant stakeholders.
Ensuring adequate funding and field staffing	SFDs	In the near term, states should examine existing budget resources to determine if enough is being allocated for FFPM. CAMPA offers a potential source of funding. In the longer term, states should seek to increase funding by increasing productivity of forests and thereby the revenue generated from the sector. A top priority is for SFDs to fill vacancies for field staff and community firewatchers in fire-prone areas. Boots on the ground are essential for all aspects of FFPM, including prevention, detection, and timely response to fires.
Training in fire suppression (prevention, detection, and post-fire reporting) for field staff	DFE (training curriculum) to be rolled out in coordination with SFDs	There is a real need for this, and this activity must begin immediately with the development of a curriculum for all forest guards and other field-level officers in the SFDs.
Provision of equipment for field staff	SFDs in coordination with FRI	There is a real need for this, and this activity must begin immediately. The focus

(continued)

Table 9.16 (continued)

Recommendation	Lead Implementer	Priorities and Timing
		should be on basic hand tools, safety gear and other equipment for ground crews that are appropriate and suited to local needs and conditions.
Establishment of coordination mechanism, at national, state, and district levels, between forest departments and disaster management agencies	MoEFCC at the national level, and SFDs at the state and district level, working with relevant disaster management agencies	This process should also begin immediately, both to define the coordination mechanism and also to establish it. MOEFCC and NDMA should take the lead and provide guidance for the state-level mechanisms
Development and deployment of fire danger rating system	FSI with SFDs	FSI to continue the development of FDRS in collaboration with SFDs, with the recognition that this is a long-term process. The immediate priority is to formalize this process and create a mechanism for SFDs to provide input to the FDRS and field data/feedback for testing the FDRS.
Continued improvement of satellitebased fire detection system	FSI with SFDs	FSI has a well-functioning nationwide satellite-based fire detection system in place. This system can be refined as new technologies and detection algorithms come available, and both FSI and SFDs should work toward this. The immediate priority is to improve twoway communication between FSI and SFDs and strengthen the process by which field-level forest officers provide feedback to both SFDs and FSI on the accuracy of the alerts.
National Policy or action plan (which would also clarify role of other agencies)	MoEFCC	Core group with director general of Forest and representatives from SFDs, NDMA, NGOs, and research institutes to be established immediately to initiate a consultative process for the development of the national policy and action

(continued)

Table 9.16 (continued)

Recommendation	Lead Implementer	Priorities and Timing
		plan over the course of one to two years.
Incentivizing communities	SFDs working with Communities and local NGOs	There is a real need for this, and this activity must begin immediately, although it will entail a longer-term process
Standard operating procedures	SFDs in consultation with relevant agencies	SFDs to begin development once MoEFCC issues guidelines.
Defining a national research agenda (with funding)	ICFRE	ICFRE, as part of its mandate, has developed a National Forestry Research Plan for 2000–2020. FFPM research needs can be defined as part of this ongoing process.
Formal mechanism for knowledge sharing between states	MoEFCC	MoEFCC organizes annual meetings of PCCFs and one of these meetings can focus on forest fires.
National Forest Fire Information Database	FSI	While such a database will serve many needs, it can be developed over the coming years once the underlying processes to collect the necessary data have been established.
National Center of excellence	ICFRE in coordination with FSI	While these are needed for such a Center of Excellence, this too can be developed over the coming years, once the underlying processes have been established.

Acronyms: *DFE* Directorate of Forest Education; *FDRS* fire danger rating system; *FRI* Forest Research Institute; *FSI* Forest Survey of India; *ICFRE* Indian Council for Forestry Research and Education; *MoEFCC* Ministry of Environment, Forest and Climate Change; *NDMA* National Disaster Management Authority; *NGO* non-governmental organization; *SFD* state forest department. Source: Adopted from Dogra et al. 2018

some of the endangered species might have been extinct (The University of Sydney 2020; Harvey 2020; Readfern 2020a) Air pollution reached its peak hazadous level and the estimated cost of reconstruction are expected to exceed A$ 4.4billion (Butler 2020). Tourism revenue is also impacted badly and had a down fall below A$1 billion (Reuters 2020). Only in May 2020 the fire is extinguished all over the country.

Lessons learned

Many people primarily blamed climate change for this incident. According to them faulty government climate policies are to be blamed for prolonged dry spells and extreme hot summer (The Australian 2020). However, later on later on the blame was on the lack of prescribed burning and fire break management, though experts have explained that prescribed burning has increased than earlier years following the recommendation of the 2009 Black Saturday Royal Commission (Readfern 2019). Experts also explained that since the weather condition has shifted towards the warmer and drier conditions it is actually difficult to achieve the prescribed level of burning (Readfern 2019). Further experts are skeptical about the effectiveness of the fuel reduction treatment to stop the bush fire where the climate and weather condition primarily controls the bush fire (Gibbons n.d.; Readfern 2020b; Gibbons 2012). Therefore, it is clear now that the old methodologies of precautions from fire may not be effective in recent times. Therefore, new solutions have to be found to cope up with the changing climate solution and here comes the role of climate change adaptation and mitigation to control forest fire.

9.5 Climate Change Adaptation and Forest Fire Mitigation: A Way Forward

From the above discussion it is clear that we need to have adaptation and mitigation strategies, which are the two sides of a coin, to fight climate change driven triggering of forest fire. Studies those are targeting to achieve adaptation strategies are assessing impacts and vulnerability but that is not leading to any better strategies. Multi-disciplinary approaches are coming up with the ideas of merging traditional knowledge with modern socioeconomic and behavioural achieve better management strategies (Keenan 2015).

Normally forest ecosystems are resilient and capable of adapting to the new weather condition and many species have already adapted (Lucier et al. 2009). In fact in some some of the places some tree species have shown positive response with the changing climate. But it is possible that it will not be able to keep pace with the fast changing climate and that may lead to reduced carbon stock and carbon sequestration capacity which will ultimately work as a positive feedback process and work as a catalyst for the climate change (Seppälä et al. 2009).

Adapting forest management would include monitoring and anticipating changes on the basis of previous data and proper action to avoid the negative impact or making potential use of the changing situation (Levina and Tirpak 2006). Adaptation approaches may either aim to reduce vulnerability or may be reactive to the changing condition or can go for long term transformational changes. Forest thinning is one of such ideas for exclusively the places where possibility of temperature change is higher. Varying species composition of harvested trees are one of the most effective ways for increasing the forest age and promoting climatically suited species

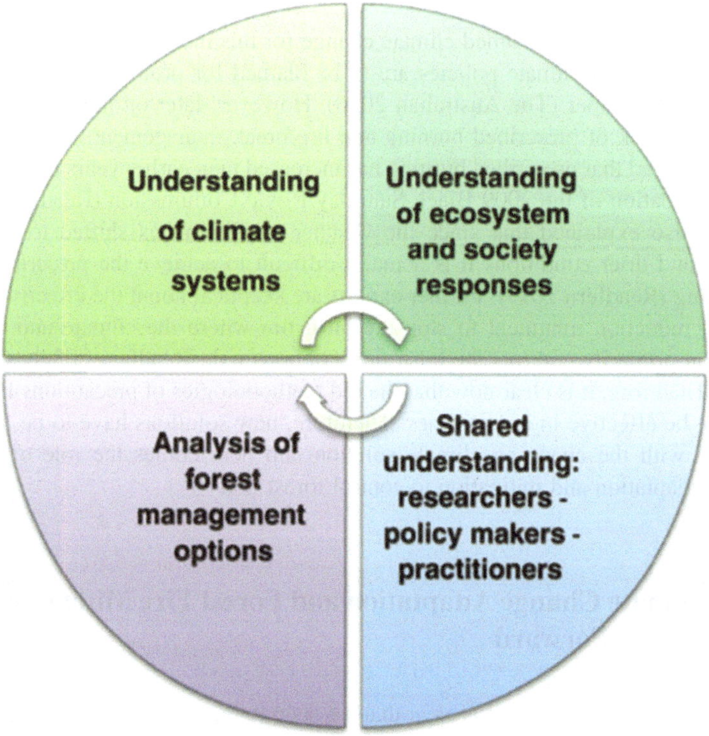

Fig. 9.13 Components of climate smart forest management (Source: Keenan 2015)

(Steenberg et al. 2011). the future trajectory of climate change and the interaction of the ecosystem to that changed climate is too uncertain (Wilby and Dessai 2010) to make any long term and costly decisions. Therefore for effective adaptation one has to be very particular about the objective of the adaptation strategy. The following diagram gives an idea of how the climate adaptive forest policies and management should be (Fig. 9.13).

We need to clearly identify whether to facilitate ecosystem adaptation by changing the species composition or build resilience through engineering resistance (Joyce et al. 2008). Adaptation is a social learning process, which needs the understanding of a particular place, capacity for individual and society to consider the potential climate change and the implication of that to their life. Policy makers need to give a greater diversity of input for decision making (Konkin and Hopkins 2009). Decisions should only made after close consultation with environmental scientists, other related organizations, communities etc. There is dire need for understanding the response of individual species or ecosystem as a whole to the climate change and the community attitudes towards the climate change and forest management.

References

Abatzoglou JT, Kolden CA (2013) Relationships between climate and macroscale area burned in the western United States. Int J Wildland Fire 22:1003–1020. https://doi.org/10.1071/WF13019

ABC News. Australia (2020) Victorian bushfires death toll rises as authorities confirm contractor's death was fire-related. Retrieved 15 January 2020

Aggarwal M, Mehta N (2016) Understanding the Uttarakhand forest fire, livemint

Ahmad F, Goparaju L (2018) Climate change and its impact on Forest fire in the state of Himachal Pradesh and Uttarakhand states of India: remote sensing and GIS analysis *Contemp*. Trends Geosci 7(2):229–246. https://doi.org/10.2478/ctg-2018-0016

An et al. (2015): An, Zhisheng, Wu Guoxiong, Li JIanping, Sun Youbin, Liu Yimin, Zhou Weijian, Cai Yanjun, Duan Anmin, Li, Mao Jiangyu, Cheng Hai, Shi Zhengguo, Tan Liangcheng, Yan Hong, Ao Hong, Chang Hong, and Feng Juan. 2015. "Global Monsoon Dynamics and Climate Change." Annu. Rev. Earth Planet. Sci. 43: 2.1–2.49.

Anmin, Li, Mao Jiangyu, Cheng Hai, Shi Zhengguo, Tan Liangcheng, Yan Hong, Ao Hong, Chang Hong, and Feng

Aragao LEOC, Anderson LO, Fonseca MG, Rosan TM, Vedovato LB, Wagner FH, Silva CVJ, Silva CHL, Junior AE, Aguiar AP, Barlow J, Berenguer E, Deeter MN, Domingues LG, Gatti L, Gloor M, Marengo JA, Miller JB, Phillips OL, Saatchi S (2018) 21st century drought related fires counteract the decline of amazon deforestation carbon emissions. Nat Commun 9:536. https://doi.org/10.1038/S41467-017-02771-Y

Bahuguna VK (2018) Changing dimensions of forests in India: impact of climate change and deforestation MOJ. Ecol Environ Sci 3(5)

Burton J (2020) 'It was a line of fire coming at us': Firefighters return home. Busselton-Dunsborough Mail. Retrieved 11 Feb 2020

Butler B (2020) Economic impact of Australia's bushfires set to exceed $4.4bn cost of Black Saturday. The Guardian. Retrieved 18 Jan 2020

Dalei NN (2016) Forest fires in Indian state of Uttarakhand. Eurasia Review

Doerr SH, Santin C (2016) Global trends in wildfire and its impacts: perceptions versus realities in a changing world. Phil Trans R Soc B 371:20150345. https://doi.org/10.1098/rstb.2015.0345

Dogra P, Andrew MM, Urvashi N, Christopher S, Ross S, Shraddha S (2018) Strengthening forest fire management in India. World Bank, Washington DC

Flannigan M, Stocks BJ, Turretsky M, Wotton M (2008) Impacts of climate change on fire activity and fire management in the circumboreal forest. Glob Chang Biol 15(3):549–560

Földi L, Kuti R (2016) Characteristics of Forest fires and their impact on the environment. AARMS 15(1):5–17

FSI (1999): F.S.I. (2000) State of Forest Report (1999) Forest Survey of India (Ministry of Environment & Forest, India), Dehradun, p. 113.

Gibbons P (2012) Land management practices associated with house loss in wildfires. Plos One 7(1):e29212. https://doi.org/10.1371/journal.pone.0029212

Gibbons P (n.d.) What is the evidence behind hazard-reduction burning?. The Canberra Times. Retrieved 6 Jan 2020

Giglio L, Randerson JT, Van Der Werf GR (2013) Analysis of daily, monthly, and annual burned area using the fourth-generation global fire emissions database (GFED4). J Geophys Res Biogeosci 118:317–328. https://doi.org/10.1002/jgrg.20042

Green M (2020) Australia's massive fires could become routine, climate scientists warn. Reuters. Retrieved 14 Jan 2020

Guha-Sapir D, Below R, Hoyois P (2015) EM-DAT: international disaster database. University Cathol Louvain, Brussels, Belgium. See http://www.emdat.be

Harvey J (2020) Number of animals feared dead in Australia's wildfires soars to over 1 billion. HuffPost. Retrieved 7 Jan 2020

Henriques-Gomes L (2020) Bushfires death toll rises to 33 after body found in burnt out house near Moruya. The Guardian. Retrieved 25 Jan 2020

India.com (2020) 2020 Is so cruel: heartbreaking visuals of Uttarakhand forest fires emerge on Twitter, Netizens Appeal for Help. https://www.india.com/viral/2020-is-so-cruel-heartbreaking-visuals-of-uttarakhand-forest-fires-emerge-on-twitter-netizens-appeal-for-help-4040773/

Joyce LA, Blate GM, Littell JS, McNulty SG, Millar CI, Moser SC, Neilson RP, O'Halloran K, Peterson DL (2008) National forests. Preliminary review of adaptation options for climate-sensitive ecosystems and resources. A report by the U.S. In: Climate change science program and the subcommittee on global change research. US Environmental Protection Agency, Washington, DC

Keenan RJ (2015) Climate change impacts and adaptation in forest management: a review. Ann For Sci 72:145–167

Kodandapani N, Parks SA (2019) Effects of drought on wildfires in forest landscapes of the Western Ghats, India. Int J Wildland Fire 28:431–444. https://doi.org/10.1071/WF18188

Konkin D, Hopkins K (2009) Learning to deal with climate change and catastrophic forest disturbances. Unasylva 60:17–23

Lucier A, Ayres M, Karnosky D, Thompson I, Loehle C, Percy K, Sohngen B (2009) Forest responses and vulnerabilities to recent climate change. In: Seppälä R, Buck A, Katila P (eds) Adaptation of forests and people to climate change: a global assessment report., World Series, vol 22. IUFRO Helsinki, pp 29–52

Levina E, Tirpak D (2006) Adaptation to climate change: key terms. OECD/IEA, Paris

Liu Y, Stanturf J, Goodrich S (2010) Trends in global wildfire potential in a changing climate. For Ecol Manag 259:685–697. https://doi.org/10.1016/J.FORECO.2009.09.002

Mukhopadhyay D (2009) Climate change: global risks, challenges and decisions IOP Publishing IOP Conf. Series: Earth and Environmental Science 6 (2009) 382027 https://doi.org/10.1088/1755-1307/6/8/382027

NIFC (2016) National interagency fire center statistics. See https://www.nifc.gov/fireInfo/nfn.htm

Paliath S (2018) P38.10 impact of climate change on forest ecosystem and forest fire in India Forest fires increased 158% in 6 years; warming a factor: experts. IndiaSpend. https://www.indiaspend.com/forest-fires-increased-158-in-6-years-warming-a-factor-experts-53125/

Pandey K (2019) Forest fires in India tripled in the last four months, down to earth. https://www.downtoearth.org.in/news/natural-disasters/forest-fires-in-india-tripled-in-the-last-four-months-63388

Readfern G (2019) Factcheck: is there really a green conspiracy to stop bushfire hazard reduction? Guardian Australia. Retrieved 3 Jan 2020

Readfern G (2020a) 'Silent death': Australia's bushfires push countless species to extinction. Guardian Australia. Retrieved 8 Jan 2020

Readfern G (2020b) Explainer: how effective is bushfire hazard reduction on Australia's fires? Guardian Australia

Reuters (2020) Australian tourism industry seeks urgent help as cost of bushfires grows. Retrieved 18 Jan 2020

Riley KL, Abatzoglou JT, Grenfell IC, Klene AE, Heinsch FA (2013) The relationship of large fire occurrence with drought and fire danger indices in the western USA, 1984–2008: the role of temporal scale. Int J Wildland Fire 22:894–909. https://doi.org/10.1071/WF12149

San-Miguel-Ayanz J, Moreno JM, Camia A (2013) Analysis of large fires in European Mediterranean landscapes: lessons learned and perspectives. For Ecol Manag 294:11–22. https://doi.org/10.1016/j.foreco.2012.10.050

SBS News (2020) The numbers behind Australia's catastropic bushfire season. Retrieved 8 Jan 2020

Seppälä R, Buck A, Katila P (2009) Adaptation of forests and people to climate change: a global assessment report, vol World Series, vol 22. IUFRO, Helsinki

Smith R (2017) Fire danger rating systems. Background note to World Bank assessment of forest fire prevention and management in India, New Delhi

Srivastava, K. (2020) Most forest fires in India on account of human activity, Mongabay. https://india.mongabay.com/2020/01/most-forest-fires-in-india-on-account-of-human-activity/
State of Forest Report (2017) Forest survey of India. https://www.fsi.nic.in
State of Forest Report (2019) Forest survey of India. https://www.fsi.nic.in
Steenberg JWN, Duinker PN, Bush PG (2011) Exploring adaptation to climate change in the forests of Central Nova Scotia, Canada. For Ecol Manag 262:2316–2327. https://doi.org/10.1016/j.foreco.2011.08.027
The Australian (2020) Melbourne climate change protest not appropriate: Daniel Andrews
The Sydney Morning Herald (2020) NSW bushfires: body found in burnt house on NSW coast. Retrieved 24 Jan 2020
The University of Sydney (2020) More than one billion animals impacted in Australian bushfires. Retrieved 13 Jan 2020
Tiwari BK, Tynsong H, Lynrah MM, Lapasam E, Deb S, Sharma D (2013) Institutional arrangement and typology of community forests of Meghalaya, Mizoram, and Nagaland of NorthEast India. J For Res 24(1):179–186
Upadhyay K (2016a) Uttarakhand battles fire crisis. The Hindu. https://www.thehindu.com/news/national/other-states/forest-fire-claims-five-lives-ndrf-forces-deployed-to-control-flames/article8538875.ece
Upadhyay K (2016b) Uttarakhand battles fire crisis. The Hindu
Wilby RL, Dessai S (2010) Robust adaptation to climate change. Weather 65:180–185. https://doi.org/10.1002/wea.543
Wohlgemuth T, Brigger A, Gerold P, Laranjeiro L, Moretti M, Moser B, Conedera M et al (2012) Leben mit Waldbrand am Beispiel von Leuk (VS) 2003. Vierteljahrsschrift der Naturforschenden Gesellschaft in Zürich 157(3–4):97–106. https://www.wsl.ch/en/forest/diseases-pests-and-disturbances/forest-fires.html
World Bank (2016) The cost of fire: an economic analysis of Indonesia's 2015 fire crisis. Indonesia sustainable landscapes knowledge, note no. 1, http://documents.worldbank.org/curated/en/776101467990969768/The-cost-of-fire-an-economic-analysis-of-Indonesia-s-2015-fire-crisis
WWF (2020) Forest landscape restoration. https://forestsolutions.panda.org/approach/forest-landscape-restoration
Zhisheng A, Guoxiong W, JIanping L, Youbin S, Yimin L, Weijian Z, Yanjun C, Anmin D, Jiangyu M, Hai C, Zhengguo S, Liangcheng T, Hong Y, Hong A, Hong C, Juan F (2015) Global monsoon dynamics and climate change. *Annu Rev Earth Planet Sci* 43:2.1–2.49

Dr. Sweta Baidya Das , has done her PhD in Oceanography from National Institute of Oceanography. She has worked on the Paleo-climate and monsoon shift from Last Glacial Maximum to Holocene. Later she was working on Glaciology in Jawaharlal Nehru University and Ministry of Earth Sciences. She was also teaching Disaster Management in Jamia Millia Islamia and now working in National Institute of Disaster Management and is currently associated with NDMA (Govt. of India) as a Sr Consultant.

Dr. Dhawan Vijay Kumar is retired as Scientist-B from Forest Research Institute, Dehradun in August 2017. Presently, he is working as 'Forestry Expert' in Silviculture and Forest Management Division, FRI, Dehradun. Mr. Vijay has 30 years' experience of Research and Teaching in "Forestry" especially in the field of Silviculture including Plant Propagation, Tree Improvement, Nursery and Plantation Management, Forest Fire Management and Climate change.

Part II
Thematic and Cross-cutting Issues

Chapter 10
Climate Resilient Infrastructure in Developing Countries

Divya Sharma and Anil K. Gupta

Abstract Infrastructure is imperative for development and links closely to achieving the desired GDP-growth of countries and regions. Both social and economic infrastructure developments are required in the changing phase of human development where urbanization and rural to urban transition is rapid. Increasing population and changing political economic realities of urban areas are stretching cities and city regions to build more and more of infrastructure in terms of energy reproduced below telecommunication, transport, and housing and provide services like water and sanitation, health and education. However, these cities and regions are also the ones that are at huge risks of climate change impacts due to financial and structural constraints, low levels of development, high vulnerability and climate-vulnerable geographies. Cities are increasingly becoming centres of environmental disasters, while also being the centre stage for economic development. To achieve growth targets, it is imperative that we build strong and resilient infrastructure that can withstand changing climate and projected increase in extreme events.

This chapter views infrastructure development in the perspective of urban centres in developing countries and need for building safe and climate resilient infrastructure. It also highlights through various statistics how cities are vulnerable to climate impacts and why there is a need to think about building safe and resilient infrastructure. It draws up policy recommendations to achieve a climate resilient urban development accompanied by safe and resilient infrastructure and makes a case for infrastructure to be resilient in terms of design, finance, regulations and standards, policy, mainstreaming.

Keywords Urban development · Resilience · Infrastructure · Developing countries · Climate safe · Disasters

---

D. Sharma (✉)
Climate Group, New Delhi, India
e-mail: DSharma@theclimategroup.org

A. K. Gupta
ECDRM Division, International Cooperation, CAP-RES DST Project, National Institute of Disaster Management (Ministry of Home Affairs), New Delhi, India

A. K. Gupta et al. (eds.), *Disaster Risk and Management Under Climate Change*, Disaster Resilience and Green Growth,
https://doi.org/10.1007/978-981-99-4105-6_10

10.1 Introduction

10.1.1 Infrastructure Is Imperative for Development

Inadequate infrastructure is widely recognized to be holding back development in fast-growing regions of Latin America, Asia and Africa, lowering the quality of life of its citizens. The developing world looks at infrastructure investment as a priority to achieve growth (World Economic Forum 2015). Communications infrastructure like telecom, transport infrastructure like road, rail, port; energy, housing, water and sanitation, health and education are considered to be the topmost requirement in terms of infrastructure provision in developing world. Asia needs an annual commitment of about 6.5% of its GDP towards infrastructure provision by 2020, of which, about half is for energy infrastructure, a third for transport, 13% for ITC, and 3% for W&S (Antonio E et .al, 2012). Infrastructure growth is a key component of the 2030 Development Agenda. The same is being highlighted in the Sustainable Development Goals;

(i) SDG 6-Ensure availability and sustainable management of water and sanitation for all;
(ii) SDG 7-Ensure access to affordable, reliable, sustainable and modern energy for all;
(iii) SDG 9- Build resilient infrastructure, promote inclusive and sustainable industrialization and foster innovation
(iv) It is also to be noted that infrastructure provision will have distributional effect on the basis of type of infrastructure and the area. For example, provision of roads in Bangladesh has contributed to reduction of poverty; transportation infrastructure in Africa has benefitted small and remote cities and infrastructure rehabilitation has provided positive benefits to the poor in Vietnam (Gurara et al. 2017). Effects of infrastructure provision are regional and any deficiency in infrastructure leads to regional impacts too. For a long time in India for example, providing matching infrastructure in rural areas has been an agenda favoured to contain migration and improve quality of life in rural areas. However, across the developing world, the changing dimensions of urbanization have influenced the way infrastructure provision is seen and attended to.

10.1.2 Cities Engines of Economic Growth or Centres of Environmental Disasters

Cities are an important centre for economic development and growth. The narrative around economic development is increasingly bringing in issues and concerns around urbanization and raising question on how to achieve urban development that induces more and more of economic development. The best cities in the world are centres of vitality, diversity and centres for growth for all, where the ideal of common growth can be realised (David and Sheridan 2016). However, with growing urbanization and number of cities, this ideal face of cities is diminishing quickly. Most of the cities, particularly in the developing world are chaotic, un-livable and far from being sustainable centres of economic growth. They lack infrastructure, basic services and innovative governance mechanisms and have weak local governance where the common good is easily violated. Growth of these cities is impeded due to failed planning and lack of knowledge of environmental sustainability, climate resilience and disaster risk reduction (Therrien et al. 2018).

It is also to be noted that most of the efforts on cities are driven towards economic development only. Economic success or the aspiration for the same has not brought forth processes that induce sustainability, climate resilience, risk reduction and disaster managed cities. These have rather been two different narratives dealt with in silos. Instead, the economic development pathways that have been adopted and the resultant consumption patterns have made cities and city regions centres of risks and inequality.

10.2 Growing Urban Population and Rapid Increase in Number of Urban Centres Are Driving Infrastructure Development

Box 10.1 The World is Moving to Cities
Close to half of the world's urban dwellers reside in settlements with fewer than 500,000 inhabitants, while around one in eight live in 33 megacities with more than ten million inhabitants. By 2030, the world is projected to have 43 megacities, most of them in developing regions.

Source: World Urbanization Prospects: Revision 2018. https://esa.un.org/unpd/wup/publications/Files/WUP2018-KeyFacts.pdf

According to United Nations, 55% of the world lives in urban areas (UN 2014). By the end of this century the population of the world is projected to grow by 6.9 billion to 13.1 billion, and the percentage of people residing in 101 largest cities will grow from 15% to 23% (Hooernweg and Pope 2017). Asia has the highest number of people living in urban areas, being 48% urbanised and house to 53% of world's population (UNHABITAT 2016). Besides Asia, Africa shows extreme urban growth as well, displaying an urban growth rate that is 11 times more rapid than the growth rate in Europe.

Box 10.2 Asia Leads on Urban Population

The urban population of the world has grown rapidly since 1950, having increased from 751 million to 4.2 billion in 2018. Asia, despite being less urbanized than most other regions today, is home to 54% of the world's urban population, followed by Europe and Africa (13% each).

Source: World Urbanization Prospects: Revision 2018. https://esa.un.org/unpd/wup/publications/Files/WUP2018-KeyFacts.pdf

It is also estimated that Sub Saharan Africa will have the most dramatic rise in the number of large cities growing from 7 in 2007 to 20 in 2050 and 30 by the year 2100 (Daniel Hoornweg et.al, 2017). Both Africa and Asia's rapid urbanization is driven mainly by natural increase, rural–urban migration, spatial expansion of urban settlements and reclassification of rural areas. The growth rates of these regions pose major challenge to their resource base and the need to build and sustain infrastructure and basic services for the growing population.

It is projected that by 2030 the urban population within these regions and in particular developing countries will double, while the area covered by cities would triple. Currently, the top 600 cities that contain a fifth of the world's population and generate 60% of global GDP consist mainly of cities in developed countries. However, it is predicted that 21 out of the global megacities will be found in Asia by the year 2025, increasing the contribution to the league of Asian cities that produce around 80% of the GDP of their region, use up 60–80% of the regions energy and create 75% of the region's carbon emission (Asian Development Bank 2015a, b).

Closely linked to the issue in the fast growing cities of Asia and Africa is the challenge of providing adequate basic services and infrastructure. This would mean that a lot of infrastructure is yet to be built to support the ever growing population and number of cities. For example a study suggests that almost 70% of India's infrastructure is yet to be built (McKinsey 2010). Similar are the cases for other developing and middle income country cities. It is to be noted that Asia alone puts in an estimated $881 billion each year into infrastructure and this is required to be doubled to match the population increase, somewhere to the range of $22.6 trillion on infrastructure by 2030 (QUARTZ 2017). Approximately 1.2 billion Asians are predicted to move to cities over the next 35 years. This will require the construction of hundreds of new homes, roads, and water and electricity networks besides other infrastructure to support this volume of population (Asian Development Bank 2015a, b).

In India for example, over the past decade, there has been a huge increase in the number of urban centres from 5161 in 2001 to 7935 in 2011(census 2011). This is mainly because of reclassification of rural settlements to urban. These areas are significantly lacking in the institutional framework provided by a municipality. This calls for significant attention on provision of infrastructure and the resolve to build better and not to repeat the mistakes of most of the Indian cities. This is equally important in the context of developing countries where transitioning cities and infrastructure provision will be major problems for the times to come. This takes us back to the regional dimensions of the infrastructure needed in the rural, urban and urban areas, to strengthen regional linkages and create the desired social and economic impact.

According to UN Habitat, cities consume 78% of the world's energy and produce more than 60% of greenhouse gas emissions in spite of the fat that they occupy only 2% of the world's area.[1]Built environment, transport emissions, emissions from industries, fossil fuel burning all have significant share of GHG emissions and attributed to urban centers more than anything else.

10.3 Vulnerabilities and Impacts

10.3.1 Growing Cities and Also at Growing Risks to Climate Change

While the city and State Governments think about urban development and ways and means to cater to the rapid increase in urban population and number of cities, climate change impacts have taken a toll on urban areas where these people, resources, and infrastructure are concentrated. An estimated 46 million people living in cities are at risk yearly from flooding from storm surges in the East Asia Region alone.

Some of Asia's fastest growing cities are susceptible to climate impacts like sea level rise. 341 Million Asian cities will be at risk of climate change induced flooding both inland and coastal by the year 2025 (Asian Development Bank 2015a, b). About half of world's largest urban centres are located in coastal areas and around 25 of them are located in the seismic zone (Hooernweg and Pope 2017).

In 2017, more than 1000 people died and 45 million people had to bear huge losses in terms of loss of livelihood, homes and services when severe floods hit Southeast Asian cities, including Dhaka in Bangladesh and Mumbai and Chennai in India. In 2017 itself, California's suburbs and Rio de Janeiro in Brazil have experienced the fury of floods, drought, wildfires and heavy rains. Cape Town in South Africa has been known to endure extreme drought since 2015, this year the city had to be declared at zero level in terms of water availability (Bai et al. 2018).

In 2018, extreme rains and floods have affected Kerala one of the coastal states in India, with a loss of lives of 385 people leaving around 0.314 million homeless and counting. The floods, the worst in last 100 years have caused a damage and loss worth INR 80 Billion in the last four days of fury (Economic Times 2018a, b). In 2019, the tourist city of Shimla in Himachal Pradesh the mountain state of India, had to be declared at zero water stage and tourists were requested not to come to visit the city during summers, the peak time for tourists' activity in the city. This affected the

[1] https://www.un.org/en/climatechange/cities-pollution.shtml

economy of the city very badly while the city dwellers were already dealing water shortage of extreme kind. 2020 caught the world unaware with COVID19 outbreak and then India was hit by cyclone Aphan causing huge loss and damage to property, infrastructure and lives.

By 2030, millions of people and US$4 trillion of assets will be at risk from such events (Bai et al. 2018). Critical infrastructure is the worst hit due to these events leaving hundreds and thousands of people strangled in dangerous situation. The cyclone HUDHUD in the year 2013 destroyed the airport of the Visakhapatnam city and more than 70% of the critical infrastructure in the city was affected (Business Standard 2014). Very recently, the airport at the Kochi city in the coastal state of Kerala state had to be shut down for over a week due to flooding[2]. The South Asia flooding in 2017 that effected India Nepal and Bangladesh had impacts on 41 million people due to heavy rains and floods, over 950,000 houses were destroyed due to the floods and over 1200 lives were lost. Tropical cyclones led flooding in Vietnam affected 233,271 houses, damaged 1782 kilometres of roads, 585 bridges or culverts and 60 kilometres of dikes leading to an overall economic loss of VND 10,520 billion- approximately US$460 million (World Bank 2017). The State of Uttarakhand experienced an unprecedented high rainfall between June 15 and 17 in the year 2013 leading to extreme flash floods and landslides within the State. A total of 580 human lives were lost, more than 4000 persons went missing. This event is said to have affected over 900,000 people in Uttarakhand that year (World Bank and Asian Development Bank 2013).

10.3.2 Climate Change Will Severely Impact Urban Infrastructure

Many of the world's largest and fastest growing cities are vulnerable to sea-level rise having located on the coast. They are equally exposed to windstorms, tsunamis and floods the occurrence of which is increasingly attributed to climate change.

It is important therefore to understand infrastructure vulnerability and ways to build resilience within planning and decision making that can enhance the disaster resilience of cities and their vital and critical infrastructure systems (Chang et al. 2014). Studies indicate that climate impacts on infrastructures will depend on their geophysical risk exposure, the existing adaptive capacity and resilience, and the level of regional economic development. Climate change will also affect the environmental and social systems around infrastructure assets and will greatly influence the functions and interaction of these infrastructures with the environment and their interactions with these systems. Understanding full range of climate change impacts

[2]https://www.indiatoday.in/india/story/kochi-airport-damage-rs-250-crore-kerala-floods-massive-repair-work-on-1320191-2018-08-22

on infrastructure will often require complex local and regional analysis of these interactions.

Vulnerability to climate change impacts is also strongly linked to technology used for construction and operation of an infrastructure. These impacts are much more pronounced in congested built-up areas that create their own micro-climates like that of urban heat islands. In addition to this, and particularly in many developing country cities, climate impacts on infrastructure are much more severe due to the development pattern these cities have adopted. These cities characterise degradation of natural protection, deforestation and building on floodplains, poor-quality housing construction on exposed slopes, and extensive ground coverage of concrete without adequate drainage. Even a day of heavy rains in these regions often result in intense, flash floods, such as Mumbai, July 2005 and Chennai in 2017 in India.

This highlights the importance of making sure that existing and future infrastructure are built to cope with climate impacts and extreme events. The complexity involved around city systems and how they are developed requires acting in an integrated, cross-sector way on climate risks and resilience. There are three logical layers to infrastructure resilience:

1. Infrastructure that is resilient and sustains functional in the event of climatic shock or event,
2. The infrastructure development in city or town doesn't create new or aggravate risk or vulnerability to climatic change impacts / disaster impacts, and
3. Interdependence of infrastructure and looking infrastructure as systems and not only as static entities.

10.4 Addressing Climate Vulnerability to Urban Infrastructure

Climate impacts can not only destroy infrastructure during extreme events but will also affect functioning and operations of the infrastructure due to gradual climate change. The Sustainable Development Gaols also recognise the need for risk reducing infrastructure in SDG no 6.

It is to be noted that both physical as well as social infrastructure is critical to be addressed for climate vulnerability. However, within the scope of his paper we are only looking at the physical infrastructure. Addressing impacts of climate change on infrastructure will involve a threefold solution:

(i) Working on early warning systems and communications for extreme events
(ii) Building infrastructure that could be used to evacuate and restore during extreme event for example bridges, roads, shelter homes, health centres
(iii) Making the infrastructure resilient to climate impacts Ex Ante and Ex Post

As per Tanner et al. (2015), investing in DRM strategies' offers a triple dividend. These dividends hold good for investing in resilient infrastructure where it would have the capacity to avoid the losses in the event of disasters. This will lead to reduced loss of lives and damage to infrastructure. Climate resilient infrastructure will also unlock the potential for economic growth by catalysing investment, innovation and entrepreneurship to these urban areas despite their risks to climate impacts. Besides this, resilient infrastructure will generate development co benefit's that will go a long way in making the system robust and reliable.

In broad terms, for cities to be resilient, it is important that its roads utilities, bridges etc. are designed to continue functioning during a disaster, while institutions both private and public are prepared with up to date information on extreme event and are effectively connected with the communities such that they can work together during a calamity (Chang et al. 2014). Infrastructure resilience should characterize specific vulnerabilities and resilience within existing infrastructure and foster future infrastructure design and functioning to be resilient. Much of the literature also suggests that land use planning and location specific decisions of infrastructure as per land topography is also an important part of building resilience.

10.5 Resilience of Urban Infrastructure

There are varied definitions and interpretations for resilience in the urban context and with growing knowledge the term resilience is being used in deeper sense of institutional, social and structural responses to calamities and adversity. According to the United Nations International Strategy for Disaster Reduction resilience (now UNDRR) is:

"The ability of a system, community or society exposed to hazards to resist, absorb, accommodate, and recover from the effects of a hazard in a timely and efficient manner, including through the preservation and restoration of its essential basic structures and functions". (UNISDR 2009).

In terms of defining resilience of urban infrastructure the resilience enhancing characteristics as proposed by Tyler and Moench (Tyler and Moench 2012) seem to bring in the most critical:

(i) Flexibility and diversity of infrastructure systems to perform essential tasks under variety of circumstances be able to convert /modify to perform functions in different possible ways.
(ii) Redundancy and modularity: Multiple pathways and variety of options for service delivery and components that can replace one another in case one fails.
(iii) Safe failure: Ability to absorb sudden shock and adaptability for slow onset changes such that a catastrophic failure is avoided.

Some researchers have framed resilience as the ability of infrastructures to withstand structural shocks brought about by climate change (Chirisa et al. 2016). A physical infrastructure is considered resilient if it can withstand, adapt and recover from external disruptions those caused by natural forces such as extreme events and man-made forces such as pressure of the growing population, inadequate O & M and degradation .

It is to be understood that in the case of infrastructure resilience, merely installing adaptive systems in silos might not help much (Fraser et al. 2016). It is to be looked at in the light of intertwined and interdependent systems that can function well at times of a disaster and make up for safe failure if it has to by doing so.

10.6 A Discussion on Best Practices and Implementation Examples for Resilient Infrastructure

Within the Asian Ministerial Conference on Disaster Risk Reduction (AMCDRR) in the year 2016, the Prime Minister of India had announced that India will work with partner countries and stakeholders to build a coalition for promoting disaster resilient infrastructure. This is in response to the future investment needed in building infrastructure in the country and recognizing the fact that investment in new infrastructure will be a an important step for attaining sustainable development and climate resilient future. The proposed coalition was built with an objective of generating new knowledge on hazard risk assessment, disaster resilient technologies and mechanisms for integrating risk reduction in infrastructure financing (NIPFP 2017).

The coalition identifies 4 broad thematic areas for the work of the coalition that include:

(i) Development of disaster risk assessment methodologies, metrics and indicators of risk to ensure sustainability for different infrastructure classes;
(ii) Revising standards, design and regulation for infrastructure development and also operations and maintenance of infrastructure;
(iii) Financing for disaster resilient infrastructure including risk transfer mechanisms; and
(iv) Reconstruction and recovery of key infrastructure sectors after disasters.

In terms of successful programs on resilience, the Asian Cities Climate Change Resilience Network (ACCCRN) supported by the Rockefeller Foundation initiated the very discussion on implementing resilience in cities in Asia way back in 2007. Starting off with 10 cities in 4 countries in Asia- India, Indonesia, Thailand, and Vietnam and extending to two more countries that included Bangladesh and Philippines (ACCCRN 2008). The initiative through a league of like-minded local partners in each country started the mission that led to over 30 cities conducting their hazard and risk assessment to climate change and preparing their own climate resilience strategies. Most of the successful cities under the program were led by city's political leadership and municipal commissioners and the in depth involvement of vulnerable communities, private and public stakeholders (Sharma et al. 2017). Through the

Rockefeller foundation's initiative like ACCCRN and 100 resilient cities, more than 30 Indian cities have started planning for resilience strategies to impacts of climate change. The cities of Surat, Indore and Gorakhpur were the foremost in conducting risk and vulnerability analysis and extensive stakeholder communications and shared learning dialogues to prepare a workable city resilience strategy.

While Gorakhpur resilience strategy follows an integrated approach that addresses the institutional, behavioural, social and technical interventions (The Energy and Resources Institute 2011); Surat and Indore resilience strategies are structured on the principles of building on existing initiatives, preparing pilot and demonstration projects and multi-sectoral synergies and integration (Rajasekar et al. 2012: 5) As part of the resilience strategy, Surat Municipal Corporation set up a trust fund for implementation resilience action in the city, while Gorakhpur resilience strategy looked at restoring and reviving the main water body- the Ramgarh Taal and implemented resilience projects in the peri-urban areas. These projects were primarily on restoration of agricultural lands, improving drainage and solid waste management in the poor localities in the peri-urban areas (ACCCRN 2014). It is to be noted though that the resilience strategies, although being a holistic document were more focussed upon policy and regulatory changes but not detailed enough to take some of the strategies to the level of implementation, particularly for the infrastructure sector. The major challenges were that of missing mandates and budget lines to support implementation of projects. Some of the challenges were that of capacity and the fact that cities were not fully equipped with scientific knowledge and tools to plan for resilience even if they wanted to.

Box 10.3 Smart Cities for Climate Resilience

There is little explicit focus on climate change and flood resilience in the Smart Cities Mission guidelines and the smart city plans of most of the selected smart cities. However, some of the cities by virtue of their exposure to extreme events frequently have designed projects within their smart city plans. For example:

Vishakhapatnam Smart City Plan suggests for a shore protection plan to counter tsunamis and prevent beach erosion.

Bhubaneswar city plans to set up an early warning system for flood and cyclones.

Chennai city demands for a sensor-based water level monitoring, along with a surveillance system to forecast and generate warnings for floods and tsunamis.

Guwahati plans for a hydrological information system (HIS) to be installed for generating real time data for flood forecasting.

Source: https://indiaclimatedialogue.net/2017/10/11/climate-disconnect-indias-smart-cities-mission/

After increase in the frequency of extreme events like flooding and heavy rainfall, cities in India are beginning to realise the importance of building resilient infrastructure and climate proofing the existing ones. Flagship national schemes like Smart Cities (Smart City Mission 2016a) and Atal Mission for Rejuvenation and Urban

Transformation (AMRUT 2016b) are looked at with hope for deliberately bringing in resilience measures to all new infrastructures that will be built under these schemes. However, only few have envisioned environment sustainability, climate resilience and disaster risk reduction as part of their mission towards building their cities as Smart Cities. Cities of Dharamsala, Pune, Guwahati, and Visakhapatnam are some of them (please refer to Box 10.3 for details). Yet there is a long way to achieve desired level of understanding and mission oriented measures to address resilience in infrastructure sector.

The Ministry of Housing and Urban Affairs is currently preparing to launch a competition based certification system for smart cities on climate smart urban development. Key thematic areas are being identified and indicators based framework is being developed at present. One effort worth mentioning is the establishment of Urban Climate Change resilience Trust Fund (UCCRTF) instituted by ADB and supported by The Rockefeller Foundation, Government of UK, Switzerland and USA to foster infrastructure resilience in cities in Asia, India being one of them (Asian Development Bank 2013).

The Rockefeller Foundation created the 100 Resilient Cities program in the year 2013. The program selected 100 prospective cities from around the world including those from the developing world to receive support from the initiative. The City Resilience Strategies that were prepared under the program were one of the core tools that propelled 100 Resilient Cities member cities to implement resilience initiatives. The strategy—product of a six-to-nine month process came up with crucial new solutions so that cities could collectively act on their resilience challenges. The focus areas for each city are spelt out and measures to address those issues have been out in place. For example, Pune city is the only city where Chief Resilience Officers (CRO) has been successfully institutionalised within the Pune Municipal Corporation. The city views resilience in terms of shocks as well as stresses and considers mobility, population pressure and increasing number of private vehicles as major concerns for resilience building initiatives in the city. The city is establishing Urban Mobility Labs within the city that is supported by Niti Ayog's Light House Challenge. Similarly the Preliminary Assessment Report (PAR) of Chennai city has identified the following areas/sectors for intervention:

(i) Water systems
(ii) Informal settlements
(iii) Planned and healthy urbanization
(iv) Urban resilience

The city is learning from International examples like' room for the river' project in Netherland to address its problems of concurrent floods. As part of another innovative solution, the 100 RC program is helping Chennai to tweak the design of its storm water drainage. The new design will not have usual concrete drainages but systems built out of permeable material that allows water to permeate to ground level- a solution that serves the dual purpose of addressing flood protection and manage drought. Besides, Greater Chennai Corporation, Parks division is working with the University of British Columbia on building green infrastructure within the city, besides creating vegetable gardens on flat urban roofs. Some of the successful pilots have been adopted on the roofs of municipal schools. The produce from these 'urban farms' is used to feed children through the mid-day meal program. The green roofs have a cooling effect. Drip irrigation system is being introduced for these urban farms.

As per the 100 Resilient Cities programs' philosophy, "the Resilience Strategies are more than a milestone—they are a roadmap, a call to action" (100 Resilient Cities 2016). Some of the member cities include Chennai, Bangalore, Jaipur and Surat in India; Bangkok Thailand; Adis Ababa, Ethiopia; Belgrade, Serbia; Can Tho and Danang, Vietnam; Accra, Ghana; Melaka, Malaysia; Jakarta and Semarang, Indonesia. All these cities and others that are part of the initiative receive the following support from the initiative and its global partners (100 Resilient Cities 2016):

(i) Financial and logistical guidance for establishing an innovative new position in city government, a Chief Resilience Officer, who will lead the city's resilience efforts
(ii) Expert support for development of a robust Resilience Strategy
(iii) Access to solutions, service providers, and partners from the private, public and NGO sectors who can help them develop and implement their Resilience Strategies
(iv) Membership of a global network of member cities who can learn from and help each other.

Japan has a previous history of Tsunamis and earthquakes of extreme impacts. The city of Sendai has been known as one of the top cities that have built climate resilient infrastructure as a response to recurring earthquakes and tsunamis. The city also displays one of the best examples of decentralised planning efforts where the local government built its own infrastructural assets that included telecommunications, roads and railways. The decentralised nature of infrastructure development not only accelerated the implementation process through improved governance but also prevented corruption in service delivery. Many other Japanese cities have been known to collaborate with private-sector players through public–private partnerships arrangements that have worked very well in building resilience of large infrastructure in Japanese cities (Chirisa et al. 2016).

Similar to these, Many Latin American cities such as Mexico, Sao Paulo and Paraguay have built resilient infrastructure through integrating them with the urban development plans (Chirisa et al. 2016). My Neighbourhood Programme (*Programa Barrio Mio*) in the Lima City Council in Peru is a settlement upgrading initiative with an overall budget of a $30 million. The programme involves construction of physical assets that foster resilience like stairways and defence walls; and social infrastructure such as sports, cultural and community centres. These are often combined with reforestation and urban reorganisation to mitigate risk and prevent disasters.

10.7 Challenges in Building Resilient Infrastructure

While some of the cities in the developing world, particularly in Asia, have shown leadership and innovation in building resilience of their urban infrastructure, most of the African cities are known to face formidable challenges of building and maintaining resilient infrastructure. Innocent (Chirisa et al. 2016) in her paper attributes it to civil disturbances and stagnating urban economies that have not only led to urban poverty but also has led the governments to ignore considerations for resilient infrastructure development and climate change given the priorities of service delivery to ever increasing urban population.

Chang et al. (2014) in his paper attributes the challenges to fostering resilience in infrastructure sector to three noteworthy considerations;

1. There is often a conflict of interests between infrastructure providers (builders) and infrastructure managers. While the managers (city government that is responsible for maintaining and operating infrastructure) are more concerned about minimising investments, repair costs and revenue losses, the infrastructure providers do not have many incentives to think and plan for future disruptions of the infrastructure they are building or even the impact of disruption of infrastructure on other dependent infrastructure. Besides this, retrofitting existing infrastructure to become resilient is not considered as a cost effective strategy in spite of its social value.
2. There are many barriers to information sharing when it comes to large infrastructure or system of infrastructure as a whole due to various reasons. Most of the western developed world would restrict data sharing for security concerns, while many small and medium sized cities in the developing world would not have enough capacity and expertise to collect and maintain dynamic data that is useful for understanding risks and building resilience.
3. One of the challenges to building resilience comes from the fact that most of the infrastructure managers do not have direct experience of a disaster, hence in their limited capacity, most of the time, infrastructure are neither constructed to be resilient nor are their operations attuned to being resilient to short or long term climate disruptions. One of the problems is also that most of the individual infrastructure providers do not have a clear understanding of impacts of the infrastructure to the regional or system of infrastructure within the influence zone of infrastructure they build.

All these lead to under investment in building resilience. Another barrier in resilient infrastructure planning is lack of capacity as well as lack of integrated planning practices where land use planning, economic development planning and disaster resilience planning are interventions most often taken up in silos. This leads to pockets of good and bad infrastructure, more often, networks of incomplete infrastructure that lead to a system failure in cities. One of the examples is the frequent flash floods in most of the large Indian cities during monsoon.

10.8 Conclusion and Way Forward

10.8.1 Policy Implications and Conclusions

A number of strategies can be used in promoting resilience in cities. Improving urban infrastructure is one of the foremost priorities in improving the city systems-that constitute a city's structure and control or foster the city's functions. In most of the cases in systems failures in urban areas in the developing country context, the causes are often related to substandard construction, lack of maintenance and aging infrastructure within an already weak network. To foster this it is pertinent to create effective, pro-poor structures of governance, while also building the capacity of communities to address these challenges (Dodman 2009).

The critical success factors would essentially include;

1. Strengthening livelihoods (natural resource management; provision of basic services; infrastructure development),
2. Good urban governance (regulatory frameworks; planning for growth),
3. Financial tools including credits, insurance, and incentives for building resilience into infrastructure planning and design that are linked to financial assistance for infrastructure.
4. Eco- system management (protected areas; payments for ecosystem services) and
5. Community- based risk reduction approaches

Box 10.4 Smart Cities for Climate Resilience

Smart City Mission is a great opportunity to fix the problems of urban India and build functional economically thriving, environmentally sustainable and climate resilient cities that offer great quality of life to its citizens and is just safe and liveable.

There is a need to make use of the State Action Plan on climate change that draw out state and district level risks and vulnerabilities which could be used as a starting point to plan resilience for city regions.

Integration of disaster risk reduction and climate resilience (including adaptation and mitigation) into the city development plans and city land use plans should be made mandatory.

Institutional synergies and systems to achieve the same have to be created afresh so that the challenges and sometimes the pretension for inaction can be avoided and reduced.

Urban infrastructure projects need to be climate proofed and looked at through a climate lens before planning. Resilience measures both structural and institutional have to be integrated in infrastructure planning and city planning in Smart Cities.

A good reference point for the purpose of this paper could be The European Commission's guidelines on resilient infrastructure building, which European cities have to comply with in order to boost their resilience. The guidelines provide a range of measures classified in the White Paper entitled 'Adapting to Climate Change: Towards a European Framework for Action' (European Commission 2009). The white paper strategizes infrastructure resilience as a function of:

1. Building 'grey infrastructure' that is related to physical interventions or construction measures and using engineering services to make buildings and infrastructure that can withstanding extreme events,
2. Build 'green infrastructure' that increases ecosystems resilience, reduces biodiversity loss, waste of water and degradation of ecosystem and,
3. Build 'soft measures' that consist of policies, plans, programs and procedures implemented for ascertaining mainstreaming of infrastructure resilience and integrated urban planning practices. This also calls for influencing behavioural changes that can contribute to increase adaptive capacity in the wake of uncertain future.

As reflected by the literature on this subject and subsequent discussions made in this chapter, some of the important considerations for policies on infrastructure resilience would include:

(i) Looking at peri –urban and urban ecosystems that help in reducing vulnerability of city regions by acting as buffers and sponges to some of the extreme events like floods. Peri Urban areas are described as *'transition or interaction zone, where urban and rural activities are juxtaposed, and landscape features are subject to rapid modifications, inducing by human activities*' (Douglas 2006). The better we maintain these buffers and allow for development of these buffers the better equipped we will be to avoid extreme risks.

(ii) Infrastructure of any kind cannot be looked at in isolation. There are regional impacts and reach of infrastructure like roads, telecommunication and even energy. Resilience in infrastructure would therefore mean looking beyond the boundaries of cities and understand these regional non local linkages to maintain functionality of infrastructure and to build robustness to the network of systems.

(iii) Urban transformation has been rapid and organic in most of the developing world's cities. Changing land use and landscape of urban areas is therefore critical to building infrastructure resilience while updating old infrastructure or creating new ones. Location of a structurally robust infrastructure in a risky region may not help in the long run as functionality may be affected negatively during extreme events.

(iv) Structural resilience and robustness is important to contain the casualty at the time of calamity, is helpful in evacuation, and aids safe failure if nothing works. Therefore design and material aspect of infrastructure should be one of the considerations. Building codes, structural standards could help a great deal in tweaking site specific and infrastructure specific design parameters to respond to climate risk factors.

(v) More importantly functionality of infrastructure during a calamity is the core concern rather than only looking at structural safety. Hence a broader framework that allows for functionality of critical infrastructure has to be built. This not only includes emergency services and evacuation plans, but also links to avoiding system failures like telecommunication and water availability.

(vi) The Ministry of Housing and Urban Affairs, Government of India is setting up Climate Smart Cities Alliance and has come up with sets of robust indicators that smart cities will use to assess their progress towards developing climate smart. These include both mitigation and adaptation measures. This is a welcome step to create lighthouse cities that other cities can learn from as well as will build resilience in the long term within the fast-growing urban centres from climate impacts. These indicators also touch upon resilience of urban infrastructure and systems in detail.

Urban development in India, is a State Subject. Therefore, state level policies and interventions go a long way in creating sustainable climate resilient cities and ensuring that particularly large and critical infrastructure is designed, located and built to mitigate climate risks. Then there are standards and bye laws present as a rule book to follow but implementation is weak at local level. This is an institutional issue and needs to be tackled through strong penalties and monitoring. In other parts of the world models for risk insurance of large infrastructure have been designed and applied. It is high time Indian authorities start looking at risk insurance in the infrastructure sector. SAPCC have been criticised for not accommodating sustainable habitats and urban centres into its policies adequately. However, in the recent past, string steps such as the climate smart cities initiative of MoHUA instil some hope.

References

ACCCRN (2008) Connecting s. Retrieved from: https://www.acccrn.net/about-acccrn/history

ACCCRN (2014) Climate change in the context of India. Retrieved from: https://www.acccrn.net/country/india

Asian Development Bank (2013) Urban climate change resilience trust fund. Retrieved from: https://www.adb.org/site/funds/funds/urban-climate-change-resilience-trust-fund

Asian Development Bank (2015a) Asia's Booming cities most at risk at risk from climate change. Retrieved from https://www.adb.org/news/features/asias-booming-cities-most-risk-climate-change

Asian Development Bank (2015b) Climate change resilience in Asia's cities. Retrieved form: https://www.adb.org/news/infographics/climate-change-resilience-asias-cities

Bai X et al (2018) Six research priorities for cities and climate change. Nature 555(23–25). https://doi.org/10.1038/d41586-018-02409-z

Business Standard (2014) Cyclone Hudhud damage in AP estimated at Rs. 8000 cr; death toll rises to 24. Business Standard. Retrieved from: https://www.business-standard.com/article/current-affairs/cyclone-hudhud-damage-estimated-at-rs-8-000-cr-114101301099_1.html

Chang SE, McDaniels T, Fox J, Dhariwal R, Longstaff H (2014) Towards disaster resilient cities: characterising resilience of infrastructure systems by expert judgement. Risk Analysis 34(3): 416–434. https://doi.org/10.1111/risa.12133

Chirisa I, Bandauko E, Mazhindu E, Kwangwama NA, Chikowore G (2016) Building resilient infrastructure in the face of climate change in African cities: Scope, potentiality and challenges. Development Southern Africa 33:1113–1127. https://doi.org/10.1080/0376835X.2015.1113122

David S, Sheridan B (2016) Urbanization, development and the sustainable development goals. In: Sheridan B, David S (eds) Cities on a finite planet–towards transformative responses to climate change. Earthscan, Routledge, pp 1–16

Dodman D (2009) Building urban resilience in the least developed countries. International Institute for Environment and Development, London

Douglas I (2006) Peri-urban ecosystems and societies transitional zones and contrasting values. In: McGregor D, Simon D, Thompson D (eds) Peri-urban interface: approaches to sustainable natural and human resource use. Earthscan Publications Ltd., London, UK, pp 18–29

EC (European Commission) (2009) White paper. Adapting to climate change: towards a European framework for action, (Brussels). http://eur-lex.europa.eu/LexUriServ/LexUriServ.do?uri=COM:2009:0147:FIN:EN:PDF Accessed 18 Aug 2018.

Fraser A et al (2016) Understating risk in the context of urban development- definitions, concepts and pathways. In: Bartlett S, Satterthwaite D (eds) Cities on a finite planet–towards transforming responses to climate change. Earthscan, pp 17–40

Gurara D, Klyuev V, Mwase N, Presbitero A, Xin Cindy X, Bannister G. (2017) Trends and challenges in infrastructure investment in low-income developing countries. IMF Working Paper. Strategy, Policy, and Review Department

Hooernweg D, Pope K (2017) Population predictions for the world's largest cities in the twenty-first century. Environment and Urbanization 29:195–216

McKinsey (2010) India's Urban Awakening: Building Inclusive cities, sustaining economic growth. Available at https://www.mckinsey.com/~/media/McKinsey/Featured%20Insights/Urbanization/Urban%20awakening%20in%20India/MGI_Indias_urban_awakening_executive_summary.ashx

Ministry of Housing and Urban Affairs, Government of India (2016a) Smart city mission. Retrieved from: http://smartcities.gov.in/content/

Ministry of Housing and Urban Affairs, Government of India (2016b) AMRUT. Retrieved from: http://amrut.gov.in/content/

National Institute of Public Finance and Policy (2017) Disaster resilience infrastructure. Retrieved from: https://nipfp.org.in/disaster_resilience_infrastructure/

PTI (2018a) Kerala flood live updates: death toll reaches 370; No heavy rains expected for next 5 days. The 1economic Times. Retrieved from: https://economictimes.indiatimes.com/news/politics-and-nation/kerala-floods-live-death-toll-reached-164-rescue-operations-underway/articleshow/65418310.cms

PTI (2018b) Kerala rains: Rs 40,000 crore loss as per initial estimate. The Economic Times. Retrieved from: https://economictimes.indiatimes.com/news/politics-and-nation/kerala-rains-rs-40000-crore-loss-as-per-initial-estimate/articleshow/65787288.cms

QUARTZ (2017) Map: Asia's infrastructure boom is coming to the cities most threatened by climate change. Retrieved from: https://qz.com/1034066/map-asias-infrastructure-boom-is-coming-to-the-cities-most-threatened-by-climate-change/

Rajasekar U, Bhat GK, Karanth A (2012) Tale of two cities: developing city resilience strategies under climate change scenarios for Indore and Surat, India. http://acccrn.net/sites/default/files/publication/attach/TaleofTwoCities_TARU_0.pdf. Accessed 18 August 2018

Resilient Cities (2016). Retrieved from: http://www.100resilientcities.org/about-us/

Sharma D, Singh R, Singh R (2017) Approaches to resilience planning in Indian cities: The ACCCRN experience. In: Archer D, Colenbrander S, Dodman D (eds) Chapter paper in Responding to climate change in Asian cities- Governance for a more resilient urban future. Routledge, pp 89–107

Tanner T, Surminski S, Wilkinson E, Reid R, Rentschler J, Rajput S. (2015). The triple dividend of resilience-realising development goals through the multiple benefits of disaster risk management. Overseas Development Institute (ODI), International Bank for Reconstruction and Development / International Development Association or The World Bank.

The Energy and Resources Institute (2011) Mainstreaming urban resilience planning in Indian cities: a policy perspective. TERI, Delhi

Therrien M, et al. (2018) Towards urban resilience: synthesizing the strategies that enable and factors that impede implementation (Social Sciences and Humanities Research Council of Canada Working Paper). Retrieved from Governance of Urban Resilience website: http://cite-id.com/documents/synth%C3%A8se-de-connaissances/Scoping-review-Final-version-siteweb.pdf

Tyler S, Moench M (2012) A framework for urban climate resilience. Climate and Development 4(4):311–326

UN HABITAT (2016) Urbanization and development–emerging futures. World Cities Report. Retrieved from: http://wcr.unhabitat.org/

UNISDR (2009) Terminology on disaster risk reduction. Available at https://www.unisdr.org/files/7817_UNISDRTerminologyEnglish.pdf. Accessed 7 Sep 2018

United Nations (2014) World urbanization prospects. The 2014 Revision, Population Division, Department of Economic and Social Affairs, New York. Retrieved from https://www.un.org/en/development/desa/publications/2014-revision-world-urbanization-prospects.html

World Bank (2017) Rapid flood damage and needs assessment–Vietnam. Available at https://www.gfdrr.org/sites/default/files/publication/Vietnam%20Rapid%20Damage_FinalWebv3.pdf

World Bank; Asian Development Bank (2013) India, Uttarakand disaster, June 2013: joint rapid damage and needs assessment report. New Delhi. © World Bank https://openknowledge.worldbank.org/handle/10986/16759

World Economic Forum (2015) What kind of infrastructure does developing world need. Retrieved from: https://www.weforum.org/agenda/2015/08/what-kind-of-infrastructure-does-the-developing-world-need/. Accessed 7 Sep 2018

Divya Sharma is the ED of Climate Action Group, India Office. She is an urban policy and climate resilience expert with extensive experience on capacity building, policy and city strategy formulation on smart cities, climate and disaster resilience and pro-poor development. Her career spans more than 15 years where she has worked very closely with decision makers and practitioners in the Government, international research community and academia.

Gupta Anil K , Professor of Policy & Strategies, Integrated Resilience. Head of the ECDRM Division, International Cooperation, Programme Director—Centre for Excellence on Climate Resilience, Nodal Officer for Central Ministries at National Institute of Disaster Management, (Govt. of India), New Delhi. He is also a member of the expert Team for WMO Climate Statement and Core Group Member of IUCN-Council of Ecosystem Management.

Chapter 11
Climate Change Adaptation in Industrial Areas for Disaster Resilience

Anil K. Gupta, N. Raghu Babu, Florian B. Lux, and Shruti Bansal

Abstract Climate is changing, and effects can be seen across the world. The climatic disturbances triggered due to earth's heating is putting industrial areas at risk. The effect of this can be seen in the form of decreased industrial productivity, decreased labor productivity and increased risk of natural disasters. India has witnessed many disasters triggered due to Climate Change. In 2015 there were nearly 3000 deaths in Andhra Pradesh and Telangana because of heat waves. Urban Heat Island effect has been observed in Gujarat, resulting in the rise of local temperature and posing severe risk in agricultural sector. Adaptation to these climatic events is the challenge for a more sustainable and safe future. The present paper focuses on industrial risks associated with climatic hazards, their possible impacts on industrial areas and explains the possible adaptation strategies to reduce hazards vulnerability by taking examples of three states viz. Gujarat, Andhra Pradesh, and Telangana. High exposure to hazards, improper planning of the areas, increased population and urbanization are some of the causes associated with increased risks in these industrial areas. Designing industrial areas based on the principal of Eco-Industrial Parks ensures sustainability with integration of social, economic and environemntal quality aspects.

Keywords Climate change adaptation · Disaster risk reduction · Industrial parks

A. K. Gupta (✉)
ECDRM Division, International Cooperation, CAP-RES DST Project, National Institute of Disaster Management (Ministry of Home Affairs), New Delhi, India

N. R. Babu
GIZ Germany, New Delhi, India

F. B. Lux
GIZ Ifanos Concept and Planning, Nürnberg, Germany

S. Bansal
Environment Resource Foundation, Gurgaon, India

A. K. Gupta et al. (eds.), *Disaster Risk and Management Under Climate Change*, Disaster Resilience and Green Growth,
https://doi.org/10.1007/978-981-99-4105-6_11

11.1 Introduction

Climate Change is posing severe risks for industrial areas. Changing climate is leading to raise in temperature, variation in precipitation rates, changes in wind direction and velocity and an increased frequency of cyclones. These changes are driving serious risks on industrial buildings, industrial production, businesses and employees.

There have been many evidences, which show the impacts of Climate Changes on industrial areas. The 21st century has seen slowing of growth rate of industrial production in Finland from 6% (from 1925 to 2000) to 3% (SYKE n.d.). As per Climate Change Risk Assessment 2017 Evidence Report, UK heat wave of 2003 has caused business losses of £400–500 million. Climate Change is inducing severe risks for industrial sector in India also. India has witnessed many disasters triggered due to Climate Change. Heat waves lead to nearly 2,500 deaths in Telangana (Heat Wave Action Plan 2018, Government of Telangana) and Andhra Pradesh in 2003 and nearly 3,000 deaths in 2015. Drought of 2015 in Telangana has affected 10 out of 13 districts and resulted in reduction in water level to 50 % lower than long-term means in major reservoirs (EPTRI, Sate Action Plan on Climate Change for Telangana State 2015). As per reports, highly erratic rainfall pattern has been observed in Gujarat since last couple of years and has posed pressure in the productivity of agricultural crops (TERI and GIZ 2014). Studies suggests that this state is expected to lose 40% of coconut yield because of rising temperature (MoEF 2010). In Tamil Nadu, Chennai's devastating flood of 2015 had taken away the lives of around 500 people, displaced over 18 lakhs of people and damaged approximately 3.8 lakhs hectares of agricultural land (Narasimhan et al. 2016). Catastrophic flood of 2018 in Kerala that started in July and continues till August is said to be the worst calamity in the last 100 years, and has caused damages of around $3 billion (Anand 2018). Reports suggested that many Indian states are vulnerable to cyclones, storm surges, floods, drought and climate extremity. Approximately, 44% of the Andhra Pradesh State is vulnerable to cyclonic storms (NDMA n.d.). it has been observed in Andhra Pradesh that due to rise in rainfall intensity, problems of floods, spilling of smaller rivers and drainage problems in delta areas are being observed (EPTRI, State Action Plan on Climate Change for Telangana State 2015). Adaptation to these changes and risks is essential to combat serious risks associated with these events. Climate Change Adaptation has the potential to deal with these changes by reducing the vulnerability associated with climatic events.

The chapter ***Climate Change Adaptation for Disaster Resilience in Industrial Areas*** illustrates some of the risks associated with Climate Change to Industrial Park, their possible impacts, possible Climate Change Adaptation strategies to address Climate Change by considering case studies of Andhra Pradesh, Telangana, and Gujarat.

11.2 Risks for Industries and Industrial Parks from Climate Change

11.2.1 Availability of Natural Resources

Climate Change has added stress on availability of natural resources. Due to Climate Change, a shift in weather pattern and climate has been observed. On top of this, Climate Change has increased the frequency of natural disasters including drought, forest fire, pests, and increased the invasive species risks. Rise in temperature has shifted the distribution of trees species (Long 2014). Due to the decrease in availability of natural resources, the raw materials and supply chains to industries get severely affected posing threats to the survival of the industry itself.

11.2.2 Work Productivity

Global temperature rise increases outside and indoor heat stress. This increased heat stress can cause health and productivity problems. Excessive temperature reduces the person's ability to work, and for India about 4–6% of loss of working hour in a year is expected by 2025 (Seelig et al. 2018). High body temperature causes dehydration, exhaustion, and in some cases, it can even cause person's death. To avoid this, it is advised to reduce worker's working hours (Kjellstrom et al. 2016). However, this solution may not be adequate for the affected industries and may require other adapation measures to overcome the risks associated.

11.2.3 Buildings and Infrastructure

Climate Change related induced disasters, e.g. floods, heat waves and high wind velocities can severely affect design life and performance of buildings. These extreme events demand for modified building designs, new construction material and techniques, so that the industrial area get adapted to these risks without compromising on overall productivity and building design life.

11.2.4 Energy Requirement

Temperature rise causes increase in energy demand due to increased air-conditioning requirements and increase in cooling requirements of water. Extreme climate also increases power outage, as change in summer and winter durations can decrease network reliability (Finley and Schuchard 2011). This calls for designing energy-efficient building and utilization of clean energy as the energy source for the industrial areas.

11.2.5 Transportation and Logistics Operation

Increase frequency and intensity of extreme weather events can cause disruption in industrial logistics operation. The events like heat waves, increased precipitation rates which can cause delays in logistics by causing failure in land-based, air, and marine transportation systems. These events also reduce the life expectancy of roads (EPA 2017).

11.2.6 Risk from Natural Disasters

Industries will be exposed to disasters from floods, cyclones and winds. These disasters would affect factory buildings, roads, electricity installations and could also lead to pollution risks due to washing-off of waste dumps, chemical spills, damage to chemicals storage tanks etc. There is a need to take adaptation measures. Climate Change Adaptation (CCA) is a series of multiple steps and has various possibilities, to choose the best adaptation option it is essential to assess risk.

11.3 Risk Assessment

Risk Assessment is the process of analyzing the impacts, occurrence frequency and response to the consequences of Climate Change events. The key steps involved in risk assessment are:

- Define the system of interest – industry and/or industrial park (IP). The generic information about the industry and/ or industrial park including name, zone, area, type of the industry and/ or IP, population etc. need to be collected during this phase
- Identify prevailing climatic hazards in the chosen system of interest, for example in Telangana the climatic hazards are heat waves, droughts and heavy rains
- Collect information on past exposure to climatic hazards including its severity (e.g. very severe, severe, moderate, slight, very slight), frequency, change/ increase in frequency observed. Based on the collected information, and combination of spatial and temporal exposure, classification of IP's exposure has to be done in the classes (very low)-low – medium – high – (very high). Refer Table 11.1, to understand the classification of exposure based on temporal and spatial exposure matrix:

Table 11.1 Exposure classification matrix

Exposure		Spatial dimension: Low	Medium	High
Temporal dimension	Low	Low	Low	Medium
	Medium	Low	Medium	High
	High	Medium	High	High

Table 11.2 Impact classification matrix

Expected impact		Susceptibility: Low	Medium	High
Exposure	Low	Low	Low	Medium
	Medium	Low	Medium	High
	High	Medium	High	High

Table 11.3 Vulnerability classification matrix

Vulnerability		Resilience: Low	Medium	High
Impact	Low	Low	Low	Medium
	Medium	Low	Medium	High
	High	Medium	High	High

- Collect information on impacts observed in the past on humans, infrastructure and buildings, water/energy supply and waste management services, operations and supply chain
- Collect information on existing resilience of the industry/IP including economic status of the IP and preliminary information on its capability to adapt to climate change. The output of the process is the classification of resilience into classes: low, medium, high
- Collect information on susceptibility of components of industries/IPs (e.g., buildings, roads, drainage systems, energy and water supply, greenery, industrial community at site, workforce, operations etc.) and categorize the susceptibility in the classes: low, medium, high
- On the basis of exposure and susceptibility, the classification of expected impacts can be done as explained in the Table 11.2:

- Combine the information of impact and resilience to determine the vulnerability, as also explained in Table 11.3

- Analyse the risk by combining the vulnerability, with the probability of events occurrence and expected impacts. Figure 11.1 summarizes the factors on which industrial risks depends.

Fig. 11.1 Factors affecting industrial risks (Source: Welle 2011)

11.4 Eco-Industrial Park

An Eco Industrial Park can be defined as a dedicated area for industrial use at a suitable site that ensures sustainability with integration of social, economic and environemntal quality aspects into its siting, planning, management and operaitons. The Deutsche Gesellschaft für Internationale Zusammenarbeit (GIZ) GmbH, the World Bank and UNIDO have brought out International Framework for Eco Industrial Parksthat has various performance parameters, which are to be intergated into industrial aprks to qualify them as Eco Industrial Parks. CCA aspects are intergral part of this document.

Eco-Industrial Parks should follow the principals of green buildings. Green buildings promote low carbon material and techniques, requiring significantly less amount of energy and relatively decrease energy cost. Some of the aspects that could be covered include:

11.4.1 Passive Designs for Industrial Parks

Passive designing is the way of designing the buildings, which leads to the less consumption of energy by taking advantage of technological innovation including natural movement of heat and air, passive solar gain, better insulation etc (van Dam and van Huet 2015). Passive building designs takes benefit of five principals that include: **orientation, overhangs and shadings, insulation, double or triple glazing,** and **thermal mass** (ESRU 2010). Industrial Parks to be designed with passive housing techniques require a well-insulated building envelope with a ventilation system that can recover heat from the exhaust air. The section below, takes reference from ESRU (2010), and explains some of the design principles, which should be kept in mind while designing the Industrial Park.

11.4.1.1 Orientation

Correct orientation of facade of industrial areas is necessary for summer cooling conditions. The basic principle here is the orientation of southern façade of the building towards the equator in the northern hemisphere and northern façade of the building towards the north in the southern hemisphere. The longer dimension of the building should be in east/ west direction for maximizing solar gain. This orientation is also advantageous for summer cooling conditions because it minimizes the east-west façades to morning and afternoon sunlight (ESRU 2010).

11.4.1.2 Overhangs and Shading

Overhangs and shading are important components of passive architecture. Proper sizing of overhangs and shading is required for reducing heating during summers and reduces heating requirement during winters. Some general recommendations have been listed just below:

- Limit the number of east and west windows because they are very difficult to shade compared to the south side
- North façades receive very less solar radiation hence can be avoided for shading
- Interior shading devices, such as Venetian blinds or vertical louvers, might be used to control glare, however exterior shadings must be included since the interior ones have already admitted the solar gain in.

11.4.1.3 Insulations

Well-insulated buildings minimize heat loss during winters and keeps interior cool in summers. Insulators being bad conductor of heat forms a barrier between exterior and interior spaces. Use of low emissivity roofing paints, high performing glazing, and strategic plantation are some of the ways of doing it (Snow and Prasad 2011). Heat loss w.r.t. insulation is represented by U-value, which represents how much heat in Watts is lost per m^2 at a standard temperature difference (ESRU 2010). Insulation can be done in walls, floors and roofs of the building in the Industrial Parks (IP). R-value determines the resistance to heat flow; higher the value higher the resistance.

11.4.1.4 Windows

Glazed windows are important component in passive architecture as they store heat and light and provide natural ventilation. Windows can also be used for maximizing daylighting, helping in using daylight as the primary source of lightning for industries resulting in less consumption of energy. Positing of windows is important in this regard. Maximizing the windows in the southern façade helps in achieving the maximum solar gain and minimizing in the northern façade helps against cold during winters. Type of glazing is of importance for insulation purpose. Double/triple glazed windows can be used depending on the requirement and site location.

11.4.1.5 Thermal Mass

Thermal mass takes benefit of solid or liquid mass which absorbs and store warmth and releases it when is needed. For this, the following can be used:

- Brick, tile or thick concrete floors
- Large brick or stone internal fireplace or an interior wall made of adobe or brick
- Masonry or concrete wall or water filled containers (called a Tube Wall).

11.4.2 Infrastructure of Industrial Parks

The support infrastructure of the industrial area should be well planned. The infrastructural components designing is well explained in MoEFCC and EPTRI (2015). The infrastructural components of Industrial Parks generally include transportation, parking, energy and water supplies etc.

11.4.2.1 Transportation

Public transport mode should be encouraged to the Industrial Park. The pedestrian paths should be integrated with the green pathways to provide a natural, attractive and safe option of travel. Provision of amenities to encourage pedestrian movement, such as benches, street trees, waste receptacles, street lighting, and shelter at public areas should be included.

11.4.2.2 Parking

Provisions for proper parking, based on hierarchy, purpose, safety accessibility and space standard should be included. To avoid truck parking and busses on-street in a haphazard manner, organized parking to be allocated in the industrial area (overnight stay), zone/cluster level (temporary stay) and plot level (loading/unloading).

11.4.2.3 Common Infrastructure Services

These includes administrative buildings, banks, post office, kiosks, and other common infrastructure services in the industrial park as support services to be planned and implemented.

11.4.2.4 Energy/Power Supply

Energy efficient and renewable energy can be used to improve the efficiency of the power in industrial areas. Crucial would be resilient systems of energy supply i.e. backdrop solutions in case main power supply will fail which is important to bridge for cooling, communication and human health. Utilization of solar parks, solar benches, and solar panels can be done to optimize the use of renewable energy.

11.4.2.5 Water Management

Industrial Park should have storm water collection system to avoid excessive discharge and water contamination. Decentralized flood management system shall be there, so that planning can be proposed depending on the site location, topography and slope. Lined surface/ treated porous surface can be used for collection purpose. Collected water shall be utilized (after treatment if required) for industrial purposes depending on the water quality.

11.4.2.6 Plantation and Landscaping

Green corridor should be planned within industrial areas. Wherever possible native species of trees and shrubs should be planted and a maintenance system installed for proper management of the greeneries.

11.5 Climate Change Adaptation for Industries

Climate Change Adaption (CCA) is the process of responding to Climate Change impacts to reduce risk and vulnerability associated (VCCCAR 2018). Adaptation planning refers to identifying measures and planning implementation of the

Table 11.4 Basic profile of studied states

Parameters		Andhra Pradesh	Telangana	Gujarat
Location	Latitude	12°41’ to 22° N	15°46’ to 19°47’ N	20°6’ N to 24°42’ N
	Longitude	77° and 84°40’ E	77°16’ to 81°43’	68°10’ E to 74°28’ E
Geographical Area (km^2)		275,045	114,840	196,024
Climate	Avg. Rainfall (mm)	906	906	1107
	Temperature	Winters: 14–22 °C Summers:30–47 °C	Winters: 13–17 °C Summers: 40–43 °C	Winters: 12–29 °C Summers: 30–49 °C
	Coastline	974 km	–	1663 km
Landform		Hilly, plateau, highlands, Coastline areas	Plateau, hills	alluvial plains, hilly areas, highlands, desert areas, coastline areas
Prevailing Climatic Hazards		Floods, cyclones, droughts, heat stress	Floods, cyclones, droughts, heat stress	Floods, cyclones, droughts, heat stress

Source: (EPTRI, State Action Plan on Climate Change for Andhra Pradesh 2012; EPTRI, State Action Plan on Climate Change for Telangana state 2015; TERI and GIZ 2014)

identified measures to reduce Climate Change related risks and vulnerabilities (Kabischet et al. 2015).

11.5.1 Case Studies

This chapter ***Climate Change Adaptation for Disaster Resilience in Industrial Areas*** attempts to explain Climate Change Adaptation methodology for Industrial Parks considering three industrial states, which are Andhra Pradesh, Telangana and Gujarat.

Andhra Pradesh and Telangana are Indian states located in the southern part of India. These states have tropical climate with moderate to subtropical weather. Gujarat is in the western part of India. Table 11.4 explains the state profile of Andhra Pradesh, Telangana and Gujarat.

11.5.1.1 Case Study 1: Andhra Pradesh and Telangana

Andhra Pradesh and Telangana are vulnerable to multiple hazards including cyclones, storms, floods, droughts, heat waves etc. Andhra Pradesh’s districts situated near coastal area are highly prone to cyclones. As per reports, more than

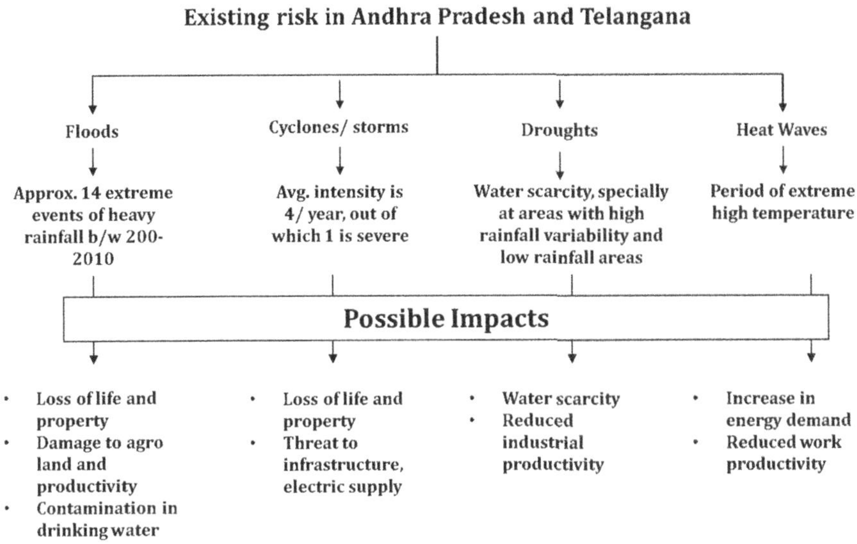

Fig. 11.2 Existing hazards risks and their possible impacts (EPTRI, State Action Plan on Climate Change for Andhra Pradesh 2012; EPTRI State Action Plan on Climate Change for Telangana State 2015)

100 cyclones have hit the State of which 31 were reported severe. The frequency and intensity of their occurrence will rise in future (EPTRI, State Action Plan on Climate Change for Andhra Pradesh 2012). The report also states that the 44% of its area is vulnerable to tropical storms. The uneven distribution of rainfall in Andhra Pradesh and Telangana makes some states highly vulnerable with heavy rainfall. Heavy rainfall triggers flood events, and if drainage system is not intelligently designed, these events can cause heavy loss to life and property. The delta region of Krishna River and Godavari River are more prone to these events (Kabisch et al. 2015). Droughts are also frequent here in these States. Heat waves are also common, with a temperature maximum of 48.8 °C is recorded in 2002 (Kabisch et al. 2015). Figure 11.2 explains the existing hazards in Andhra Pradesh and Telangana, risks, and their possible impacts.

Inadequately planned areas will have comparatively very high risks associated with climatic change. For example, poorly designed drainage system will not be adequate in accommodating water accumulated due to heavy rainfall and this will lead to infrastructural damages. Timely adaptation to these risks can help industries in reducing the impacts associated with these events.

At the state level, SAPCC of Telangana and Andhra Pradesh identifies the industrial sector impacted by climate change. However, the focus is primarily on mitigation activities and not on adaptation activities. This chapter highlights the CCA strategies that can be adopted for disaster resilience in Industrial Parks.

The fundamental action of choosing CCA strategies needed is to reduce the risk associated with the hazardous events. Adaptation options are well defined in **CCA for Sustainable Industrial Development** by GIZ (Kabisch et al. 2015). Section 11.5.3, CCA Measures of this chapter covers the variety of option that can be considered for building Climate Resilient Industrial Park

11.5.1.2 Case Study 2: Gujarat

Naroda is in the northeast of central part of Ahmedabad, Gujarat. Naroda industrial area was established in 1980s, accommodating around 1200 industries. Based on the risk profile and vulnerability the prevailing hazards in Gujarat includes floods, heat waves and increasing temperature (refer Fig. 11.3). Studies have shown that temperature of industrial areas and dense urban area is much higher than the suburb areas of Ahmedabad (Joshi et al. 2015).

As can be seen from the above illustration (Fig. 11.3), that the primary causes for flooding and heat islands is the inadequate infrastructure and improper planning. Vehicular load increase has also been observed over last couple of years (Shodhganga 2010), and contributes to an increasing heat stress in Gujarat as shown in Fig. 11.4. The climatic risks can have serious impacts on industrial areas, some of them are explained below:

- Analysis results of Kabisch et al. (2015) shows that approx. 50 industries are at risk due to floods. The number can go up to 300 in case of severe flooding

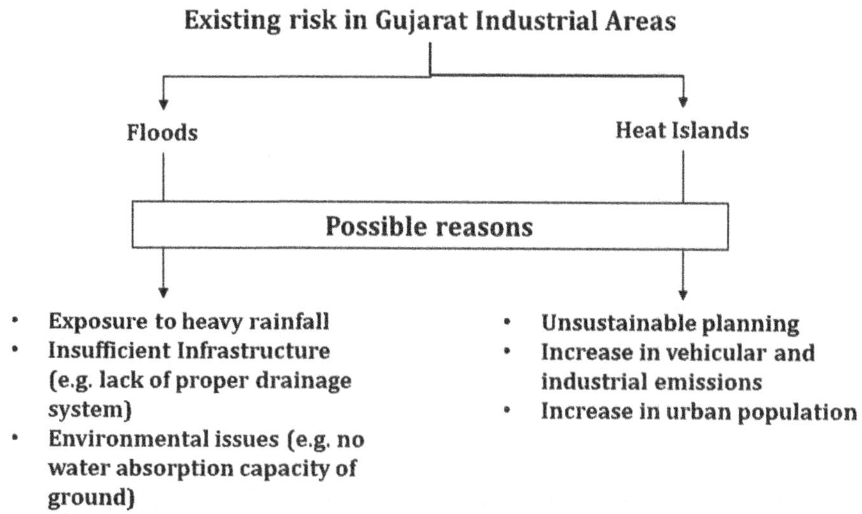

Fig. 11.3 Concept prepared using information from Joshi et al. (2015) and Snow and Prasad (2011)

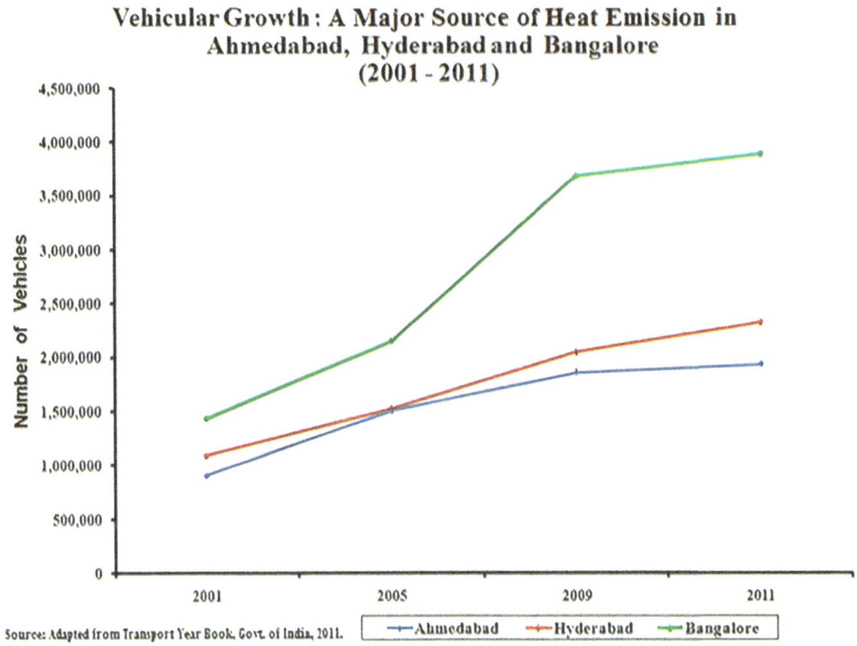

Fig. 11.4 Increase in vehicle count in Ahmedabad city (Source: Shodhganga 2010)

- Temperature rise can labour productivity by adversely impacting the general comfort of workers

The coming section explains the overall CCA approach, which can be followed to build climate resilient Industrial Parks.

11.5.2 Adaptation Planning

Adaptation planning for existing industrial parks involves the following steps:

- Collect information (e.g. the location, type of hazard the Industrial Park is exposed to, return period of exposed hazard, resilience of Industrial Parks etc.)
- Assessment of Climate Change risks (e.g. the expected impact of hazard and the risk it can cause to Industrial Parks)
- Identification of adaptation measures and their prioritisation (explained in the coming section)
- Cost-benefit analysis
- Implementation planning including infrastructure delivery mechanisms, identification of funding sources and their mobilisation, preparation of an implementation plan

- Initiate implementation
- Operation and management, monitoring.

Adaptation planning for new industrial parks involves the following steps:

- Detailed Climate Risk Assessment (CRA)
- Costing and financing of concrete CCA Measures
- Formulation and selection of concrete CCA measures
- Communicate resilience as Unique Selling Point
- Include CRA as part of EIA
- Realization of CCA measures include CCA in organizational set-up and statutes.

A wide range of adaptation measures can be there depending on the industrial type and location. Adaptation generally involves inclusion of combination of structural (physical) or non-structural (non-physical) options (Kabisch et al. 2015). Physical strategies involve enhancement in the building/infrastructural designs, modifying operational aspects, and choosing site location wisely. Non-physical aspects on the other hand includes improvement in the management/monitoring services, emergency response programs, training staff, available insurance services etc. The adaptation measures are required to be integrated into a site master plan. The CCA parameters for inclusion in the site master plan are given below:

11.5.2.1 High Wind Resistance

- Appropriate location for new IPs
- High wind resistant foundation design
- Growing vegetation in buffer zones
- Effective external cladding material
- Orientation and shape of buildings
- Anchoring of roofs
- High-wind responsive building design
- Cyclonic wind resistant doors and windows
- Construction of impact resistant structure
- Emergency service

11.5.2.2 Protection of Infrastructure

- Locating and planning of buildings and facilities
- Development of public notification plan
- Lightning detection system
- Lightning resilient building design and material
- Lightning protection system
- Protection to electrical facilities

- Installation and distribution of surge protection device
- Planting and protection of tree

11.5.2.3 Planning Storm Surge Responsiveness

- Appropriate location for new IPs
- Storm-surge preventive building design
- Protective materials for walls and floors
- Planning an adequate building layout
- Storm-surge proof doors and window
- Access safety and road protection
- Construction of coastal protection measures
- Developing soft coastal protection measures

11.5.2.4 Reduction of Soil Erosion

- Demarcation of soil disturbance areas
- Construction of defences to control erosion
- Landscaping features to control erosion
- Erosion proof structural system
- Protection of drainage facilities
- Interception of toxic wastes and pollutants
- Stabilization and development of topsoil
- Planning Drainage Facilities

11.5.2.5 Prevention from Flood Water

- Suitable IP site selection and flood preventive layout
- Prevention of site using flood barrier
- Prevention of building using flood barrier
- Access/egress safety routes, signage and posts
- Protection and maintenance of roads
- Materials for water infiltration into ground
- Protection measures for electrical facilities
- Adaptive building design
- Sturdy foundation design
- Raising ground floor level
- Water-force resistant
- Structures
- Flood-proofing the walls and floors

- Flood-proofing doors and windows
- Providing means of evacuation

11.5.2.6 Ensuring Appropriate Site for Industrial Parks

- Locating IP at an appropriate site
- Storm-surge preventive building design
- Construction of coastal protection measures
- Protective materials for walls and floors
- Planning an adequate building layout
- Storm-surge proof doors and window
- Access safety and road protection
- Developing soft coastal protection measures

11.5.2.7 Cooling the Urban Realm

- Compact buildings
- Light coloured and reflective paving materials
- Open water and water features
- Trees and vegetation to provide shading
- Building to building shading
- Access to overheating relief
- Artificial shading
- Orientation of buildings to prevailing winds
- Green ventilation corridors

11.5.2.8 Cooling of Buildings

- Building insulation
- Thermal mass
- Renewably powered cooling systems
- Shading elements
- Reflective roofs and facades
- High performance glass
- Cross-ventilationn
- Evaporative cooling
- Earth cooling
- Solar chimney
- Labour management

11.5.2.9 Management of Water Stress

- Education relating to demand reduction measures
- Drip irrigation
- Rainwater storage and harvesting
- Waste water reuse technologies
- On-site storage
- Smart water metering
- Water efficient fittings
- Level-controlledvalves

11.5.2.10 Planning for Resilient Infrastructure

- Modernization of transmission system
- Electricity storage
- Heat resilient energy technology
- Energy security, additional power capacity/supply
- Electricity Mix Diversification
- Cold storage facilities

11.5.3 CCA Measures

This section covers various adaptation options that can be adopted for prevailing hazards in Andhra Pradesh, Telangana and Gujarat, which are floods, cyclones, droughts and heat waves. The coming section is adopted from ***Engineering Measures for Planning and Resilience Measures for Climate Change Adaptation in Industrial Parks*** (Kabisch, et al. 2016), which gives the detailed overview of adaptation measures according to disaster type.

11.5.3.1 Adaptation Measures for Floods

Suitable Site Selection for Industrial Parks and Their Overall Layout

This is an important parameter in designing resilient IPs. This directly reduces the exposure of hazards and facilitate cost reduction for making disaster resilient industrial areas. The parameter to be considered includes contours map, slope, drainage pattern, existing water bodies. These are the following design parameters, which can be considered here:

- Apply sequential approach, where medium to high risk areas can be used for recreational purposes, amenities, non-emergency roads and other similar uses; and low flood risk areas can be used for industries development.
- Land raising at sufficient height to prevent the impact of flood. If not possible adequate flood routes and additional storage areas need to be designed.
- *Flood Barrier Construction w.r.t to Landscape and Vegetation Cover.* Designing the site periphery in cut and mound manner helps in diverting the water flow away from IP site.
- Low cost earth-embankments can also be created to serve as a barrier against floodwater.
- Shallow ditches can also be constructed, facilitating water transportation through the pipes.
- Green spaces reduce run-off and promotes interception of floodwater, hence more amount of green spaces must be developed around roadside, recreational areas and open areas.

Constructed Barriers

Floodwalls can be constructed at the site periphery, providing a permanent wall against floodwater forces and inundation. 3–4 feet floodwalls are generally provided, which can raise up-to 20 m depending on the flood risk and vulnerability.

Road Maintenance

As discussed in the previous sections, transportation network suffers considerable disruption. Transportation network must be considered, and adequate protection measures should be applied while designing climate resilient Industrial Park to avoid delay in emergency services and essential supplies at the time of disastrous events. Flowing are the recommendation, which can be followed to keep the roads in operating conditions:

- Increase depressions/pits for preventing water accumulation towards industries
- Replace fixed pathways to removable ones, so reduce flood level
- Use permeable material for constructing roadways to facilitate interception of floodwater into the ground
- Fix frequency of roads inspection and prioritize repairing work wherever needed.

Permeable Surface for Water Infiltration

Water percolation into ground (or diverted towards storage tanks) can significantly reduce the intensity and impact of floods. Design specification are listed here:

- Use permeable material (of permeability of at least 0.3m level) around the building for water percolation

- The permeable bed should be lined with non-permeable material (e.g. clay) to avoid water entering into the foundation zone
- Permeated water should be diverted into adequately dimensioned storage tanks to prevent upward pressure on bottom of the building.

Electrical Facilities

Electrical facilities should be operational to avoid hindrances in rescue services during emergencies. Following are important design recommendations:

- Essential facilities including (fire stations, power plants, water treatment plant etc.) should be located at low risk areas. Also, power plants should be located at higher level
- Position external electrical transformer at least 2m higher than expected flood level
- Insulated conduits should be used for laying cables, should be located at height higher than expected flood level
- Position of electrical panels, water supply and gas lines should be correctly marked and known, for shutting them off whenever needed.

Drainage Facilities

Drainage system needs to be adequately designed to accommodate excess floodwater, sewer water etc. Here are the specifications:

- Drainage systems should be designed either to store the floodwater or should facilitate diversion of water towards the storage tank. These systems should be placed near infrastructure, buildings, roads and pavements
- Wider water and sewer system pipes can be installed to increase drainage capacity. Screens can be used to avoid solid waste/ debris coming into the drain
- Check valve can be used to control the flow.

Buildings

There are multiple options to make buildings adaptable to the flooding impacts, some of which are explained:

- Flood walls or embankments can be constructed to prevent storm water from entering inside
- Use basement as non-living or 'non-productive' space at the sites with high flood risk
- Use higher building for providing emergency shelters at upper floors

- Adopt pile foundation to resist all base flood condition to avoid water accumulation
- For single/double buildings raft foundation can be adopted
- Provide ground slabs at higher elevation, to avoid water entry inside. The slabs should be heavy enough to resist buoyancy forces due to flooding conditions
- RCC shear walls may be provided depending on the flood risk
- For external cladding engineering bricks, such as fireclay (along with cement or lime plaster paste as binding element) can be used because of their high-water penetration and drying ability
- For internal cladding, cement plaster cement: sand/ lime ratio of 1:6:1 could be used because of rapid drying ability.

Doors and Windows

Doors and windows are highly susceptible to floods. If broken by floods, it would result in flooding of water directly into the building. Here are the recommendations, which can be followed to make them more resistant to floods:

- Install overhangs around doors/ windows to prevent infiltration
- Use PVC sealant stoppers at doors bottom edge; silicone sealants may be used around doors and windows
- Timber doors should be avoided. Fiberglass plastic or other waterproof material can be used for doors and windows.

Evacuation

Evacuation requires emergency relocation of the people away from the threat area. To provide safe evacuation following processes can be implemented:

- Evacuation plans must be provided in every individual building
- Audible sirens/ boards should be installed. The sirens installed should be automatically triggered by flood water
- Provisions for inflatable boats for transportation use during flooding.

11.5.3.2 Adaptation Measures for Cyclones/Storms

Suitable Site Selection for Industrial Parks and Their Overall Layout

Here are some recommendations for selecting location for Industrial Parks:

- Areas near to water bodies, mud flats, salt flats should be avoided. If not possible, the site with the elevation higher than the possible, storm level should be chosen. If higher elevation areas are not available, the buildings should be constructed on stilts or on raised earthen mounds

- To decrease wind load, rough terrain, or flat terrain with obstructions can be used
- Ridges should not be chosen, because they experience accentuation of wind velocity. On the other hand, valleys experience lower velocity.

Vegetation Cover

Vegetation acts as natural barriers against cyclones. The inclusion of following recommendations while designing the Industrial Parks may be suitable:

- Loosely dense and low height vegetation should be in the front position seaward side. The medium dense and moderate height plants should be in the middle position and finally the densest and tallest trees should be planned after the medium height mangrove row
- Multiple rows of trees, in three to four layers planted upwind act as an effective shield against high winds. The influence of such a shield will be over a limited distance, equivalent to 8 – 10 times the height of the trees. Hence distance of tree from the buildings, must be kept 1.5 times the height of the tree to avoid building damage due to broken trees
- Trees with trunk lesser than 6 inches of diameter are recommended to be planted near site. As larger diameter trees can cause significant damage to the buildings.

Orientation and Shape

Shape of the building affects the wind pressure and the effective applied load to the surfaces. The orientation of building should be in direction opposite to the wind flow. The following recommendations may be considered while deciding the orientation and shape of the building:

- Most recommended layout is circular, hexagonal, or octagonal. In case of square shaped, few cantilevers can be included. Corners should be rounded for better aerodynamic properties
- The best option is to have length and width ratio as 3:1
- Non-load-bearing walls and door and window frames should be designed in line with the principles of rain-screen to minimize damages.

Buildings

- Taller buildings experiences higher wind forces and velocity. Therefore, for 2-3 story building optimum height of not more than 4m per story is preferable
- Walls should have vertical and horizontal reinforcing and filing to resist wind loads
- Type of foundation recommended includes: slab/ raft foundation, stepped foundation, short bored pile foundation, pad foundation
- Connections between roof and wall panels should be designed with adequate uplift load resistance to prevent the wall panels from collapsing.

Doors and Windows

- Openings in load bearing walls should not be within a distance of h/6 from inner corner for the purpose of providing lateral support to cross walls, where 'h' is the story height up to eave level
- Doors operable outward would be preferred as they would provide more resistance to the wind flow compared to the doors operable inward
- Leakage can occur; therefore, window design should be done considering water infiltration issues. Sealants could be used as the secondary line of defence against water infiltration.

Evacuation

- Providing at least two means of site egress would be necessary. If one route becomes blocked by trees or other debris, or by floodwaters, the other access route would still be available
- If multiple buildings are occupied during a storm, it is recommended that enclosed walkways be designed to connect the buildings
- Emergency and standby generators should be placed inside wind-borne-debris resistant buildings.

11.5.4 Adaptation Measures for Droughts and Heat Waves

11.5.4.1 Green Corridor, Vegetation and Artificial Shading

Green corridors help in mitigating heat island effect by improving urban ventilation, transporting colder and fresh air from outside into built up areas. Green ventilation corridors designed along major prevailing wind directions will allow effective air movements. Here are the recommendations to design green corridors:

- Green ventilation corridors should be continuous green spaces creating ecological connectivity and improving biodiversity and animal species dispersal
- The design should offer multifunctional spaces deriving the maximum benefits for the costs
- Local species of plants should be preferred
- If trees in public spaces are not applicable (e.g. due to ground/structural considerations), artificial shading can be used such as pavilions, tents or sun umbrellas.

11.5.4.2 Orientation

To assure fresh air supply within Industrial Parks, both street pattern and buildings should be oriented to prevailing winds. Likewise, the design of buildings should avoid blocking prevailing winds. Here are suggestions that can be followed:

- Main streets should be aligned in parallel or up to 30 degrees to wind direction to maximize the penetration of winds
- The alignment of the longer frontage of plots should be in parallel to the wind direction to increase the ventilation
- Roads should be wide enough to allow for good ventilation
- Site design should also allow for north south orientation of buildings for minimum heat gain.

11.5.4.3 Building to Building Shading

The height of buildings and the dimensions between buildings will have shading effect, which will help to reduce temperatures and increase the thermal comfort. Here is some technical detailing:

- The building height to street width ratio (H/W-Ratio) gives orientation. The higher the ratio, the more effective is the building layout to minimize heat gains
- The exact dimensioning of the urban geometry can however only be determined by considering the solar geometry in combination with street and buildings orientation.

11.5.4.4 Open Water and Water Features

Water reduces the local temperature through evaporation. Water features can reduce air temperature as well as create pleasant outdoor spaces.

11.5.4.5 Light Coloured and Reflective Paving Materials

Both color and type of material affect its temperature. Following are some of the recommendations:

- Porous asphalt for roads and light-coloured brick should be considered for pavements
- Coatings and grass paving are suitable pavement materials
- Areas that expect high traffic turning forces and heavy goods vehicle traffic are not suited for its application.

11.5.4.6 Compact Building

Compact buildings lead to a low S/V ratio. The lower the S/V ratio the higher the compactness of buildings and the lower the heat or cooling loss.

11.5.5 Cost-Benefit Analysis

After all adaptation measures are identified, next step is the selection of the best measure. The cost benefit analysis is the appropriate method to select the most optimized option, by quantifying all the option, which involves calculating, and comparing all the costs and benefits of an adaptation measure (Kabisch et al. 2016). The analysis is explained by taking example of retrofitting existing toilets with water efficient. Flush tank replacement needs initial investment. The replacement of flush tank also decreases water consumption and hence the water cost of the Industrial Park, as well as also makes it more resilient towards water scarcity. Let us take a hypothetical scenario. The present information about the building is explained in Box 11.1 and proposed water efficient scenario is explained in Box 11.2.

Box 11.1 Existing Flush Tank System

Office building area=1800 m^2

Total no. of employees = 300

Working days in a year= 239

Visitor's/day = 80

Total annual water consumption = 1116 m^3

Water consumption in toilets = 61% of 1116= 681 m^3

Water supply and wastewater tariff = 120 INR/m^3 = 81720 INR

Average flow of existing flush = 9 L per Flush

Total no. of toilets = 10

Box 11.2 Proposed Water Efficient Flush System

Average flow of water efficient flush = 4 L per Flush

Lifetime of new fitting = 5 years

Investment cost of replacing one flush tank = 1350 INR

Total cost of replacing 10 flush tanks= 13500 INR

Discount Rate = 9%

Percentage of saving per fitting = 55.6%

Average annual water savings = 378.3 m^3

Average annual savings = 45400 INR

Discounted benefits over life time* = 192483.28 INR, *calculated by adding the discounted value for each year (for 5 years with a discount rate of 9%)*

Benefit-Cost ratio = 14:3 *Discounted benefits over life time/Total cost of retrofitting*

Payback period = 3.57 months *Total cost of retrofitting/Average annual savings*12*

As clearly mentioned in the above two boxes retrofitting the existing flush system with the water efficient one would be effective measure to reduce water consumption and to reduce water tariff. The investment incurred will be paid back in 4 months after the replacement. The high benefit- cost ratio suggest the suitability if the proposed system.

11.5.6 Funding of Engineering Measures

The next step is the determination of funding sources once all costs are determined. Several sources are available to finance adaptation measures for industrial parks in India as mentioned in (Kabisch et al. 2016) which are:

- Funds, directly financing programs or measures
- Governmental Incentives or compensation
- Privately financed measures.

11.5.7 Raising Awareness on Adaptation

The success of Climate Resilient Industrial Parks requires appropriate implementation of CCA Measures. The important step is the raising hazard risk awareness among all the industrial stakeholders, to make them responsive of the potential impacts of hazards induced due to Climate Change. This can be done by:

- Internal training to understand Climate Change impacts
- Conducting awareness-raising campaigns.

This can be achieved through publishing all relevant reports, documents and case examples on web sites, conducting awareness campaigns, meetings, workshops and conferences and publicizing through media. For the training there is a need to design target group tailored modules which includes methods of assessment, methods of adaptation and financing tools and mechanisms for implementation and the required changes in management or the adaptation processes.

11.5.8 Monitoring and Evaluation

Monitoring and evaluation are the vital steps to check the efficiency and effectiveness of the Climate. There are multiple ways of doing this, which involves:

- Enforcement of law and order, so that the industry owners follow the CCA approaches
- Post-building inspection to check for decay and deterioration.
- Scheduling of repair work/ replacement work if needed
- Special inspection immediately during early warning signals

11.6 Role of NAPCC and SAPCC in Building Climate Resilient Industrial Park

National Action Plan on Climate Change (NAPCC) was released on 30th June 2008 and has eight missions on water, renewable energy, energy efficiency agriculture etc. The main agenda of this plan is to reduce country's stress from Climate Change without compromising on economic development through its eight National Mission and a series of adaptation measures like promoting use of green and renewable energy, developing smart cities, encouraging use of energy efficient devices, conserving biodiversity etc. Four new national mission are also under discussion for the NAPCC. Out of which waste-to energy would promote Industries to use their large amount of waste to generate electricity, which will eventually decrease the energy depend on fossil fuel. However, NAPCC has not supported much on CCA activities for Industrial Park.

The State government involvement is required to achieve the balance between strategies and actions as most of the events including drought, floods, deforestation, and water scarcity are experienced at state level. With this motive, State Action Plans for Climate Change (SAPCC) are developed to address existing risks in the states, vulnerabilities and impacts (EPTRI, State Action Plan on Climate Change for Andhra Pradesh 2012). SAPCC for Gujarat, Andhra Pradesh and Telangana is well established, with industries also identified as key sectors. Table 11.5 depicts the concerning areas, strategies taken by concerned SAPCC and highlight their strengths, weakness and recommendation for enhanced CCA for industries.

11.7 Conclusion

Adaptation to Climate Change for industrial areas is the need of the hour. Impacts on industrial areas are already noticeable including increasing vulnerability of the industries to the disasters, reduced work productivity, fight for natural resources etc. To develop Climate Resilient Industrial Parks, it is important to identify climatic hazards, the industries/industrial parks are exposed to, the probability or frequency of their occurrence and the industrial resilience to climatic hazards. A wide range of adaptation measures are available based on industrial type, climatic hazards, and location. Eco-industrial park concept looks very promising as it results in

Table 11.5 Key issues and strategies for CCA (EPTRI, State Action Plan on Climate Change for Andhra Pradesh 2012; EPTRI, State Action Plan on Climate Change for Telangana state 2015; TERI and GIZ 2014) and own interpretation

Key issues	Strategies	Strengths	Weakness	Recommendation
Andhra Pradesh and Telangana				
Increasing fossil fuel consumption and GHG emissions from the industries	• Clean energy use • Enhance the share of public transport in the total transportation mix • Enhance the share of low emission/fuel-efficient vehicles and vehicles that run on alternate fuels	Implemented strategies will result in: • Less Carbon emissions technologies • Less waste generation • Better infrastructure • Improvement in air quality	• Focused mainly on Climate Change Mitigation. Climate Change Adaptation is taken for existing industries only. For newly planned industries adaptive strategies are not clearly explained • Building recommendation for industries to cope with adverse effect of Climate Change can be more elaborated	• Design code and recommendation for new industries should be included • Capacity buildings programs for Industrial Adaptation to CC should be included. Mentioning in brief what industrial process best fits, designing green buildings, utilizing resources efficiently etc. • Concept of eco-industrial park should be included
Increase in industries count and waste generation	Waste minimization			
Surface water depletion	Drainage improvement			
Air quality degradation due to mining	Policy enforcement to minimize environmental damage			
Industrial risks to CC	Adaptive capacity enhancement			

(continued)

Table 11.5 (continued)

Key issues	Strategies	Strengths	Weakness	Recommendation
Gujarat				
High energy consumption	• Clean energy use • Incentivize to energy efficient industries • Installation of solar power plants	Implemented strategies will result in: • Less Carbon emissions technologies • Less waste generation • Better infrastructure • Improvement in air quality • Green Employment generation	• Not focused on developing building resilience to Climate Change • Building recommendation for industries to cope with adverse effect of Climate Change can be more elaborated	• Design code and recommendation for new industries should be included • Capacity buildings programs for Industrial Adaptation to CC should be included • Concept of eco-industrial park should be included • Guidelines to retrofit existing buildings adapting current/future Climate Change scenarios
Increase in industries count and waste generation	• Waste recycling and reuse			
Increase in industries, causing high vehicular load, depletion in air quality, rise in temperature	Creation of alternate routes to reduce vehicular load between 2 points			
Industrial pollution is putting marine ecosystem at risk	Mangrove plantation in coastal areas			
Water depletion	Groundwater recharge through rainwater harvesting			
GHG emissions from agriculture, agricultural productivity losses	Sustainable agriculture, installation of Automatic Weather stations (AWS), crop insurance			

minimization of energy and resources. CCA strategies for eco-industrial parks should be consider. Designing building based on passive architecture supports less consumption of energy and resources. Industrial areas should follow these principals for managing risks associated with changing climate. Recommendations/guidelines for developing climate resilient industrial parks should be a part of each SAPCCs and need to be enforced for both newly planned and existing industries.

References

Anand N (2018) The mounting economic toll of Kerala's once-in-a-century floods. Quartz India. https://qz.com/india/1368531/kerala-floods-automobile-it-rubber-sectors-worst-hit/. Accessed 27 Aug 2018

EPA (2017) Climate change impacts on transportation. Climate Change Impacts. https://19january2017snapshot.epa.gov/climate-impacts/climate-impacts-transportation_.html. Accessed 19 Jan 2017

EPTRI (2012) State action plan on climate change for Andhra Pradesh

EPTRI (2015) State action plan on climate change for Telangana state. MoEF&CC. http://www.moef.gov.in/sites/default/files/Telangana.pdf

ESRU (2010) Passive design. http://www.esru.strath.ac.uk/EandE/Web_sites/09-10/Hybrid_systems/passivedesign.pdf

Finley T, Schuchard R (2011) Adapting to climate change: a guide for the energy and utility industry. BSR. https://www.bsr.org/reports/BSR_Climate_Adaptation_Issue_Brief_Energy_Utilities.pdf

GLISA (n.d.) Climate change impacts on natural resources. Huron River Watershed Council

Government of Telangana (2018) Heatwave action plan—2018 revenue (disaster management). Department Government of Telangana

Jain S (2015) National Action Plan on Climate Change (NAPCC). Arthapedia. http://arthapedia.in/index.php?title=National_Action_Plan_on_Climate_Change_(NAPCC). Accessed 4 Aug 2015

Joshi R, Raval H, Pathak M, Prajapati S, Patel A, Singh V, Kalubarme MH (2015) Urban heat island characterization and isotherm mapping using geo-informatics technology in Ahmedabad City, Gujarat State, India. Int J Geosci 6:274–285

Kabisch S, Bollwein T, Bank P, Brulez D, Varaprasad S, Mahadev R, Ganta R (2015) Climate change adaptation for sustainable industrial development. INTEGRATION Environment & Energy Climate Change Adaptation Project (CCA), Hyderabad

Kabisch S, Seelig S, Ulguner O, Banerjee U, Banerjee D, Bank P et al (2016) Manual 2: engineering measures for planning and resilience measures for climate change adaptation in industrial parks. INTEGRATION Environment & Energy, Telangana

Kjellstrom T, Otto M, Lemke B, Hyat O, Briggs D, Freyberg C (2016) Climate change and labour: impacts of heat in the workplace. United Nation Development Programme

Long T (2014) Climate change and its effects on natural resources. Michigan State University Extension

MoEF (2010) Climate change & India: a 4x4 assessment. Indian Network for Climate Change Assessment, INCCA

MoEFCC, EPTRI (2015) State of environment report, Telangana—2015, Chapter 9: INDUSTRIAL DEVELOPMENT. Govt. of Telangana. http://www.eptri.com/wp-content/uploads/2018/04/SoERTS-Chapter-9-Industrial-Development.pdf

Narasimhan B, Bhallamudi SM, Mondal A, Ghosh S (2016) Chennai floods 2015: a rapid assessment. Interdisciplinary Centre for Water Research, Bangalore

NDMA (n.d.) NDMA. NDMA, Andhra Pradesh. https://ndma.gov.in/en/andhra-pradesh-sdma-office

Raghu NB (2018) Climate change adaptation for industrial areas. GIZ. http://cca.urban-industrial.in/live/hrdpmp/hrdpmaster/igep/content/e66188/e66197/e66326/e66383/ClimateChangeAdaptationforIndustrialAreas_final.pdf

Seelig S, Ülgüner Ö, Happold B, Banerjee U, Banerjee D, Bemmerlein-Lux F et al (2018) Climate risk assessment for the adaptation and increasing resilience of industrial parks to the impacts of climate change. GIZ

Shodhganga (2010) Causes and consequences of urban heat island and remedial measures

Snow M, Prasad D (2011) Climate change adaptation for building designers: an introduction. Environment Design Guide. http://environmentdesignguide.com.au/media/misc%20notes/EDG_66_MSa.pdf

SYKE (n.d.) Impacts of climate change on industrial production. Climate Guide. https://ilmasto-opas.fi/en/ilmastonmuutos/vaikutukset/-/artikkeli/79840ec2-4723-442b-a6b3-5a2ebc46f6da/teollisuus.html. Accessed 5 Sept 2018

TERI, GIZ (2014) State action on climate change. Government of Gujarat

UNIDO (2015) Promoting climate resilient industry

van Dam D, van Huet M (2015) GSDR 2015 brief passive housing. Wageningen University and Research Centre

VCCCAR (2018) Climate change adaptation definitions. VCCAR. http://www.vcccar.org.au/climate-change-adaptation-definitions. Accessed 19 Oct 2018

Welle D (2011) Eco parks offer firms a low-carbon road to growth. Deutsche Welle. https://www.dw.com/en/eco-parks-offer-firms-a-low-carbon-road-to-growth/a-6532982

Gupta Anil K, Professor of Policy & Strategies, Integrated Resilience. Head of the ECDRM Division, International Cooperation, Programme Director - Centre for Excellence on Climate Resilience, Nodal Officer for Central Ministries at National Institute of Disaster Management, (Govt. of India), New Delhi. He is also a member of the expert Team for WMO Climate Statement and Core Group Member of IUCN-Council of Ecosystem Management.

Raghu Babu Nukala is an engineer-planner by qualification with over 29 years of experience. He has extensive experience in the areas of International [Bilateral Development] Cooperation, Sustainable Development [eco industrial parks, green industrial parks, planning new industrial parks and investment zones, retrofitting old industrial parks, Eco-cities] etc. He is currently working as Project Director at the Deutsche Gesellschaft für Internationale Zusammenarbeit (GIZ) GmbH (public enterprise of the German Government).

Florian Bemmerlein-Lux is a graduated ecologist (Germany and USA) and since 1985 partner of ifanos concept & planning (Institute for applied ecological studies, Germany). He worked in the TU-Berlin, Institute for Ecology for the research programme "Digital Geo-Scientific Maps" and applied these skills in nature conservation, natural resource management as well as environmental impact assessments spatial planning and Climate Change Adaptation/disaster management.

Shruti Bansal is working as a consultant at Environment Resource Foundation. She has also worked as a Research Associate at National Institute of Disaster Management (NIDM). She is a graduate in Civil Engineering from Aligarh Muslim University, and a post graduate in Environmental Engineering & Management from IIT Kanpur with over 5 years of work experience. She has contributed extensively in the areas of Climate Change, Disaster Management, Carbon Footprints, and Environmental Engineering etc. She is also experienced in geospatial technologies including GIS and Remote Sensing.

Chapter 12
Managing Disaster Waste in the Aftermath of Emergencies: Addressing Future Climate Risk-Integrating Adaptation

Bindu Aggarwal and Anil K. Gupta

Abstract Sendai Framework identifies Climate Change as one of the drivers of disasters. Disasters are responsible for causing substantial damage globally. This, in turn, results in a severe environmental and economic burden on living conditions as well as management of additional waste created by disasters. Disaster Waste Management has been identified as a critical issue while responding to disasters. It burdens existing waste management systems and severely impacts post-disaster response and recovery. If not managed properly Disaster Waste entails significant environmental and public health impacts leading to environmental emergencies. The paper presents that strategies, issues and challenges related to Disaster Waste Management vary depending upon type of disaster, lntensity and location. Paper also reflects Environmental Emergency Response System/tools available internationally while same kind of system and tools are currently non- existent in India.

Keywords Climate risk · Climate change adaptation · Disasters · Disaster waste (DW) · Environmental emergency (EE) · EE Response systems/tools

12.1 Introduction

Natural Disasters now mostly influenced by climate variabilities manifest as floods, hurricanes, landslides, droughts, forest fires and earthquakes and anthropogenic disasters may be industrial accidents involving chemical release and oil spills. In the present scenario we do not have adequate statistics on disaster related waste generation and management in India whereas the intensity and frequency of hydro meteorological disasters are known to cause $ 80 billion loss in India. As per 'Report

B. Aggarwal (✉)
Indian Red Cross Society, New Delhi, India
e-mail: binduaggarwal123@rediffmail.com

A. K. Gupta
ECDRM Division, International Cooperation, CAP-RES DST Project, National Institute of Disaster Management (Ministry of Home Affairs), New Delhi, India

A. K. Gupta et al. (eds.), *Disaster Risk and Management Under Climate Change*, Disaster Resilience and Green Growth,
https://doi.org/10.1007/978-981-99-4105-6_12

on Economic Losses, Poverty and Disasters for 1998–2017', India ranks among the top five countries that have suffered losses due to disasters (UNISDR 2018). These disasters are directly related to impact of climate change. So this chapter has been developed based on global knowledge base incorporating lessons and case studies from various places. The contextual inferences have been drawn for their use in Indian Scenario.

A natural disaster such as an earthquake can have a huge humanitarian impact – deaths and injuries, houses ruined, crops and livestock destroyed. In addition to humanitarian impacts disasters entail serious environmental ramifications. While disasters like floods and landslides imperil the life of survivors and rescuers uniformly, debris and waste generated during disaster threaten the environment and human health. Access routes may be blocked, delaying the arrival of food and supplies, and slowing the economic recovery of affected areas in the longer term (Joint UNEP/OCHA Environment Unit 2009).

Disaster waste is far more complicated than ordinary waste – toxins, inert and infectious material which are difficult to segregate and hence disposal is a time-consuming affair (Kobayashi 1995). Due to the toxicity of disaster waste, it is a critical health and environment hazard. Thus, ineffective management of disaster waste lay the foundations for serious environmental and economic problems in the country (Karunasena et al. 2009). The inert waste generated from building and infrastructure collapse can lead to the spread of vector diseases by blocking water bodies and drainage system. "Inerts" means wastes which are not bio-degradable, recyclable or combustible street sweeping or dust and silt removed from the surface drains Such conditions promote water logging and an ideal situation for the rapid growth of vector as well as foul odour (Perera 2003; Kobayashi 1995). Under this situation, construction waste carries the significance, capable of being salvaged and reused in the reconstruction process thereby reducing waste load and stabilizing the economy.

12.2 Disaster Waste and Climate Change

Waste management is identified as the third largest emitter of anthropogenic methane in the world. It represents 3% of overall green-house gas emitted across world. In India, SWM is the second largest anthropogenic methane emitter and the largest green-house gas emitter among activities which do not add to the economic growth of the country.

As per second biennial updated report to UNFCC, 3% of total GHG emissions in India were contributed by waste sector. Presently, an insignificant fraction of methane emitted from solid waste dumpsites is captured in India, rest of it is left into the atmosphere, not captured and unused. Control of GHGs from SWM is considered an achievable goal in the short term, among many other efforts to avert climate change.

Methane is produced as final gaseous product from anaerobic digestion of organic wastes under anoxic conditions inside landfills. Due to anaerobic reactions, landfills emit methane throughout their life time and also for several years after closure. Methane has high energy content and is a potential renewable energy source if captured economically.

Methane has 21 times more global warming potential as compared to CO2, methane emissions from organic wastes mean 'net positive emissions'. One ton of methane equals 21 tons of CO2 equivalents over a long period of time. In short time periods, CH4 is much more potent than CO2. During the first year of release, CH4 is 71 times more potent than CO2. Therefore, net positive emissions of GHGs in the form of methane warm the planet faster and contribute to global warming and in turn climate change (Annepu 2012).

12.2.1 What Is Disaster Waste (DW)?

A combination of all kinds of solid and liquid waste resulting from disaster represents disaster waste. Disaster waste ranges from collapsed building and infrastructure rubble like concrete, steel, wires, wood to household furniture and articles like electronic gadgets, wooden furniture, almirah, kitchen material, gas cylinders, cleaning materials, vehicles, agricultural tools and chemicals like fertilizers, pesticides, hospital wastes like surgical waste, bandage gloves, coats and marine vessels like wrecked ships.

Disaster Waste is the end product of not only disaster itself but relief and rehabilitation activities also generates waste significantly in large volumes in the form of packaging material, empty bottles, used diapers, used sanitary napkins and excreta. Public health risks can arise from: direct contact with waste accumulated in the streets, hazardous wastes such as asbestos, pesticides, oils and solvents, and indirectly from vectors such as flies and rodents and post disaster collapse of unstable structures.

12.3 Vulnerability and Impacts

It is a known fact that mostly communities do not have waste management plans in place even for normal conditions and in such cases when disaster strikes and add disaster waste communities are overburdened with new load and face challenge of waste disposal of an extraordinary level.

The inability to consider environmental concerns while planning humanitarian response leads to deterioration of the environment in various ways during disaster at different countries. Poor management of disaster waste post-earthquake in Haiti immensely contaminated the environment and caused Cholera outbreak at a big scale. These examples provide evidence that the high environmental footprint of humanitarian action exacerbates impacts on ecosystems (Environment and Humanitarian Action Network).

Table 12.1 Typical disaster waste issues and their impacts

Issue	Typical human and environmental impacts
Uncollected building rubble from Damaged buildings	Impeded access and constrained rehabilitation & reconstruction activities. Waste tends to attract more waste since the site is already considered a dumping site.
Scattered waste dumping	If the waste is dumped close to the human settlement it poses potential human health and injury risks. Loss of valuable land due to waste encroachment. Infiltration of waste in drinking water supplies and fresh water that ultimately creates a harmful living condition for aquatic animals. High chances of outbreak of disease vectors (flies, mosquitoes, rats, etc.). Risk of catching fires. Risk of cuts and injury from sharp materials i.e. syringes, broken glasses etc. to waste handlers.
Breakdown of municipal solid waste services	Unavailability of collection service and uncontrolled dumping.
Open discard/ dumping of health care waste	Serious health risks to local populations. High risk of spread of disease and infection.
Exposure of Asbestos sheet in collapsed structures or re-use of asbestos for reconstruction	Respiration related health risks.

(Source: Joint UNEP/OCHA Environment Unit 2013)

12.3.1 *Impacts of Disaster Debris*

As mentioned earlier disaster waste contains sufficient quantities of toxic substances in the form of chemicals and heavy metals which can impact the environment critically by way of contaminating waterways, agricultural lands and subsequently communities. Physical obstruction of waterways can also occur (Tables 12.1 and 12.2).

12.3.2 *Common Risks by Disaster Waste Hazard Type*

12.3.2.1 Chemical Risks

The chemical risks associated with disaster waste are as follows:

Direct contact of skin with contaminants such as pesticides, oils and acids

Inhalation of:

Table 12.2 Hazard types and their waste characteristics

Earthquakes	Structures collapse 'in-situ', i.e. floor slabs collapse on top of each other, trapping waste within damaged buildings and structures. This can lead to challenges in sorting out hazardous waste (e.g. asbestos) from non-hazardous (e.g. general building rubble). Collapsed buildings impede commute and commuters of streets and roads as debris extends over roads making access difficult for the search and rescue and relief operations. Waste accumulation is high in comparison with all the disasters.
Flooding	Initial damage depends on the structural integrity of infrastructure, while building contents are normally damaged extensively. Based on severity and longevity of flood, moulding of building material and edible materials may exist and rotting of timber may take place. Waste is often mixed with unhygienic and hazardous materials such as household cleaning products, electronic goods, used healthcare products etc. Flooding is likely to bring mud, clay and gravel to affected areas.
Tsunami	Strong tsunamis cause grievous damage to infrastructure, spreading debris over large areas. Debris is often mixed up with soils, trees, bushes and other loose objects such as vehicles.
Volcanoes	Dispersal of ash and pumice stone along with a stream of molten rock (lava) is the major aftermath scenes of volcanic eruptions. Fine particles of ash cause strain on mechanical and hydraulic equipment as well as electronic devices.
Hurricanes, typhoons, cyclones	Strong winds can tear the roof off buildings, collapse poorly constructed houses and huts. Waste and debris are spread over the open area and small items and dust are carried by the wind. This may cause serious problems where asbestos is present. Ships and boats are often thrown back at the shore and destroyed. Electrical and telephone grids, transformers and PCBs may be destroyed.
Conflict (Short Term)	Intense, short-term conflicts involving arm and lethal weapons such as rockets, missiles and bombs which, combined with land combat, result in huge loss of economy (destruction of infrastructure, industrial and residential areas) and human lives. Bombed and burnt buildings result into massive debris of concrete, bricks etc. This condition calls for the use of heavy machinery i.e. excavators and bulldozers. Waste collection vehicles may be damaged. The possible presence of Unexploded ordnance (UXO) including undetonated landmines in the waste.

(Source: Joint UNEP/OCHA Environment Unit 2013)

- Hazardous chemicals or products like pesticides
- Chemicals like dioxins/furans, polyaromatic hydrocarbons released from partial burning of waste and concentrate of heavy metals from unrestricted combustion of waste.
- Suspended Particulate matter comprised of particles of dust, including small particulate matter (PM10)and asbestos fibers

- Intake of contaminated surface or groundwater. Leachate from waste is the main culprit contaminant. Contaminated water contains high levels of heavy metals, organics, polychlorinated biphenyls and volatile organic compounds.
- Annoyance from bad smell emanating from chemicals residing in waste or from the decomposition of waste.

12.3.2.2 Biological Risks

The following are examples of biological risks:

Contaminates food webs and disrupts detritus food chain, Generates fungal spores which act as bio pollutants causing asthama and COPD, Bio magnification of organic heavy metals in bio- geosphere.

- Direct contact of skin with waste/ intake of excreta and body fluids through contaminated water and exposure of waste handlers to infectious hospital/ healthcare waste.
- Breeding of disease causing vector animals near waste dump : Faeces of rats are a potential source of hanta virus, diseases like leptospirosis, plague and scrub typhus. Cases of leptospirosis were recently visible in Kerala after floods in August 2018.
- Appearance of Malaria and dengue-causing mosquitoes.
- Presence of flies in general depicts unhealthy conditions and may lead to bacterial infection.
- Annoyance from various waste feeding species – Birds, insects and rodents.

12.3.2.3 Physical Risks

- The collapse of buildings and other constructions as well as waste piles, such as large piles
- of rubble that have been pushed to the side of a road
- Injuries and scratches from sharp objects in waste like metals, broken glasses, needles and surgical knives.
- Sudden fires at stocks in waste
- Accidents of vehicles transporting huge quantities of hazardous waste from one place to another. Accidents can also occur during loading and unloading of waste.
- Annoyance from windblown litter and columns of smoke rising from waste burning.
- Disrupts structure and function of Ecosystem may terrestrial and aquatic ecosystems.

12.3.2.4 Local Environmental Risks

The immediate environment of disaster waste stocks can face adverse impacts due to following:

- Longer stay of hazardous waste on land contaminates the soil below and land becomes unfit for agricultural practices as well as for habitation of the population.
- When waste lies on land for long the leachates find their way into ground water sources leading to contamination of ground water causing Soil, surface and ground water contamination
- Disaster waste while decomposing on open landfill release obnoxious gases making the air unhealthy for humans and animals.
- Outbreak of insects and rodents who feed on waste.
- Litter gets blown with air, mixes with air and travels to other places.

12.4 Mitigation and Management

12.4.1 Post Disaster Waste Management

DW if managed judiciously presents the option to extract and salvage valuable material from waste like steel and concrete can be reused for reconstruction while wood chips and organic material can be subjected to composting for preparing nutrient rich manure. This activity reduces transportation cost from disaster hit area to landfill, reduces the requirement of land for landfill, reduces waste load, reduces the burden on existing resources and income can be earned by persons involved in extracting waste for reuse.

Currently local authorities/communities are not having means/ trained to manage disaster waste and when disaster strikes there is inaction from the community because they do not know how to use their available resources and capacities to deal with waste. They are unaware of various steps of disaster waste management namely segregation, recycling, composting, combustion and final disposal and unable to opt anyone of these. Previous experience has taught that it is inadequate to bank upon only one of the available disaster waste management options while handling immense quantities of disaster waste. Communities, if trained can help in identification of initial staging areas for waste and for later storage of same where it can be segregated and processed before it is subject to scientific handling and disposal as per its characteristics (EPA 2008).

Post 2001 earthquake in Bhuj which occurred in the month of January it was found that disaster debris was still lying in some areas till October 2001 though the task of debris removal was assigned to multiple agencies. About 109 lakh MT of disaster waste was removed by various agencies (Table 12.3).

There was a loss of INR 74.4 crores to Kerala State post August 2018 floods, 45% of which was attributed to cleaning at household level, 31% spent on disposing dead

Table 12.3 Removal of debris by different agencies in Bhuj

Agency involved	Disaster waste removed in lakh MT
Roads & Buildings Deptt	20.48
Irrigation Deptt	18.72
Kandla Port Trust	4.06
Government buildings	4.66
Drought relief work	20.48
Village committee/village *panchayat*	40.41

remains of animals and 24% on cleaning of public areas by Urban and Rural administrative bodies.

Hurricane Katrina is known to generate as high as 90.3 million $m^3$of disaster waste in America. US Army Corps of Engineers (USACE) was assigned the job of waste management and they exhibited swift results and removed more than 15.1 million m^3 of disaster waste from 16 southern counties. (Brandon et al. 2011).

Post Katrina Dell Computer and Best Buy (American multinational electronics retailer) partnered with EPA, the states, and local governments to increase collection and safe recycling of computers and related electronic equipment, including monitors, printers, computer peripherals, multifunction machines, fax machines, scanners, DVD players, Notepads, Laptops, keyboards, Mouse, radios, mobiles, disks and speakers were wrecked by impacts of Hurricane Katrina. This effort resulted in the recycling of more than 220,000 pounds of electronics debris (EPA 2008).

San Diego County experienced two wildfires in October 2003 which caused destruction of 6000 structures, destroyed 4000 vehicles and burned 400,000 acres of land. It generated as high as 128,000 tons of debris. Out of this 128,000 tons 60% of waste in the form of concrete, metal and vegetative debris amounting to 74,000 tons was recycled. Recycling lowered the demand for landfill site and consequently 185,000 cubic yards of landfill space was preserved (County of San Diego Debris Removal and Recycling Programs for the 2003 Cedar and Paradise Fires Final Report 2005).

Another situation is from Escambia County, Florida which was hit by Hurricane Ivan on September 15, 2004. Within 12 hours ten million cubic yards of disaster waste was generated and required immediate handling. Escambia County successfully diverted more than half of the debris from disposal in landfills and recovered more than 90% of the displaced beach sand. Escambia county opted to dispose of vegetative waste in a manner which generated revenue through export to Italy, helped in land reclamation, used as a raw material in paper mill and generated energy through incineration and preserved 6.5 million cubic yards of landfill space. Almost 60% of vegetative waste was exported to Italy to be used as biomass for energy generation, 15% of it was sold to paper mills; 15% used as landfill cover; and 10% incinerated on site in air curtain incinerators (EPA 2008).

Post earthquake in January 1994, the city of Los Angeles depended on recycling for managing disaster debris. City staff recognized the fact that recycling would reduce the requirement of the landfill area and stabilize the economy through recycling. Within six months 56% of the waste generated by an earthquake which amounted to 1.5 million tons was recycled (EPA 2008).

12.5 Case Study

12.5.1 26 December 2004: Indian Ocean Tsunami

Registering 9.0 on the Richter scale, the earthquake that struck just off the coast of northern Sumatra on Sunday 26 December was one of the strongest ever recorded. The tsunami hit Banda Aceh on the north Sumatra coast of Indonesia, the coast of Thailand, India and Sri Lanka. The Maldive Islands felt the devastating impact 1.5 hours later, with all 199 inhabited islands inundated.

In all three countries, disaster waste management emerged as a major concern. The tsunami itself created a wave of debris, mixing together waste from destroyed buildings, domestic waste dumping sites, and sometimes more-hazardous materials. In Sri Lanka, UNEP made the simple calculation that almost 100,000 homes, with an average weight of 3,000 kg each, were destroyed. Combined with other debris from vehicles, boats, damaged houses and shops, they estimated a total exceeding 500 million kilograms of rubble and waste material that needed to be cleared.

In Banda Aceh, Indonesia, **domestic waste and debris** (including refrigerators, cars, furniture and plastics) was mixed with oil, chemicals, contaminated water and sewage. In the Maldives, the tsunami exacerbated already poor waste management practices. As no facilities existed for disinfecting medical waste, it was mixed with household waste and dumped. The tsunami effectively distributed this mixed waste over the islands, combining it with demolition debris that included asbestos cement roofing sheets.

In context of India Disaster waste management data regarding 2004 Tsunami is lacking. Had there been some system in place some figures could have been established. Now the Union Ministry of Environment, Forests and Climate Change (MOEFCC, 2016) recently notified the new Solid Waste Management Rules (SWM), 2016. These have replaced the Municipal Solid Wastes (Management and Handling) Rules, 2000, which have been in place for the past 16 years.

These rules are the sixth category of waste management rules brought out by the ministry, as it has earlier notified plastic waste management, e-waste management, biomedical waste management, hazardous waste management and construction and demolition waste management rules.

In Sri Lanka, at Galle District, the UNDAC team environmental experts introduced a model for waste removal that engaged thousands of displaced people in a pay-for-work scheme to clear land safely and systematically. Based on this experience, a similar waste management program was implemented in Banda Aceh, Indonesia, in cooperation with the United Nations Development Program. The Joint Environment Unit's initial reports into the situation in Indonesia and Sri

Lanka highlighted the need for donors to provide experts in waste management as well as such heavy specialized equipment as aggregate crushers, wood chippers and compactors, to support the waste management and recycling processes. Guidelines were also needed for the large number of organizations working in the field (JEU 2009).

12.5.2 Solid Waste Management in the Affected Area of Thailand Post Tsunami 2004

The methodology to discard waste in Tsunami hit area of Thailand led to categorizing the solid waste under three heads in Thailand. They were MSW generated by the people living in the area, infectious waste from dead human/animal bodies and medical activities and the debris from demolished construction works. According to a field survey, the amount of solid wastes being collected after the tsunami disaster was more than 76,250 tons on-land area and 252 tons from the sea.

12.6 Management of Municipal Solid Wastes (MSW)

Presence of pre installed waste disposal facilities like incinerators and identification of landfill areas helps in early waste recovery and treatment post disaster. Local authorities managed demolition waste in different ways depending upon their capacities and availability of resources. Availability of private companies in Pathong municipality in Phuket province and PP island in Krabi province enabled authorities in these areas to hire private companies for transportation and disposal of waste in private lands. In some other area, the recyclable materials were separated and reclaimed by scavengers before the disposal of the remaining wastes by responsible local authorities.

12.6.1 Infectious Wastes Management

This waste was generated during care of patients, handling of dead bodies, disposal of used medical equipment like syringes, bandages, blood bags, sheets, and clothes of patients, gloves of medical and paramedical staff. Almost 104 tons of such waste was subjected to incineration. Phuket had its own incineration plant which used to treat 10–11 tons of infectious waste per month before disaster, however, quantities treated rose to 80.3 tons post Tsunami in the month of January 2005.

Ta Kua pa municipality too had its own incinerator though of smaller capacity of 500 kg/day. This was also used to incinerate infectious waste after disaster amounting to 2.3 to 4.6 tons per month. Around 5000 baht/ton of cost was involved in getting rid of infectious waste through incinerator.

12.6.2 Recovery of Ship Vessels and Debris from the Ocean

Massive quantities of waste from construction debris and ship wreckage had entered the ocean waters after Tsunami disaster. Almost 102 tons of such waste was salvaged from sea shores and coastline by hand picking and collection.

12.6.3 Recycling Activities

Informal sector represented by waste pickers and scavengers play a significant role in removal of waste and recycling practice in Thailand. After the disaster event, some re-sellable components were separated from the wastes for reuse by some groups of people.

12.6.4 Cost Involved in Solid Waste Management

In Thailand the estimated cost of incinerating infectious wastes was about 5,000 baht/ton, totaling a loss of Bath 0.4 million (1 USD approx. 38 Baht) and MSW incineration was about 200 baht/ton. All of these marine recovery operations required Thai Baht 9.4 million. However, the largest financial loss of Baht 110 million incurred for handling and disposing building debris. The direct costs then work out to be a total of Baht 120,313,123 (Basnayake et al. 2006).

12.6.5 Post Tsunami 2011 Disaster Waste Management at Japan

On 11th March 2011 a massive earthquake occurred off the pacific coast of Japan. A snapshot of damages occurred is reflected in Box 12.1.

Box 12.1 Japan Earthquake: Vital Statistics

Epicentre	38∘19' 19.2" N, 142∘22'8.4"E
Earthquake magnitude	Mw 9.0*
Peak Acceleration	3g
Aftershocks	1235
Casualties	15,854 deaths**, 3155 missing 26,992 injures
Building damage	129,225(fully collapsed)** 254,204 (half collapsed) 69,177(partially damaged)
Economic damage estimate	USD 210 billion (¥16,800)***

Sources: * US Geological Survey ** National Police Agency, Japan *** The Economist

Initial findings indicated that huge amount of Disaster waste was accumulated at temporary storage sites and situation demanded large scale incineration and recycling before disposal of these wastes. Volumes of Disaster Waste generated during this event is indicated in Fig. 12.1.

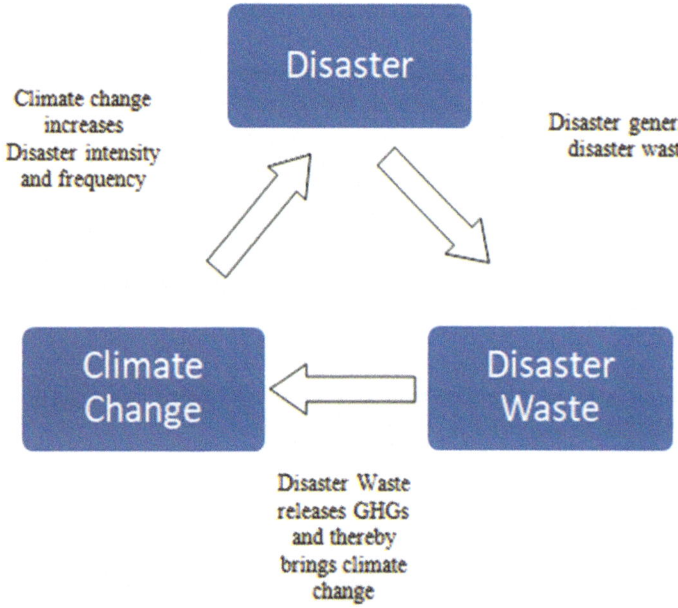

Fig. 12.1 Relationship between disaster waste and climate change

Prefecture	Local Government	Amount of waste (tons)	Prefecture	Local Government	Amount of waste (tons)	Prefecture	Local Government	Amount of waste (tons)
Iwate	Hirono	15,000	Miyagi	Sendai	1,352,000	Fukushima	Iwaki	880,000
	Kuji	96,000		Ishinomaki	6,163,000		Soma	217,000
	Noda	140,000		Shiogama	251,000		Minami Soma	640,000
	Fudai	19,000		Kesen-Numa	1,367,000		Shinchi	167,000
	Tanohata	86,000		Natori	636,000		Hirono	25,000
	Iwaizumi	42,000		Tagajyo	550,000		Naraha	58,000
	Miyako	751,000		Iwanuma	520,000		Tomioka	49,000
	Yamada	399,000		Higashi-Matsushima	1,657,000		Ohkuma	37,000
	Ohtsuchi	709,000		Watari	1,267,000		Futaba	60,000
	Kamaishi	702,000		Yamamoto	533,000		Namie	147,000
	Ofunato	752,000		Matsushima	43,000			
	Rikuzen	1,016,000		Shichigahama	333,000			
				Rifu	15,000			
				Onagawa	444,000			
				Minami Sanriku	56,000			
	Total	**4,755,000**		**Total**	**15,691,000**		**Total**	**2,280,000**

Source: Sendai City Waste Management Guidelines for Great East Japan Earthquake, Environmental Bureau, City of Sendai, January 2012

Fig. 12.2 Waste volumes in selected impacted cities

Immediately after Great East Japan Earthquake and Tsunami a master plan was created for Disaster Waste Management. It urged 3 prefectures Iwate, Miyagi and Fukushima to draw their own disaster waste management plans in consultation with various stakeholders namely – prefectural government, municipal government, Central government and representation from various related industries.

Realizing the seriousness of disaster waste management issue a task force to manage disaster waste was constituted immediately after Tsunami by Japan's Ministry of Environment. The disaster waste guidelines from the ministry, together with the technical experts deployed in the field, essentially formed the foundation for the post-disaster debris management operation and recommended approaches for managing disaster waste as depicted in Table 15.1. Procedure followed for Disaster Waste Management in Japan are shown in Fig. 12.2.

Key considerations in Disaster waste include – **health and safety** of staff involved in debris management, **management of stakeholders** like affected Community, informal rag pickers, private service providers, NGOs, Hospital, local municipalities, practitioners, administrators and donors as well as **communication and coordination** among various agencies (Table 12.4).

Table 12.4 Approaches recommended for Disaster Waste Management, Government of Japan UNEP (2012)

S. N.	Debris category	Suggested approach
1	Combustible Waste	After shredding use for cement calcination process and power generation wherever possible
2	Waste wood	Expose the wood to rain to wash out the salt to meet user requirements. Mainly use for making multipurpose wooden boards and as fuel for boilers and power generation Copper chrome arsenate (CCA) treated wood should be incinerated at waste treatment facilities
3	Non- combustible waste	Separate from combustible waste and dispose of at landfill sites
4	Scrap metal	Should be recycled after separating ferrous and non- ferrous metals wherever possible
5	Waste Concrete	Should be separated into asphalts, concrete, stone and other materials Should preferably be used as materials for reconstruction in the impacted areas
6	Home appliances and automobiles	Items should be separated to the extent possible (televisions, air-conditioners, washing machines, dryer and refrigerators) Should be managed under relevant Designated Home Appliance Recycling Act 1998 Automobiles should be delivered to collection companies for recycling pursuant to the End - of- Life Vehicle Recycling law 2002
7	Watercraft	Ships should be dismantled after removing fuel and batteries Scrap metal should be recycled Waste plastic and wood must be incinerated and used for power generation, to the extent possible Parts of vessels containing asbestos should be disposed of according to the specified procedures for asbestos- contaminated waste
8	Hazardous waste	Waste containing asbestos, polychlorinated biphenyls (PCBs) and other hazardous substances should be separated from other waste, treated as a discreet category of specially-controlled waste, and disposed of according to its properties.
9	Tsunami sediments	Materials containing toxic substances (i.e. heavy metals), perishable combustible materials and sediments containing oil should be used as raw materials of cement or subjected to incineration for landfill Other materials with similar properties to water botton sand should be segregated from foreign matter and used as backfill in ground subsidence areas, recycled into civil engineering materials or placed into the ocean
10	Waste at Post- fire sites	At sites affected by fires, ash should be segregated from scrap metal and waste concrete Ash, along with tsunami sediments mixed with ash, should be molten or disposed of in landfills at final disposal sites, deemed suitable based on the detected dioxin levels.

Source: Guidelines (Master Plan) for Disaster Waste Management after the Great East Japan Earthquake, Ministry of the Environment (2011)

12.6.6 *International Environmental Emergency Response System/Tools*

Currently numbers of resources and tools are available for response to International Environmental emergencies as mentioned below:

Joint UNEP/OCHA Environment Unit (JEU): In order to exclusively address environmental emergencies and develop special international tools to respond to Environmental Emergency JEU was established in 1993 which is essentially cooperation between UNEP technical expertise and OCHA (Joint UNEP/OCHA Environment Unit 2009) (Fig. 12.3).

Advisory Group on Environmental Emergencies (AGEE): In the year 1995, Advisory Group on Environmental Emergencies was established to bring together environmental experts from across the globe for the purpose of sharing information, expertise and lessons learned for improved response to Environmental Emergency globally.

National Focal Points: National focal points for response to Environmental Emergencies have been designated by JEU to serve as bridge between national government and AGEE and it provides relevant information during emergencies.

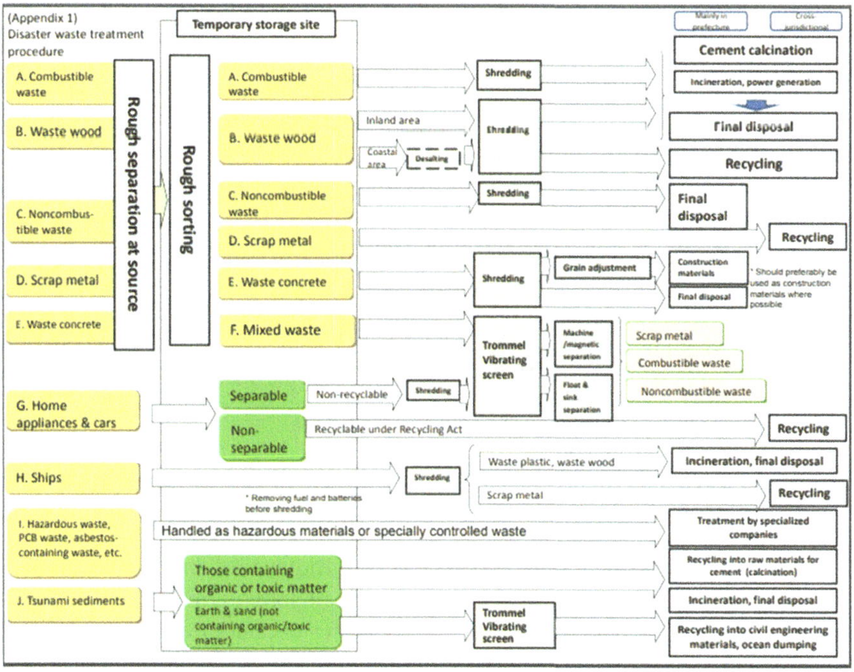

Fig. 12.3 Procedures followed for Disaster Waste Management in Japan is shown. (Source: Guidelines (Master Plan) for Disaster Waste Management after the Great East Japan Earthquake, Ministry of Environment Japan, May 2011)

Virtual On-Site Operations Coordination Centre (Virtual OSOCC): The idea to develop Virtual On-site Operations Coordination Centre was mooted by OCHA. It has multiple benefits like facilitating Decision Support System (DSS) through real time information exchange among various stakeholders of the International Emergency Response Community, facilitate training & Workshop.

International Environmental Response Resources: Joint Environmental Unit has a wide network of donors and has developed a pool of International experts with technical background. These skillful and competent experts can be deployed to affected location as and when need arises.

Online: Environmental Emergencies Centre: The Environmental Emergencies Centre (EEC) (www.eecentre.org) is one of the International initiatives aimed to act as online preparedness tool and capacitate National responders.

Online: Humanitarian Response: OCHA, an international organization provides unique and exclusive service of humanitarian response through website (www.humanitarianresponse.info) which is a dedicated website for management of information.

Flash Environmental Assessment Tool (FEAT): FEAT methodology enables a non- expert to map locations of affected area with an evident risk of technological accidents and corresponding chemical release.

Awareness and Preparedness for Emergencies at Local Level (APELL) program: While the main focus of the program is on environmental emergencies related to industrial activities with potential for fire, explosion or toxic release, it is also relevant to natural disaster preparedness.

Hazard Identification Tool (HIT): It provides the emergency responder with a list of facilities with potential secondary environmental impacts after a natural disaster and thus serves as a basis for on-site verification and further assessments. However, at the same time HIT is capable to recognize only massive and tangible entities and associated secondary threats.

Disaster Waste Management Guidelines (2013): The objective of these guidelines is to improve management of Disaster waste in a safe and cost effective manner by providing pragmatic guidance to National and International stakeholders uniformly. DWM guidelines have incorporated best contemporary wisdom through consultative process involving National and International stakeholders.

12.7 Conclusion

Current waste management rules (MSW, E-waste, Biomedical, hazardous, building and construction) under EPA fail to incentivise and impose a strict penalty in case of poor implementation. The rules have not pushed for decentralised management of waste but have encouraged centralised treatment such as waste to energy, the present state of which is not good in the country. Also, the informal sector has been considerably neglected in the new rules.

According to statement made by Sh. Prakash Javedkar, Union Minister of State for Environment, Forests and Climate Change, in 2016, 62 million tonnes of waste was generated annually in the country, out of which 5.6 million tonnes was plastic waste, 0.17 million tonnes was biomedical waste, hazardous waste generation was 7.90 million tonnes per annum and 15 lakh tonnes was e-waste. He added that only about 75–80 per cent of the municipal waste gets collected and only 22–28 per cent of this waste is processed and treated.

12.8 Way Forward

Mishandling of Disaster Waste or delayed disaster waste management post any disaster leads to Environmental Emergency. It is also established that 77% of direct economic losses amounting to $ 2.2 trillion are caused by climate-related disasters (UNISDR 2018). These findings lead us to following suggestions:

- The humanitarian and development communities are urged to **include the disaster waste issue into all phases of an emergency response,** including needs assessment, strategic planning, cluster response, preparedness, and monitoring and evaluation.
- It is urged national and local authorities and partners engaged in disaster waste management conduct **pre-disaster planning**, including the mapping of waste management facilities in the country and establish and regularly update the roster of technical experts on specific areas of waste management, and to **incorporate disaster waste management into emergency preparedness plans**.
- Within Disaster waste management is embedded the economic opportunity which can only be exploited fully when prioritized. Recycling disaster waste and debris offers benefits, by preserving natural resources, by saving money and space through landfill disposal, by generating energy and reducing pollution instead of open burning, and many more.

Disaster Waste management Plan be strengthened with appropriate budgets to ensure rapid restoration of public services and humanitarian access routes.

References

Annepu RK (2012) Sustainable solid waste management in India. Columbia University, New York

Basnayake BFA, Chiemchaisri C, Visvanathan C (2006) Wastelands-clearing up after the tsunami in Sri Lanka and Thailand. Waste Management World, (R), 31

Brandon, et al (2011) A case history study of the recycling efforts from the United States Army Corps of Engineers Hurricane Katrina debris removal mission in Mississippi. Adv Civil Eng

County of San Diego Debris Removal and Recycling Programs for the 2003 Cedar & Paradise Fires Final Report (2005). http://www.ciwmb.ca.gov/disaster/Fires2003

Environment and Humanitarian Action Network (n.d.) Environment and Humanitarian Action in the age of global reform agendas. Background Document to the nexus dialogue. http://www.eecentre.org/wp-content/uploads/2017/10/Backgroundconcept.pdf

Environmental Protection Agency (2008) Planning for natural disaster debris. http://wwwepagov/CDmaterials/pubs/pnddpdf

http://www.eecentre.org/about-eec/
http://www.eecentre.org/wp-content/uploads/2017/11/2-Joint-Statement-Feb2017.pdf
https://pib.gov.in/newsite/PrintRelease.aspx?relid=138591
https://www.downtoearth.org.in/news/waste/asia-witnessed-63-per-cent-jump-in-e-waste-generation-in-five-years-56811
Joint UNEP/OCHA Environment Unit (2009) Environmental emergencies learning from multilateral response to disasters. https://www.unocha.org/sites/unocha/files/EnvEm_LearningFromMultilateralResponse.pdf
Joint UNEP/OCHA Environment Unit (2013) Disaster waste management guidelines. https://www.msb.se/RibData/Filer/pdf/26599.pdf
Karunasena G, Amaratunga D, Haigh R, Lill I (2009) Post disaster waste management strategies in developing countries: case of Sri Lanka. Int J Strateg Prop Manag 13(2):171–190
Kobayashi Y (1995) "Disasters and the problems of wastes". International Symposium on Earthquake Waste, 12–13 June, Osaka Shiga: Japan, pp 6–13
Ministry of Environment, Forest and Climate Change (2016) Solid waste management rules, 2016. http://www.moef.nic.in/sites/default/files/SWM%202016_0.pdf
Ministry of Environment Japan (2011) Guidelines (master plan) for disaster waste management after the Great East Japan Earthquake. https://www.env.go.jp/jishin/attach/haiki_masterplan-en.pdf
Perera KLS (2003) "An overview of the issue of solid waste management of Sri Lanka" 3rd international conference on environment and health, 15–17 December 2003, India. Chennai: University of Madras, pp 46–352
UNEP (2012) Managing post-disaster debris: the Japan experience report of the International Expert Mission to Japan
UNISDR (2018) Economic losses, poverty and disaster 1998–2017

Ms. Bindu has been working as Programme Manager, Flood Early Warming, for Indian Red Cross Society (NHQ) in the National Capital Region since 2014. In her current position her work profile includes designing project related to Integrated Risk Management, Climate change adaptation and Disaster resilience as well as Monitoring and Evaluation of the project being implemented in various states of India.

Gupta Anil K , Professor of Policy & Strategies, Integrated Resilience. Head of the ECDRM Division, International Cooperation, Programme Director - Centre for Excellence on Climate Resilience, Nodal Officer for Central Ministries at National Institute of Disaster Management, (Govt. of India), New Delhi. He is also a member of the expert Team for WMO Climate Statement and Core Group Member of IUCN-Council of Ecosystem Management.

Chapter 13
Climate Resilient Healthcare System in India

Anjali Barwal, Atisha Sood, and Anil K. Gupta

Abstract Climate change is known to have deleterious effects on human health, environment and wellbeing. The consequences of public health of natural and man-made disasters are complex and varying. With the diverse agro-climatic systems, India faces a variety of climate risks such as temperature extremes, precipitation extremes, cyclones, dust and hailstorms, frost, and a variety of secondary disasters such as forest fires, landslides, pest attacks, disease outbreak epidemics, pandemics and so on. It's becoming clear that environmental and climate influences play a key role in spreading various diseases around the world, affecting public health responses and systems. Poor air quality also appears to be associated with more severe health outcomes and the ongoing COVID-19 pandemic has added emergent new dimensions to public health crisis. The current healthcare system, facilities, and services seem to be insufficient and unprepared to cope with the evolving and multiplying risk scenarios. Disaster health implications are best addressed by enhancing health systems resilience through a well-designed systemic mechanism. The National Health Adaptation Plan for disaster of India related illnesses has been developed involving a systemic, consultative and science-policy integrated approach, aiming for the resilient healthcare system by prioritizing budget allocation and strengthening the existing gap areas. This is a model approach that will help with the integration of health dimensions into the existing response, development and disaster management plans.

A. Barwal (✉)
Environment, Climate and Disaster Risk Management Division, National Institute of Disaster Management, New Delhi, India

A. Sood
Environment, Climate and Disaster Risk Management Division, National Institute of Disaster Management, New Delhi, India

A. K. Gupta
ECDRM Division, International Cooperation, CAP-RES DST Project, National Institute of Disaster Management (Ministry of Home Affairs), New Delhi, India

A. K. Gupta et al. (eds.), *Disaster Risk and Management Under Climate Change*, Disaster Resilience and Green Growth,
https://doi.org/10.1007/978-981-99-4105-6_13

Keywords Climate change · Public health · Health system resilience · Disaster management · Adaptation plan

13.1 Introduction

Over the past few years, interest in climate change and health research has grown substantially among the public health community which has led to better understanding of the overlapping domains of climate change and health, sensitisation on specific health risks and finding sustainable solutions keeping in mind the mitigation costs. A number of organisations are working together to meet the commitments made for climate change negotiation at the World Health Assembly including the United Nations agencies, government and non governmental bodies (WHO 2016).

Climate change is a global issue as it affects agriculture, forestry, biodiversity, water resources, sanitation, transportation, communication, health, life style, etc. Currently, climate change gets high attention globally as its effect challenges both developed as well as developing nations. But still developing countries are more vulnerable to climate change effects as they have less capacity to adapt to climate change (Ravindranath and Sathaye 2002; Mertz et al. 2009; Azadi et al. 2019). If we are not in a position to tackle climate change and its effect in integrated and cooperative manner, it might ruin the social, environmental, economic and political achievements gained so far. Climate Resilient Green Economy (CRGE) and many other studies revealed that health sector is one of the top three vulnerable sectors to climate change related adverse effects (National Health Adaptation Plan to Climate Change 2018). *Moreover,* health is a key element of the Sendai Framework for Disaster Risk Reduction 2015–2030 and it is also a key point of convergence across global and national policy frameworks (Reifels et al. 2018).

India, being a developing country with high population density and varied temperature and climatic profile, it might experience myriad human health effects because of climate change. These effects could include vector-borne and water-borne infectious diseases such as malaria, chikungunya, cholera, diarrhoea etc. Increases in extreme weather events due to climate change can have a number of health implications as depicted in Fig. 13.1. Displacement due to a lack of accommodation, poverty, hunger and accidents are only some of the population's negative consequences.

The main goal of developing the National Health Adaptation Plan (HAP) of India is to make sure the health system is climate change related disaster resilient. There is a need to develop and implement the HAP by mainstreaming it to various programmes and by strengthening its partnership with relevant ministries, departments and other stakeholders. The HAP shall be implemented with full participation of communities at grass root level through health awareness programmes using healthcare professionals and workers. The implementation of the plan should be supported and monitored at all levels so that it will be successful in building resilient healthcare infrastructure.

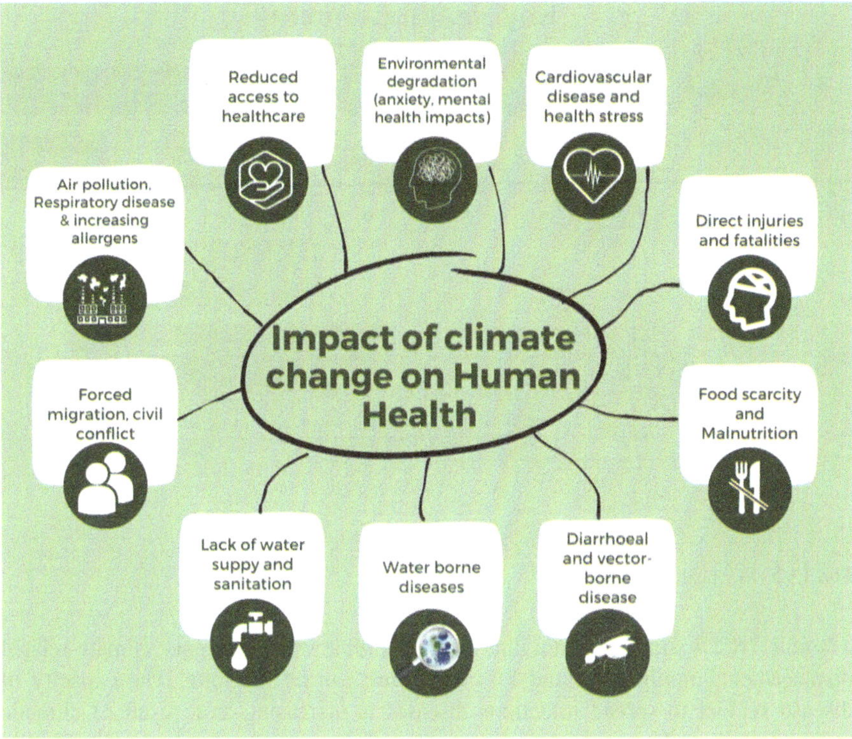

Fig. 13.1 Impact of climate change oh human health (*Source: Author*)

13.2 Health Sector Vulnerability

Global warming trends are continuing worldwide, accompanied by increasing numbers of extreme weather events, rising by 46% between 2000 and 2013 (IPCC 2018; Watts et al. 2018). A changing, more variable climate is now recognized as the most likely, highest-impact global risk to society as a whole and which presents a clear and present danger to health security (WMO 2017). Climate change can affect human health both directly and indirectly. The direct health impacts include physiological effects of exposure to higher temperatures, increasing incidences of Non-Communicable Diseases (NCDs) such as respiratory and cardiovascular disease and injuries and death due to extreme weather events such as droughts, floods, heat waves, storms, wildfires etc. Whereas, indirect effects on health are due to ecological changes, such as food and water insecurity and the spread of climate-sensitive infectious diseases, and also to societal responses to climate change, such as population displacement and reduced access to health services (Smith et al. 2014). As indirect effects of climate change may result from long causal pathways, they are particularly difficult to anticipate. The effects may be short or long-term and direct or indirect, sometimes with life-long consequences for health and well-being. For

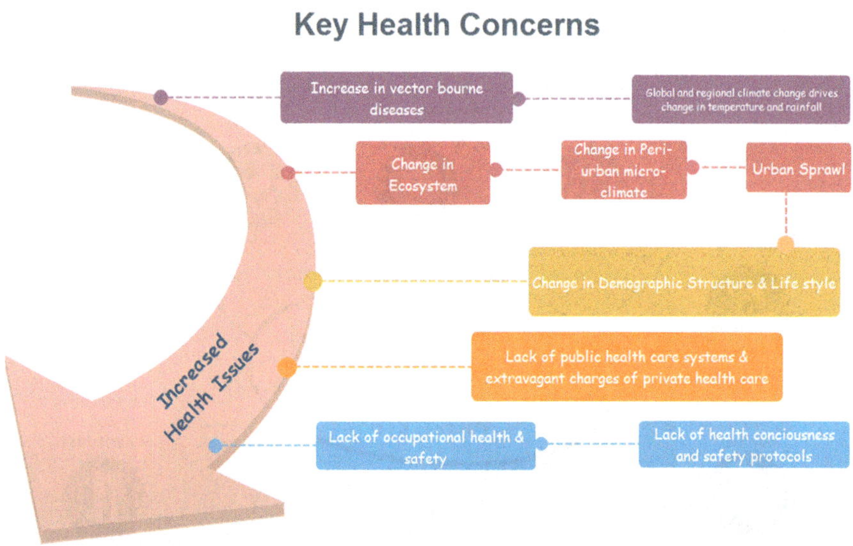

Fig. 13.2 Key health concerns

example, NCDs such as mental illness after extreme weather events, climate-related displacement, immigration and loss of culture can be life-long. The capacity of disease vectors to spread infectious diseases is increasing as a result of climatic shifts; for example, the vectorial capacity of the mosquitoes that are primarily responsible for the transmission of dengue fever has risen by approximately 10% since the 1950s (Watts et al. 2018). Ecological shifts as a result of climate changes may have further health effects, by affecting water and sanitation and causing food insecurity and malnutrition (Meybeck et al. 2018). Malnutrition is anticipated to be one of the greatest threats to health resulting from climate change, and the young and the elderly will be affected particularly. Climate variation and extremes are among the leading causes of severe food crises, and the cumulative effect is undermining all dimensions of food security, including availability, access, use and stability. Rising temperatures, floods and droughts also affect food safety; for example, rising temperatures can increase the levels of pathogens in food sources (such as ciguatera in fish) and in food, and flooding increases the risk that pathogens will spread from livestock.

The changing climate is heavily impacting the health sector and aggravating several communicable and non-communicable diseases such as malaria, dengue, yellow fever, meningitis, asthma, diarrhea, heat stroke, heart and lung diseases, etc. The burden to the health services would increase as climate change aggravates natural disasters that would cause social instability, physical damage, morbidity and mortality. More flooding events have started taking place because of increased rain intensity, which would destruct health infrastructures unless otherwise preventive measures taken ahead. Some of the key health concerns with respect to the climate change related disasters are shown in Fig. 13.2.

As a consequence of climate change extreme weather events are likely to become more frequent. Extreme weather events can have catastrophic effects on humanity. Historically many disasters, epidemics and famines have been triggered by floods and droughts. Weather can affect the infectious diseases pattern which is primarily spread through water and food contamination. Water borne diseases outbreaks like cholera, typhoid etc. can occur after floods when water is infected with faecal matter (Kovats 2000; Sood et al. 2020).

13.3 Developing National Health Adaptation Plan

The major objectives of HAP is to create enabling environment, build the capacity of climate resilient healthcare system; help institutions strengthen their preparedness, enhance the health system resilience in terms of universal health coverage; early warning, surveillance and community mobilization in the context of health emergency risk management.

The primary intervention areas to implement HAP are strengthening and expanding health infrastructure, identifying and assessing health hazards and vulnerabilities associated with climate change related disasters, building human resource capacity through policy planning, advocacy, research, training, education and knowledge management, promoting climate resilient water and sanitation facilities, revising building codes of health facilities, promoting community health insurance scheme, advocating and creating awareness on climate change and health as well as encouraging research and development programmes on health and climate change (Barwal et al. 2022).

Strengthening the mitigation of climate-sensitive health risks and building adaptive capacity to absorb the changing, increasing risks posed by climate change are the immediate public health activities required to address the challenge of climate change. Climate change adds further pressure on health, which is strongly influenced by effects on other sectors, such as food, water and sanitation. As a result, a multisectoral response is needed, based on existing strengths. Thus, the provision of health care and public health and work in sectors such as water and sanitation, food systems and energy provision, should be integrated with the additional functions and capacities required to build climate resilience (Chaturvedi and Siwan 2020).

Coronavirus disease 2019 (COVID-19) has changed the entire world forever. It has written new pages in world history. Worldwide countries are responding quite differently to the COVID-19 outbreak. While COVID-19 may be devastating in every aspect of life, we should not lose the opportunity to learn and grow. This is an excellent time to develop such policies that consider the health of the population, as well as the healthcare system and infrastructure. The countries' healthcare systems must be strong and reliable. The general public should also be encouraged to report and inform about infected people.

Fig. 13.3 Pathways for adaptation, mitigation and building resilience in health system

The need to plan for infectious disease outbreaks, whether naturally occurring or caused by bioterrorism, is now more than ever. Any such outbreak constitutes a threat to national and international security, with a potential to cause a health disaster. Legal frameworks, protocols and strategic HAP need to be developed and adapted to handle such public health crisis and to build disaster resilient healthcare system (Fig. 13.3). This can delineate the scope of the healthcare providers and government's responses to the public health emergencies (Sood et al. 2020).

Both healthcare practitioners and managers at all levels should receive capacity-building training on the health effects of climate change and the adaptation measures that should be taken based on the context, with a focus on regions or areas that are frequently impacted by climate-related disasters and diseases. It is very essential to improve the health management information system so that it can deliver accurate and high-quality data for decision-making. There is also a pressing need to integrate the systems in order to create a shared database that can be used for policy and decision-making. Food, air, and water quality, vector breeding, malnutrition, and other risk factors and behaviours that render the population more susceptible to the consequences of climate change-related disasters must all be monitored by health facilities to strengthen their surveillance systems (Barwal et al. 2020).

References

Azadi Y, Yazdanpanah M, Mahmoudi H (2019) Understanding smallholder farmers' adaptation behaviours through climate change beliefs, risk perception, trust, and psychological distance: Evidence from wheat growers in Iran. J Environ Manage 250:109456

Barwal A, Sood A, Gupta AK (2022) India's health adaptation plan: strategic tool for minimizing disaster related losses and damage. In: Goyal MK, Gupta AK, Gupta A (eds) Hydro-meteorological extremes and disasters. Disaster resilience and green growth. Springer, Singapore

Barwal A, Sood A, Gupta AK, Goyal MK (2020) Lessons from trans-domain assessment of COVID 19 outbreak. In: Integrated risk of pandemic: Covid-19 impacts, resilience and recommendations. Springer, Singapore, pp 481–496

Chaturvedi M, Siwan RM (2020) Resilience of healthcare system to outbreaks. In: Integrated risk of pandemic: Covid-19 impacts, resilience and recommendations. Springer, Singapore, pp 397–412

IPCC (2018) Intergovernmental panel on climate change special report on global warming of 1.5 °C Switzerland. Popu Dev Rev 45(1):251–252

Kovats RS (2000) El Niño and human health. Bull World Health Organ 78:1127–1135

Mertz O, Halsnæs K, Olesen JE, Rasmussen K (2009) Adaptation to climate change in developing countries. Environ Manage 43(5):743–752

Meybeck A, Laval E, Lévesque R, Parent G (2018) Food security and nutrition in the age of climate change. FAO, Rome

National Health Adaptation Plan to Climate Change (2018) Federal Ministry of Health (2018–2020). https://www.who.int/globalchange/resources/wash-toolkit/national-health-adaptation-plan-to-climate-change.pdf

Ravindranath NH, Sathaye JA (2002) Climate change and developing countries. In: Climate change and developing countries. Springer, Dordrecht, pp 247–265

Reifels L, Arbon P, Capon A, Handmer J, Humphrey A, Murray V, Spencer C (2018) Health and disaster risk reduction regarding the Sendai Framework. Aust J Emerg Manage, The 33(1):23–24

Smith K, Woodward A, Campbell-Lendrum D, Chadee D, Honda Y, Liu Q et al (2014) Human health: impacts, adaptation, and co-benefits. In: Climate Change 2014: impacts, adaptation, and vulnerability. Part A: global and sectoral aspects. Contribution of Working Group II to the fifth assessment report of the Intergovernmental Panel on Climate Change. Cambridge University Press, pp 709–754

Sood A, Barwal A, Gupta AK, Goyal MK (2020) Introduction to virus outbreaks. In: Integrated risk of pandemic: COVID-19 impacts, resilience and recommendations. Springer, Singapore, pp 3–20

Sood A, Barwal A, Gupta AK, Kishore J (2020) Novel coronavirus epidemic: new dimension for disaster management and health resilience. Curr Sci 118(8):1149–1150

Watts N, Amann M, Ayeb-Karlsson S, Belesova K, Bouley T, Boykoff M, Byass P et al (2018) The Lancet countdown on health and climate change: from 25 years of inaction to a global transformation for public health. Lancet 391(10120):581–630

WMO (2017) World Meteorological Organization. WMO statement on the state of the global climate in 2016. World Meteorological Organization (WMO)

World Health Organization (WHO) (2016) Second global conference: health and climate, conference conclusions and action agenda

Anjali Barwal, She is an experienced professional with a history of working in the environmental services and disaster mangement sector. With over a decade of experience in the sector, she has successfully completed many projects on environmental awareness, sustainability, climate change, public health resilience, capacity building and preparation of Disaster Management Plans of the Central Ministries and Departments of the Government of India. She had been associated with National Institute of Disaster Management, Ministry of Home Affairs, for developing the National Disaster Management Plan for the Department of Chemicals and Petrochemicals and the National Health Adaptation Plan for the Ministry of Health & Family Welfare, GoI.

Atisha Sood, Currently working as a Research Associate looking after the Centre for Disaster & Health at NIDM and development of the Disaster Management Plan of Ministry of Ayush. Have also contributed in the preparation of National Chemical and Petrochemicals Disaster Management Plan Project supported by DCPC, Ministry of Chemicals and Petrochemicals and worked previously in 'Health Resilience and Capacity Building' Project supported by World Health Organization at National Institute of Disaster Management, Ministry of Home Affairs, Government of India.

Anil K. Gupta, Professor of Policy & Strategies, Integrated Resilience. Head of the ECDRM Division, International Cooperation, Programme Director - Centre for Excellence on Climate Resilience, Nodal Officer for Central Ministries at National Institute of Disaster Management, (Govt. of India), New Delhi. He is also a member of the expert Team for WMO Climate Statement and Core Group Member of IUCN-Council of Ecosystem Management.

Chapter 14
WATSAN and Public Health in Hydro-Climatic Disasters

Jugal Kishore, Mukesh Kumar, and Anil K. Gupta

We shall not defeat any of the infectious diseases that plague the developing world until we have also won the battle for safe drinking water, sanitation and basic health care.
—Kofi Annan

Abstract Water, sanitation and hygiene are vital components for the health and development of any community. They can be attributed as fundamental aspects of Right to Health. Whenever a disaster occurs the availability of safe and potable water and a dependable sanitation system becomes priority for that community. Many water-borne, and vector borne diseases like cholera, typhoid, malaria, dengue and chikungunya can easily be countered with the simple measures of using safe water and having good sanitation facilities available during disasters. In today's world of scientific and technological superiority the onus of provision of such necessities lies on the authorities for safeguarding the health interest of the community. Burden of diseases spreading due to unavailability of safe water and unsafe sanitation practices increases the impact of any disasters with more problems in the vulnerable groups of the community. Phenomenon of climate change is also affecting the occurrence of hydroclimatic disasters. An outline of this chapter consists of types of communicable diseases, response to such problems and to manage various diseases spreading in such circumstances. Impact over health facilities will also be discussed and hydroclimatic disasters will be looked through public health perspective as well. Certain suggestions in order to get the better mitigation will also be investigated during the discussion.

J. Kishore · M. Kumar
Department Community Medicine, Vardhman Mahavir Medical College & Safdarjung Hospital, New Delhi, India

A. K. Gupta (✉)
ECDRM Division, International Cooperation, CAP-RES DST Project, National Institute of Disaster Management (Ministry of Home Affairs), New Delhi, India

A. K. Gupta et al. (eds.), *Disaster Risk and Management Under Climate Change*, Disaster Resilience and Green Growth,
https://doi.org/10.1007/978-981-99-4105-6_14

Keywords Disaster · Hydroclimatic · Water · Sanitation · Hygiene

14.1 Introduction

Water is one of the vital components of life in true sense and sanitation and hygiene can be categorized among basic pillars for the strategy to achieve the goal of health for all. Therefore, these are the fundamental aspects of Right to Health. For any community, water supply and sewerage system are vital for its health and development. Water, being a daily requirement for many purposes like drinking, cooking, washing clothes and for household waste disposal including human wastes, becomes indispensable in modern human life. Sanitation services on the other hand utilize the water systems and provide a barrier for transmission of diseases from animal and inanimate sources to various hosts including humans.

In a world where more than a billion people are denied safe drinking water while more than double of this number is not using safe sanitation practices; where the developing nations are still fighting with water borne and vector borne diseases (UNICEF Media Team 2017), an urgent need is felt for educating people regarding importance of safe drinking water and sanitation facilities. In order to expand international cooperation and to have capacity building on water and sanitation activities, commitment of international community has been requested under the Sustainable Development Goals. This commitment is to be utilized to support local communities. Through its Goal 6, the countries of the world have resolved to achieve universal access to safe drinking water and adequate sanitation and hygiene to all in the next fifteen years (United Nations in India n.d.).

In India, though we saw improved water sources in households rising from 68% in 1992–93 to 89.9% in 2015–16. But 63.3% of rural households and 19.7% of urban households were still not having improved sanitation facilities as in 2015–16. With government initiatives like Swachh Bharat Abhiyan, the National Rural Drinking Water Programme, and Namami Gange, the problem is expected to reduce in future (United Nations in India n.d.).

In the event of disaster especially hydroclimatic disasters like flood, drought, cyclone etc the availability of safe and potable water and sanitation facilities become of highest priority. A disaster is a burden over a community and its health system. It is something which a community cannot tackle itself and hence require outside support. In such circumstances, availability of safe potable drinking water and sanitation facilities can help to take away the misery of a suffering community to a great extent and help them to recover sooner from the suffering rendered after disaster.

Disasters like floods contaminate the sources of drinking water and renders the sewerage and sanitation system ineffective while disasters like earthquake, heat wave etc increase the demand for potable water. Most effects of various disasters are negative for water supply and sanitation system.

Here, we will try to look into the problems related to health conditions seen during the events of any natural disaster specifically hydroclimatic disasters towards a community or a country, how things like climate change have made this situation more severe and how health system of any country can gets affected with such disasters and how it can prepare for future events.

14.1.1 Climate Change and Hydroclimatic Disasters

Earth's climate has changed at a pace which is beyond any forecast made till date. With such rapidly changing climate the occurrence of disasters is also seen more frequently. Nowadays we experience extreme weather in many parts of the world. Be it the glaciers of Arctic and Antarctica or the deserts of Sahara, Africa or the plains of Siberia, we have been observing extreme weather conditions in every terrain (Blunden 2018). These conditions have resulted in unexpected occurrences of events with potential of disaster and the way people behave towards it. It changes how people anticipate the impending disaster. The resources at original places are not for managing unforeseen events and thus increase the impact of disaster.

14.1.2 Public Health and Hydroclimatic Disasters

Public health professionals have been really concerned with the natural disasters and their human consequences for decades. Several reasons are attributed to this concern. First, major disasters involving international aid are reported several times a year; and with better reporting by the mass-media, these disasters appear to occur more frequently, and as the population density increases in the disaster-prone areas these disasters become more devastating. Second, there is a strong feeling that nowadays disasters should be amenable to management to prevent them or mitigate their impact, or at least to improve rescue and relief.

Now, hydro-climatic disasters involve public health professionals because of the impacts of these disasters and their effects which can easily be prevented with timely intervention and adequate planning. Such interventions and preparations can lead to decreased sense of apprehension among lay public as well as special quarters of the society who have a keen watch over the occurrences rates of hydroclimatic disasters.

Recently, it's being realized that the effects of natural disasters on the health of populations are amenable to study by epidemiological methods. One can calculate death rates for different types of disaster. The burden of diseases occurring after the disaster can also be assessed and proper mitigation measures can be designed keeping in mind all such rates and indicators.

There can be variability in mortality rates as seen in the natural circumstances, as it depends upon many factors ranging from type of hydroclimatic disaster, affected population density along with its distribution, environmental conditions, preparedness of the population as well as authorities and any opportunity for warning. Death rates may be of lesser use as an indicator for planning of relief and rescue of an impending disaster, but they prove to be of great use when one evaluates the effectiveness of preventive measures for mitigating the effects of various hydroclimatic disasters. There may be differences in the distribution of damages and the loss of lives caused because of hydroclimatic disasters in different parts of the country as various factors contributing to the final impact of disaster may also differ in those areas.

In certain hydroclimatic disasters such as floods, the extent of casualties may depend on early recognition of the impending event of flooding as well as an appropriate warning system with which one can timely and effectively warn the population giving them enough time to leave the area and seek refuge. This timing of the warning becomes very important as whether authorities want to be on the safe side and may run the risk of issuing a false alarm or they want to wait to have definite signs of disaster, which may lead to a late warning and can affect the outcome of the disaster. Such type of dilemma can be compared to the one which epidemiologist have while deciding a cut off value of a screening test keeping in mind the sensitivity and specificity of the test to get the best out of a test for case detection, there by showing the responsibilities a public health professional has with the mitigating methods for the hydroclimatic disasters.

In certain circumstances, the number of deaths may be used to assess the adequacy of the criteria adopted for issuing the warning of the hydroclimatic disasters. Even after receiving warning, how people of different groups react to it can be assessed by looking at various age specific mortality rates which give interesting insights into such situations. Thus, it can get age-specific mortality rates and age group which is the most vulnerable to the disaster and also the type of education needed to be given.

Box 14.1 Control of Cholera Epidemic in a Camp (Source: Wisner and Adams 2002)

If there is an epidemic of cholera post disaster following measures may be undertaken:

Control Measures:

Establish emergency treatment facility.
Visitors to the patients should be limited.
Chlorine concentration should be at least 0.2 mg of free residual chlorine per litre of drinking water.
For washing and cleaning walls and floors chlorine concentration should be maintained at 0.5 g per litre and 2 g per litre respectively.
And for contaminated bedding and clothes, for latrines chlorine concentration should be 10 g per litre of water.

(continued)

Box 14.1 (continued)
Public-health Measures

Wells in the affected area should be treated and try to cover these wells if possible.
Case detection should be active by the health workers.
Any type of mass gatherings should be restricted.
There should be measures to reduce contamination of food articles sold in the market.
Water samples and stool samples should be tested at laboratory for contamination and if possible, for presence of cholera.
The status of epidemic can be assessed if there has been good record keeping done at clinics and treatment centers.
For better assessment, spot maps can be used to depict certain disease outbreak.
And, if resources permit, patient's home disinfection should be undertaken.

There is also considerable interest at present in prevention, especially in the field of earthquake engineering, and such indices will provide essential tools for evaluating the effectiveness of various structural designs and building regulations in reducing death and injuries. The effectiveness of various types of assistance and the long-term effect of aid on the restoration of the pre-disaster situation could be assessed as well, if adequate pre-disaster information were available.

During most of the disasters number of casualties after a disaster is usually low in relation to the number of deaths except in earthquakes, and by studying the distribution and types of injuries one can help in planning the amounts and types of relief supplies and rescue personnel required.

In disasters such as floods, the number of deaths depends largely upon early recognition of the impending disaster and on an appropriate warning system that may permit enough time to leave the area or seek refuge.

Morbidity because of hydroclimatic disasters is another cause of concern for better planning and management of disasters. Usually the level of morbidity is low as compared to number of deaths in most of the hydroclimatic disasters. For example: in floods, about 0.2–2% of the people require first aid for various injuries and illnesses and among them also most of them required for skin infections and ulcerations. However, general level of morbidity in an area is also affected by disasters because of either interruption of normal health care services or other disease control measures.

Though not many of the people require specific medical care after hydroclimatic disasters, still the type of illnesses and injuries reported give an idea about the types of services needed after the disaster. There is always a need to monitor such reports to properly plan medical care and to provide resources needed by the health personnel, ranging from generalized care to the specific care of specialized personnel.

Therefore, monitoring of the use of health facilities, once the disaster has passed, is important. Indices like, bed occupancy, duration of stay and delineation of catchment area prove to be of great use for such monitoring. This is especially important for better preparedness for future disasters.

Incidence and prevalence of communicable diseases is also affected and forms important aspect of epidemiology of disasters. As the control system is collapsed after the occurrence of disasters there may be increased transmission of diseases after that and this aspect requires the implementation of a disaster-adapted epidemiological surveillance system with appropriate indices and trends. These consequences of the disruption of routine health services about mortality and morbidity are often overlooked. Recently, there have been cases where mortality from ischemic heart disease, renal failure, and possibly from obstetric causes, has apparently been affected by the unavailability of normal health services (Lechat 1979).

Nutrition is another aspect which requires attention in post disaster conditions. There are many methods of nutritional assessment but the indirect effects of famine on the community are much beyond the simple statistics and broad socio-epidemiological understanding is required. The nutritional indices such as weight-for-height, weight for age, etc., are considered only one facet of the problem. Much more study is needed to develop appropriate measurements of the effects of malnutrition and to evaluate the appropriateness of various forms of food aid and supply in post disaster period.

Mental health after the occurrence of disasters is among common problems that requires further investigation and management. But indices of mental health are hard to design as there occurs differences in cultural patterns found in a disaster-struck community and ignorance about what should indeed be considered an adjusted reaction to disasters in exposed individuals.

But such type of studies is not possible mainly due to paucity of data, particularly concerning the health situation immediately after the impact. Such type of data could be collected either by conducting case studies of past disasters or by collecting information during relief operations; this lack of on-the-spot assessments has been a weak spot in the whole process of disaster management. Epidemiological surveys require the organization of assessment teams that can be activated rapidly after the disaster and, wherever possible, the teams should be composed of people from disaster-prone countries who are familiar with the area and the local customs. The organization of such stand-by epidemiological assessment teams seems to be one area in which international cooperation could yield immense benefits.

14.2 WATSAN and Disaster Mitigation

It is not possible to have a system which is completely safe from disasters, but it is upon the organizations both public and private to be capable of resolving any problem arising out of any kind of disaster in a best possible fashion and in the shortest possible time. With past experiences with us, there is always a need to prepare for such contingencies.

Special care is required for those with weak or compromised immune system, children, pregnant females and geriatric population as they should be provided with the facilities at highest priority.

14.2.1 Water, Sanitation and Hygiene (WASH) Interventions in Emergency Context

An event of disaster can affect people's health and lives along with the infrastructure supporting the survival. The various environmental health problems seen after disasters can be attributed to their effects on the physical, biological and social environment. All these problems are a serious threat to human health, its well-being and to survival.

Prevention of water borne diseases remains priority after a disaster. Various causes which can be attributed are malfunctioning water supply, pollution of water resources, and lack of sanitation facilities. Provision of enough quantity of safe water, availability of basic sanitation facilities, and a good hygiene behaviour are top priorities for immediate response.

The authorities managing water supply and sanitation system are supposed to be well equipped and ready to deal with any sudden occurrence of an event with potential of disaster. Given the importance of water supply and sanitation system, the authorities managing these should work in coordination to effectively manage any such event. Various preparatory methods which can be investigated may include an infrastructure which can avoid and untoward incident triggering or manipulating the events with potential of disasters, the proper training of the task force involved in the management of water supply and sanitation system and an always available blue print of action plan which can be used in events of disasters among other measures.

Various institutions, be it social, health or of any kind should always be prepared for such emergencies. We need to incorporate disaster scenarios while planning the infrastructure. Also, institutional, community, and societal response is a critical step for better management of risk in order to minimize vulnerability of population during and after a disaster.

The authorities must ensure adequate quantity of water to be used by the affected community. Along with quantity we must also ensure quality of water (both drinking and domestic use). In the situation after a disaster the availability of safe defecation facilities present at the relief site along with proper excreta disposal and management becomes another important aspect of the interventions during disaster emergency.

Apart from basic water and sanitation need, there is also requirement of proper solid waste management at or in the vicinity of the relief facility. There should be proper facilities for waste water and drainage from the relief facility to the disposal or treatment site.

Vector control and personal protection measures from vector borne diseases are another aspect of these interventions to be done during emergency. These have been discussed earlier also. One must also ensure improvement of hygiene through safe hygienic practices and behaviour change communication during and after a disaster has struck.

14.2.2 Water & Sanitation for Emergency Shelters

While designing and establishing relief shelters for victims of disaster, following steps for planning and implementation of water and sanitation program require (Practical Action 2019; National Disaster Management Authority 2009):

- Assess immediate needs of the affected population and various supplies available.
- All surface water is to be treated as polluted. And all upstream water supplies and wells should be protected.
- For all protected sources, basic collection, storage and treatment facilities are to be provided.
- Proper facilities should be provided to prevent indiscriminate defecation.
- There should be provision for safe disposal of excreta and refuse, and methods to control rodents and pests should also be employed.
- Foundations for long-term infrastructure should always be considered and it should also be ensured that such activities in future will not be impeded.
- Local water resources which may be scarce should be ensured to be safe in future also.

Also, recent advances in information management, with advent of technological resources such as geographic information systems (GIS), should also be used for emergency and disaster management to have the best possible information available for effective decision-making.

Total commitment and support of key decision makers is needed, so that the necessary material, human and logistical resources are made available for an effective and appropriate response. There is a need for proper planning along with appropriate organizational development for a speedy response to emergency by any agency whether government or private.

In recent years, urgent need for disaster reduction planning by water supply and sewerage systems administrators has been stressed repeatedly. It is important to make sure that efforts of such responses will have a long-lasting effect and that they will be reinforced with the knowledge and involvement of a significant number of the agency's or company's technicians and professionals.

14.3 Vulnerability & Impacts

14.3.1 Health Conditions Associated with Disasters

Disasters have a great adverse impact over health of the community. These health conditions arising after a disaster may either be due to direct impact of the disaster or due to indirect impacts of a disaster. Theses may include various physical morbidities arising after the disaster, the mental morbidities resulting from the loss of loved ones of survivors as well as due to displacement of the native population. These morbidities leave a great impact over the suffering community. The physical morbidities any community suffers after any disaster may include trauma due to disaster itself (traumatic injury, burns injuries due to electricity wire, snake or animal bites, drowning etc.) and may also result due to various diseases spreading after disaster because of other added factors, which will be discussed later.

Certain diseases leading to physical morbidities among the suffering population may also result from their spread aided by various vectors, contamination of food and water after the disaster and diseases spreading due to change in physical environment affecting the living conditions of the population. Generally, communicable diseases are having spread in such a scenario as all these factors favour the spread of these diseases.

14.3.2 Hydroclimatic Disasters and Communicable Diseases

Occurrence of diseases after disaster specifically hydroclimatic disasters has always been a topic of discussion among various authorities. Although it is generally perceived that piles of dead bodies seen after a disaster may result in outbreak of certain communicable diseases, it is not always true. Baring few situations where death following a disaster has occurred due to disease like cholera (Sack and Siddique 1998) or due to haemorrhagic fever (Boumandouki et al. 2005), it can be said that remains of human bodies have no risk for any kind of outbreaks. But one needs to dispose of human remains in any kind of hydroclimatic disaster. This should be done with full dignity. Points to keep in mind for management of dead bodies are given in Box 14.3. During a disaster, generally there is mass displacement of the affected population. Such a displacement has always been a cause of concern for the health and civic authorities among others as such a large-scale displacement are always having a risk of outbreaks of communicable diseases. Such a risk is always associated with the size of displacement and the characteristics of the population which has been displaced. The characteristics which generally determine the risk include the nutrition status of the community, the vaccination status, sanitation facilities in the community and availability of health facilities (Noji 1997).

Box 14.2 Floods of Uttarakhand (Source: Prakash 2015)
On the night of August 3rd 2012, there was very heavy rainfall and cloudburst which along with tree logs and boulders blocking the drainage lead to transient lakes formation in tributaries of main rivers like river Bhagirathi and river Ganga. These lakes overflowed into the main rivers taking all debris with them. The flash floods resulting from these chains of events swept away many lives with them also causing heavy destruction of infrastructure, roads, bridges, houses and even some hydro-electric projects. The devastation of that night and following days was followed by something more disastrous.

Many people including tourists were left stranded with little or no support for food, drinking water, sanitation and healthcare facilities. There was immediate response from district administration as well as support from state and central government level. The actions included evacuation of the survivors trapped in this situation. The next task was identification of shelters for those who had lost their homes in this disaster. Provision of food, medicine, potable water and other facilities was next task for the authorities. Many governments as well as non-governmental authorities came forward to help people suffered in the flash floods with commodities like tirpal, lamps, ration, sanitary pads and water purification tablets (chlorine tablets). With sustained and associated efforts of various agencies these aftermaths were dealt with great care to help people rebuild what they had lost on that critical night.

Communicable diseases associated with hydroclimatic and other natural disasters can be categorized under following headings:

- Communicable diseases related with water
- Crowding related communicable diseases
- Vector borne diseases
- Other diseases

Box 14.3 Management of Dead Bodies
Always prefer burial of dead bodies over cremation.

Try to identify all dead bodies and allow family of every diseased to have funeral appropriate and accepted to their cultures.

Try to locate alternate facilities for cremation and burial if existing ones are proving to be inadequate.

Avoid mass burial whenever possible.

Every worker who is handling the dead bodies should always follow universal safety precautions and should use proper disinfection and disposal facilities available.

During most of the disaster the affected community faces the problem of scarcity of safe and potable water. Hydroclimatic disasters like floods are usually associated with such outbreaks and the displacements occurring after such disasters is also associated with outbreaks of diseases like diarrhoea seen after Bangladesh floods in 2004 (Qadri et al. 2005), typhoid during hurricane Allison and Katrina in United States of America, hepatitis A, E after Aceh Tsunami of December 2004 (World Health Organization 2005) and leptospirosis as seen after Mumbai flooding in 2000 (Karande et al. 2003).

Displacement of any community after a disaster commonly leads to the problem of crowding in that population. Crowding facilitates transmission of diseases like measles (Marin et al. 2006), Acute respiratory infections (ARI) and Neisseria meningitidis meningitis etc. Such transmission was seen during Aceh Tsunami in 2004 (World Health Organization 2005).

Transmission of vector borne diseases depends to a great extent upon vector-breeding sites in and around the area where the disaster has occurred. In cases of hydroclimatic disasters one can see spurt in the presence of breeding sites for many vectors responsible for transmission of diseases like malaria whose outbreaks are frequently reported with periodic flooding of EL-Nino-Southern oscillation (Gagnon et al. 2002). Though outbreaks of malaria are reported with hydroclimatic disasters, dengue transmission generally depends upon other climatic factors like temperature, humidity etc and its increased transmission during any disaster is generally because of occurrence of disaster during rainy or favourable season for breeding as well as transmission of dengue vector and virus respectively.

Many diseases which are not transmitted person to person, but sometimes their outbreaks are reported after some disasters. These diseases are generally caused by some pathogens or their products and growth or survival of these pathogens is aided by the conditions prevailing in the community after the disasters. Tetanus caused by *Clostridium tetani* and coccidioidomycosis a fungal infection are examples of such diseases. Diseases like tetanus are also reported more in the communities with low immunization coverage against it. Such outbreaks were reported in Pakistan after earthquakes of 2005 (World Health Organization 2006a, 2006b).

14.3.3 Other Health Conditions

Besides the health conditions discussed above, there are certain other health conditions which are seen to have increased incidence during and after the vent of a disaster. These conditions may be classified to occur when general services to the community are interrupted following a disaster.

After a hydroclimatic disaster it is very common to see episodes of power failures in any community. In today's world, the electricity has been into each aspect of one's life in some or the other way years after its invention. Now after a disaster if electricity failure occurs it throws off other essential services too off their shelves. There is increased risk of water borne diseases as the water treatment and supply plant don't function optimally after electricity failure which may hamper the quality of water. Also, since food preservation is becoming a part of our life and this majorly depends upon electricity, the chances of diarrheal diseases increase in such scenario.

Power failures during and after a hydroclimatic disaster also renders health care facilities functioning ineffectively. An improper power supply to health care facilities may affect various aspects of a health care facility. Services like basic health care may not be affected much in absence of continuous power supply but at secondary and tertiary levels this may become a necessity as many life-saving procedures nowadays depend upon electricity. Essential services like vaccination are also affected in the absence of a sound functioning of a storage facility which may not function in cases of interrupted power supply.

14.4 Mitigation & Management

14.4.1 Prevention of Communicable Diseases

Prevention is always said to be better than cure. So, in case of communicable diseases after disaster, preventing them beforehand is the best way to counter them. There should always be an attempt to assess the risk of communicable diseases following an event of disaster. One should try to evaluate the endemic diseases prevalent in that locality. Living conditions should also be assessed meticulously as nutritional status, an independent risk for outbreak also depends upon it. We can further understand preventive measures for communicable diseases after an event of disaster under following two categories:

14.4.2 Pre-Disaster Phase

It's always better to be prepared for the unforeseen disasters. The pre-disaster phase aims at reducing the vulnerabilities of the population towards the disease and risk factors. It refers to assessment of risk factors and identify the most important one [which is many a times population displacement (Watson et al. 2007)], so that a detailed plan and preparations can be done accordingly. Many other factors which accelerate spread of communicable diseases include overcrowding, poor sanitation & waste management, poverty, food shortages leading to malnutrition among others (Connolly et al. 2004; Jensen et al. 2010; Ivers and Ryan 2006; Wilder-Smith 2005; Lashley 2003).

So, one needs to plan in a way so that minimal loss is there in situation when a disaster struck. Apart from factors discussed above non-availability of health facilities post disaster is another reason which increases the impact of communicable diseases (Loghmani et al. 2008). Also, the authorities need to strengthen surveillance system for early warning and thus minimizing casualties and impact due to an outbreak of communicable diseases and other health conditions. There is also a need to sensitize and increasing awareness among the population which is at most risk regarding the does and don'ts in the event of a disaster and after that (Wisner and Adams 2002; Tierney et al. 2001; Brown et al. 2008; Aldrich and Benson 2008).

14.4.3 Post-Disaster Phase

Post disaster is a phase of emergency response. It includes provision of medical care to the victims, provision of shelter, availability of safe water and sanitation facilities, proper nutrition, care of health of humanitarian workers and health education among others.

Some measures which can prove to be useful to reduce the impact of disaster include (World Health Organization 2006a, 2006b):

- Site planning and provision of safe water and sanitation
- Mass immunization
- Early warning system or effective surveillance system
- Services for primary health care
- Prevention and control of vector borne diseases

Provision of safe drinking water and good sanitation facilities should always be of utmost importance as these basic things can help in containment of spread of many communicable diseases like cholera, typhoid etc and can reduce the mortality and morbidities post disasters to a great extent. Also, there should be provision of safe and nutritious food available to the population displaced during a disaster.

Mass immunization with certain vaccines like measles along with vitamin A is generally recommended as soon as possible for all children below 15 years of age. Among them priority is given to those below 5 years of age and rest are covered if the resources allow. Other vaccines like typhoid vaccine, hepatitis A vaccine and cholera vaccine are not recommended for mass immunization due to cost constraints but vaccine like typhoid may prove to contain an outbreak at local level if the resources allow.

After a disaster it is necessary to identify rapidly the early signs of an impending outbreak. Health workers should be trained to identify all possible signs by which one can detect an outbreak in a given population. Special care should be taken among susceptible age groups and population. Also, all resources like adequate training to health workers as well as logistics for sample collection and transportation shall also be maintained.

Primary health care facilities can be very helpful to reduce the impact of a disaster as early diagnosis and treatment is critical for many diseases. These facilities should ensure proper wound care whenever needed and provision of primary treatment wherever required. It should be ensured that basic vaccination like tetanus toxoid is given at these facilities. There should be early diagnosis and treatment for basic URI, diarrhea, malaria and common fever. Basic rehydration facilities should be there, and basic and essential medication should also be available. Health education is another aspect to be covered at primary health care facilities for the community as well as health care workers.

Dengue and malaria are most prevalent vector borne diseases. Specific prevention of these diseases differs from each other. For malaria, prevention depends upon assessment of local area for prevalent risk factors. As flooding is followed by decrease in number of mosquitoes, such disaster gives health workers some time to adopt measures like indoor residual spraying, distribution of Long-lasting Insecticidal Nets (LLINs) etc. Early detection is mostly dependent upon effective local surveillance mechanism. Proper and complete treatment for each type of malaria helps in decreasing the burden of the disease. For dengue, prevention is manly limited to vector control. This may be achieved through source reduction, community mobilization and health education.

An example may be seen in Box 14.1 for control a communicable disease (Cholera in the given example) in post disaster scenario.

14.4.4 Health Facilities During Disasters

Health systems are among the most vulnerable to natural disasters. As seen after the 2004 Indian Ocean tsunami, when many health institutions were damaged. The devastating tsunami didn't spare hospitals, drug stores, cold rooms, preventive health care offices, health staff accommodation facilities and even district health offices. There was a loss of manpower also including medical officers, nurses, midwives and support staff among which many health staff were injured, traumatized or displaced by the event, hence were unable to assist the affected.

Other examples of disasters affecting health facilities severely include flood in Srinagar of 2001 where government medical college was submerged and collapse of Children's hospital during Kashmir earthquake of 2005 among others.

A well-documented and tested disaster management plan (DMP) is to be in place in every hospital. To increase their preparedness for mass casualties, hospitals must expand their focus to include both institutional and community-level planning. A meticulous plan at the institution level may prove to be beneficial in managing the emergency arising out of disaster apart from reducing the effects of disaster.

National Disaster Management Guidelines on Hospital Safety have been developed by National Disaster Management Authority with a vision of structural and functional safety of all hospitals from disasters in order, such that the risks to human life and infrastructure are minimized (National Disaster Management Authority Govt of India 2016).

The guidelines introduce the need and importance of hospital safety. Various objectives addressed in these guidelines include:

- hospital safety through a multi-hazard and inter-disciplinary approach
- structural safety of hospitals

- the day to day operation of hospitals are prepared to respond to disasters involving all professionals
- that every hospital in the country has a fully functional and regularly tested Hospital Disaster Management Plan

These guidelines address hospital preparedness and response focuses upon provisions required to be put in place to ensure functional safety of hospitals/health facilities in disaster situations.

Structural and design safety elements as well as fire safety are focused in the plan. Elements pertaining to regular maintenance and inspection for hospitals are also highlighted.

The guidelines states that conscious efforts are needed for a 'safer and functional hospitals' and hence a focused strategy, 'National Action Framework for Hospital Safety' is also discussed.

There should also be a plan for collaboration with outside organizations for planning; establishment of alternate care sites; training of medical personnel in the management of diseases which can be of potential of biological warfare; drills on aspects of the response plans; and equipment and bed capacity available at the hospital.

The most important external agencies for collaboration would be state and local public health departments, emergency medical services, fire departments and law enforcing agencies among others.

14.5 Conclusion and Way Forward

Disasters have been there for ages. Mankind has been braving enough to cope with the adversities of various disasters. Disasters, specifically hydroclimatic disasters have been seen to occur with diseases like water borne diseases, vector borne diseases. Prevention of these diseases requires certain measures to reduce the vulnerabilities of the population towards the disease and risk factors. There is a need to work on various preventive measures using available resources. Post disaster phase also require measure to contain the effect of disasters. Health facilities also require a good disaster management plan to prepare for the causalities and effects of disasters. More studies are required to explore various epidemiological indices which can be used to assess the impact of disasters and can be used in developing different preventive measures.

To sum up as said by Petra Nemcova, we cannot stop natural disasters, but we can arm ourselves with knowledge: so many lives wouldn't have to be lost if there was enough disaster preparedness.

References

Aldrich N, Benson W (2008) Disaster preparedness and the chronic disease needs of vulnerable older adults. Prev Chronic Dis 5:A27

Blunden JDS (2018) State of the climate in 2017. Bull Amer Meteor Soc 99:Si–S310

Boumandouki P, Formenty P, Epelboin A, Campbell P, Atsangandoko C, Allarangar Y (2005) Clinical management of patients and deceased during the Ebola outbreak from October to December 2003 in Republic of Congo. Bull Soc Pathol Exot 98:218–223

Brown D, Young S, Engelgau M, Mensah G (2008) Evidence-based approach for disaster preparedness authorities to inform the contents of repositories for prescription medications for chronic disease management and control. Prehosp Disaster Med 23:447–457

Connolly M, Gayer M, Ryan M, Salama P, Spiegel P, Heymann D (2004) Communicable diseases in complex emergencies: impact and challenges. Lancet 364:1974–1983

Gagnon A, Smoyer-Tomic K, Bush A (2002) The El Nino southern oscillation and malaria epidemics in South America. Int J Biometeorol 46:81–89

Ivers L, Ryan E (2006) Infectious diseases of severe weather-related and flood-related natural disasters. Curr Opin Infect Dis 19:408–414

Jensen P, Meyrowitsch D, Konradsen F (2010) [Water and sanitation in disaster situations]. Ugeskr Laeger 172:109–112

Karande S, Bhatt M, Kelkar A, Kulkarni M, De A, Varaiya A (2003) An observational study to detect leptospirosis in Mumbai, India, 2000. Arch Dis Child 88:1070–1075

Lashley F (2003) Factors contributing to the occurrence of emerging infectious diseases. Biol Res Nurs 4:258–267

Lechat M (1979) Disasters and public health. Bull World Health Organ 57:11–17

Loghmani A, Jafari N, Memarzadeh M (2008) Determining the field hospital setting in earthquake: Using RAND/UCLA appropriateness method. Iran Red Crescent Med J 10:184–192

Marin M, Nguyen H, Langidrik J, Edwards R, Brian M (2006) Measles transmission and vaccine effectiveness during a large outbreak on a densely populated island: implications for vaccination policy. Clin Infect Dis 42:315–319

National Disaster Management Authority (2009) National policy on disaster management 2009. National Disaster Management Authority, New Delhi

National Disaster Management Authority Govt of India (2016) National disaster management guidelines-hospital safety. National Disaster Management Authority, New Delhi

Noji E (ed) (1997) Public health consequences of disasters. Oxford University Press, New York

Practical Action (2019, March 9) Water & sanitation for emergency shelters. Retrieved from Practical Action: https://practicalaction.org/emergency-shelter-with-water-sanitation

Prakash DS (2015) A study on flash floods and landslides disaster on 3rd August 2012 along Bhagirathi Valley in Uttarkashi District, Uttarakhand. New Delhi. National Institute of Disaster Management (NIDM), Ministry of Home Affairs, Govt of India

Qadri F, Khan A, Faruque A, Begum Y, Chowdhury F, Nair G (2005) Enterotoxigenic Escherichia coli and Vibrio cholerae diarrhea, Bangladesh. Emerg Infect Dis 11:1104–1107

Sack R, Siddique A (1998) Corpes and the spread of cholera. Lancet 352:1570

Tierney K, Lindell M, Perry R (2001) Facing the unexpected: Disaster preparedness and response in the United States. Joseph Henry Press, Washington DC

UNICEF Media Team (2017, July 12) 2.1 billion people lack safe drinking water at home, more than twice as many lack safe sanitation. Retrieved from UNICEF: https://www.unicef.org/press-releases/21-billion-people-lack-safe-drinking-water-home-more-twice-many-lack-safe-sanitation

United Nations in India (n.d.) SDG 6: Clean water and sanitation. Retrieved from United Nations in India: http://in.one.un.org/page/sustainable-development-goals/sdg-6/

Watson J, Gayer M, Connolly M (2007) Epidemics after natural disasters. Emerg Infect Dis 13:1–5

Wilder-Smith A (2005) Tsunami in South Asia: what is the risk of post-disaster infectious disease outbreaks? Ann Acad Med Singap 34:625–631

Wisner B, Adams J (2002) Environmental health in emergencies and disasters: a practical guide. World Health Organization, Geneva

World Health Organization (2005) Epidemic-prone disease surveillance and response after the tsunami in Aceh Province, Indonesia. Wkly Epidemiol Rec 80:160–164

World Health Organization (2006a) Communicable diseases following natural disasters: Risk assessment and priority interventions. NTD Information Resource Centre, Geneva

World Health Organization (2006b, August 10) Morbidity and mortality report. Retrieved from Humanitarian Health Action: http://www.who.int/hac/crises/international/pakistan_earthquake/sitrep/Pakistan_WMMR_VOL23_03052006.pdf

Jugal Kishore is a medical educator and public health expert. He is part of development of various departments particularly Community Medicine in BP Koirala Institute of Health Sciences in Nepal and the Centre for Occupational & Environmental Health at Lok Nayak Hospital, New Delhi. His more than 200 research papers are published in journals, in magazines, and in dailies. He is authored two dozen books and many chapters and pamphlets on public health, national programs, social reform, and rationality. He is member of many expert groups of health planning and implementation of the government. He is the recipient of "Rashtrya Ekta Puruskar", "Dr. S Radhakrishnan Memorial Teacher Award 2007", President Appreciation Award in 2009 and "Chikitsa Ratan" in 2016. His scientific and rational outlook attracts international body and made him Hon. Director of Centre for Inquiry Delhi, founder of Sabasva foundation and KISHORE Foundation and Secretary General of Indian Association for Adolescent Health. His mission is to liberate the man through education and development of public health.

Mukesh Kumar has done MD (Community Medicine) from VMMC & Safdarjung Hospital, New Delhi (GGSIPU) in 2018. Presently working as a Senior Resident in the Dept of Community Medicine, VMMC & Safdarjung Hospital, New Delhi, India. Has presented oral papers and posters at various national and international conferences. Has 5 papers in various indexed and peer reviewed scientific journals. Has special interest in Adolescent Health, Mental Health, Non-communicable diseases, Immunization and Maternal Health.

Gupta Anil K. Professor of Policy & Strategies, Integrated Resilience. Head of the ECDRM Division, International Cooperation, Programme Director - Centre for Excellence on Climate Resilience, Nodal Officer for Central Ministries at National Institute of Disaster Management, (Govt. of India), New Delhi. He is also a member of the expert Team for WMO Climate Statement and Core Group Member of IUCN-Council of Ecosystem Management.

Chapter 15
Understanding Water-Energy-Food Nexus in Urban Ecosystem for Resilience to Climate Risks

Swati Singh and Shresth Tayal

Abstract Growing population, urbanization, changing climate, increasing prosperity and change in dietary habit is putting unprecedented challenge on Water, Energy and Food (WEF) resources of India. WEF nexus has been gaining global momentum today and developing synergies between them for optimization of resource use is the demand of the hour. This paper is an attempt to analyze the importance of nexus approach for India and develop an interrelationship between Water, Energy and Food from consumption point of view for better management of natural resources through literature review and analysis of available secondary data. The finding comes up with few recommendations like integrated development of cities, concept of sustainable food consumption, integrated policy approach, green economy approach and ecosystem based approach for optimal use of resources. The connections between Water, Energy and Food is so robust that improving efficiency in one process either through technological interventions or through regulatory approach will have impact on the overall nexus elements and thus helps in achieving the goals of Sustainable Urban Development.

Keywords Water-energy-food nexus · Climate change · Urbanization · Sustainable development

15.1 Introduction

Ecosystems are dynamic complexes of living communities and their non-living environment interacting as functional unit. Ecosystems are basis of life and livelihoods and provide essential services for existence and socio-economic well being of people living in the surroundings. Urban ecosystem represents the built

S. Singh (✉)
UNDP, New Delhi, India

S. Tayal
The Energy and Research Institute, New Delhi, India

A. K. Gupta et al. (eds.), *Disaster Risk and Management Under Climate Change*, Disaster Resilience and Green Growth,
https://doi.org/10.1007/978-981-99-4105-6_15

environments with high human density (MEA 2005) which is highly managed by humans to satisfy their needs and demands. Ecologically, they are complex systems that show distinctive ecological characters, patterns, processes, disturbances and subtle impacts (Mariana 2008). The relationship between ecosystems and cities are interlinked and is often a two-way process. Ecosystems provide a multitude of physical and environmental services to cities and their residents which also help in enhancing city's resilience. However, the cities, which are rapidly urbanising and experiencing unplanned development is leading to a threatening decline in ecosystems. This is impacting the resilience of several cities across the world.

The world is under sustained process of urbanization which has led to major socio-economic and environmental changes. 60 years back the world urban population was 29% (UNDESA 2006), today it is 50% and as per the projections made, by 2050 it will grow by 72% (UNDESA 2012). The numbers of megacities are increasing across the globe and more than half of the world's megacities i.e. 28 out of 36 are found in Asia-Pacific (UNDESA 2018), three of which are in India. The point of concern is that, most of this growing population will be absorbed into rapidly growing cities, notably in Asia's fast developing economies (Bentham 2011). These huge urban agglomerations exceed the carrying capacity (Mathur and Sharma 2016) of the region and require enormous amount of water, energy and food within a small area to support them.

Though the concept is in use for ages, water-energy-food (WEF) nexus is being promoted as a management tool to address the issue of resource scarcity (World Economic Forum 2009; Hoff 2011; Middleton and Allen 2014) and develop synergy with economic growth, socio-ecological linkages and path of resilience (Hellegers et al. 2008; Rasul 2014). Thus, this approach has been used by many experts, researchers and organizations like FAO, ICIMOD, UNEP, Global Water Partnership and many more in order to develop synergies between water-energy-food system and aims to provide a framework for optimal use of resources to manage trade-offs between them (Hellegers et al. 2008; Rasul 2014). The WEF concept has been gaining global momentum over last few years and has become centre point for Sustainable Development discussions (Biggs et al. 2015). The SDGs 2 and 12 talks about zero hunger and responsible consumption and production respectively, SDG 6 is related to clean water and sanitation whereas SDG 7 lays emphasis on affordable and clean energy. Apart from these goals that are directly addressing the water, energy and food resources, there are certain goals which are cross cutting across the nexus sector. SDG 11 have implications on urban ecosystem and emphasizes on making cities inclusive, safe, resilient and sustainable by implementing integrated policies and plans for resource use efficiency and adaptation to climate change. SDG 13 is related to climate action and SDG 15 talks about life on land that lays emphasis on integrating ecosystem based approach in local planning and developmental processes (UN 2016). The WEF relationship is continuously evolving with changing resource conditions, development priorities and technological advancements and thus provides an interface to explore synergies for most positive outputs. Thus, an integrated approach towards harmonizing WEF nexus can play a crucial role in developing countries like India which can lead to smarter and more resilient development solutions.

15.2 Vulnerability and Impacts

The growing population and speed of urbanization in India puts unprecedented challenge on its environment and natural resources. Currently, 400 million people in India live in towns and cities; which accounts for 33% of its total population, and it is likely to increase in the coming years putting stress on its resources if not managed sustainably. The report by Planning Commission (now NITI Aayog) projects that within 20 to 25 years another 300 million people will migrate to various cities and towns (The Planning Commission 2012). Rise in population, rampant urbanization, growing prosperity and transition in lifestyle demands for more water and energy along with increase in the demand of food. Urban development, growing population density and the infrastructure development like housing societies, mall and markets, highways, office premises etc., all will have an obvious impact on the energy efficiency, water use and availability of food to support this growing population. As stated by Hoff (2011), urban living promotes more resource intensive lifestyles and concentrates consumption and waste production and thus escalating the scarcity of natural resources base.

History is the evidence that no nation can survive for long if natural resources of a state are exhausted or polluted. Meadows, Meadows et al. (1974) came up with an excellent work on "Limits to Growth" where it has been illustrated that natural resources depletion may jeopardized the economic growth of the nation (Meadows et al. 1974). However, the work further stated that global equilibrium could be designed to alter the negative growth trend and set up a harmony between economy and ecological stability that is sustainable for future as well. Thus, developing synergies and interrelationship between water, energy and food security for optimization of resource use efficiency to sustain is one of the solutions to the above statement and it is also the demand of time. The concept has gained a global momentum over the past few years. The nexus plays a crucial role in developing countries like India which can lead to smarter and more resilient development solutions.

Recognizing the importance and relevance of nexus approach in assuring sustainable economic growth of the country, India organized its first event, the "India Water Week" on this very theme of "Water, Energy and food Security: Call for solutions" in April 2012. The event recognized the importance of inter linkages among different sectors (water, energy, food and environment) as well as emphasized on strengthening the institutional framework for the same. Steps have already been taken in the direction; however, research is required to understand and strengthen the nexus. This paper is an attempt to analyze India's understanding of the nexus approach and develop an interrelationship between Water, Energy and

Food from the consumption point of view for better management of natural resources, using literature review and analysis of available secondary data.

15.2.1 Drivers to Integrate Water, Energy and Food: Key Challenges

Amongst many stresses and challenges being faced by our country today, key drivers are population growth, urbanization, increasing prosperity, change in the dietary habit, environmental pollution and climate change. It is important to understand the connections between these drivers and nexus elements.

15.2.1.1 Rising Population

India stands second in the world only after China in terms of population and supports 17.5% of total world's population with just 2% of geographical land (Ministry of Agriculture 2011). As per the recent projections made by United Nations, India is on the way to be the most populous country by 2022, surpassing China within seven years. Population pressure on land and water resources is very high to meet its food and developmental priorities. Comparing the data of last 50 years, India's population has grown and doubled from 541 million in 1971 to 1210 million in 2011. To support its large and growing population, India made significant progress in the field of food grains, doubling its production from 108 million tons in 1970's to 218 million tons in 2011 (Ministry of Agriculture 2011) and reached a stage of self-sufficiency to surplus agricultural produce (Swaminathan and Bhavni 2013; Ahmad and Haseen 2012). The "Green Revolution" that started in late sixties and made India from a country of food grain importer to food grain exporter is to be credited for such progress.

However, to reach this stage of food sufficiency water resources were exploited manifold and have reached a stage of water stressed condition today. With only four percent of total global water available, agriculture remains the biggest use, accounting for almost 80% of water withdrawal. The demand for water will continue to rise in future due to rapid population increase and economic growth.

According to Rao (2002) and projections made by National Commission for Integrated Water Resources Development Plan (NCIWRDP), the requirement of water for irrigation in India will grow by more than 50% by 2050. The per capita water availability has declined from 5200 m^3 in 1951 to 1588 m^3 in 2010, showing a decrease of 69% which is more than double. As per the projections made it will decrease further to 1191m^3 by 2050 (http://india-wris.nrsc.gov.) almost reaching a stage of water scarce condition (Fig. 15.1).

Water is used in the production of almost all kind of energy and demand of electrical energy for agricultural usage is showing an increasing trend in India

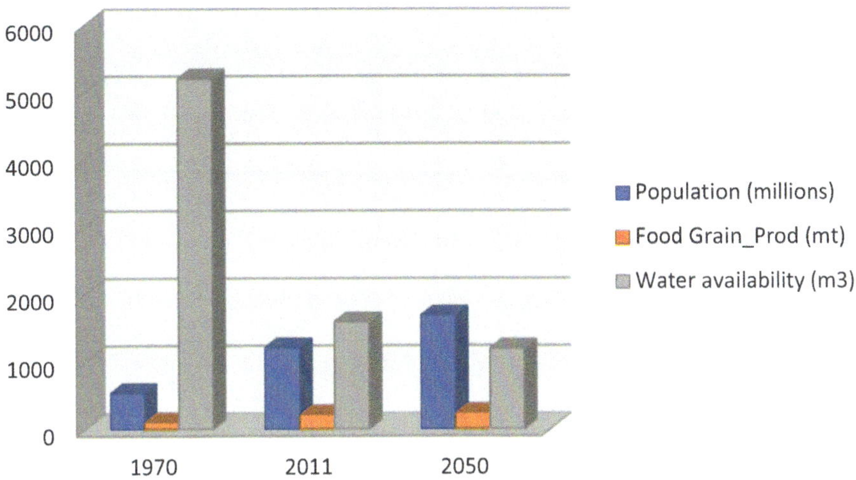

Fig. 15.1 Graph projecting the declining trend of water resources with increasing population & food production

(Swarnkar and Singh 2013). Mechanization of various agricultural operations in the supply chain including farm irrigation, harvesting, processing etc. accounts for the energy demand in agricultural sector. The energy consumption in terms of electric utility in our country has increased manifold during the last five decades. It has increased from 56 billion kwh in 1971 to 768 billion kwh today; which is 13 times more. According to United Nations, 90% of global electricity requires water either to rotate turbines or as a coolant in thermal power plants (Lundy and Bowdish 2013). Population will continue to rise and as per the projections made by various organizations (Planning Commission of India) by 2050, India's population will be 1690 million which implies that we need to increase the rate of production with limited land and available water resources. Thus water, energy and food needs are expected to increase in near future, in order to meet the demand of the growing population.

15.2.1.2 Urbanization

As urbanization continues, half of the world's population is living in cities, largely seeking increased economic opportunity to sustain them. An analysis of Indian Census data of last 50 years shows that number of urban agglomeration of towns and cities has grown more than three times from 1970 to 2011. Population residing in these urban areas has almost doubled and is likely to increase in coming years. If the current trend continues, it has been projected by United Nations that India's urban population will increase to 50% by 2050; showing an increase of almost 20% from current years (Fig. 15.2); intensifying the stress on its resources if not managed sustainably.

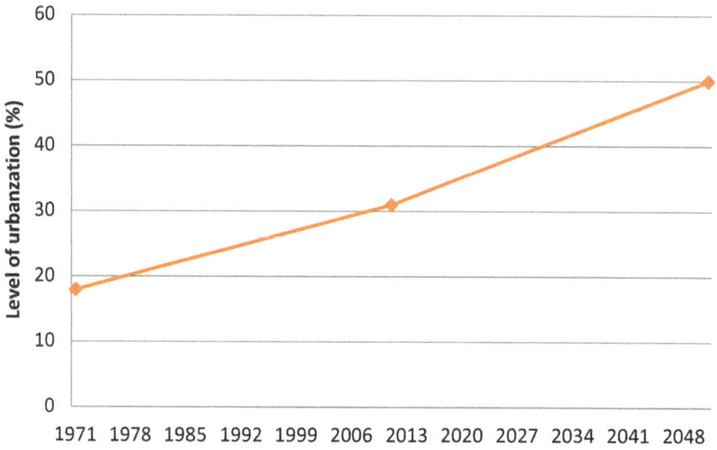

Fig. 15.2 Line chart projecting level of urbanization in India

As stated by Dutta (2006), India's urbanization has been termed as "over-urbanization" or "pseudo-urbanization" because of inordinately large population size in big cities leading to virtual collapse in the urban services followed by basic problems in the field of housing, sanitation, slums, water, infrastructure as well as quality of life. The current pattern of urbanization is actually taking place on the fringes of cities, wherein, putting pressures on Municipal Corporations for the supply of basic amenities. In urban water supply and municipality, sanitation, wastewater treatment and drinking water supply, energy is required at every step and forms the main operational cost component. In most cities, water supply is sourced from long distances and the length of pipeline determines the cost including the cost of pumping water. The very process of collecting, transporting and distributing water is energy intensive and cost of operation is very high. As per the Planning Commission of India (now NITI Aayog), 75% of urban households are covered with piped water; 40–50% of water is lost in the distribution system and approximate 30–50% electricity is spent to pump water in most Indian cities (Planning Commission 2011). This rapid and accelerating economic growth, growing population, burgeoning middle class, expanding cities and urbanization are increasing the demand for water, energy and food.

15.2.1.3 Increasing Prosperity and Change in Dietary Habit

India's per capita income has grown ten times in the past six decades. As per the report of Rangrajan Committee on poverty status in India (Planning Commission 2014) there has been a significant decline in the Indian poverty ratio-which fell down from 38.2% to 29.5%, lifting almost 10% of population out of poverty line. This implies that more and more people have migrated to cities and burgeoning middle class and thus the purchasing power of individual has increased substantially. This is

Table 15.1 Water and energy input for processes foods (Source: *IME (2013), **Pimental (2011))

S. No	Food stuff	Water consumed (litres)*	Energy consumed (Kcal/kg)**
1	Chocolate (Kg)	17,196	18,948
2	Meat (kg)	10,412	1206
3	Milk (l)	1022	354
4	Wine (l)	436	830
5	Coke (l)	600	1425

indeed a positive development but it will be accompanied by increase in the demand and supply of basic amenities, intensifying the environmental stresses and putting pressures on natural resources.

Food basket has diversified significantly with the rising incomes of the country. The economic growth, urbanization and taste preferences have resulted shift in food consumption pattern deviating from traditional food commodities (Meenakshi 1996; Murty 2000; Amarsinghe et al. 2007; Pathak et al. 2010). Indian diet is diversifying with fruits/vegetables and animal based food share increasing and cereal and pulses declining (NCEAR 2014). Increase in income has also changed the dietary habit of affluent urban people where consumption of processed foods and ready to eat foods has gone up (Vepa 2004). There has also been an increase in the nutritional intake in terms of calories and it is projected to increase to 3000 calories in 2050 from 2495 calories at present.

According to the NSSO report of 2013, there is around 33% increase in the consumption expenditure from 66th round to 68th round (Sinha 2014) where expenditure on beverages, refreshment and processed food is highest and forms the significant amount to drive the food sustainability issues. Energy is used in the production, processing as well as delivery of food, pumping of water for irrigation, transport of food, its distribution and storage. The full food production and supply chain is responsible for around 30% of total global energy demand (Hoff 2011). Food processing consumes substantial amount of energy, which is comparatively high in bread, animal food products and sugar processing (UNIDO 1995). The Table 15.1 gives water and energy inputs for some of the processed foods most commonly used these days.

Government policies are predominantly focused on the production side to increase the food availability in order to feed billion populations. But at present, around one-third of all food produced is lost along the food supply chain and considerable amount of food gets wasted at consumption level. It has been estimated by FAO (2011) that global food loss and waste is approximately 1.3 billion tons annually and occurs at point throughout the supply chain and across the socio-economic spectrum (Moomaw et al. 2012). In India, the domestic food loss has been estimated to be 30–40% of total food supply (Emerson Report 2013). A significant share of total energy and water inputs are embedded in these losses. Therefore, there is an urgent need to develop holistic and effective strategies for promoting sustainable consumption as well as improving the food supply chain vis-a-vis efficient production, which will address the nexus issues.

15.3 Case Study of Gurugram: A Study at Micro Level

Gurugram is one of the largest cities of Haryana, which is urbanizing and expanding at a very fast rate. With the advent of economic reforms and modernization in 1990s, the cultural and ecological landscape of the city has undergone a rapid change. Apart from real estate development, the city has emerged as one of the prominent business hub in North India because of the favourable tax policy of the Haryana State Government, rapid development of infrastructure as well as good connectivity with the International Airport. Many automobile and software multinationals have found their way into the city premises due to the availability of high level of infrastructure of airways, railways, highways, medical and educational institutions in its close proximity.

Despite its planned and gleaming infrastructure, the government has not been able to keep at pace with the growing population and increasing level of urbanization in the city. The areas that are dilapidated and poorly connected, lacks adequate water and power supply with no proper sewage handling mechanisms. The city regularly faces water scarcity and water quality problems are acute during summer season. The city meets its water supply demand from Yamuna Nagar Canal and groundwater. Groundwater is used to supplement the water need that constitutes the main source of water for construction sector. The unsustainable developmental pattern has led to some obvious consequences with wells running dry and city turning into a concrete desert.

The population of the city (as per 2011 census) is 1.5 million, where male and female constitutes 54% and 46% of the population respectively (GoI 2011). The total water demand of the city as per the city agencies is 184 MLD. Water required for cooking is approximately 11 MLD, which is 6% of the total water demand. This water requirement includes water for cooking as well as washing utensils. The city needs more than 1 lakh kg of energy in form of LPG for cooking raw food whereas, 854.64 Mwh of electricity in consumption of food other than cooking. This includes reheating, processing and storage of foods using electrical appliances. In addition the city needs 114.6 Mwh of energy in the form of electricity for pumping water into the storage tanks which is further used for cooking along with other purposes like washing clothes, sanitation and domestic use.

In the past two decades, Gurugram city has experienced urban growth at a phenomenal rate. The rapid rate of population growth and increasing urbanization resulted in fast deterioration in the quality of life for the mass living in the city. Prosperity and growth of a region depends upon the availability of infrastructure and services. Water and energy is the most crucial among these. Advance planning for development of such resources are necessary, taking into account the projected population of the region. Hence, in such scenarios the water energy food nexus provides a synergy and must be looked upon for optimization of resources to meet the future demands.

15.4 Mitigation and Management

The key mitigation and management strategies for optimizing the WEF nexus are:

15.4.1 From Master Plan to Integrated Development of Cities

India is growing very fast in terms of population as per the recent projections made by United Nations. Therefore, proper planning should be done by taking the cognizance of projections made for the future; otherwise the country will face unprecedented level of urbanization and unmanageable concentration of population. Master plans of the country have been unable to cope up with the pace of growth of Indian cities. Inefficient regional planning approach has led to proliferation of slums in cities. As per the 12th Five Year Plan of India very few Indian cities have 2030 master plans that take into account basic services like water, sanitation, food, transportation, roads etc. It is time to develop an integrated development of "smart cities" which aims at developing the urban ecosystem by strengthening institutional, physical, social and economic infrastructure. Ministry of Housing and Urban Affairs (MoHUA) has already taken a step by releasing a new Urban and Regional Development Plan Formulation and Implementation Guidelines (URDPFI) in 2015. The plan came with the objective of replacing the existing 1996 guidelines of formulating master plans in all cities of the country. Thus cities and towns need to be designed to be compact and connected, with energy efficient transport, green buildings, easy and secure access to water and food.

15.4.2 Sustainable Food Consumption

The food consumption pattern of India's urban and burgeoning middle class is changing with change in lifestyle and increasing prosperity. Research shows that there is increasing trend towards fast food and ready to eat food with more shares of animal products, sugar and fats. The change in food habit is not only causing health related issues like obesity and hypertension but significant amount of water and energy gets wasted at consumption level. According to FAO and UNEP, Sustainable Food System Program, which was adopted at the Rio+20 Conference in 2012, consumers exert strong influences through the ways they buy, transport, conserve, cook and consume their food. Food availability, food accessibility, food choice, geography, demography, socio-economic status, urbanization, culture, marketing and consumer attitudes etc. should be considered while developing holistic and effective strategies for promoting sustainable consumption as well as improving the food supply chain vis-a-vis efficient production which will address the nexus issues.

15.4.3 Governance, Institutions and Integrated Policy Approach

Urban planning and management in India is a state subject. With the enactment of the 74th Amendment Act of India in 1992, Urban Local Bodies (ULBs) like Municipal Corporations and Municipal Councils were constituted and entrusted with the task of town planning as well as social and economic development of the urban areas. Cities are growing beyond the municipal boundaries, hence, ULBs needs to be strengthened for the proper management of cities. Proper institutional mechanisms need to be relooked so that ULBs have their defined roles and responsibilities. Currently, India does not have policies that look at food, water and energy sectors in an integrated manner. Policies in the energy, water and food can only be successful if they are interlinked with each other, inclusive of climate change and environment policies.

15.4.4 Green Economy Approach

"Green economy" and "Green Growth concept" are new policy approaches promoted in many international conferences and seminars to address the WEF nexus. The green economy promotes low carbon, efficient utilization of resources and social inclusion as drivers for sustainable development. Investing more on renewable energy, increasing efficiency in irrigation system, sustainable agriculture, reducing food wastage, checking water wastage, increasing fuel efficiency, etc. will reduce the carbon footprint along with addressing the nexus issues. India has already taken steps towards developing a road map for transition towards green economies; however, more investment is needed on research and development focusing on green solutions.

15.4.5 Ecosystem Approach

Ecosystem-of which we are an integral part-provides us with many goods and services (regulating, provisioning, supporting and cultural services) that are required to sustain our livelihood. Organizations like International Centre for Integrated Mountain Development (ICIMOD) and United Nations Environment Program (UNEP) have promoted ecosystem based approaches in addressing WEF nexus policy dialogues. The ecosystems must be protected and enhanced to ensure the resilience of water, energy and food sectors.

15.5 Conclusion and Way Forward

This study reveals that increase in population, urbanization and economic growth will strain the demand-supply balance of water, energy and food resources and estimates show that by 2031 the average supply of water and energy per person will drop significantly putting stress on the natural resource base of the country.

According to Fifth Assessment Report of IPCC, climate change and anthropogenic greenhouse gas emissions are the functions of increasing population size, unsustainable economic activity, prosperity and change in lifestyle, intensive energy use, land use patterns, technology and climate policy of a nation (IPCC 2014). Thus, it is imperative to develop future adaptation strategies in order to reduce the risks and vulnerability of climate change impacts and contribute to the Goals of Sustainable Development for the well-being of human beings. Developing adaptation framework is very context specific (IPCC 2014; IPCC 2007) and needs to be integrated into development plans, policy and decision making to promote synergies across the sectors. This framework can be very well aligned with India's National Mission on water, solar energy, enhanced energy efficiency and sustainable habitat.

Energy and water is required at all stages of food production as well as consumption. The interrelationship is so sturdy and strong, that improving the efficiency in one process will have impact on the overall nexus elements. These interventions will also help to achieve the targets envisaged in Sustainable Development Goals by 2030 along with achieving the targets of National Action Plan on Climate Cshange. India needs to work on several areas to manage its developmental goal along with key challenges like urbanization and rising population.

Henceforth, the nexus approach identifies synergies between social, institutional, environmental and economic pillars of the society and mainstream the concept of "sustainable development" in planning at global, regional and national level.

References

Ahmad F, Haseen S (2012) The performance of India's food grains production: a pre and post reform assessment. Int J Sci Res Publ 2(3):1–15. http://www.ijsrp.org/research_paper_mar2012/ijsrp-Mar-2012-64.pdf. Accessed 27 July 2015

Amarsinghe UA, Shah T, Singh OP (2007) Changing consumption patterns: implications on food and water demand in India. http://nrlp.iwmi.org/PDocs/DReports/Phase_01/03.%20Consumption%20pattern%20changes-%20Amarasinghe%20et%20al.pdf

Bentham J (2011) Water, energy and food security in urban context. Published speech on World Water Week-Stockholm. http://s07.staticshell.com/content/dam/shell/static/aboutshell/downloads/our-strategy/shell-global-scenarios/speech-jeremy-benthamworldwaterwater24082011.pdf

Biggs EM, Bruce E, Boruff B, Duncan JMA, Horseley J, Pauli N, McNeill K, Neef A, Ogtrop FV, Curnow J, Haworth B, Duce S, Imanari Y (2015) Sustainable development and the water-energy-food nexus: a perspective on livelihoods. Environ Sci Policy 54:389–397

Dutta P (2006) Urbanization in India. http://www.infostat.sk/vdc/epc2006/papers/epc2006s60134.pdf

Emerson Report (2013) The food wastage and cold storage infrastructure relationship in India: developing realistic solutions. http://www.emerson.com/SiteCollectionDocuments/India%20Cold%20Storage%20Report%202013/Report_layout_Reduced.pdf. Accessed 28 July 2015
Govt. of India (2011) Size, growth rate and distribution of population. http://censusindia.gov.in/2011-prov-results/data_files/india/Final_PPT_2011_chapter3.pdf
Hellegers P, Zilberman D, Steduto P, McCornik P (2008) Interactions between water, energy, food and environment: evolving perspectives and policy issues. Water Policy 10(1):1–10
Hoff H (2011) Understanding the Nexus. Background paper for the Bonn 2011 conference: the water, energy and food security Nexus. Stockholm Environment Institute, Stockholm
IME (2013) Global food report. http://www.campaignforrealfarming.org/wp-content/uploads/2013/01/IME-Global-Food-Report.pdf
IPCC (2007) Contribution of working groups I, II and III to the fourth assessment report of the Intergovernmental Panel on Climate Change. Cambridge University Press, Cambridge
IPCC (2014) Climate change 2014: synthesis report. Contribution of working groups I, II and III to the fifth assessment report of the Intergovernmental Panel on Climate Change [Core Writing Team, R.K. Pachauri and L.A. Meyer (eds.)]. IPCC, Geneva, Switzerland, 151pp
Lundy J, Bowdish L (2013) The energy-water-food nexus: insights for the business community. US Chamber of Commerce Foundation
Mariana A (2008) Advances in urban ecology: integrating humans and ecological processes in urban ecosystems. http://link.springer.com/chapter/10.1007/978-0-387-75510-6_1. Accessed 14 April 2016
Mathur M, Sharma K (2016) Modelling urban carrying capacity and measuring quality of life using system dynamics. TERI, New Delhi
MEA (2005) Ecosystems and human well-being: synthesis. Island Press, Washington, DC
Meadows DH, Meadows DL, Randers J, Behrens WW (1974) The limits to growth: a report for the club of Rome's project on the predicament of mankind. Universe Books, New York
Meenakshi JV (1996) How important are changes in taste? A state-level analysis of food demand. Econ Polit Wkly:3265–3269
Middleton C, Allen S (2014) The (re)discovery of "the Nexus": political economies and dynamic sustainabilities of water, energy and food security in Southeast Asia. Paper presented at the Asia Pacific Sociological Association (APSA) conference "Transforming Societies: Contestations and Convergences in Asia and the Pacific", 15–16 February 2014, Chiang Mai, Thailand
Ministry of Agriculture, GoI (2011) Agricultural statistics at a glance. http://eands.dacnet.nic.in/latest_20011.htm. Accessed 27 July 2015
Moomaw W, Griffin T, Kurczak K, Lomax J (2012) The critical role of global food consumption patterns in achieving sustainable food systems and food for all. A UNEP Discussion Paper. United Nations Environment Programme, Division of Technology, Industry and Economics, Paris, France
Murty KN (2000) Changes in taste and demand patten for cereals: implication for food security in semi-arid tropical India. Agric Econ Res Rev 13(1):25–51
NCEAR (2014) An analysis of changing food consumption pattern in India. A research paper prepared under the project agricultural outlook and situation analysis reports Impact of globalization on food consumption pattern of urban India. http://www.fao.org/3/a-y5736e/y5736e02.pdf
Pathak H, Jain N, Bhatia A, Patel J, Aggarwal PK (2010) Carbon footprints of Indian food items. Agric Ecosyst Environ 139:66–73
Pimental D (2011) Food for thought: a review of the role of energy in current and evolving agriculture. Crit Rev Plant Sci 30(1-2):34–44
Planning Commission (2011) Report of the working group on urban and industrial water supply and sanitation for the twelfth five-year plan (2012–2017). http://planningcommission.nic.in/aboutus/committee/wrkgrp12/wr/wg_indu_sani.pdf
Planning Commission, GoI (2014) Report of the expert group to review the methodology for measurement of poverty. http://planningcommission.nic.in/reports/genrep/pov_rep0707.pdf

Rao CH (2002) Sustainable use of water for irrigation in Indian agriculture. Econ Polit Wkly 37(18):1742–1745

Rasul G (2014) Food, water and energy security in South Asia: a nexus perspective from the Hindu Kush Himalayan Region. Environ Sci Policy 39:35–48

Sinha G (2014) Linkages between food consumption pattern, food security and sustainable food system. A synopsis submitted for Doctor of Philosophy in Economics, Dyalbagh Educational Institute, Agra. http://shodh.inflibnet.ac.in:8080/jspui/bitstream/123456789/2060/1/synopsis.pdf

Swaminathan MS, Bhavni RV (2013) Food production and availability. Essential prerequisites for sustainable food security. Indian J Med Res 138:383–391

Swarnkar KN, Singh SN (2013) Analysis of electrical energy consumption of agricultural sector in Uttarakhand state. Int J Emerg Technol Adv Eng 3(3):343–347. http://www.ijetae.com/files/Conference%20ICERTSD-2013/IJETAE_ICERTSD_0213_53.pdf

The Planning Commission, GoI (2012) The challenges of urbanization in India: approach to 12th Five Year Plan. http://12thplan.gov.in/12fyp_docs/17.pdf

UN (2016) Sustainable cities: why they matter. http://www.un.org/sustainabledevelopment/wp-content/uploads/2016/08/16-00055K_Why-it-Matters_Goal-11_Cities_2p.pdf

UNDESA (2006) "World urbanization prospects: the 2005 revision". United Nations Department of Economics and Social Affairs- Population Division, New York. http://www.un.org/esa/population/publications/WUP2005/2005wup.htm

UNDESA (2012) World urbanization prospects: the 2011 revision. Department of Social and Economic Affairs, Population Division. United Nations, New York. http://esa.un.org/unup/pdf/wup2011_highlights.pdf

UNDESA (2018) The world's cities in 2018: data booklet (ST/ESA/ SER.A/417). https://www.un.org/en/events/citiesday/assets/pdf/the_worlds_cities_in_2018_data_booklet.pdf

UNIDO (1995) Food processing industry. Output of a seminar on energy conservation in food processing industry. https://www.unido.org/fileadmin/import/userfiles/puffk/food.pdf

Vepa SS (2004) Impact of globalization on food consumption of urban India. http://www.fao.org/3/a-y5736e/y5736e02.pdf

World Economic Forum (2009) Thirsty energy: water and energy in the 21st century. World Economic Forum in partnership with Cambridge Energy Research Associates

Dr. Swati Singh, She has more than nine years of professional experience in the area of water resource management, climate change adaptation, environmental laws and disaster management with proven contributions to policy planning, research, knowledge building, advocacy and capacity building. She has worked with National Institute of Disaster Management, GIZ, Development Alternatives and Sir Ratan Tata Trust-CInI in various capacities. She obtained her PhD from TERI School of Advanced Studies (TSAS) and is also an alumnus of Young Researcher's School, United Nations University, Japan. She is currently working with UNDP as an Adaptation Analyst.

Dr Shresth Tayal, working as Senior Fellow with The Energy and Resources Institute (TERI), New Delhi, is acting as Area Convenor of its Centre for Himalayan Ecology and responsible for coordinating integrated research at high altitudes, for adaptation of local communities in Himalayas. He obtained his doctoral degree from Jawaharlal Nehru University, New Delhi, on 'Glacier Hydrological Systems'. With his 17 years of experience in the field of hydrology, he is involved in a number of projects related to hydrological modelling, integrated water resources management (IWRM), and analysing the impacts of climate change on water resources of the region. He has been involved in various hydrological, hydrochemical, and meteorological research in glacierized watersheds, and vulnerability assessment of water resources using hydrological modelling tools at the watershed level.

Chapter 16
Water Governance Transitions Pathway: Adaptive Water Governance

Chitresh Saraswat and Anil K. Gupta

Abstract Climate change is adversely impacting all three pillars of sustainability (Environment, Economic, and Social). Arguably, the most impacted natural resource and through which its negative impacts are strongly felt is 'water' (in the form of water insecurity). In developing countries, climate change, rapid urbanization, unplanned development, deforestation, increasing pollution and other factors resulted in increased uncertainties in managing water. Academic research showed that improving water governance is a key in dealing with uncertainties of water management and achieving water security. India's national water policy also acknowledges that most of the challenges faced by the country's water sector are governance related issues rather than technical and are not addressed adequately. In this background, this study emphasis on exploring the new forms of water governance: which are adaptive and equipped in dealing with the uncertainties and the challenges of Indian water sector. This chapter argues that adaptive water governance (AWG) is a desirable alternative in Indian context, as it presents foundation and strategies to transition towards achieving adaptation and resiliency. The study analysed the current inefficiencies of the water governance system in India to meet future demands and presented adaptive system as a transition pathway to achieve the sustainable water supply employing adaptive business practices and demand management by enhancing capabilities, integrating innovative solutions in Indian water sector.

Keywords Adaptive water governance · Disaster risk reduction · Water security · Institutional innovation · Sustainability

C. Saraswat (✉)
Fenner School of Environment & Society, Australian National University, Canberra, ACT, Australia

A. K. Gupta
ECDRM Division, International Cooperation, CAP-RES DST Project, National Institute of Disaster Management (Ministry of Home Affairs), New Delhi, India

A. K. Gupta et al. (eds.), *Disaster Risk and Management Under Climate Change*, Disaster Resilience and Green Growth,
https://doi.org/10.1007/978-981-99-4105-6_16

16.1 Introduction

Climate Change is adversely impacting the economic, social and environment, the pillars of sustainability (Hansmann et al. 2012) and presenting significant challenges in achieving sustainable development in Global South (World Bank 2014). Arguably, the most impacted natural resource and through which its impacts are strongly felt is "Water" (IPCC 2014b). Even, the Intergovernmental Panel on Climate Change (IPCC), in its fifth assessment report mentioned that directly or indirectly, more than 90% of climate change impacts will be felt on water sector (IPCC 2014a, 2014b) worldwide. In the developing economies the climate change along with other external factors such as rapid urbanization, growing population, economic growth, unplanned development, unsustainable lifestyle, pollution, and others, resulted in increased uncertainties, which increase the water insecurity (Qureshi et al. 2013; Hanjra and Qureshi 2010). Water is a basic necessity for the ecosystem and regarded as the foundation of life (UN-Water 2012; UNESCO and United 2009), which is gradually transforming into a crisis due to an ever-growing water demands, urban growth, pollution, ineffective governance, landscape changes, lack of awareness, excessive water withdrawal and available water depletion, mainly because of the anthropogenic activities (Bakker 2012; Piao et al. 2010) and climate change is worsening the situation. Magnitude of a situation is dependent on local condition but changes such as urbanization (changes in runoff patterns, pollution, and infiltrations), deforestation (retaining and recharging of water), precipitation frequency, global warming, wetlands reduction, and others have substantial impacts on water resources (Falkenmark and Rockström 2010; Mishra et al. 2017; Rockstrom et al. 2007; Saraswat et al. 2016; UN-Water 2012).

The General Assembly of the United Nations (UNGA) recognized the importance of challenges faced by the water sector (Pahl-Wostl et al. 2013, 2013; UN-Water 2012). The adopted resolution "Transforming our world: the 2030 Agenda for Sustainable Development" (Agenda 2030) have a water-specific goal (SDG 6). 2030 Agenda represents the today's most relevant globally negotiated normative framework on sustainability and involves 17 Sustainable Development Goals (SDGs) and 169 targets (Loewe and Rippin 2015; Saraswat et al. 2017; Sengupta 2016; UN 2015). In which, SDG 6 is focused on ensuring universal access to safe and affordable drinking water for all by 2030, this requires a considerable investment in water infrastructure and sanitation facilities (Hutton and Varughese 2016). Achievements regarding water specific goals are also highly relevant for other SDGs such as making communities and cities more resilient, safe, and sustainable (SDG 11) and ensuring a healthy life and promote well-being for everyone (SDG 3), as there are multiple synergies and trade-offs between the SDGs (Group et al. 2015). Water governance plays an essential role in achieving SDG6. Previous academic research shows that improving water governance is a key to deal with uncertainties of water management to achieve transitions towards sustainability (Araral and Wang 2013).

To achieve water security in India, it is essential to encourage the water efficiency (Bogardi et al. 2012; Vörösmarty et al. 2010), integration of new technology and promoting effective water governance at national and state level (Meene et al. 2011). India's national water policy 2012, acknowledges that most of the challenges faced by the country's water sector are governance related rather than technical and are not addressed adequately (Ministry of Water Resources, River Development, and Ganga Rejuvenation 2018; Ministry of Water Resources NWP 2012; Ministry of Water Resources NWP 2002). In contrast to developed countries, water consumption in India is more into agriculture sector (Mekonnen and Hoekstra 2016; Ministry of Water Resources 2012), more than 85%–87% of available water used for agricultural use, 8–10% for industrial use and rest for domestic purposes. The sustainable use of water is approaching or exceeding the limits (Gleick and Palaniappan 2010; Mishra et al. 2017) and the competition for water in agriculture and industrial sector is intensifying as the size of cities are expanding as well as their political influence (Bahri 2012; Mishra et al. 2017). Here we argue that Indian water sector is not very efficient with the water use in agricultural and industrial developments and ineffective governance plays a significant role in inefficiency (ADB 2013; Hanasaki et al. 2013; Qadir et al. 2010).

According to Indian government data, between 2001–2011 the average annual per capita available water fell 15%, and it will keep falling by another 13% and 15% by 2025 and 2050, respectively (Ministry of Water Resources 2012). This means that in the next 30–35 years, which the currently available water per capita consumption will increase to around 170 liters per day. In that case, India will be the most water scarce and highest water demanding country by 2050 (Biswas and Tortajada 2019). Water crisis and water wars: Myths and realities. Water scarcity leads to unsustainable ecosystems, affects flora and fauna, deteriorates health conditions, destroys livelihoods and inflicts unnecessary suffering for poor (Ejaz Qureshi et al. 2013; Hanjra and Qureshi 2010). Achieving water security and transitions towards sustainability, therefore, one of the biggest challenges our generation is facing (Bakker 2012; Biswas and Seetharam 2008; Hanjra and Qureshi 2010; IPCC 2014a, 2014b). The transitions towards sustainability comprises of developing clean potable water, managing wastewater, use water efficiently and providing necessary sanitation facilities (Wouters 2008).

Against this background, this study emphasis on exploring the new forms of water governance in India, which are based on adaptive business practices and capable of integrating innovative solution to transition towards sustainability. Here, we discussed whether the current water governance system is equipped to provide adequate responses to uncertainties and water security challenges and make a case of water transitions from centralised predict and control approach with an emphasis on technical and engineering solutions towards an adaptive and integrated water management approach, which focuses on social learning and adaptive capacity. The study presented adaptive system as a transition pathway to achieve the sustainable water supply employing adaptive business practices and demand management by enhancing capabilities, integrating innovative solutions in the Indian water sector.

16.1.1 Water Governance Transition: Adaptive Systems

Amidst of increasing uncertainty posed by changing climate to meet future water demands and sustainable water supply, the current water governance system must transition towards adaptive management practices and effective institutional arrangement (Araral and Wang 2013; UN-Water 2012). The current water governance system is facing challenges of inefficient management and ineffective coordination between different water institutions (Araral and Ratra 2016; Araral and Wang 2013). The water governance transition pathways aim to achieve sustainable water supply and demand management, which requires a transformation from supply-only oriented practices towards adaptive water management to adapt the challenges of climate change and external pressures such as urbanization, pollution, and land-use changes. Water-related institutions encouraging adaptive practices are also crucial to address the ineffective coordination and inefficiencies of management practices, which hinders the water sustainability.

The United Nations Development Program (UNDP), Water Governance Facility, described the water governance as a system of institutions dealing with water service allocation, effective and equitable water distribution, social issues, applying integrated water management approaches, focusing on water conservation by balancing the water use for ecosystem and socioeconomic activities along with the strong political will (UNDP, SIWI,, and Facility 2009; Wouters 2008). This clearly defined the roles of civil society, government institutions, management, and ownership of the private sector. This definition covers most of the critical agenda of water challenges in the Indian context and issues described in the last section but fall shorts on the details of the implementation level. Based on the review of water governance practices, Ararl and Wang, 2013, concluded that current government practices and definition of water governance are not entirely fit and there is need for exploring pathways to transitions towards water sustainability that pays attention to the incentives-based system and multidisciplinary with clear policy implications (Araral and Wang 2013). This gives an exciting insight, but the importance of enhancing capacities of system, integrating, and adaptive business practices were not discussed.

Academic researches recommended multiple pathways to achieve a transition towards sustainable practices and recognized the importance of effective water governance systems in achieving water security in developing countries (Araral and Ratra 2016; Araral and Wang 2013; Bakker and Cook 2011; Cook and Bakker 2012). In India, the urban water system is facing external and internal pressures, changing climate, and new challenges, which require a significant transformation from traditional (fixed rule-based) to adaptive and participatory practices. The authors, Briscoe and Malik 2008, pointed out that India is at more risks in comparison to the other developed countries due to its high population growth, rapid urbanization, and its inability to handle new challenges due to weak institutional arrangements (Briscoe and Malik 2008). The study is presenting the adaptive water governance system as a pathway to achieve sustainable water supply and demand management in the Indian water sector. This pathway aims to achieve transition by enhancing and developing the capabilities of the current water governance system to adapt changes, integrating innovative solutions, and removing barriers. This study argues that the water governance transitions pathway towards adaptive water governance systems are necessary to predict, manage and cope with the uncertainties of the water sector at the national level, which will reflect regional/state level in India.

Adaptive water governance system is considered as an essential driver to address uncertainty implementing good governance, sustainability, and resiliency in the water sector by enhancing the capacity to learn while managing (Akamani 2016; Rogers et al. 2003). Adaptive management of the water sector increases the flexibility of system and ability to predict and manage uncertainty by employing management programs to improving the efficiency of water-related institutions, public-private participation (Rola et al. 2016; Pahl-Wostl et al. 2008). The authors Godden et al. in 2011 argued that the need for an adaptive system in water management by concluding that concluded that well-structured water governance is vital in developing the develop the adaptive capacity of a water system (Godden et al. 2011; Kashyap 2004). IPCC, in its fifth assessment report in 2014, noted the need for innovation, infrastructures, well-structured governance, and economic wealth in adaptation strategies (IPCC 2014b). It argues the good governing principles and efficient institutions enhance the adaptive capacity of the system. The institutions help in shaping the national/local water policies, pave paths for innovation, and spatial planning (Meene et al. 2011).

To conclude, water governance is a range of the social, economic, environmental, political, and administrative systems to manage an equitable distribution of water (Wouters 2008). A well-balanced structure of water governance deals with the negative impacts of climate change and builds adaptive capacity by providing required feedback to build the framework in local context. This can have different organizational settings under public or private ownership and discourse.

16.2 Operationalization & Key Features of Adaptive Water Governance

The importance of enhancing the adaptive capacity of the system is of utmost importance not only in the water management but disaster risk reduction, building resilience strategies, and ecosystem resource management (Dietz et al. 2008; Munaretto et al. 2014). The studies concluded that the focus-shift from traditional (fixed rule-based institutions) to dynamic, participatory and flexible management is essential to enhance the adaptive capacity of the system (Pahl-Wostl and Knieper 2014; Pahl-Wostl et al. 2012; Pereira and Ruysenaar 2012). The authors reasoned that increasing the adaptive capability of the system is a cyclic process of understanding problems and formulating policy to implementing it along with monitoring and providing feedback to re-evaluate the problem (Ayre and Nettle 2017), as shown in Fig. 13.2. The learning-based approach, responsiveness, and user-oriented policymaking are essential key factors in enhancing the adaptive capacity of the water system (Pahl-Wostl et al. 2012; Rola et al. 2016). As shown in Fig. 13.2, the adaptive water governance frameworks rely on the interaction between the different set of stakeholders in multi-layered institutions operating at different scales and adaptive capabilities in its core (Dietz et al. 2008; Jordan et al. 2015; Olsson 2004; Ostrom 2014). The adaptive water governance builds the system capacity to learn from previous experiences, evolve to be capable of responding to the uncertainty (Godden et al. 2011; Shinn 2016; Walch 2018). Under this background, it is understood that the adaptive water governance systems results in being resilient due to its ability to handle uncertainties exerted by climate change and deal with complexities of the water sector (Huitema et al. 2009). The chapter argues that collective actions in the form of stakeholder involvement in decision-making along enhancing adaptive capacity of the system and effective institutional arrangements reforms can play a significant role in building resilience against climate change in India (showed in Fig. 16.1).

The Indian water governance system is facing multilevel challenges at the national and state level. The current governance structure prioritizes supply-side management, augmenting, and maintaining the water supplies to the citizens besides managing the demand effectively. The inability of the system to focusing on demand-side management is due to the limited capabilities of centralized structure and command-control bureaucracy, which is responsible for administering water at the national and state level. The ineffective institutional arrangement discourages the transparency and limits the access to water information for the researchers and stakeholders. Under the current institutional structure, water management and development of water resources are entirely state subjects. There is a need for effective coordination between national and state water institutions as the national government provides financial resources for the project of national interests, and state government responsibility is to manage and maintain. State water institutions are regulatory authorities, water department, public works, and irrigation departments.

National water institutions are coordinating with state-level institutions for policy implementation, financial resources, and administration, but the state has full authority over water in the state's boundary. Mostly at the state level, water is managed by municipality and district level bodies, and they are responsible for policy formulation, implementation and water service provisioning, infrastructure maintenance, and other activities. The water resource department implements this in municipality or public water utility. Generally, water utilities are underperforming, and a massive improvement in infrastructure and capacity is required. The authors argued that the reason behind such lag is the complexity of the decision-making system, conflicts of water rights between states, poor coordination and implementation of policies, and lack of water expertise among water planners (Araral and Ratra 2016).

The chapter argues that complex interplays in the Indian water governance system are well suited to benefit from the implementation of adaptive water system. As fundamental principles of the adaptive system is that the environmental changes are becoming more complex, so the water management system has to develop the

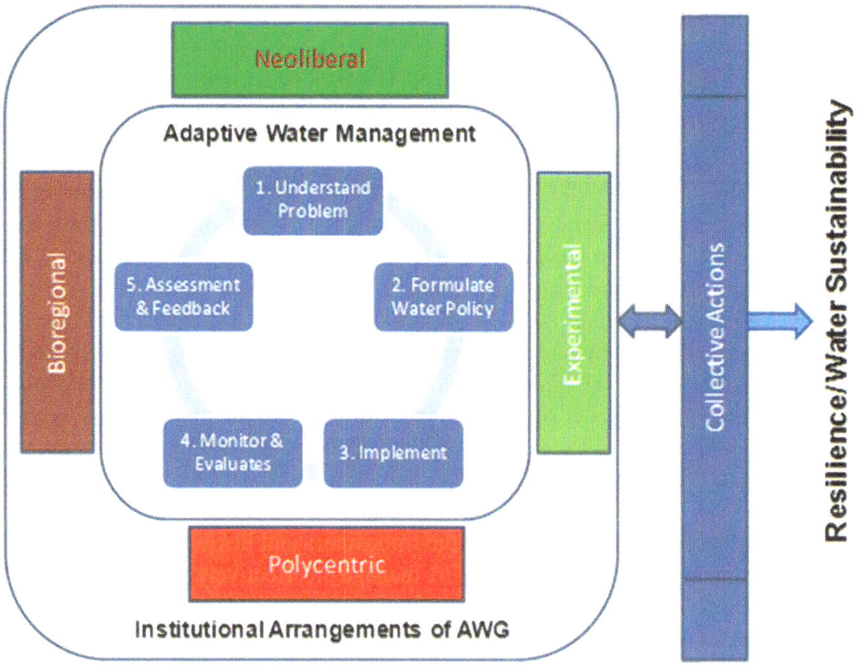

Fig. 16.1 Synthesis of adaptive water governance processes, institutional arrangements, and resilience to climate change (Adapted from Ayre and Nettle 2017; Huitema et al. 2009)

ability to learn to adapt changes (Cosens 2015; Huitema et al. 2009). To develop this ability it is essential to incorporate the understanding of water-ecosystem dynamics into water management and planning, the utilization of an ecological knowledge system able to interpret and respond to ecosystem feedback, and the monitoring of external drivers to deal with uncertainty and surprise (Folke 2006) in Indian water governance system. The key feature of adaptive system is based on the collaboration of different stakeholders operating at a different level (from local level organizations to international level) and the governance frameworks relies on this collaboration between multilevel institution and organization to learn to adapt (Jordan et al. 2015; Olsson 2004; Pahl-Wostl 2009). As explained, Indian water governance is complicated due to multilevel interventions, and collaboration are ineffective in such conditions. To improve that, the water utilities need to play an essential role by providing the leadership and direction (Jensen 2017; Olsson 2004) in India.

16.3 Management Strategies

16.3.1 Characteristics of Adaptive Water Governance: Indian Context

The literature categorized the multiple characteristics and evaluating criteria of the adaptive water governance system, but this chapter focuses on the most relevant characteristics suitable in the Indian water governance context. To enhance to the adaptive capacity of water system the governance system should be 'receptive and reactive' (Diaz and Hurlbert 2013), it is the ability of institution (water utilities and others) to respond on time and appropriately to climatic variability and uncertainty (Dietz et al. 2008; Hatfield-Dodds 2006). As explained, focus Indian water planning is towards augmenting water supply; this focus must be shifted to managing demand effectively and using it sustainably. Enhancing the reactive and receptive approach in the Indian water governance system will lead to more integrated & holistic planning.

Another characteristic is to develop the 'robustness and flexibility of the system,' which are the policies of the system to manage it effectively during periods of changes and challenges. Also, it creates flexibility in the system to identify gaps and adapt timely (Huntjens et al. 2012; Huntjens et al. 2010; Mollenkamp and Kasten 2009). Implementing this characteristic in policymaking positively encourages the Indian water governance system to reformulate current policies to focus on local context and avoiding centralized planning to enhance the robustness and flexibility in the system. The third critical characteristic of adaptive system is 'transparency and openness,',' which is that the system allows access to water data and information to researchers and stakeholders smoothly (Gupta et al. 2013). The lack of transparency in Indian water governance exacerbated the quality and conflict situation. Accessing required information in a form that is used to primary stakeholders and openness to discuss the water solution is essential in India.

Another critical characteristic of adaptive system is 'sectoral integration,' which is required to integrate the sectoral responses to crisis situation so that similar institutional arrangement can be adjusted to each other's (Mollenkamp and Kasten 2009) to be useful In India, water issues are divided between different sectors such as surface, ground, domestic water and other, which leads to communication gaps. Strong coordination is required to integrate the different water sectors. Another significant characteristic is 'feedback and learn,' the system should remember the experiences, and the routines should be improved based on it (Armitage 2005; Dietz et al. 2008; Huntjens et al. 2012; Olsson 2004). Indian water governance system is considered uni-dimension and disciple; the focus is narrowed to engineering solutions for water problems such as constructing the dam and extraction. The feedback from previous solutions/crisis is essential to learn and prepare for the next policy change.

The adaptive system should be 'participatory and collective,' there should be participation from non-state actors and enhance the involvement from the stakeholders involved in decision making regarding adaptation policies, and it needs to be regularly monitored (Folke 2006; Huntjens et al. 2012). As mentioned in the previous section, there is an ineffective coordination between national and state-level water institutions in India and bureaucracy, which limits the ability to respond promptly. To achieve the objective, the similar kind of institutional structure on both level should be encouraged. Another vital characteristic is 'trustworthy and open to experiment,' the system should be an institutional pattern to promote mutual trust, and policy experiments should allow feedback loops (Huntjens et al. 2012). In India, the trust deficit among stakeholder and community towards water institution is relatively higher in comparison to the other sector. There is strong need of social interaction from water utilities to explain and build confidence in community. The essential characteristic of the adaptive system is 'legitimacy and equitable,' the governance regime must be perceived as legitimate and accountable, as well as fair in its process and impact (Gupta et al. 2010; Huntjens et al. 2012; Ostrom 2014). Another important key feature of the adaptive system, it should be 'innovative,' it needs to be engaged in research and development to explore the further option for adaptation (Bakker and Cook 2011; Kallis et al. 2009). Indian governance system is focused on equitable access to everyone, as government policy, but this needs to be combined with the awareness to end case based inequality and laws to curb the water distribution inequality based on economic condition. To achieve the objectives, innovation is the area where the Indian water sector has to focus. Implementing an adaptive system can act as a catalyst to achieve social, product, infrastructure, management, and administration innovation in Indian water governance.

At the same time, the governance regime should seek and have 'strong political support,' as responding changing climate is the long-term process and required visionary policies that require reliable political support for more extended plans (Mollenkamp and Kasten 2009). Finally, few others characteristics, which are worth to mention is, 'build capacity' to deal with the changes, 'informative,' need to be updated and able to 'clearly defined the boundaries' in water conflicts (Gupta et al. 2010; Mollinga and Tucker 2010). Such critical characteristics of an adaptive water system based on the idea that what abilities adaptive governance systems should have to adapt new changes and integrate to build a governance system capable of responding effectively to the uncertainties and water challenges in the Indian water sector. Adaptive water governance should not be understood only in terms of managing water but also in terms of innovation, organizational efficiency, performance, reducing disaster-related risk, increasing awareness, improve social cohesion and opportunities for participation in adaptation processes (Diaz and Hurlbert 2013; Shinn 2016; Walch 2018). The chapter analysed that implementing adaptive system will benefit country's water governance by enhancing key characteristics and provide a pathway for transition towards sustainability.

16.3.2 Institutional Arrangement of Adaptive Water Governance

Institutional arrangement in water governance decides how quickly and effectively the system can make decisions and shape the water policies on various levels, which drive the behaviour of water services and overall organizational level efficiency (Huitema et al. 2009). It is crucial to understand the institutional arrangements of adaptive water governance System. Huitema et al. 2009, highlighted the four kinds of institutional arrangement or approaches for the adaptive management system, all apply to water management. The first is polycentric governance system, second is neoliberal approaches in water management, third is an experimental approach to water (resource) management, and the final one is management at the bioregional scale (Huitema et al. 2009). In this section, we discuss the four institutional arrangements for adaptive water management and analyse its suitability for the Indian water sector. India water system is complex in nature and to analyse that it is well-suited for adaptive water governance system is based on a range of factors, but here our focus is the crucial feature which is significant to enhance the adaptive capacity of the Indian water sector.

The pointed out the five requirements of adaptive system as information sharing, conflict resolution, encouragement of adaptation measures, following rules and infrastructure development (Diaz and Hurlbert 2013; Dietz et al. 2008). These requirements of adaptive system encourage the focus on new institutional arrangements. One of which is polycentric institutions, the concept based on Ostrom's work (Ostrom 2014). The definition of polycentricity suggests that "it implies many centres of decision-making which are formally independent of each other" (Araral and Hartley 2013; Carlisle and Gruby 2017; Jordan et al. 2018; Jordan et al. 2015; Marshall 2015; Ostrom 2014; Vaas et al. 2017). This provides an opportunity to speculate any policy subsystem that is recognized as having polycentric features, i.e., a policy subsystem with many independent centers of decision-making (Araral and Hartley 2013; Pahl-Wostl et al. 2012; Tan-kim-yong et al. 2003). This approach to governance embraces complexity and use it to make a decision (Bakker and Morinville 2013) because of this it is considered more resilient than other monocentric systems and it can respond effectively under uncertainty (Bakker and Morinville 2013; Bogardi et al. 2012; Jordan et al. 2015; Huitema et al. 2009; Pahl-Wostl 2009).

The first mode of governance to discuss is the polycentric governance system, which have multiple independent centres of decision-making and authority is dispersed in overlapping jurisdictions (Araral and Hartley 2013; Ostrom 2014; Pacheco-Vega 2013). It is considered desirable in governance literature theoretically, however there is little literature available providing empirical evidence to prove the fact that polycentricity enhance the adaptive capacity of system and performance. As in India, Indian water resources are divided between multiple states and jurisdictions over the country provides a platform to look polycentric water governance as a desirable solution from the Indian perspective.

Another is neoliberal approaches in water or public-private partnership, which is the collaboration between governmental and non-governmental stakeholders (Jensen 2017). In India, water governance is historically are governed under public service provisions (Wu et al. 2016). Neoliberal water governance systems often hailed and promoted by an international organization such as the World Bank and International Monetary Fund (IMF). India recently started experimenting with these types of governance, the government of India, also hails its first experiment as success in Nagpur city, India but there are controversies around this approach, as the exploitation of environmental resources and the displacement of locals. There are many critiques of neoliberal water reforms (Bakker 2012), but to enhance the adaptive capacity of the water system, this could be a success based on strong regulation, accountability, transparency, and innovativeness. Also, this could be only the success if the process is well defined for stakeholders, which are supposed to be the part and correctly reflect the concerns of stakeholders (Huitema et al. 2009; Lieberherr and Truffer 2015). This seems most relevant and more comfortable to implement as the first step as adaptive water governance, with provision/factor contributing transparency, equity, and effectiveness.

The third one, suggested by Huitema et al. 2009, is "Experimentation" approach, which is comprised of two methods, one is as a research methodology, which recognizes our limited knowledge or as an approach to management that emphasizes learning from experience (Huitema et al. 2009). According to analysis, Huitema et al. 2009, the experimentation as a research methodology are hypothetically effective, but governance scholars considered it more appropriate in technical issues than for goal reflection. But this as an approach to management found hypothetically sound, but scholars strongly prefer small-scale experiments due to high risk and cost involved (Huitema et al. 2009; Jordan et al. 2018). This makes it difficult in implementation in Indian perspective. The bioregional approach or river-basin approach is seen as an opportunity to improve the governance system in the ecosystem. It includes collaboration at river/basin level, the scale of the ecosystems, and the governance system. The effectiveness of this system depends on a range of factors such as availability of resource and stable sources of funding and seems not very fit for Indian perspective (Huntjens et al. 2010; Huitema et al. 2009).

16.4 Conclusion

16.4.1 Water Transition in India: Towards an Adaptive Water Governance System

Currently, Indian water governance is facing numerous challenges such as rapid urbanization, unplanned development, growing population, pollution, and others, which results in uncertainties in managing water. The impacts of changing climate are also adversely affecting the water availability, hampering service provision and resulting in water insecurity in India. Arguably, the most impacted natural resource in India is water, in the form of water scarcity. The IPCC fifth assessment mentioned that the most challenging issue world is facing is climate change.. In particular, India is more vulnerable to the impacts of climate change due to the limited capacity to cope up with disasters, limited ability to recover, and high risk of water crisis. The transformation towards sustainability required effective water governance. As discussed, Indian water governance lags and currently is not capable of transitions. The reason of lag is complexity of Indian water governance, conflict, poor coordination and ineffective management. To achieve sustainable development, in general, water security, there is a need for strengthening the water governance system in India.

The adaptive system principles are considered as a significant catalyst to improve the water governance capabilities to address the uncertainties and enhancing resiliency by learning from feedback. The adaptive system also increase the flexibility of system by enhancing the ability to predict and manage uncertainty by employing management programmes to improving the efficiency of water-related institutions, public-private participation. Previous academic research shows that improving water governance is a key to deal with uncertainties of water management to achieve water security. India's national water policy also acknowledges that the most of the challenges faced by the Indian water sector are governance related and there is a need of exploring new form of water governance for sustainability and dealing with uncertainties. This presents a strong case for water transitions in India from centralised predict and control approaches with an emphasis on technical and engineering solutions towards an adaptive and integrated water management approach focusing on social learning and enhancing the adaptive capacity of the water sector. Overcoming the water security challenges in India requires water transitions and its emphasis on exploring the new forms of water governance, which are adaptive and robust enough to deal with sustainability and uncertainties in the Indian water sector.

Indian water sector dependences on centralised predict and controlled, command and control, i.e., monocentric governance, which limits the opportunity to effectively address the uncertainties and risks. The focus of reform of the Indian water sector is more focused on technical and engineering solution, which is also crucial for a developing country such as India but to achieve sustainability, there is a need to do more. This study focused on exploring the adaptive and integrated water management approach with an emphasis on social learning and adaptive capacity. According to the existing research and analysis that in comparison to monocentric governance, the adaptive governance is more resilient to uncertainties and threats posed to water security. Adaptive management/governance can tackle the uncertainties effectively. Recently, internationally, there is an increase in the research agenda of adaptive water governance, but there are very few researches available on implementation level (Diaz and Hurlbert 2013; Walch 2018). Also, there are minimal examples of functioning adaptive water governance, mostly in developing countries (Diaz and Hurlbert 2013; Huitema et al. 2009). So there is an important question that is it possible that adaptive water governance can work in complexities of India, where the state has limited resources (Cosens and Chaffin 2016; Karpouzoglou et al. 2016). Majority of studies have evaluated the adaptive water governance are in the area of river basin management, trans-boundary water resource management, and IWRM (Rouillard 2012). Also, discussed that which type of institutional arrangements is well-suited in Indian context, and concluded that polycentric water governance and neoliberal approaches present the better deal to explore further.

In this chapter, we explored the relevance of adaptive governance system to enhance the adaptive capacity of Indian water management to face uncertainty, risks, and increase resiliency. The chapter aims to accelerate the discussion on water security challenges in India and to highlight the current water governance system adequacy in providing responses to water challenges. For the same, the adaptive water governance's key feature, characteristics, and institutional arrangement were discussed to understand that to what extent the current water governance system in India are capable of enhancing the adaptive capacity of the water sector. Instead of debating the institutional strengths and weakness, this study focused on the reviewing institutional arrangement that can enhancing the adaptive capacity of system. The research presented the adaptive water governance key features, characteristics, and the institutional arrangement to offer an alternative to current inefficient Indian water governance system. The chapter contributes to the policy discourse of the adaptive governance in enhancing adaptive capacity, and water security in India, in support of the United 'Nation's sustainable development goals (SDGs).

16.5 Way Forward

The study asserts focusing on adaptive water governance to achieve sustainability transitions and building resilience against climate change in India. The adaptive system is capable of addressing the need for implementing strategies to increase water usage efficiency in agriculture, encouraging rainwater harvesting policies, preventing the unregulated abstraction of groundwater, encouraging recycling and treatment in Indian context. Another aspect is integrating innovation in water management to minimize the loss of water along with promoting awareness on water conservation in India. These are the essential features mentioned in Indian water policy (Ministry of Water Resources NWP 2012; Saraswat et al. 2016), enhancing the scope of national water mission under national action plan on climate change and national water plan is utmost important.

1. This encourages the policy and decision making to design measures such as increasing water storage capacity by focusing on stormwater management, catchment level rainwater harvesting schemes, and adapting using ponds and tanks to store water at the local level will help to deal with water uncertainty and transformation towards sustainability.

2. Controlling water flow and flood by canal system and strong water distribution network, strict measures to deal with any kind of water pollution is needed to be incorporated in Indian national water policy.
3. Transitioning from centralized approach towards adaptive learning is recommended for transformation towards sustainability.
4. Establishment of water institution deals with overall governance and water policy and work directly under Ministry of water resources with nodal offices, which have strong presence and understanding of major river basin in India.
5. There is acute lack of capabilities due the leadership is trained to apply engineering solutions of most of the problem. Multi-disciplinary skills are required in water institutions.
6. Investment in innovation and research to develop capabilities to understand mechanism of effective water governance.
7. Strengthening the capabilities of local level water institution and utility by managing partnership and collaboration with local academic institutes and practioners in water management.
8. Developing transparent and online accessible data management system on water, which primary stakeholders can access for knowledge/ research purposes.
9. Strengthening the local capabilities by grooming a new generation of water managers are keys to building a resilient and secure water future for India. Also, there are many Indian working outside the country in the field of water management, inviting them to work in India could be a good start to bring knowledge and develop local capabilities in the country.
10. Another important recommendation is to focus on neoliberal approaches in the water sector, Public-Private-Partnerships (PPP), generally, viewed as innovative and efficient and India can surely get the advantage if the strong regulations are in place, which focuses on improving water governance system rather than profiting specific individuals and companies. With strict rules and effective water policies, PPP could bring efficiency and innovativeness to Indian water management and involving the private sector, public sector along with other stakeholders, civil societies and NGOs will have a positive impact on sustainability transitions of water governance.

The chapter argued that integrating adaptive system principles current governance system of India will transitions towards sustainability. Another important topic to discuss is that, what are conditions that encourage governments to shift from conventional public provision of water management to adaptive water management (Walch 2018). Successful implementation of adaptive water governance in India depends on the enhanced awareness, political willingness and capacity of decision and policy-makers.

References

ADB (2013) Asian water development outlook 2013: measuring water security in Asia and Pacific. Asian Development Bank, Philippines. https://doi.org/10.1080/07900620701760556

Akamani K (2016) Adaptive water governance: integrating the human dimensions into water resource governance. J Contemp Water Res Educ 158(1):2–18

Araral E, Hartley C (2013) Polycentric governance for a new environmental regime: theoretical frontiers in policy reform and public administration. International conference on public policy, panel: polycentric policy and the environment, pp 1–31

Araral E, Ratra S (2016) Water governance in India and China: comparison of water law, policy and administration. Water Policy 18:14–31. https://doi.org/10.2166/wp.2016.102

Araral E, Wang Y (2013) Water governance 2.0: a review and second generation research agenda. Water Resources Management 27(11):3945–3957. https://doi.org/10.1007/s11269-013-0389-x

Armitage D (2005) Adaptive capacity and community-based natural resource management. Environ Manage. https://doi.org/10.1007/s00267-004-0076-z

Ayre ML, Nettle RA (2017) Enacting resilience for adaptive water governance: a case study of irrigation modernization in an Australian catchment. Ecol Soc 22(3). https://doi.org/10.5751/ES-09256-220301

Bahri A (2012) Integrated urban water management. TEC Background Papers. https://doi.org/10.1080/09614520701469427

Bakker K (2012) Water security: research challenges and opportunities. Science 337:23–24. https://doi.org/10.1126/science.1226337

Bakker K, Cook C (2011) Water governance in Canada: innovation and fragmentation. Int J Water Resour Dev 27(2):275–289. https://doi.org/10.1080/07900627.2011.564969

Bakker K, Morinville C (2013) The governance dimensions of water security: a review. Philos Trans Royal Soc A. https://doi.org/10.1098/rsta.2013.0116

Biswas AK, Seetharam KE (2008) Achieving water security for Asia. Int J Water Resour Dev. https://doi.org/10.1080/07900620701760556

Biswas AK, Tortajada C (2019) Water crisis and water wars: myths and realities. Int J Water Resour Dev 35:727

Bogardi JJ, Dudgeon D, Lawford R, Flinkerbusch E, Meyn A, Pahl-Wostl C et al (2012) Water security for a planet under pressure: interconnected challenges of a changing world call for sustainable solutions. Curr Opin Environ Sustain. https://doi.org/10.1016/j.cosust.2011.12.002

Briscoe J, Malik RP (2008) India's water economy: bracing for a turbulent future. Water. The World Bank, Agriculture and Rural Development UnitSouth Asia Region

Carlisle K, Gruby RL (2017) Polycentric systems of governance: a theoretical model for the commons. Policy Stud J 47(4):1–26. https://doi.org/10.1111/psj.12212

Cook C, Bakker K (2012) Water security: debating an emerging paradigm. Glob Environ Change 22(1):94–102. https://doi.org/10.1016/j.gloenvcha.2011.10.011

Cosens B (2015) Application of the adaptive water governance project to the management of the Lake Eyre Basin and its connections to the Great Artesian Basin. The Goyder Institute for Water Research

Cosens B, Chaffin BC (2016) Adaptive governance of water resources shared with indigenous peoples: the role of law. Water 8(3):1–15. https://doi.org/10.3390/w8030097

Diaz H, Hurlbert M (2013) Water governance in Chile and Canada: a comparison of adaptive characteristics. https://doi.org/10.1007/978-3-642-29831-8_11

Dietz T, Ostrom E, Stern P, C. (2008) The struggle to govern the commons. In: Urban ecology. Springer, Boston, MA. https://doi.org/10.1007/978-0-387-73412-5_40

Falkenmark M, Rockström J (2010) Building water resilience in the face of global change: from a blue-only to a green-blue water approach to land-water management. J Water Resour Plan Manag. https://doi.org/10.1061/(ASCE)WR.1943-5452.0000118

Folke C (2006) Resilience: the emergence of a perspective for social-ecological systems analyses. Glob Environ Change. https://doi.org/10.1016/j.gloenvcha.2006.04.002

Gleick PH, Palaniappan M (2010) Peak water limits to freshwater withdrawal and use. Proc Natl Acad Sci U S A. https://doi.org/10.1073/pnas.1004812107

Godden L, Ison RL, Wallis PJ (2011) Water governance in a climate change world: appraising systemic and adaptive effectiveness. Water Resour Manag. https://doi.org/10.1007/s11269-011-9902-2

Group OW, Goals SD, Group OW, Goals SD, Goals D, & Goals SD (2015) Sustainable development goals and targets. United Nations

Gupta J, Akhmouch A, Cosgrove W, Hurwitz Z, Maestu J, Ünver O (2013) Policymakers' reflections on water governance issues. Ecol Soc 18(1):35

Gupta J, Termeer C, Klostermann J, Meijerink S, van den Brink M, Jong P et al (2010) The adaptive capacity wheel: a method to assess the inherent characteristics of institutions to enable the adaptive capacity of society. Environ Sci Policy. https://doi.org/10.1016/j.envsci.2010.05.006

Hanasaki N, Fujimori S, Yamamoto T, Yoshikawa S, Masaki Y, Hijioka Y et al (2013) A global water scarcity assessment under shared socio-economic pathways - part 2: water availability and scarcity. Hydrol Earth Syst Sci. https://doi.org/10.5194/hess-17-2393-2013

Hanjra MA, Qureshi ME (2010) Global water crisis and future food security in an era of climate change. Food Policy. https://doi.org/10.1016/j.foodpol.2010.05.006

Hansmann R, Mieg HA, Frischknecht P (2012) Principal sustainability components: empirical analysis of synergies between the three pillars of sustainability. Int J Sustain Dev World Ecol 19(5):451–459. https://doi.org/10.1080/13504509.2012.696220

Hatfield-Dodds S (2006) The catchment care principle: a new equity principle for environmental policy, with advantages for efficiency and adaptive governance. Ecol Econ. https://doi.org/10.1016/j.ecolecon.2005.09.015

Huitema D, Mostert E, Egas W, Moellenkamp S, Pahl-Wostl C, Yalcin R (2009) Adaptive water governance: assessing the institutional prescriptions of adaptive (co-)management from a governance perspective and defining a research agenda. Ecol Soc. https://doi.org/10.5751/ES-02827-140126

Huntjens P, Lebel L, Pahl-Wostl C, Camkin J, Schulze R, Kranz N (2012) Institutional design propositions for the governance of adaptation to climate change in the water sector. Glob Environ Change. https://doi.org/10.1016/j.gloenvcha.2011.09.015

Huntjens P, Pahl-Wostl C, Grin J (2010) Climate change adaptation in European river basins. Reg Environ Change. https://doi.org/10.1007/s10113-009-0108-6

Hutton G, Varughese M (2016) The costs of meeting the 2030 sustainable development goal targets on drinking water, sanitation, and hygiene. World Bank, Water and Sanitation Program: Technical Paper. https://doi.org/10.1596/K8543

IPCC (2014a) Climate change 2014: mitigation of climate change. Working group III contribution to the fifth assessment report of the Intergovernmental Panel on Climate Change. https://doi.org/10.1017/CBO9781107415416

IPCC (2014b) Climate change 2014 synthesis report summary chapter for policymakers. IPCC. https://doi.org/10.1017/CBO9781107415324

Jensen O (2017) Public–private partnerships for water in Asia: a review of two decades of experience. Int J Water Resour Dev. https://doi.org/10.1080/07900627.2015.1121136

Jordan AJ, Huitema D, Hildén M, Asselt VH, Rayner TJ, Schoenefeld JJ et al (2015) Emergence of polycentric climate governance and its future prospects. Nat Clim Change 5(11):977–982. https://doi.org/10.1038/nclimate2725

Jordan A, Huitema D, Schoenefeld J, Asselt VH, Forster J (2018) Governing climate change polycentrically. In: Governing climate change. Cambridge University Press, Cambridge, pp 3–26. https://doi.org/10.1017/9781108284646.002

Kallis G, Kiparsky M, Norgaard R (2009) Collaborative governance and adaptive management: lessons from California's CALFED Water Program. Environ Sci Policy. https://doi.org/10.1016/j.envsci.2009.07.002

Karpouzoglou T, Dewulf A, Clark J (2016) Advancing adaptive governance of social-ecological systems through theoretical multiplicity. Environ Sci Policy. https://doi.org/10.1016/j.envsci.2015.11.011
Kashyap A (2004) Water governance: learning by developing adaptive capacity to incorporate climate variability and change. Water Sci Technol 49:141
Lieberherr E, Truffer B (2015) The impact of privatization on sustainability transitions: a comparative analysis of dynamic capabilities in three water utilities. Environ Innov Soc Transit. https://doi.org/10.1016/j.eist.2013.12.002
Loewe M, Rippin N (2015) The sustainable development goals of the post-2015 agenda: comments on the OWG and SDSN proposals. SSRN. https://doi.org/10.2139/ssrn.2567302
Marshall GR (2015) Polycentricity and adaptive governance (January), pp 1–19
Meene VSD, Brown R, Farrelly M (2011) Towards understanding governance for sustainable urban water management. Glob Environ Change. https://doi.org/10.1016/j.gloenvcha.2011.04.003
Mekonnen MM, Hoekstra AY (2016) Sustainability: Four billion people facing severe water scarcity. Sci Adv. https://doi.org/10.1126/sciadv.1500323
Ministry of Water Resources (2002) National water policy. Government of India, New Delhi. http://mowr.gov.in/sites/default/files/nwp20025617515534_1.pdf
Ministry of Water Resources (2012) National water policy. Government of India, New Delhi. http://mowr.gov.in/sites/default/files/NWP2012Eng6495132651_1.pdf
Ministry of Water Resources, River Development & Ganga Rejuvenation (2018) National water policy-review of national water policy. Government of India, New Delhi. http://mowr.gov.in/policies-guideline/policies/national-water-policy
Mishra BK, Regmi RK, Masago Y, Fukushi K, Kumar P, Saraswat C (2017) Assessment of Bagmati river pollution in Kathmandu Valley: scenario-based modeling and analysis for sustainable urban development. Sustain Water Qual Ecol 9-10:67–77. https://doi.org/10.1016/j.swaqe.2017.06.001
Mollenkamp S, Kasten B (2009) Institutional adaptation to climate change: the current status and future strategies in the Elbe Basin, Germany. Climate change adaptation in the water sector
Mollinga P, Tucker SP (2010) Changing water governance in India: taking the longer view. SAWAS 1(3):i–vi
Munaretto S, Siciliano G, Turvani ME (2014) Integrating adaptive governance and participatory multicriteria methods: a framework for climate adaptation governance. Ecol Soc. https://doi.org/10.5751/ES-06381-190274
Olsson AR (2004) Electronic democracy and power. In Electronic government: proceedings of the 3rd [IFIP WG 8.5] international conference, EGOV 2004
Ostrom E (2014) A polycentric approach to climate change. Ann Econ Finance 15(1):71–108. https://doi.org/10.1596/1813-9450-5095
Pacheco-Vega R (2013) Polycentric water governance in Mexico: beyond the governing-by-river-basin-council model. 2013 Meeting of the Latin American Studies Association (LASA), pp 1–30
Pahl-Wostl C (2009) A conceptual framework for analysing adaptive capacity and multi-level learning processes in resource governance regimes. Glob Environ Change. https://doi.org/10.1016/j.gloenvcha.2009.06.001
Pahl-Wostl C, Arthington A, Bogardi J, Bunn SE, Hoff H, Lebel L et al (2013) Environmental flows and water governance: managing sustainable water uses. Curr Opin Environ Sustain 5(3–4):341–351. https://doi.org/10.1016/j.cosust.2013.06.009
Pahl-Wostl C, Knieper C (2014) The capacity of water governance to deal with the climate change adaptation challenge: using fuzzy set qualitative comparative analysis to distinguish between polycentric, fragmented and centralized regimes. Glob Environ Change 29:139–154. https://doi.org/10.1016/j.gloenvcha.2014.09.003
Pahl-Wostl C, Lebel L, Knieper C, Nikitina E (2012) From applying panaceas to mastering complexity: toward adaptive water governance in river basins. Environ Sci Policy 23:24–34. https://doi.org/10.1016/j.envsci.2012.07.014

Pahl-Wostl C, Möltgen J, Ebenhoeh E, Holtz G (2008) The NeWater management and transition framework—state and development process. In: Adaptive and integrated water management. Springer, Berlin, Heidelberg, pp 75–96

Pahl-Wostl C, Vörösmarty C, Bhaduri A, Bogardi J, Rockström J, Alcamo J (2013) Towards a sustainable water future: shaping the next decade of global water research. Curr Opin Environ Sustain. https://doi.org/10.1016/j.cosust.2013.10.012

Pereira LM, Ruysenaar S (2012) Moving from traditional government to new adaptive governance: the changing face of food security responses in South Africa. Food Secur. https://doi.org/10.1007/s12571-012-0164-5

Piao S, Ciais P, Huang Y, Shen Z, Peng S, Li J et al (2010) The impacts of climate change on water resources and agriculture in China. Nature. https://doi.org/10.1038/nature09364

Qadir M, Wichelns D, Raschid-Sally L, McCornick PG, Drechsel P, Bahri A, Minhas PS (2010) The challenges of wastewater irrigation in developing countries. Agric Water Manag. https://doi.org/10.1016/j.agwat.2008.11.004

Qureshi EM, Hanjra MA, Ward J (2013) Impact of water scarcity in Australia on global food security in an era of climate change. Food Policy. https://doi.org/10.1016/j.foodpol.2012.11.003

Rockstrom J, Lannerstad M, Falkenmark M (2007) Assessing the water challenge of a new green revolution in developing countries. Proc Natl Acad Sci U S A. https://doi.org/10.1073/pnas.0605737104

Rogers P, Hall AW, Meene VD, Brown R, Farrelly M (2003) Effective water governance global water partnership technical committee (TEC). Glob Environ Change. https://doi.org/10.1016/j.gloenvcha.2011.04.003

Rola AC, Abansi CL, Arcala-Hall R, Lizada JC, Siason IML, Araral EK (2016) Drivers of water governance reforms in the Philippines. Int J Water Resour Dev 32(1):135–152. https://doi.org/10.1080/07900627.2015.1060196

Rouillard J (2012) Adaptive water governance: flood management and the policy process in Scotland, p 299

Saraswat C, Kumar P, Mishra BK (2016) Assessment of stormwater runoff management practices and governance under climate change and urbanization: an analysis of Bangkok, Hanoi and Tokyo. Environ Sci Policy 64. https://doi.org/10.1016/j.envsci.2016.06.018

Saraswat C, Mishra BK, Kumar P (2017) Integrated urban water management scenario modeling for sustainable water governance in Kathmandu Valley, Nepal. Sustain Sci 12(6):1037–1053. https://doi.org/10.1007/s11625-017-0471-z

Sengupta M (2016) The sustainable development goals: an assessment of ambition. Global Policy J

Shinn JE (2016) Adaptive environmental governance of changing social-ecological systems: empirical insights from the Okavango Delta, Botswana. Glob Environ Change. https://doi.org/10.1016/j.gloenvcha.2016.06.011

Tan-kim-yong U, Bruns PC, Bruns BR (2003) The emergence of polycentric water governance in Northern Thailand. Workshop "Asian Irrigation in Transition - Responding to the Challenges Ahead," pp 1–21

UN (2015) Transforming our world by 2030: a new agenda for global action. In UN summit to adopt the post-2015 development agenda. https://doi.org/10.1017/S1368980015002529

UNDP, SIWI, Facility WG (2009) Issue sheet 4. Water and sanitation governance. Governance an International Journal of Policy and Administration

UNESCO, UNITED T (2009) Water in a changing world. World Water. https://doi.org/10.3390/w3020618

UN-Water (2012) WWDR4: managing water under uncertainty and risk. The United Nations World Water Development Report 4

Vaas J, Driessen PJ, Giezen M, Laerhoven VF, Wassen MJ (2017) Who's in charge here anyway? Polycentric governance configurations and the development of policy on invasive alien species in the Semisovereign Caribbean. Ecol Soc 22(4). https://doi.org/10.5751/ES-09487-220401
Vörösmarty CJ, McIntyre PB, Gessner MO, Dudgeon D, Prusevich A, Green P et al (2010) Global threats to human water security and river biodiversity. Nature. https://doi.org/10.1038/nature09440
Walch C (2018) Adaptive governance in the developing world: disaster risk reduction in the State of Odisha, India. Clim Dev 11:238. https://doi.org/10.1080/17565529.2018.1442794
World Bank (2014) Climate change is a challenge for sustainable development. http://www.worldbank.org/en/news/speech/2014/01/15/climate-change-is-challenge-for-sustainable-development. Accessed 16 Sept 2018
Wouters P (2008) Global water governance through many lenses. Global governance
Wu X, House RS, Peri R (2016) Public-private partnerships (PPPs) in water and sanitation in India: lessons from China. Water Policy 18(December):153–176. https://doi.org/10.2166/wp.2016.010

Chitresh Saraswat is a computer scientist turned water policy researcher with interdisciplinary research interests at the intersection of water security, governance, and institutional analysis. Currently, he is working as an individual environmental consultant focusing on water security and environmental governance and as a researcher at the United Nations University Institute for the Advance study of Sustainability (UNU-IAS), Tokyo, Japan.

Gupta Anil K , Professor of Policy & Strategies, Integrated Resilience. Head of the ECDRM Division, International Cooperation, Programme Director - Centre for Excellence on Climate Resilience, Nodal Officer for Central Ministries at National Institute of Disaster Management, (Govt. of India), New Delhi. He is also a member of the expert Team for WMO Climate Statement and Core Group Member of IUCN-Council of Ecosystem Management.

Chapter 17
NbS Interventions as Tool for Urban Climate Resilience: A Case Study of Peri-Urban Ecosystem in Noida

Harsha Nath, Pritha Acharya, and Anil K. Gupta

Abstract Rapid urbanization presents one of the most urgent challenges of our times. Cities are the engines of growth and markers of development. Climate change may undermine all of this by worsening resource scarcity and putting vulnerable communities at risk. Cities must cope with poor air quality, heat island effects, increased flood risk and the frequency/severity of extreme events. Adaptation and mitigation to climate change, as well as sustainable development, are critical challenges for urban cities. What must be developed therefore is a robust, evidence-base reference framework that takes into account nature-based solutions (NBS). NbS are measures that are dependent on and inspired from features and processes of natural ecosystems and have potentials to enhance inclusive urban resilience. In recent years, Nature-based solutions (NBS) are becoming more relevant to cope with climate change. It is a cost-effective approach that offers environmental, social and economic benefits at the same time and helps build disaster resilience. This study tries to analyze whether NBS could be a feasible option for preserving the stability of Noida as an alternative to conventional engineering solutions. The objective of this study is to map the existing ecosystem based adaptations/practices and identify the scope for NbS interventions as effective adaptation strategies for climate change. Global recognition for Climate change adaptation through eco-DRR is evident but the mindset is undergoing a transformation and emphasizing nature-based solutions (NBS) as an effective tools to address the climate resilience of urban cities.

Keywords Urbanization · Climate change · Climate resilience · Nature based solutions (NbS) · Ecosystem based approaches (EbA) · Noida

---

H. Nath
Corporate Sustainability, PPAP Automotive Limited, Noida, India

P. Acharya (✉)
ECDRM Division, CAP-RES DST Project, Centre for Excellence on Climate Resilience, National Institute of Disaster Management (Ministry of Home Affairs), New Delhi, India

A. K. Gupta
ECDRM Division, International Cooperation, CAP-RES DST Project, National Institute of Disaster Management (Ministry of Home Affairs), New Delhi, India

A. K. Gupta et al. (eds.), *Disaster Risk and Management Under Climate Change*, Disaster Resilience and Green Growth,
https://doi.org/10.1007/978-981-99-4105-6_17

17.1 Introduction

17.1.1 Global Scenario

Several ecosystem-based approaches have been devised over the past few decades to re-nature urban environments in an attempt to improve the resilience of urban systems (Haase et al. 2014; Sarabi et al. 2019). The term "ecosystem-based approach" is based on the description provided by the Convention on Biological Diversity (CBD), which states that "the ecosystem approach is a strategy for the integrated management of loand, water, and living resources that promotes conservation and sustainable use in an equitable way." It has been identified as a critical strategy for reducing disaster risk. Ecosystem-based disaster risk reduction (Eco-DRR) is described as "sustainable ecosystem management, conservation, and restoration with the goal of achieving sustainable and resilient development" (Estrella and Saalismaa 2013).

Nature-based strategies have emerged as a concept for combining ecosystem-based approaches to solve a variety of societal problems. Increased urban resilience is specifically addressed and contributed to by nature- based solutions (Busha and Doyon 2019). The introduction of urban resilience and nature-based solutions (NBS) concepts over the last decade has provided the opportunity to address current problems in cities in a sustainable and resource-efficient manner. One of the major challenges we face globally is that cities are rapidly expanding, with the majority of this development taking place in peri-urban areas (Maheshwari et al. 2016). Building resilience is a critical task for governments and organizations (Sayaas 2016), as evidenced by global long-term agendas like the Paris Climate Agreement, the New Urban Agenda 2030, and the Sendai Framework for Disaster Risk Reduction 2015–2030. NBS are critical to achieving the goals of the United Nations Agenda 2030 for Sustainable Development (UN 2017; UNESCO 2018) and other long-term global agendas, such as the Paris Agreement on Climate Change (UN 2019). The introduction of the NBS has been promoted in European cities in recent years as a potential solution to urban problems including climate change, urban degeneration, and ageing infrastructures (Frantzeskaki 2019). In addition to a variety of other advantages, NBS harnesses natural resources to provide ecosystem services to resolve key social, environmental and economic challenges. Some examples of the current trends that would aid the establishment of a common NBS impact evaluation framework are green roofs and covering shelters, green barriers, agroforestry (rural land management), peri-urban parks, natural and constructed wetlands etc. (Somarakis et al. 2019). However, given the diversity of ecosystem services, their multi- functionality, and the trade-offs between functions, as well as across temporal and spatial scales, implementing nature-based solutions is inherently difficult (Raymond et al. 2017).

17.2 Background

The New Okhla Industrial Development Authority (NOIDA) is situated at 28.5N 77. E, in the Gautam Buddha Nagar District of Uttar Pradesh State, in northern India. It is bound by the Yamuna River to the west and south-west, the city of Delhi to the north and north-west, the districts of Delhi and Ghaziabad to the north-east, and the Hindon River to the north-east, east and south-east. Noida has a population of about 0.64 million and is spread over an area of 203 km^2. For most of the year, Noida has a hot and humid climate. During summers, the weather stays hot, i.e. from March to June, and temperatures range from maximum 48 °C to minimum 28 °C. With an annual rainfall of 93.2 cm (36.7 in.), the Monsoon season prevails between mid-June and mid-September, but occasionally regular heavy rain triggers floods. At the winter peak, temperatures decline to as low as 3 to 4 °C and experiences fog and smog (Fig. 17.1).

The 3 selected study sites are as follows:

1. Okhla Bird Sanctuary (Protected Area)

   The Okhla Bird Sanctuary (OBS) is about 4 square kilometers in size and is located near the entrance of Noida in Uttar Pradesh's Gautam Budhh Nagar district. It is located where the Yamuna River reaches the state of Uttar Pradesh after leaving Delhi's territory (Gupta et al. 2017). It is one of the state's 15 bird sanctuaries. Because of its unique location, the sanctuary attracts migratory bird species. The construction of the Okhla Barrage resulted in the creation of this wetland. In 1990, the Uttar Pradesh government declared this area as a sanctuary.

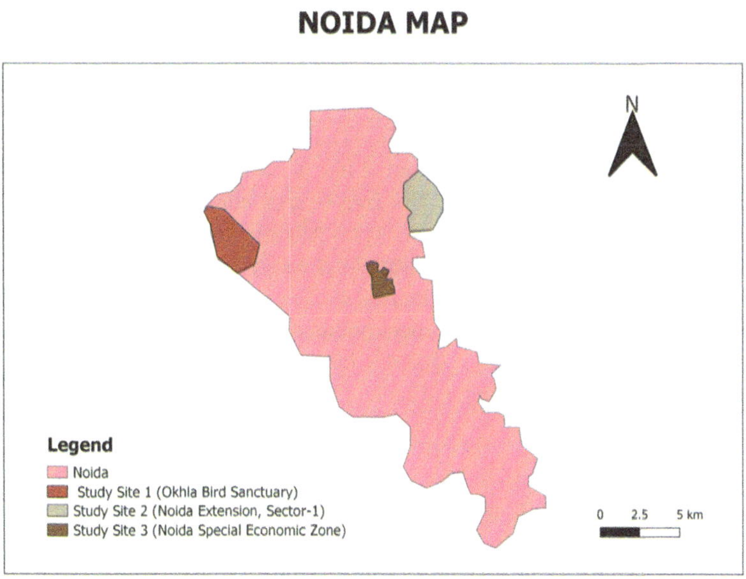

Fig. 17.1 Map of study area

Despite the fact that the Uttar Pradesh government claims to be doing everything possible to protect the Okhla Bird Sanctuary, the forest is deteriorating. Each year, the Okhla Bird Sanctuary sees a record low turnout. The Okhla Bird Sanctuary, once a flourishing habitat for flamingos, sarus, and pochards, is now devoid of most of its winged tourists. Small terrestrial birds such as stonechats, warblers, and paradise flycatchers have also declined in numbers. Experts say that industrial complexes, the DND expressway, flyovers, and even the Ambedkar Park have ruined the bird sanctuary's feeding grounds. Birds steer clear of the sanctuary because of the 400 kV lines. It has now been designated as one of India's 466 IBAs (Important Bird Areas) (obs-up).

2. Noida Special Economic zone (Industrial Area)

   NSEZ, the only Central Government SEZ in northern India, was built in 1985 on a 310-acre plot of land in Noida Phase-II, under the governance of the Development Commissioner. The Indian government has so far invested Rs. 11,149.60 million in its growth. For the thrust areas of exports such as gem and jewellery and electronics applications, NSEZ offers excellent infrastructure, supportive services, and sector-specific facilities. It is optimal for setting up jewellery and software development units because of its proximity to Delhi and the availability of professional and committed manpower. During the 2017–2018 fiscal year, these two industries account for approximately 75% of total export revenue.

   For many years, NSEZ has maintained a high rate of export growth. The zone's exports in 2018–2019 (up to January 2019) totaled Rs. 83,321.10 million. The number of people employed in the region has risen from 32,550 in June 2009 to 39,338 in January 2019. Over the same time span, the zone's operating units rose from 245 to 261 (NSEZ authority).

3. Sector-1, Noida Extension (Residential area)

   Sector 1, Noida Extension also known as Greater Noida West is a newly built sector and has transformed into a prime residential zone for investors. Noida Extension is surrounded by Greater Noida and Noida. Due to low property prices, the region is rapidly expanding, and it has good accessibility to Noida, Delhi, Ghaziabad, and Greater Noida. The area is well connected to Delhi and other parts of the NCR, as well as Meerut, Agra, Aligarh, and Mathura, through public transportation. The National Highway 24 is easily accessible from the area and the Boraki Halt railway station is located 19.7 km away (Gnida Authority). Noida Extension is among the top four most polluted cities in the world according to a report by (World Air quality). Bisrakh Jalalpur is the most polluted area in the Gautam Buddh Nagar district and ranks fourth in the world (Iqair)

17.2.1 Impacts of Urbanization on Noida

Noida is becoming populous mainly because of the availability of various means of earning a living, as Noida has emerged as one of the most favored industrial areas in the Delhi National Capital Region. Ecological imbalance has grown due to the rapid industrialization and urbanization and infrastructure growth in Noida (Sinha 2018). This exploitation of resources have adverse consequences in this area, such as rising temperatures contributing to urban heat island effect (UHI), reduced green cover, air pollution, deteriorating water quality, water scarcity, induced flood risks (due to poor water management strategies) (Nandy 2015).

According to World Air Quality report 2020 by iqair, there is a great deal of residential construction in Noida and its peri urban areas. The primary sources of air pollution are particulate matters and dust from construction and vehicular emissions. The reason that the area experiences patchy rains, could be a chance of affecting the average AQI but it hardly reaches a good or satisfactory category. Richa Sharma et al. (2020) study findings suggests that the microclimate in Noida has been exacerbated by incessant conurbation, causing the thermal anomalies within the city to increase. LST values range from a minimum of 25.57 °C to a maximum of 42.97 °C in 2011, with a mean value of 35.01 °C. LST values subsequently increased significantly in the year 2019, with a minimum LST of 29.13 °C and a maximum of 48.09 °C, with an average LST of 41.43 °C. In the span of eight years from 2011 to 2019, an overall rise of 6.42 °C is observed. One of the main contributory reasons for such elevated temperature values is the conversion of natural and agricultural lands into constructed ones.

A study of Gopal Krishna from IARI suggests the study area was found to have undergone rapid changes in LU/LC (Land Use/Land Cover). Change detection analysis indicates that the area under construction grew by 29 percent (from 10,090.68 hectares to 24,996.63 hectares) and the area under cultivation decreased by 22 percent (from 27,551.36 hectares to 15,840.81 hectares) between 1986 and 2011. Results of the transformation and post-classification ratio show that only two categories, i.e. built-up area and bare soil/landfill sites, have occupied 32,590.80 ha out of 53,000 ha (61 percent of total area). According to the Indian census, there were 204,302 households in the Gautam Buddha Nagar district in 2001, an increase of 111,976 making a total of 316,278 households in 2011. The LU/LC (area in hectares) for water bodies in 1986 was 2605.40 and this area came down to 1200.06 hectares in 2011. The vegetation cover has also reduced being 7833.56 in 1986 to 3368.97 in 2011. This finding also indicates a significant increase in urbanization over the last decade. Noida's urban Sprawl has had a negative effect on the quantity and quality of water in the study area. With the decline in agricultural land (and the rise in built-up area) the recharge of water to ground water aquifers has decreased. As a result, the Noida water level has decreased from 1 to 1.5 m per year over the last five years (Bhowmick 2018). Virendra Bahadur Singh and Tripathi (2016) study suggests that salinity and pollution are the main factors affecting the quality of groundwater in Noida.

17.2.2 Eco-System Based Approaches

EbA worldwide is gaining momentum. Case studies and literature have shown that EbA can provide a flexible, cost-efficient and wide-ranging approach in mitigating climate change impacts (Munang et al. 2013, 2013). EbA refers to the use of biodiversity and ecosystem services as part of a broader adaptation plan to assist people in dealing with the detrimental impacts of climate change. Healthy ecosystems help to mitigate the climate change vulnerability and disaster risk in the following ways:

a) Adding value to more costly infrastructure investments (Munang et al. 2013, 2013; Temmerman et al. 2013).
b) Providing disaster risk mitigation and adaptation strategies that are in line with national growth and adaptation priorities (WWF 2013).
c) Contributing to climate change mitigation by increasing carbon sequestration (Duarte et al. 2013), Deforestation and land loss prevention (Busch et al. 2015).
d) Engagement among citizens and communities helps build trust and accountability, (Naumann et al. 2011).

17.2.3 Existing Practices of EBA on Study Sites

Okhla Wetlands
Wetlands are one of the world's most active ecosystems that provide ecosystem services along with a supply of rare flora and fauna (MA 2005). Wetlands habitats also help to minimise disaster risk by acting as natural barriers or buffers to protect the environment and thus mitigate risks (Gupta and Nair 2012). Decline in the ecosystem increases the disaster risk by both reducing an ecosystem's ability to act as a natural buffer and reducing human resilience by reducing its livelihood bases such as food, medications, and building materials (ProAct Network 2008). The Yamuna River Floodlands, over 90 sq. kilometres, are home to forests, agricultural fields, villages, reservoirs and lakes, along with a range of wildlife and flora. It can also store 2 billion cubic metres of water. Near Okhla, where a rich ecosystem of biodiversity occurs, is found to have maximum width of the flood plain (Gupta et al. 2017). The wetlands and water supply near Okhla helps to sustain the minimum amount of water needed to operate the floodplain throughout the year. The excess water percolates during the monsoon season and helps control flooding and also maintains the moisture levels in lean times. The organic build-up in the floodplain of essential nutrients i.e. bioaccumulation often reduces pollution stress. It also offers ecosystem services including the recharge of groundwater, excess water storage, control of diseases, carbon sequestration and hot summer thermal protection.

Noida Extension Agricultural Land
There are vast agricultural lands on the outskirts of Noida extension. These agricultural lands provide a variety of ecosystem services in addition to food for cities and

farmer livelihoods. It offer various ecosystem services such as mitigating urban heat island effect. It also helps in flood prevention as during monsoon season, Hindon river overflows due to excessive water in Yamuna leading to risks of floods to the villages situated along the bank of the river. Along with that, it also aids in green space preservation and development aid in carbon sequestration and augmentation. And most importantly, it provides a source of income to the urban poor (Gupta et al. 2017).

NSEZ Park

NSEZ Park is located in Noida's NSEZ, Sector 81. It is the only park in the vicinity of the industrial zone. This urban park provides a tranquil atmosphere as well as green open space for lunch hour picnics. This city park has long been a haven for employees seeking a respite from the grey of the industrial zone's asphalt and concrete. Nature and green spaces directly contribute to public health by reducing stress and mental illnesses (Ward Thompson et al. 2012; Annerstedt et al. 2012), enhancing the impact of physical activity (Mitchell 2012), reducing health inequalities (Mitchell and Popham 2008), reducing health disparities (Mitchell and Popham 2008), and improving self-reported general health (Maas et al. 2006; Stigsdotter et al. 2010). Providing arenas and opportunities for physical activity (Coombes et al. 2010), increasing satisfaction with living environment and social interactions (Björk et al. 2008; Maas et al. 2009), and various forms of recreation all have indirect health effects (Weber and Anderson 2010).

17.3 Resilient Urban Development Through Peri-Urban Ecosystem

Urban resilience cannot be achieved in isolation as the flow of goods and services - including the natural continuous - connect the urban areas with peri-urban areas. Peri-urbanization is a dynamic process that alters land use in the peripheries of growing cities and towns (large and small), often displaying a form, structure, and interaction that is unique and geared to support the urban centre (Gupta et al. 2017). In general, these are considered transitional zones between the rural and urban regions, comprising characteristics from both (Narain et al. 2013; Mitra et al. 2015; Dutta 2012). The ecosystem services provided by transition zones are important for a variety of aspects of urban sustainability and resilience, including food security and disaster risk management. The Sustainable Development Goals of the United Nations (Goal No. 11) also stresses the need to make cities inclusive, sustainable, resilient, and robust through the implementation of integral resource-efficiency and climate change adaptation policies and plans (UN 2016) which is not possible without conservation of these peri-urban areas. However, the ecosystems on which these services rely are increasingly threatened and are frequently ignored in policy and planning (ESPA 2017). Gupta et al. 2017 suggested in his study that due to its position adjacent to the city's edges, where many dynamic socio-economic

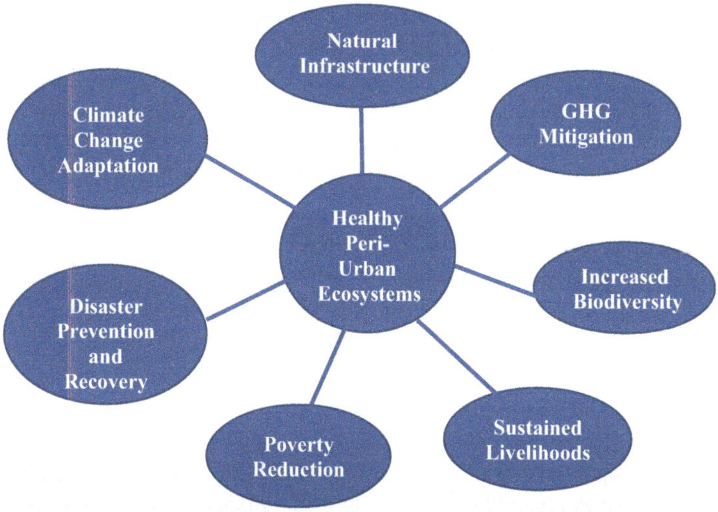

Fig. 17.2 Peri-urban ecosystems and role in resilience (Source: Gupta et al. 2017)

processes exist, a peri-urban region is an ever-changing zone of both activity and change. As a result, its potential for resilience is primarily influenced by its input-output dynamics with the nearby urban region on the one side, and with the purely rural structures on the other. Building urban and peri-urban resilience to natural hazards and climate extremes necessitates, first and foremost, a transition away from existing reactive approaches and toward the creation and implementation of successful local planning and management processes to promote higher quality of life. The resilience of urban and peri-urban areas depends on maintaining ecosystem health. To strengthen urban resilience, city-region planning and policy should consider the value of peri-urban environments (Fig. 17.2).

Case Study of Noida Extension

The Noida Extension or Greater Noida West area is located on the outskirts of the cities of Noida and Greater Noida. Land mafias (builders) are more active in this zone as a result of rapid urbanization, converting open spaces/agriculture lands into residential areas. The rapid urbanization in Noida Extension is putting a strain on natural resources and absorbing existing agricultural land on the outskirts of the city, resulting in fewer green/open spaces, a disruption in the supply of food to cities, a disruption in livelihood patterns, and a reduction in natural drainage of excess storm water. By increasing water logging and run-off, large-scale conversion of agricultural land to non-agricultural uses exacerbates climate change risks. About 54 percent of the peri-urban area designated for agricultural use in the Noida Extension Master Plan-2021 has a population of 0.1 million people, with a significant proportion of small and marginal farmers. Flooding and waterlogging, sewage dumping, rising agricultural costs, changing land use patterns, and governance issues all affect these farmers, making them socially and economically vulnerable. In addition, villages

Table 17.1 Ecosystem services of peri urban agriculture (Espa 2017; Gupta et al. 2017)

Provisioning services	Regulating services	Supporting services	Cultural services
Access to ample and nutritious food on a regular basis (food security).	Reduces the effect of the urban heat island.	Green space preservation and development aid in carbon sequestration and augmentation.	Creates urban open space at the city's outskirts and gives it a pleasing appearance.
For the urban poor, this is a significant source of income (Economic security).	Aids in the absorption of excess water during floods, preventing inundation in urban areas.		A well-kept peri-urban location, such as a bird sanctuary, can be used for eco-tourism.
The amount of transportation infrastructure required to carry food into city limits is reduced, resulting in a shorter supply chain.			

along the Hindon River's banks are vulnerable to flooding. Peri-urban areas are undergoing rapid change, displacing conventional agriculture and reducing cities' ability to adapt to non-linear change. The best way to preserve urban sustainability is to take a regional approach that integrates urban and peri-urban networks (Michael et al. 2016) (Table 17.1).

17.3.1 Nature Based Solutions for Urban Resilience

Cities' ability to adjust and adapt in the face of change determines their resilience (Alberti and Marzluff 2004; Alberti et al. 2003; Pickett et al. 2004). Responding to incremental change and persistent pressures, often of a socioeconomic sort, as well as sudden change or acute shocks, such as natural disasters, are all part of resilience (Resilient Melbourne 2016). Climate change, natural disasters, food and water protection, human health and well-being, and economic and social growth have all become societal challenges and 'Nature-based solutions' has emerged as a concept, or umbrella term, for ecosystem-based approaches to address these issues (Cohen-Shacham et al. 2016; EC 2015). Actions that are inspired by, supported by, or copied from nature are referred to as nature-based solutions (Cohen-Shacham et al. 2016; EC 2015). NBS serves as an "umbrella" concept that develops on and supports concepts such as ecological engineering and catchment systems engineering, green-blue infrastructure, natural infrastructure, ecosystem-based approach, ecosystem-based adaptation/mitigation, ecosystem services, re-naturing, and natural capital (Somarakis et al. 2019). Nature, according to NBS, is a source of inspiration, providing long-term sustainable solutions for dealing with the effects of human activities and enhancing natural capital (EC 2015). NBS serve a multi-functional

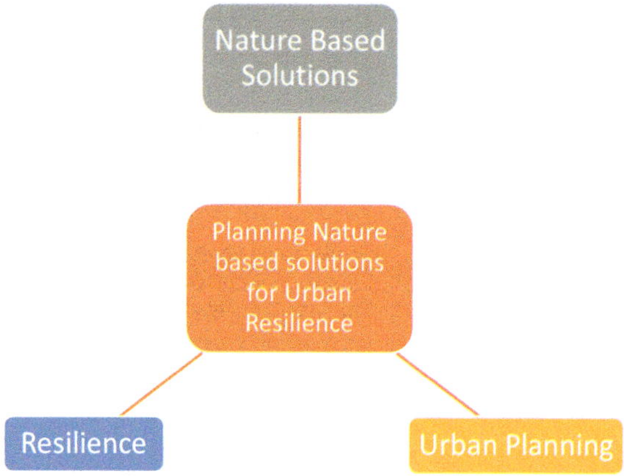

Fig. 17.3 Relationship between nature based solutions, urban planning and resilience

purpose, allowing them to address the social, environmental, and economic aspects of global challenges. NBS have been described as vital for urban regeneration and improvement, coastal resilience, multifunctional watershed management, and biodiversity conservation, increasing the sustainability of matter and energy usage, enhancing ecosystem insurance value, and increasing carbon sequestration (EC 2015; Krauze and Wagner 2019). NBS have been promoted by practitioners (in particular the International Union for Nature Conservation, IUCN) and by policy (EC) as a way for the sustainable use of nature in solving urban societal challenges (Eggermont et al. 2015) like mitigation of climate change, management of water resources, land use, and development of urban areas (Bulkeley et al. 2017). The concept of NbS was used by Gómez-Baggethun et al. (2013) to emphasize the provision of ecosystem services in urban areas that contribute to a city's resilience. Nature-based solutions function as decentralised, distributed infrastructure service delivery systems, which are typically more resilient than large, centralised grey infrastructure (Depietri and McPhearson 2017) (Fig. 17.3).

According to Final Report 2020 on 'Nature-Based Solutions and Re-naturing Cities' by European commission, the goals of NBS include:

- *Sustainable urbanisation* – urban areas host an enormous share of the world population facing multiple challenges (natural resources shortage, human wellbeing, etc.).
- *Restoration of degraded ecosystems* – various ecosystems have been severely degraded due to human interventions and activities (agriculture, industry, etc.).
- *Adaptation and mitigation of climate change* – climate change is a worldwide challenge affecting not only environment but also economy and society.
- *Risk management and resilience* – there are diverse hazards, which can result in extreme losses for both natural and societal resources without the proper preparation.

17.3.1.1 NbS and SDG Agenda 2030

The UN 2030 Agenda for Sustainable Development, which was adopted by all UN Member States, is another action plan (UN 2015). It consists of 17 Sustainable Development Goals (SDGs), which address global challenges and place strain on society, the economy, and the environment. Similarly, one of NBS's main priorities is to solve global issues that are specifically related to the Sustainable Development Goals. NBS contribute to a variety of UN Sustainable Development Goals, not just those relevant to biodiversity and ecosystems. There are campaigns using NBS in relation to different SDGs all over Europe (Fig. 17.4).

17.3.1.2 Applied Examples of NbS

Some case studies are given below, to provide a better understanding of what NBSs are and how they function.

Case Study 1: Copenhagen Strategic Flood Masterplan
Location: Copenhagen, Denmark

In Copenhagen in 2011 a cloud burst took place. The city was inundated by 150 mm of rainfall, up to 1 m in altitude. In order to counter potential floods that include climate adaptation options and current grey infrastructures, the town switched to NbS (blue and green master plan). The strategy to revise Lake Sankt Jørgens was a major difference between New and Old Masterplans. The lake level was decreased from +5.8 m to 02.8 m, resulting in a new storage area of 40,000 m^3 for cloud burst storage. Thc lakcfront development allowed people to enter an area otherwise abandoned. A tunnel with a 2.5 m diameter is designed to flow into a nearby port the water from the overflowing lake. This strategic master plan protects Copenhagen from floods and provides an urban climate of high quality (Fig. 17.5).

Case Study 2: Stuttgart Green Ventilation Corridor
Location: Stuttgart, Germany

Stuttgart's location makes it vulnerable to poor air quality due to its moderate climate, low wind speeds, and heavy industrial activity. As a result, the municipality devised a new plan focused on the city's environment and population. A significant

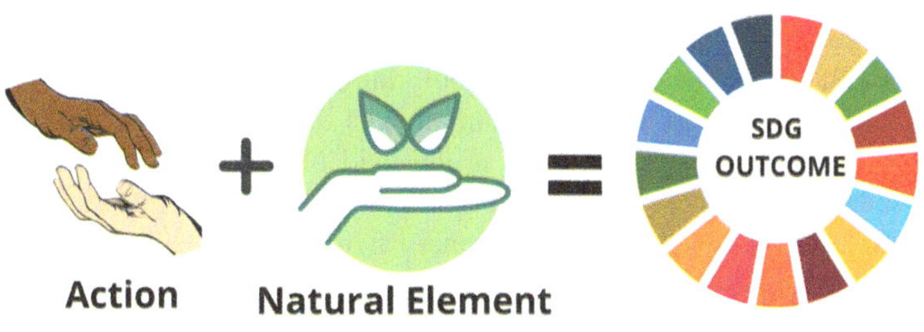

Fig. 17.4 NbS as three part equation

Fig. 17.5 Lake development for flood masterplan (Source: Asla.org)

vegetation barrier was built around the development, with connected green spaces strategically positioned to allow for air exchange. Valleys that provided clean air were cordoned off to allow for free flow of air. A preservation order was given for all trees with a trunk diameter of 80 cm or more. These nature based solutions have reduced the urban heat island effect and increased biodiversity while also sequestering carbon (Fig. 17.6).

Case Study 3: Shenzen Sponge City Transition

Shenzhen district, in China's subtropical south, experiences massive flooding during the monsoon season and water shortages during droughts. The local government has long relied on natural-based solutions to address the dual challenge, especially during the construction of Guangming New District. The People's Sports Center in Guangming, for example, has a green roof, raingardens, and permeable pavement that can capture over 60% of annual rainfall. The district was named China's first low-impact development model town in the field of storm water management in 2011. Shenzhen was chosen to participate in the national Sponge City programl, which will provide the city with an additional 1.5 billion yuan (205 million Euros) in subsidies over the next three years. The local government's action plan calls for the re-designing of an additional 256 square kilometres in 24 areas in a water-sensitive manner (Fig. 17.7).

Fig. 17.6 Vegetation barrier along the developmental area (Source: Climate ADAPT)

Fig. 17.7 Shenzen's peri-urban wetland (ICLEI 2017)

17.4 Results

17.4.1 Categorization of the Questionnaire Results (Table 17.2)

Based on the results and analysis of the questionnaire, the dominant environmental problems in study sites were Air pollution, Water Pollution, Green spaces (more trees and greenery). Other prominent topic were increasing temperature (Urban heat island effect), storm water management due to congested roads in NSEZ.

17.4.2 Addressing Urban Societal Challenges

The identified challenges that possible NbS could address in Noida Extension, NSEZ and Okhla Bird Sanctuary are Air quality, Green space management, Water management (Quality and storm water), Increasing temperatures (Urban heat island

Table 17.2 Comparison of responses from the participants of questionnaire from 3 study sites to evaluate the urban societal challenges

	Study sites		
Questions	Noida extension	Okhla Bird Sanctuary	Noida Special Economic Zone (NSEZ)
Age-group	15–30 years 30–50 years More than 50 years	15–30 years	30–50 years More than 50 years
How long have you lived/visited/worked in this area?	0–5 years	0–5 years 5–10 years	15–25 years More than 25 years
Gender	50% Men 50% Women	90% Women 10% Men	90% Men 10% Women
Are you a local resident or migrated from other area?	50% Local 50% Migrated	60% Non-local residents, 30% Local, 10% Migrated	90% Migrated 10% Local
On a scale of 1 to 5, please rate the environment of your area?	2	3	3
What is your opinion on the environment/ surroundings of your area?	80% Polluted 20% Moderate	50% Moderate 50% Polluted	50% Moderate 50% Polluted
How Urbanization has impacted your lives and the surrounding environment?	Poor air and water quality, noise pollution, high temperature, deforestation	Air and water pollution, deforestation, waste management issues	Air and water pollution, deforestation
How frequently have you faced environmental problems like poor air and water quality, increasing temperature and flood risks in your area?	Always	Sometimes	Sometimes
On a scale of 1 to 5, please rate the greenness of your area in general	2 (poor)	3 (moderate)	3 (moderate)
On a scale of 1 to 5, please rate the air quality of your area in general	2 (poor)	3 (moderate)	2 (poor)
On a scale of 1 to 5, please rate the water quality of your area in general	2 (poor)	2 (moderate)	2 (poor)

(continued)

Table 17.2 (continued)

	Study sites		
Questions	Noida extension	Okhla Bird Sanctuary	Noida Special Economic Zone (NSEZ)
Do you face any health problems due to the ongoing environmental problems in your area?	Yes (100%)	No	Yes (50%), No (50%)
If yes, how is it impacting your health and what problems have you encountered?	Lung diseases like Asthma, breathing problem, eye irritation, skin irritation due to water quality	NA	Breathing problem, skin and eye irritation
How do you think our actions are impacting the environment?	Overpopulation, Deforestation, pollution, climate change	Overpopulation, deforestation, littering waste, pollution trigger climate change	Deforestation, Pollution from vehicles, industrial pollution
Which environmental issue in your area/city do you believe the government/administration should prioritize?	Air quality and Greenness of area	Water pollution, waste disposal	Air Pollution Water Pollution, storm water management (Blockage of roads due to heavy rain)
Any suggestions or ideas that you could add to effective ways of solving the pollution problem?	Stop reckless construction, planting more trees, closing of tunnels (Nala)	Strict governmental policies, proper waste management around the sanctuary	Proper waste treatment before disposal in water body, keeping a check on air pollution caused by a company, opting ecofriendly lifestyle, planting trees

effect). The NbS identified and discussed are mainly based on Raymond et al. (2017), EC (2015) and Somarakis et al. (2019). The NbS identified may help address the following challenges in the area.

17.4.2.1 Air Quality

Green Barrier and City tree (Somarakis et al. 2019), Street trees, green walls, and the preservation and construction of urban green spaces are among the proposed NBSs to address component air quality (EC 2015; Raymond et al. 2017; Maxter 2018). The synergy between ecosystem processes that control pollutants and CO2 in the atmosphere is also exploited by NBS based on the creation, enhancement, or restoration of ecosystems in human-dominated environments. Vegetation can lower air temperatures, reducing BVOC emissions and delaying the formation of secondary pollutants such as ozone (Wang et al. 2015b; Calfapietra et al. 2013). Pollutants can be removed by vegetation in a number of ways (Xing et al. 2017; Maxter 2018). For example, vegetation, especially leaves, has been found to remove air pollutants and improve air quality by depositing and affecting dilution and physical transport of air pollutants (Salmond et al. 2013; Janhall 2015; Klingberg et al. 2017). The density of the vegetation and its height and location influence both deposition and dispersion of air pollutants. For example, low vegetation and vegetation near the surface may be diluted with clear air above the vegetation; and, due to its proximity to pollution sources, vegetation near the source permits greater deposition. Somarakis et al. (2019) stated that Green barrier and City tree could be used as NbS interventions to mitigate the poor air quality. Air pollution and noise from automobiles have an effect on the atmosphere in some situations, such as in neighborhoods near transportation facilities such as in NSEZ specially during the peak morning and evening hours. Green barriers created by creeper plant species on simple bearing structures are an easy-to-implement NBS that needs little maintenance. These green barriers block contaminants from vehicular emissions from penetrating the environment and reduce traffic noise by up to 15 decibels at a low cost and with minimal maintenance. Another proposed NbS was the installation of City Tree. Green City Solution's, City Tree is a creative mobile installation that uses mosses and controllable ventilation technology to eliminate pollutants from the air. Local air quality, soil humidity, temperature, and water quality are all monitored using integrated sensors. The installation is self-contained and needs very little maintenance. The type of trees planted also need to be taken into consideration when implementing this NBS. The air temperature can however also be reduced by vegetation (Wang et al. 2015). Because of the various co-benefits they produce and their contribution to amenity value over time, initiatives to enhance air quality by improving green infrastructure can be considered a good investment (Baró et al. 2015).

17.4.2.2 Increased Temperature/Heat Waves

Increasing the volume of vegetation, preventing the loss of vegetation, increasing the amount of vegetation including urban trees, pockets, green roofs, green walls, deculvertation and water retention and ground infiltration through green space could help to resolve and mitigate the expected increase in temperature and heat waves due to climate change. These measures reduce local temperatures by increasing shades and evapo-transpiration (EC 2015; Raymond et al. 2017; Kabisch et al. 2017). Green roofs are one of the most intriguing options for densely populated urban areas with little green space. They are a form of climate change adaptation green and blue space solution that provides multifunctional benefits. These can reduce indoor temperatures by up to 5 °C when used widely in densely populated areas (usually as part of government-supported initiatives). In comparison to intensive green roofs, extensive green roofs need less irrigation and have lower construction and maintenance costs. They are much less likely to cause structural harm if an effective growth mechanism is in place. Covering the building envelope with vegetation (i.e. green walls and roofs) could reduce air temperature at roof level by up to 12.8 °C on a daily basis, and within the urban canyon by up to 9.1 °C on a daily basis (Alexandri and Jones 2008). Since urban vegetation and trees provide climate regulation services, these can help to minimize heat-related mortality and morbidity (Chen et al. 2014). The noon-time indoor air temperature was reduced by 0.47 °C by greening the areas around buildings, thereby leveraging vegetation's ability to adjust albedo, specific heat capacity, and evaporative efficiency (Dhakal and Hanaki 2002). Direct solar radiation on walls was also decreased from 600 W/m^2 to less than 100 W/m^2 by greening the surrounding area (Papadakis et al. 2001). A green wall of 850 m^2 was built on a building in Vienna, and it was discovered that on a hot summer day, the green wall had the same cooling effect as 80 air conditioning units, each running at 3000 W for an 8-hour period (corresponding to 712 kWh) (Enzi et al. 2017). De-culverting is also said to increase evapotranspiration and lower UHI effects (Raymond et al. 2017; Gunawardena et al. 2017). The ability of water bodies or water courses to alter surrounding temperatures is determined by their intrinsic properties and interactions with surrounding climatic conditions. (Gunawardena et al. 2017). However, when blue and green spaces are combined, they have synergistic cooling and other ecosystem services (Gunawardena et al. 2017).

17.4.2.3 Water Management/Extreme Rainfall Events

The following are some examples of possible water management measures: Urban waterbodies are being reintroduced to their natural habitat (opening channels, de-culverting, increase vegetation, greening waterfronts), new vegetated surface waterbodies could be formed (ponds, drains, lakes, bio-retention cells) (EC 2015), restoring Okhla wetland in the Yamuna river basin to create storage areas for temporary flood water as existing wetlands are important biotopes that contribute to hydrological cycles, sustain a diverse ecosystem, can purify polluted water, and store significant amounts of carbon. Green roofs will cut down on storm water runoff by 17–20% (Somarakis et al. 2019). Sustainable urban drainage systems (SuDSs) and de-culverting waterbodies are viable choices for dealing with increased precipitation and severe rainfall events brought about by climate change (EC 2015; Raymond et al. 2017; Kabisch et al. 2017). Permeable surfaces, rain gardens (as well as other bioretention systems), and rainfall collection and interception using plants such as trees and green roofs are all examples of SUDs. Trees efficiently minimise storm water runoff by transpiration, promoting infiltration, canopy interception failure, and in relation to other green infrastructure technologies (Woods Ballard et al. 2015; Berland et al. 2017). Trees have been found to minimize surface runoff from asphalt by 62 percent by intercepting it and infiltrating it through their pits (Maxter 2018). Rain gardens or raised planters are another choice for addressing increased precipitation and flood risks s (Woods Ballard et al. 2015). Rain gardens were tested for their ability to minimise runoff volume and peak flow for a variety of precipitation events varying from 0.3 to 180 mm, as well as efficiently remove pollutants during simulated urban runoff events (Yang et al. 2013).

17.4.2.4 Green Space Management (Enhancing/Conserving Biodiversity)

Green walls and roofs (Benvenuti 2014), meadow-like vegetation (Benvenuti 2014; Bretzel et al. 2016), and trees and greenspaces (Whitford et al. 2001; Sandström et al. 2006), for example, offer a variety of plant habitats (Madre et al. 2014; Enzi et al. 2017) as well as habitat functions for birds and insects. The majority of vacant space in towns and cities is on building exterior walls and roofs. Integrating nature with buildings is one obvious way to instil green space in the urban environment (Xing et al. 2017). Green roofs with wildflowers, for example, attract both domestic and solitary bees, butterflies, bumblebees, and dipteras, and green roofs with wildflowers may also serve as green corridors (Benvenuti 2014),. Green roofs and walls may also be used for unique biodiversity interventions, such as replicating ground-level ecological conditions (Enzi et al. 2017). Parks, pocket parks, and other green spaces, especially larger green spaces with high corridor connectivity and a large number of smaller patches, help to promote biodiversity. The biodiversity index increases as ecosystems are distributed more evenly (Forman 1995, as referred to in Strohbach et al. 2013). Birds, small vertebrates, and insects are attracted to meadow-like vegetation because it provides food and habitat (Bretzel et al. 2016). Wildflowers have the potential to thrive in low-nutrient soils, making them ideal candidates for planting in urban areas to boost biodiversity (Maxter 2018). SUDs and blue components also provide important tools for biodiversity, such as water, food, and habitat. As with green spaces, connectivity with other ecosystems boosts biodiversity even further (Woods Ballard et al. 2015).

Green spaces are an important component of a compact city's sustainability (Jansson 2014). Ecosystem services offered by green and blue spaces in urban areas lead to urban development and resilience (Jansson 2014; Badiu et al. 2016). People are more comfortable with their neighborhood in general if there are green spaces nearby (within 300 m) (Björk et al. 2008). Green spaces are known to enhance the attractiveness of urban areas by growing and improving components related to quality of life, such as protection, social interaction, appealing work and living space, and involvement. (Jansson 2014).

17.4.3 Blue Green Infrastructure as NbS Intervention for Noida

As suggested by Peter et al., in the paper 'Blue-Green Infrastructures As Tools For The Management Of Urban Development And The Effects Of Climate Change', For cities facing climate change challenges, Blue-Green Infrastructure (BGI) is a viable, cost-effective, and valuable option. It works in conjunction with grey infrastructure and, in some cases, replaces it. BGI represents a paradigm shift in urban water management, recognizing the importance and value of including urban hydrology.

The "Blue" emphasizes the physicality of water, whereas the "Green" links urban hydrological functions to vegetation systems in urban landscape design. Blue and green infrastructures work together to improve urban habitats by evoking natural processes in man-made environments and combining the needs of clean water and storm water management with those of urban planning and urban life. As a result, such systems have a positive effect on urban natural resource metabolism (added green values) as well as the perception and behavior of people who use these infrastructures (added social values). BGI not only reduces storm-water quantity but also improves water quality. Plant roots, in conjunction with soil, absorb nutrients and purify infiltrating water, as well as increase overall water quality in urban catchment areas, reducing energy demands and costs associated with water treatment. It helps to prevent overheating and oxygen shortages caused by high temperatures of concrete materials in riverbeds. It improves on-site storm water retention, protecting important wetland areas, reducing the need for downstream flood buffer zones, and lowering the risk and effect of flooding along with recharging underlying aquifers and balancing groundwater levels. It has a tremendous potential to modulate the urban climate by reducing urban heat island impacts, balancing diurnal temperature fluctuations, and promoting natural air ventilation, in addition to direct benefits related to water and plants. It also mitigates the bioclimatic effects of land cover changes, such as urban soil desiccation and related wind-borne air pollution and dust hazards. BGI improves the adaptability and stability of urban infrastructure by controlling and modulating hydro-climatic fluctuations and weather extremes. BGI enhances rich biotopes and landscape connectivity, preserves aquatic habitats, and establishes biodiversity-rich areas, all of which increase urban biodiversity. BGI improves the physical and mental health of people by providing improved space for leisure, exercise, and social events. Public health expenses are reduced as a result of these amenities. BGI encourages social engagement and integration by encouraging people to use open spaces for events. The development of Blue-Green infrastructure enhances a city's overall attractiveness and livability, as well as the city's governmental institutions' reputation for caring about the living conditions of its people. Finally, BGI promotes biophilia, i.e. people's love of nature, by reuniting them with natural forms, elements, and processes.

17.4.3.1 Applied Example

Case Study: Nagoya: Adaptation to Climate Change Driven by Biodiversity Conservation

Nagoya's green spaces have been reduced to around 25% of the city's total area due to changes in land use associated with industrialization and the growth of residential areas (Kojima 2010). Another source of concern is Nagoya City's rising temperature as a result of climate change. The urban heat island effect and problems associated with high temperatures in Nagoya's urban areas are being exacerbated by these factors. Given these dangers, Nagoya has taken a number of steps to promote more sustainable lifestyles in the region (Local Action for Biodiversity 2008). The 2050

Nagoya Strategy for Biodiversity, which aims to boost and expand the city's green areas, is the key programme that can help the city become better adapted to climate change (in particular, high temperatures) (City of Nagoya 2008). The Water Revitalization Plan, which is part of the Biodiversity Strategy, aims to recharge ground water sources by increasing infiltration through the use of green spaces (CBD 2010). The Biodiversity Strategy's implementation in close coordination with the city's citizens, businesses, and non-governmental organizations is a critical feature.

17.4.4 Social, Economic and Environmental Co-Benefits from the Suggested Nature Based Solutions (Table 17.3)

The above-mentioned benefits can be viewed as co-benefits rather than direct benefits, depending on the goal of a project in which NBSs are intended to be used (Maxter 2018). Several social benefits derived from NBSs, such as green infrastructure and green spaces, have co-benefits not only on a community level but also on an individual level, such as stress relief, increased brain power, opportunities for social interaction (Jansson 2014), improved concentration and mental

Table 17.3 Examples of various ecosystem services and other NBS benefits (inspired by Faehnle et al. 2014; Somarakis et al. 2020)

Provisioning services	Nutrition and Food Security
	Drinking water and water resources
	Raw Material for Energy
Regulation and maintenance services	Carbon sequestration
	Local Climate regulation
	Water purification
	Air quality regulation
	Biodiversity (including genetic resources)
	Pollinators for food security and biodiversity (Maintaining population and habitats)
	Flood risk control and storm-water management
	Erosion control
Socio-cultural services	Aesthetic improvement
	Cultural heritage
	Active lifestyle
	Restoration from stress or illness
	Knowledge creation, education and awareness raising
	Social cohesion and social capital
Economic services	Eco-tourism development (Nature based tourism)
	Increased regional value
	Reduced maintenance costs and increased investments

control (Berman et al. 2008), and, on a community level, income from increased tax revenue and property development, as well as saving (Saraev 2012).

Green infrastructure (e.g., green roofs, walls, and street vegetation) (Jansson 2014) and green space have economic co-benefits in the form of green employment in construction and maintenance, as well as increased efficiency (Saraev 2012.; Enzi et al. 2017) and job satisfaction (Dravigne et al. 2008; Enzi et al. 2017). When compared to conventional drainage systems, rain gardens have been found to be a promising choice with a cost savings of about 42 percent (Vineyard et al. 2015). Green surfaces (Xing et al. 2017) and buildings can also save energy and money (Enzi et al. 2017) by lowering heating and cooling costs. Retrofitting green roofs to a university campus building in Canada, for example, lowered building energy demand by 3% as found by Berardi (2016). Green infrastructure, in general, has a good return on investment (Enzi et al. 2017).

The carbon sequestration and storage properties of vegetation and soil in varying quantities help to reduce climate change, which is one of the environmental co-benefits (Alexandri and Jones 2008; Fioretti et al. 2010; Kiss et al. 2015). A tree *Platanus hybrida*, for example, can sequester 60 kg per year, while a *Sophora japonica* can sequester 23.5 kg per year (Kiss et al. 2015). The regulating services of urban forests were quantified by Baró et al. (2014), and it was discovered that the studied region removed 19,036 tonnes of CO2eq and 305.6 tonnes of air contaminants per year. Urban runoff is a major pollutant source in urban areas (Kamali et al. 2017). Green roofs and bioretention systems, such as rain gardens, can store and remove stormwater runoff contaminants (Clar et al. 2004; Woods Ballard et al. 2015; Davis and Naumann 2017) (which pose risks to human and environmental health). Furthermore, rain gardens have been found to have a 62 percent to 98 percent lower environmental effect than conventional piping systems (Vineyard et al. 2015).

17.4.5 Technical Barriers and Knowledge Gaps for NbS Implementation

NBS projects tend to encounter several obstacles. Designing, implementing, and sustaining NBS, as well as quantifying (including economic valuation) the advantages and co-benefits of their ecosystem services, are not well understood. Furthermore, there is a lack of deep understanding among key stakeholders, as well as a lack of skills and experience at various stages of the NBS project growth. Decision-makers and practitioners also lack the knowledge and skills needed to effectively overcome potential trade-offs and make the best use of available technological solutions. Inadequate, or in most cases absent, follow-up monitoring of applied NBS obstructs the assessment of their efficacy, depriving decision-makers and practitioners of useful insights into the cost-benefit analysis, efficiency, and long-term viability of NBS.

In the field of NBS, there is a growing trend for technological innovation. The economic opportunities for implementing NBS are there for the taking, but conventional business models lack the system to allow the development of financial support for such schemes. Despite the fact that NBS are frequently more cost-effective than conventional grey infrastructure alternatives, the obstacles to their implementation are frequently more complex. These issues can be traced back to changes in management, a lack of knowledge, collaboration, and securing funding for a new and little understood industry. These variables will make it difficult to create a well-defined business case. The type of approach, the targeted resilience result, the degree of investment, the size of the actions, and the NBS's lifetime will all influence the project's economic risk. One of the most important factors to remember is how NBS are funded. NBS are typically financed by municipalities, state governments, and national governments (public stakeholders), as well as private sectors. The method of obtaining finance varies greatly between states and territories, as well as between public and private organizations. In certain instances, different types of funding are available, depending on the local context and the stakeholders' willingness to collaborate (WBCSD 2017).

In order to be mainstreamed, NBS face many challenges in their execution, necessitating constructive and creative policy interventions. The majority of policies at all levels (local, regional, global, and international) were developed without considering NBS as a viable alternative to traditional grey solutions or other similar methods. As a consequence, current policies can make it difficult or impossible for NBS projects to be considered.

17.5 Discussion

NBS are seen as a way to bridge the gap between economic development and socio-environmental issues, providing a concrete route to a sustainable economy (Maes and Jacobs 2015) and urban resilience. However, Raymond et al. (2017) highlighted that scaling up and connecting and absorbing small-scale initiatives on the ground into larger and potentially more impactful interventions is a major challenge and opportunity for NbS. In this case, stronger policy coherence may be beneficial. Future research should be conducted to provide a solid evidence base for the contribution of NbS to job growth, as well as to demonstrate the economic feasibility of NbS in relation to other types of solutions on a timeline that is reasonable. Established urban planning frameworks for nature-based solutions, however, have significant gaps and omissions that must be resolved and improved if planning is to comprehensively support their implementation. New and emerging research methods using multi-species approaches are needed for urban planning to effectively promote the implementation of nature-based solutions. The evidence for NBS benefits is inadequate, and generalizations are often made. NBSs are location specific, which may clarify why generalizations are used. In terms of the environmental value of carbon sequestration, for example, the effects of a project of this size may not be sufficient enough to meet local greenhouse gas emission reduction goals. However, if applied in larger scales and quantities, a difference can be distinguishable. For sufficient documentation and to bridge information gaps, it is recommended that the results of the implemented NBSs be monitored on a regular basis.

Somarakis et al. (2020) highlighted that Since NBSs are still in the early stages of growth, there are knowledge gaps in both research and practice. This can be a drawback because information gaps can lead to uncertainty on how to implement solutions. It does, however, open up possibilities for collaboration. Collaboration with research projects, for example, could result in a win-win situation in which the research group is able to address their research questions while the city benefits from the implementation of NBSs in their urban environments. Our aim should be to contribute to this emerging multidisciplinary area by developing a planning system for nature based solutions (NbS).

References

Alberti M, Marzluff JM (2004) Ecological resilience in urban ecosystems: linking urban patterns to human and ecological functions. Urban Ecosyst 7(3):241–265. https://doi.org/10.1023/B:UECO.0000044038.90173.c6

Alexandri E, Jones P (2008) Temperature decreases in an urban canyon due to green walls and green roofs in diverse climates. Build Environ 43(4):480–493

Annerstedt M, Ostergren P-O, Bjork J, Grahn P, Skarback E, Wahrborg P (2012) Green qualities in the neighbourhood and mental health - results from a longitudinal cohort study in Southern Sweden. BMC Public Health 12:337

Badiu DL, Iojă CI, Pătroescu M, Breuste J, Artmann M, Niţă MR, Grădinaru SR, Hossu CA, Onose DA (2016) Is urban green space per capita a valuable target to achieve cities' sustainability goals? Romania as a case study. Ecol Indic 70:53–66

Baró F, Haase D, Gómez-Baggethun E, Frantzeskaki N (2015) Mismatches between ecosystem services supply and demand in urban areas: a quantitative assessment in five European cities. Ecol Indic 55:146–158. https://doi.org/10.1016/j.ecolind.2015.03.013

Benvenuti S (2014) Wildflower green roofs for urban landscaping, ecological sustainability and biodiversity. Landsc Urban Plan 124:151–161

Berardi U (2016) The outdoor microclimate benefits and energy saving resulting from green roofs retrofits. Energy Build 121:217–229

Berland A, Shiflett SA, Shuster WD, Garmestani AS, Goddard HC, Herrmann DL, Hopton ME (2017) The role of trees in urban stormwater management. Landsc Urban Plan 162:167–177

Berman M, Jonides J, Kaplan S (2008) The cognitive benefits of interacting with nature. Psychol Sci 19(12):1207–1212

Bhowmick S (2018) Noida's groundwater level falling by 1.5 meter every year. The Times of India, Aug 24. https://timesofindia.indiatimes.com/city/gurgaon/noidas-groundwater-level-falling-by-1- 5-metres-every-year/articleshow/65522432.cms. Accessed 25 Aug 2018

Björk J, Albin M, Grahn P, Jacobsson H, Ardö J, Wadbro J, Östergren P-O, Skärbäck E (2008) Recreational values of the natural environment in relation to neighbourhood satisfaction, physical activity, obesity and wellbeing. J Epidemiol Community Health 62(4):E2–e2

Bretzel F, Vannucchi F, Romano D, Malorgio F, Benvenuti S, Pezzarossa B (2016) Wildflowers: from conserving biodiversity to urban greening—a review. Urban For Urban Green 20:428–436
Busch J, Ferretti-Gallon K, Engelmann J, Wright M, Austin KG, Stolle F, Turubanova S, Potapov PV, Margono B, Hansen MC, Baccini A (2015) Reductions in emissions from deforestation from Indonesia's moratorium on new oil palm, timber, and logging concessions. PNAS 112(5): 1328–1333. https://doi.org/10.1073/pnas.1412514112
Busha J, Doyon A (2019) Building urban resilience with nature-based solutions: How can urban planning contribute? Cities 95:102483
Calfapietra C, Fares S, Manes F, Morani A, Sgrigna G, Loreto F (2013) Role of biogenic volatile organic compounds (BVOC) emitted by urban trees on ozone concentration in cities: a review. Environ Pollut 183:71–80
CBD (2010) City of Nagoya, Japan. http://www.cbd.int/authorities/casestudies/nagoya.shtml
Chen D, Wang X, Thatcher M, Barnett G, Kachenko A, Prince R (2014) Urban vegetation for reducing heat related mortality. Environ Pollut 192:275–284
City of Nagoya (2008) Biodiversity report. Local Action for Biodiversity, and ICLEI initiative
Clar M, Barfield BJ, O'Connor T (2004) Stormwater best management practices design guide volume 1 - general considerations. EPA/600/R-04/121. U.S. Environmental Protection Agency, Washington, DC
Cohen-Shacham E, Walters G, Janzen C, Maginnis S (2016) Nature-based solutions to address global societal challenges. IUCN, Gland
Coombes E, Jones AP, Hillsdon M (2010) The relationship of physical activity and overweight to objectively measured green space accessibility and use. Soc Sci Med 70:816–822
Davis M, Naumann S (2017) Making the case for sustainable urban drainage systems as a nature-based solution to urban flooding. In: Kabisch N, Korn H, Stadler J, Bonn A (eds) Nature-based solutions to climate change adaptation in urban areas: linkages between science, policy and practice. Springer, Cham, pp 123–137
Dhakal S, Hanaki K (2002) Improvement of urban thermal environment by managing heat discharge sources and surface modification in Tokyo. Energy Build 34(1):13–23
Dravigne A, Waliczek TM, Lineberger RD, Zajicek JM (2008) Effect of live plants and window views of green spaces on employee perceptions of job satisfaction. HortScience 43(1):183–187
Duarte CM, Losada IJ, Hendriks IE, Mazarrasa I, Marbà N (2013) The role of coastal plant communities for climate change mitigation and adaptation. Nat Clim Chang 3:961–968
Enzi V, Cameron B, Dezsényi P, Gedge D, Mann G, Pitha P (2017) Nature-based solutions and buildings – the power of surfaces to help cities adapt to climate change and to deliver biodiversity. In: Kabisch N, Korn H, Stadler J, Bonn A (eds) Nature-based solutions to climate change adaptation in urban areas: linkages between science, policy and practice. Springer, Cham, pp 159–183
Estrella, M. and Saalismaa, N.(2013). Ecosystem-based DRR: An overview. The role of ecosystems in disaster risk reduction: 26–47.
Fioretti R, Palla A, Lanza LG, Principi P (2010) Green roof energy and water related performance in the Mediterranean climate. Build Environ 45(8):1890–1904
Frantzeskaki N (2019) Seven lessons for planning nature-based solutions in cities. Environ Sci Policy 93:101–111
Gunawardena KR, Wells MJ, Kershaw T (2017) Utilising green and bluespace to mitigate urban heat island intensity. Sci Total Environ 584-585:1040–1055
Gupta AK, Nair SS (eds) (2012) Ecosystem approach to disaster risk reduction. National Institute of Disaster Management, New Delhi
Gupta AK, Singh S, Wajih SA, Mani N, Singh AK (2017) Urban resilience and sustainability through peri-urban ecosystems: integrating climate change adaptation and disaster risk reduction. Gorakhpur Environmental Action Group, Gorakhpur
Haase D, Larondelle N, Andersson E, Artmann M, Borgström S, Breuste J, Gomez-Baggethun E, Gren Å, Hamstead Z, Hansen R et al (2014) A quantitative review of urban ecosystem service assessments: concepts, models, and implementation. Ambio 43:413–433

https://obs-up.com/
https://www.iqair.com/world-air-quality-ranking
Janhall S (2015) Review on urban vegetation and particle air pollution - deposition and dispersion. Atmos Environ 105:130
Jansson M (2014) Green space in compact cities: the benefits and values of urban ecosystem services in planning. Nord J Archit Res 26(2):139–160
Kamali M, Delkash M, Tajrishy M (2017) Evaluation of permeable pavement responses to urban surface runoff. J Environ Manag 187:43–53
Kiss M, Takács Á, Pogácsás R, Gulyás Á (2015) The role of ecosystem services in climate and air quality in urban areas: evaluating carbon sequestration and air pollution removal by street and park trees in Szeged (Hungary). Morav Geogr Rep 23(3):36–46
Klingberg J, Broberg M, Strandberg B, Thorsson P, Pleijel H (2017) Influence of urban vegetation on air pollution and noise exposure – A case study in Gothenburg, Sweden. Sci Total Environ 599-600:1728–1739
Kojima M (2010) Working on transforming Nagoya into a comfortable. Low-carbon city power point presentation at the Nagoya City Hall, 18 May 2010
Local Action for Biodiversity (2008) City of Nagoya, Japan. Biodiversity for well-being and sustainable livelihoods. http://www.unep.org/urban_environment/PDFs/Nagoya_Final.PDF
Maas J, van Dillen SME, Verheij RA, Groenewegen PP (2009) Social contacts as a possible mechanism behind the relation between green space and health. Health Place 15:586–595
Maas J, Verheij RA, Groenewegen PP, De Vries S, Spreeuwenberg P (2006) Green space, urbanity, and health: how strong is the relation? J Epidemiol Community Health 60:587–592
Maheshwari B, Singh VP, Thoradeniya B (2016) Balanced urban development: options and strategies for liveable cities. Water Sci Technol Library 72. https://doi.org/10.1007/978-3-319-28112-4_1
Millennium Ecosystem Assessment (2005) Ecosystems and human well-being: wetlands and water synthesis. World Resources Institute, Washington, DC
Mitchell R (2012) Is physical activity in natural environments better for mental health than physical activity in other environments? Soc Sci Med. https://doi.org/10.1016/j.socscimed.2012.04.012
Mitchell R, Popham F (2008) Effect of exposure to natural environment on health inequalities: an observational population study. Lancet 372:1655–1660
Munang R, Thiaw I, Alverson K, Liu J, Han Z (2013) The role of ecosystem services in climate change adaptation and disaster risk reduction. Curr Opin Environ Sustain 5(1):47–52
Munang R, Thiaw I, Alverson K, Mumba M et al (2013) Climate change and ecosystem-based adaptation: a new pragmatic approach to buffering climate change impacts. Curr Opin Environ Sustain 5:1–5. https://doi.org/10.1016/j.cosust.2012.12.001
Nandy SN (2015) Urbanization in India – past, present and future consequences. Urban India 35(2): 8–24
Naumann S, Anzaldua G, Berry P, Burch S, Davis M, Frelih-Larsen A, Gerdes H, Sanders M (2011) Assessment of the potential of ecosystem-based approaches to climate change adaptation and mitigation in Europe. Final report to the European Commission, DG Environment, Contract no. 070307/2010/580412/ SER/B2, Ecologic Institute and Environmental Change Institute, Oxford University Centre for the Environment
Papadakis G, Tsamis P, Kyritsis S (2001) An experimental investigation of the effect of shading with plants for solar control of buildings. Energy & Buildings 33(8):831–836
Pickett STA, Cadenasso ML, Grove JM (2004) Resilient cities: meaning, models, and metaphor for integrating the ecological, socio-economic, and planning realms. Landsc Urban Plan 69(4): 369–384. https://doi.org/10.1016/j.landurbplan.2003.10.035
ProAct Network (2008) The role of environmental management and eco-engineering in disaster risk reduction and climate change adaptation. http://www.proactnetwork.org/proactwebsite/en/resources/ecosystem-based-drr/94-ecosystem-based-drr/205-ecosystem-baseddrr-key-publications

Raymond CM, Berry P, Breil M, Nita MR, Kabisch N, de Bel M, Enzi V, Frantzeskaki N, Geneletti D, Cardinaletti M, Lovinger L, Basnou C, Monteiro A, Robrecht H, Sgrigna G, Munari L, Calfapietra C (2017) An impact evaluation framework to support planning and evaluation of nature-based solutions projects. Report prepared by the EKLIPSE Expert Working Group on Nature-based Solutions to Promote Climate Resilience in Urban Areas. Centre for Ecology & Hydrology, Wallingford, United Kingdom

Resilient Melbourne (2016) Resilient Melbourne: resilience strategy for greater Melbourne. City of Melbourne, Melbourne

Salmond JA, Williams DE, Laing G, Kingham S, Dirks K, Longley I, Henshaw GS (2013) The influence of vegetation on the horizontal and vertical distribution of pollutants in a street canyon. Sci Total Environ 443:287–298

Sandström UG, Angelstam P, Mikusiński G (2006) Ecological diversity of birds in relation to the structure of urban green space. Landsc Urban Plan 77(1):39–53

Sarabi SE, Qi Han A, Romme GL, de Vries B, Wendling L (2019) Key enablers of and barriers to the uptake and implementation of nature-based solutions in urban settings: a review. Resources 8:121. https://doi.org/10.3390/resources8030121

Saraev V (2012) Economic benefits of greenspace: a critical assessment of evidence of net economic benefits. Forestry Commission, Edinburgh

Sayaas S (2016) Resilient cities report—preliminary version. OCDE Regional Policies for Sustainable Development Division; OCDE: Paris, France

Sharma R et al (2020) Assessing urban heat islands and thermal comfort in Noida City using geospatial technology

Singh VB, Tripathi JN (2016) Identification of critical water quality parameters derived from principal component analysis: case study from NOIDA area in India. Am J Water Res 4(6): 121–129. https://doi.org/10.12691/ajwr-4-6-1

Sinha SK (2018) Urbanization in Gautam Buddha Nagar District, Uttar Pradesh

Somarakis G, Stagakis S, Chrysoulakis N (Eds) (2019) ThinkNature nature-based solutions handbook. ThinkNature project funded by the EU Horizon 2020 research and innovation programme under grant agreement no. 730338. https://doi.org/10.26225/jerv-w202

Stigsdotter UK, Ekholm O, Schipperijn J, Toftager M, Kamper-Jorgensen F, Randrup TB (2010) Health promoting outdoor environments - associations between green space, and health, health-related quality of life and stress based on a Danish national representative survey. Scand J Public Health 38:411–417

Strohbach M, Lerman S, Warren P (2013) Are small greening areas enhancing bird diversity? Insights from community-driven greening projects in Boston. Landsc Urban Plan 114(C):69–79

Temmerman S, Meire P, Bouma TJ, Herman PMJ (2013) Ecosystem-based coastal defence in the face of global change. Nature 504:79–83. https://doi.org/10.1038/nature12859

Thompson CW, Roe J, Aspinall P, Mitchell R, Clow A, Miller D (2012) More green space is linked to less stress in deprived communities: evidence from salivary cortisol patterns. Landsc Urban Plan 105(3):221–229

UN-Climate Action (2019) The nature-based solutions for climate manifesto-developed for the UN Climate Action Summit 2019. UN Secretary General Climate Action Summit. https://stgwedocs.uneporg/xmlui/bitstream/handle/20.500.11822/29705/190825NBSManifesto.pdf?sequence=1&isAllowed=y. Accessed 25 Aug 2020

UNESCO (2018) UN world water development report, nature-based solutions for water. UNESCO, Paris

United Nations General Assembly (2017) Resolution adopted by the General Assembly on 6 July 2017. 71/313 work of the Statistical Commission Pertaining to the 2030 Agenda for Sustainable Development. A/RES/71/313. https://undocs.org/A/RES/71/313. Accessed 10 July 2017

Vineyard D, Ingwersen W, Hawkins T, Xue X, Demeke B, Shuster W (2015) Comparing green and grey infrastructure using life cycle cost and environmental impact: a rain garden case study in Cincinnati, OH. J Am Water Resour Assoc 51(5):1342–1360

Wang Y, Bakker F, Groot R, Wortche H, Leemans R (2015) Effects of urban trees on local outdoor microclimate: synthesizing field measurements by numerical modelling. Urban Ecosyst 18(4): 1305–1331

WB (2017) Implementing nature-based flood protection: principles and implementation guidance, Washington DC

Weber D, Anderson D (2010) Contact with nature: recreation experience preferences in Australian parks. Ann Leis Res 13:46–69

Whitford V, Ennos AR, Handley JF (2001) "City form and natural process"—indicators for the ecological performance of urban areas and their application to Merseyside, UK. Landsc Urban Plan 57(2):91–103

Woods Ballard B, Wilson S, Udale-Clarke H, Illman S, Scott T, Ashley R, Kellhager R (2015) The SuDS manual. Ciria, London

WWF (2013) Operational framework for ecosystem-based adaptation: implementing and mainstreaming ecosystem-based adaptation responses in the Greater Mekong Sub-Region

Xing Y, Jones P, Donnison I (2017) Characterisation of nature-based solutions for the built environment. Sustainability 9:149

Yang H, Dick W, Mccoy E, Phelan P, Grewal P (2013) Field evaluation of a new biphasic rain garden for stormwater flow management and pollutant removal. Ecol Eng 54(C):22–31

Harsha Nath holds a master's degree in Environmental Sciences and has been actively working in the field of sustainability and climate change. She has been associated with reputed organizations like GICIA and NIDM (Govt of India). She is currently working as an assistant manager, corporate sustainability at PPAP Automotive Limited.

Pritha Acharya has a background in environmental sciences. She has worked in the field of sustainability, climate change adaptation and biodiversity conservation while working with IUCN India. She is currently working as a research associate and manages the CAP-RES (DST) funded project.

Gupta Anil K , Professor of Policy & Strategies, Integrated Resilience. Head of the ECDRM Division, International Cooperation, Programme Director - Centre for Excellence on Climate Resilience, Nodal Officer for Central Ministries at National Institute of Disaster Management, (Govt. of India), New Delhi. He is also a member of the expert Team for WMO Climate Statement and Core Group Member of IUCN-Council of Ecosystem Management.

Chapter 18
Equity and Fairness in Community Based Adaptation and Disaster Risk Reduction

Sivapuram Venkata Rama Krishna Prabhakar, Norkio Shimizu, and So-Young Lee

Abstract Equity and fairness have high significance in the Disaster Risk Reduction and Climate Change Adaptation as it is evident that disasters affect communities differently depending on their developmental levels, and social status. Any public policy or intervention on disaster risk reduction deals with communities that are at different 'starting lines' in terms of their disaster vulnerabilities and hence should ensure that all are able to reach an identical 'finishing line' of reduced risks as an outcome of the intervention. To this effect, the paper presents a review of pertinent literature on how and to what extent the CBA and CBDRR approaches are able to achieve equity and fairness outcomes. In India, it was observed that the disasters could trigger solidarity and altruistic motivations that could help in recovery from disaster. In Japan, equity and fairness issues were addressed through several policies. The discussion reveals that these approaches may not automatically result in equity and fairness and that there is a need for conscious efforts to ensure such outcomes even in participatory approaches. Hence, it is not merely the participation alone but also the level of participation and nature of participation that determines the equity and fairness outcomes of interventions. It is important for the practitioners and policy makers to ask questions such as who is at risk, who is vulnerable, how the vulnerabilities are assessed, who assessed the vulnerabilities, how inclusive the process is, and if the process satisfies various aspects of the equity and fairness. The paper lists important questions and indicators that help in assessing the extent to which a risk reduction intervention can achieve equity and fairness outcomes.

Keywords Community based adaptation · Disaster risk reduction · Equity · Fairness

---

S. V. R. K. Prabhakar (✉) · N. Shimizu · S.-Y. Lee
Adaptation and Water Group, Institute for Global Environmental Strategies, Hayama, Japan
e-mail: prabhakar@iges.or.jp

A. K. Gupta et al. (eds.), *Disaster Risk and Management Under Climate Change*, Disaster Resilience and Green Growth,
https://doi.org/10.1007/978-981-99-4105-6_18

18.1 Introduction

Research carried out by many organizations such as the Institute for Global Environmental Strategies (IGES), its partners and other research community has revealed that the decision-making in areas of disaster risk reduction (DRR) and climate change adaptation (CCA), in most of the Asia-Pacific countries and elsewhere is at nascent stages. Issues such as limited understanding on how the decision-making in CCA and DRR could (or should) be different from developmental decision-making, lack of tools or limited application of available tools to prioritize CCA and DRR actions, and limitations with current policy and institutional mechanisms. Despite these challenges, there has been a significant consensus among Climate Change research and development community that Community Based Adaptation (CBA) and DRR offers best solution to address Climate Change vulnerabilities for the following reasons (World Bank 2006; Prabhakar 2013):

(i) **Knowledge on natural resources:** Communities live close to natural resources, they are benefited by them, they observe the impacts of Climate Change on them, and hence they better understand the impacts of Climate Change on natural resources and in turn on their own livelihoods.
(ii) **First responders:** Disasters affect communities first, and often they are the first responders to disasters. It entails that the capacities of communities are built first to enable better response. Hence, this is where CBA comes to help by enabling building of community capacities.
(iii) **Equity:** Communities have diverse interests in natural resources and achieving a consensus on benefit sharing is an important aspect of their development.
(iv) **Capacity:** Communities often have better understanding on resources that they live in proximity than other stakeholders and they also have better understanding of the capacities that they derive from the resources they depend up on.
(v) **Cost-effectiveness:** Local management may help reduce government costs and community management of resources could provide efficient means for the governments.
(vi) **Alignment with development:** Local participation, decentralization, and subsidiarity may, in themselves, are important development objectives. Such concept has commonality with CBA in that it promotes community participation. Hence, CBA could help in achieving these developmental objectives.

From the above discussion, it could be seen that the equity and public participation are integral requirements for successful adaptation and DRR and it is important that governments and other stakeholders consider equity and public participation in the policy interventions since Climate Change Adaptation (CCA) and DRR are related to the question of public policy making requiring collective action and cognitive decision-making.

Keeping in view the importance of promoting equity and fairness in CCA and DRR interventions, this paper presents a review of available evidence and experiences in promoting equity and fairness in the CCA and DRR literature. The paper makes efforts to present the extent the CBA and CBDRR approaches are able to achieve equity and fairness outcomes.

18.2 Vulnerability and Impacts

The societal vulnerability to disasters stems from various factors. These factors could include socio-economic, institutional and policy, and environmental elements. The research indicates that the social factors such as equity and fairness play an important role in how the disasters affect people and in turn the effectiveness of the interventions (Nutters 2012; Mazepus and Leeuwen 2019). The literature clearly indicates that the communities that are subjected to inequality are highly vulnerable to the impacts of disasters and are often subjected to unfair decision-making outcomes aftermath of disasters. Hence, best CCA and DRR outcomes are possible only by effectively addressing the equity and fairness in how the risk reduction interventions are designed and implemented.

Participatory approaches such as Community Based Disaster Risk Reduction (CBDRR) and Community Based Adaptation (CBA) have become a corner stone of risk reduction interventions. It has been widely hypothesized that participatory interventions are successful and sustainable since they are able to satisfy the equity and fairness principles. However, very little evidence has been offered in the current literature in terms of how the participatory approaches such as CBA are able to achieve equity and fairness, if at all they achieved.

Keeping in view the importance of equity and fairness (E and F) in CCA and DRR decision-making, a panel discussion was organized at the Adaptation Forum (2012) by the authors of this paper where researchers and policy makers have shared their opinions on how the E and F can be ensured (APAN 2012). The messages from the session are incorporated in this paper wherever appropriate.

Rahman (2012) has stated that adaptation is an issue of E and F and that the CBA can directly lead to ensuring E and F as it ensures public participation in the interventions. Also, ensuring public participation has long been and loudly claimed by the experts and practitioners of CBA. While CBA approaches may seem to have enabled public participation, there is no clearly published evidence to prove that these approaches have promoted equitable and fair decision-making in adaptation and DRR in CBA. Hence, it is not only important to bring out such evidence but also to create enabling conditions for ensuring fairness even in CBA, an important question to explore, since CBA approaches have widely been employed in both DRR and CCA areas with wide-reaching consequences and outcomes. Though public participation and fairness are issues that have been historically the point of focus in developmental interventions, it is not very much clear from the available literature on how these experiences can be learned and applied to decision-making in CCA and DRR fields.

Keeping the above background, the current paper sets out to address the following questions:

1. What are E and F and why it is important to consider E and F issues in CBA and related approaches?
2. What approaches promote E and F in general and where community-based participatory approaches such as CBA stand among them in ensuring E and F?
3. Do participatory approaches such as CBA automatically ensure E and F or do they need to be ensured through special efforts within those participatory interventions?
4. If the answer to above questions is that one need to make special efforts to ensure E and F in CBA, then what forms of governance and tool to be employed in CBA projects and programs to provide an enabling environment for ensuring E and F. What evidence do we have from the literature to support these approaches?

The aim of this paper is to answer some or all of the above questions based on evidences from the existing participatory experiences in CCA and DRR fields. In answering the above questions, this paper defines E and Fin the context of community-based approaches such as CBA only. In this paper, equity is defined as an equitable outcome of the fair procedure. Fairness is used in a procedural dimension in this paper. Such procedure is defined as a procedure (1) which is conducted with a great care for the vulnerable and the most affected for gaining public acceptance, considering geographically unequal impacts of Climate Change and their livelihoods; (2) involves all the stakeholders in adaptation and DRR planning from the very early stage without any oppression from others; and (3) gives enough information for stakeholders to decide among the available alternatives.

Regarding CBA, as the term 'community-based' is often referred to community as the setting for interventions, this paper adopts the definition of CBA as 'to put community as the setting for adaptation and disaster risk reduction interventions through various procedures' (Merzel and DAfflitti 2003).

18.2.1 Different Definitions of Equity and Fairness

This section presents different understandings of E and F in the CCA and DRR at the local level. Though some consider E and F as the same or in an interchangeable manner, in general, fairness is an outcome achieved through equity. Fairness is being unbiased and being just for all the sections of the society or the participants of a programme or a government policy without any discrimination based on certain characteristics of the individuals. If fairness is considered as an outcome, equity, and equality, is the means through which such an outcome could be achieved. Here, fairness through equity means that different sections of the society will have different levels of access or benefits from a public policy depending on where each individual stand vis-à-vis the policy objectives. From this point of view, Pauly and Willett (1972) indicate that E and F in public policy dictates that the equals should be treated equally and that unequal should be treated unequally, and such differences should be decided fairly.

There are different descriptions and definitions of E and F in CCA at the local level, depending on literature context and author's backgrounds. While some clearly differentiated between E and F, others do not. Here are some examples of how different literature defines them differently.

The E and F have a long history since the time of Aristotle. Since then, the E and F has been one of the important research questions in many discourses involving burden sharing. In climate change context, E and F has also been discussed in many aspects: unequal impacts of climate change; equitable distribution of climate finance among developing countries, fair burden sharing of climate finance by developed countries: fairness in the level of ambition for mitigation action; and inequitable impacts between developed/developing countries, generations and species. For example, the United Nations Framework Convention on Climate Change (UNFCCC) mentions equity in the context of protecting the climate system for the benefit of the present and future generations and for mitigating GHG emissions according to the equitable contributions of countries to global efforts (UNFCCC 1992). Although there are many aspects of E and F in climate change context, this paper focuses on E and F of CCA at the local level.

Huq and Khan (2006) defined fairness by procedural justice and equity by distributing justice and described the relationship of these two concepts i.e. *'if procedural justice, meaning fairness of the process used to prioritize the activities, can be ensured…, considerations of equity in outcome can be addressed.'* While Huq and Khan (2006) did not define equity explicitly, equity was used in the context of outcome and fairness was used in the procedural context and it was assumed that the later could bring out the former.

On the other hand, there are other literatures that do not make clear distinction between fairness and equity (Adger and Nelson 2010). Adger and Nelson (2010) discussed procedural dimension of adaptation at the local level without clear distinction between 'fairness' and 'equity'.

Here we provide an elaborative illustration for better understanding of the equity and equality. Designing a flood insurance program with subsidized premium is a relevant policy question where the principles of E and F can be applied. Since not all sections of the society can afford to pay the insurance premium, government may decide to subsidize insurance to certain sections of the society. Here the policy question could be who should have access to subsidized insurance and who should not. The principle of equity dictates that the government program may allow higher insurance premium subsidies for the low-income individuals. On the other hand, the principle of equality determines that all insurance holders are equally eligible for subsidized premiums.

Figure 18.1 shows how E and F affect the risk reduction outputs and outcomes. For easily visualizing the importance of E and F in public policy making including that of risk reduction interventions, considering two contrasting situations could help us understand well the implications of E and F in risk reduction interventions. In this mental experiment, let us assume that the Place A has a social and institutional setup that is aware of and addresses the E and F issues upfront and implements programs and policies accordingly. On the contrary, the social and institutional setup at Place

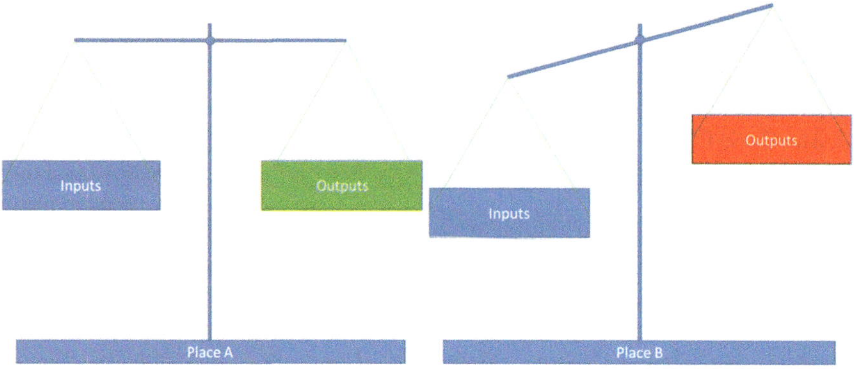

Fig. 18.1 A mental experiment for visualizing the importance of equity and fairness in risk reduction

B do not consider these as important ingredients in public policy making including designing and implementing risk reduction interventions. It can be seen from the Fig. 18.1 that the net risk reduction outputs and outcomes in the place B could be far lesser than the Place A for following reasons. (a) highly vulnerable are given priority over less vulnerable i.e. vulnerability assessments are the basis for targeting most CCA and DRR interventions, (b) presence of procedural fairness means that the Place A is able to guarantee due process in identifying the beneficiaries, (c) place A is able to ensure distributional equity i.e. interventions are targeted according to individual's risk and vulnerability characteristics, have equal access to the interventions, and that resources are allocated justly, and (d) the Place A is able to achieve process equity wherein processes are put in place for accessing resources, to express opinions and to engage with the interventions.

There are several emerging examples of employing E and F principles in the public policy. Taking the example of Japan, White Paper on Official Development Assistance (ODA) 2011 published by the Ministry of Foreign Affairs (MOFA) emphasizes 'equity'. It identified four concepts as important in order to tackle global challenges such as refugees, economic crisis, climate change, environmental problem and disasters: human security, equity, mutual support, and sustainability. Especially equity is considered to be an important concept for Japanese ODA, that there should be no one left without support (Ministry of Foreign Affairs 2012).

In addition to the ODA White Paper, there are other instances where E and F have been dealt separately in Japan. For instance, Act on Building Unit Ownership stipulates unit ownership in a single building. Article 62 (3) of the Act stipulates that share of the expenses of demolition of a building, the reconstruction of a building, and attribution of the unit ownership of the reconstructed building shall be specified in a way that does not undermine the 'equity' of each unit owner (Government of Japan 2008a). However, there are no 'fairness' words in the Act.

In another instance, discussion on burden share of medical care cost seems to be an important area for fairness consideration. In Japan, level of burden share of medical care cost by elderly persons is lower than that of the younger generation. On the other hand, on an average, disposable income of older generation is almost the same as that of the younger generation. Since the younger generation feel more burden in medical care of the elderly (Ministry of Health, Labor and Welfare 2018), the Ministry of Health, Labour and Welfare opined that the older generation has to share the burden with respective capacity in order to achieve fairness of burden between different generations (Ministry of Health, Labor and Welfare 2018).

In Japan, Omnibus Decentralization Act was enacted whereby matters related to prefectures and municipalities are dealt autonomously. In order to ensure the community participation, City Planning Act has regulated decision-making procedures of urban planning by stipulating the public consultation as a necessity to ensure E and F in urban planning (Government of Japan 2008b).

Another case for application of E and F in public policy can be found in the Republic of Korea. To respond to the growing public demands for safe and sustainable energy, the Seoul Metropolitan Government worked closely with civil society to formulate a comprehensive energy plan. It resulted in the finalization of the Comprehensive Plan for One Less Nuclear Power Plan with citizens' full participation during 16 three-way talks among the Seoul Metropolitan Government, expert advisory groups, and civil society. The goal of One Less Nuclear Power Plant was ambitious - reducing two million TOE of energy, a target equal to the production capacity of one nuclear power plant. The goal was reached 6 months ahead of schedule and achieved through citizen voluntary engagement in small but various projects i.e. the development of energy self-sufficient villages, distributions of solar PV or cooperative-sharing power plants, and other small-scale innovations. Thanks to the success, it has started the second phase activities that consist of new task - greater energy efficiency for the underprivileged through the energy-sharing community and energy welfare platform (Seoul Metropolitan Government 2014a, b, 2015; Asian Co-Benefits Partnership 2017). This success has largely been attributed to the participatory approaches employed by the government and the E and F was achieved through the citizen participation.

E and F in public policy can be complicated especially when inter-regional disparities are to be addressed in a national level policy. Such a situation could be found in Japan where the varying rates of aging in the country is causing stress on the national government to balance the financial burden in a national nursing care insurance system. The Japanese government has introduced nursing-care insurance system in 2000. Municipalities implement the system and the resource of nursing-care insurance is shared half from public money and half from the insured. Breakdown of public money resource is 25% by the central government and 12.5% by each local prefectural government and municipalities. However, this system is found to cause regional difference (Room 2018). This poses a question of fairness and equity at a country level. Background of this is that the system requires people who are over 40 years old to pay the nursing-care insurance. Hence, regions that have more population over 40 years have more budgets, but regions that have more aged people needs to spent more.

18.2.2 Significance of Equity and Fairness

In the context of climate change, E and F issues are largely limited to the international negotiations such as unequal impacts of climate change; equitable distribution of climate finance among developing countries, and fair burden sharing of climate finance by developed countries. However, E and F considerations have been less pronounced at the national and community level even though E and F is a pre-requisite for promoting better outcome, at any level and it should not be limited to the international negotiations only. Hence, this section answers the question of why fairness and equity is important in public policy and adaptation especially at the local level.

In the context of development, there are several experiences regarding fairness and equity. In 2000, the report released by the World Commissions on Dams (WCD) included discussion on the distributional inequity between groups benefiting from dam projects and those adversely affected, and equity is considered as core values and shared understanding of WCD. The Report argued that the public acceptance is essential or equitable and sustainable water and energy resources development, which emerges from recognizing rights, addressing risks, safeguarding the entitlements of all groups of affected people. Thereafter, similar arguments were introduced into policies of many countries and development agencies. In this way, experiences in the field of development has already revealed that the equitable outcome and public acceptance (i.e. a part of fair procedure according to the definition of fair procedure above) of public policies is essential.

Given that relationship between polluters and the affected is not ‘fair’, and since ‘fair procedures’ of adaptation are necessary to achieve human security; hence, the importance of E and F cannot be disregard in the context of CBA. In this context, Adger and Nelson (2010) have argued for the procedural fairness indicating that the procedural fairness is central to human security. Here, the authors have considered human security from the point of view of climate change and related impacts. This indicates one of the differences between the contexts of development and adaptation: while adaptation is often in response to threat to lives and livelihoods of people and community (i.e. human security), development is to do with continuous improvement.

Climate change has provided an opportunity for us to add new perspective to the issues of decision-making. Some literatures on community adaptation have addressed significance of E and F of adaptation at the community level. In fact, the existing literature sees the significance or necessity of fairness or equity in consideration to the vulnerable (Adger and Nelson 2010), because they are not the polluters but likely to be the most affected. Here, Adger and Nelson (2010) point out that the vulnerable find it difficult to adapt as they tend to be highly exposed, highly sensitive and have low adaptive capacity; and they have limited access to public assistance and external resources.

18.3 Mitigation and Management

This section identifies various approaches being implemented and proposed to promote E and F and discusses how the CBA is positioned among these approaches. Based on the literature, two groups of approaches can be found: (a) procedural issues of CBA, and (b) scope of information provided to stakeholders for decision-making.

18.3.1 Firstly, a Fair Procedure Is Required for Equal Outcome of CBA

What constitutes a fair procedure is a complex question. According to Huq and Khan (2006), fair procedures are realized when they vest process control or voice in those who are affected by some events. A process should be unbiased, representative of participants and stakeholders, be concerned with needs and wellbeing of stakeholders, and be flexible and ethical (Huq and Khan 2006). This argument seems to be theoretically correct but remains too conceptual. Hence, the question is what kind of procedures should be prepared to realize this in practice? Listing all the procedures is not in the scope of this paper; however, some of the key elements could be listed here:

(i) independence of a facilitator from a specific group of stakeholders;
(ii) competence and/or adequacy of a facilitator;
(iii) language used in a consultation;
(iv) appropriate format for explanation to stakeholders;
(v) whether or not the facilitator created a circumstance where those who are vulnerable or marginalized people can give their opinion without other's oppression; and
(vi) willingness to and level of reflecting stakeholder's inputs in the outcomes of the consultation.

It is important to identify and target the vulnerable populations appropriately with due consideration to all the determinants of vulnerability. Appropriate consideration of underlying factors of vulnerability, addressing factors that contributes to the full recovery, and successful implementation of mitigation and adaptation programs with due consideration of equitable and fair distribution of benefits of interventions constitute some of the prerequisite elements for achieving fairness and equity in DRR (Barrios and Battle 2018). Barrios and Battle (2018) also opined that the DRR and CCA should in fact become synonymous with the reduction of inequality and making of equity. It has been suggested that processes to identify vulnerable has to be participatory and inclusive so that there is an agreement among the communities on the people identified for targeting. In this context, important questions to be asked are: who is identifying the vulnerable, what criterion has been employed, and if all the community members had equal opportunity to participate in consultations.

Since several disadvantaged sections of the society undergo various forms of discrimination and exclusion, often it is not sufficient to target them only based on a narrow definition of vulnerability but there is a need to expand the scope to address various ways in which they are discriminated and excluded from sharing the benefits of interventions (Lovell and Masson 2014).

Of all the vulnerable, natural-resource users are often considered to be the prominent losers to climate change. However, it should be carefully examined who is really the vulnerable. 'Who is actually the vulnerable' is different depending on the communities, regions and countries and thus needs to be carefully addressed. Excessive use of natural resources can create vulnerability to climate change and can undermine the resilience and adaptive capacity of people. It is also to be noted that often the natural resource degradation, such as deforestation and related loss of biodiversity, is often due to pressures other than those who are dependent on the natural resources for their livelihoods but the dependent communities are often the losers while the wealthy mining and timber cartels often benefit from destroying the natural resources raising the issue of E and F and disproportionate impacts on the natural resource dependent communities. Greater access to natural resources, with appropriate ownership rights, is a critical entry point for enabling greater participation in decision-making since those with resources often happen to be the ones who makes decisions. Decentralization, access to information, shared learning and strengthening local governance can help promoting access to resources and can help reduce the inequality.

18.3.2 Secondly, Scope of Information Provided for Stakeholders Is Also a Key to Promote E & F

Information is power and decision-making may not be appropriate without enough and correct information, it is important that the E and F discussion cover the information imperfections that occur during participatory processes and during decision-making that relies on the information generated through participatory processes. There is ample evidence in the published literature that information equity is paramount to achieve overall developmental equity (Lievrouw and Farb 2003) and the same applies to adaptation decision-making as well. The issues of E and F emerges when the information flow is obstructed due to failure in designing proper information communication from those who have to those who need it, and how the information is generated to achieve maximum impact.

To address issues of information inequity, participatory risk reduction should work by understanding the dynamics of information generation and utilization for decision-making. The process should start by defining the stakeholders (people who affect and or are affected by the actions of an intervention), seeking their active participation in designing and implementing risk reduction activities, interpreting the information generated from participatory consultations, and deciding how to use the information to achieve intended results (Prabhakar and Parwez 2010). Hence, the process of participatory risk reduction can also be viewed as the process of equitable information generation and information flow from one stakeholder to another stakeholder.

The information flow is enabled through the 'information gradient' at the community end and this is represented by the extent of knowledge that is available with the communities and putting to use of such knowledge for beneficial purposes. This information gradient is mostly formed by the believers and potential believers who often tend to act upon the information received and put them to use thereby increasing the efficiency of the information gained. The overall success of a risk reduction depends on making sure that the gradients of information-flow do not alter the information in itself (Prabhakar and Parwez 2010).

The issue of scope of information is deeply related to the procedural issues of achieving E and F. There is a value to promote people's understanding on information equity, that providing information through participatory processes could be at times far beyond the community, and that facilitators could face challenges to understand community in an unbiased manner. To avoid this situation, it is suggested that the facilitators are chosen based on their (lack of) biases, world views, and understanding of E and F issues. It appears that contextualizing the fairness discussion into food security, energy security, water security, health security, and education securities could be an effective way of bringing this subject into the domain of communities.

From the above discussion, it is clear that public participation and fairness in decision-making will not happen automatically but rather they need to be facilitated and promoted at all levels in a conscious manner. E and F issues can be very tricky when it comes to operationalizing. That is because 'fair/equal to whom' depends on 'who is defining equity/fairness' and who is driving the discussion i.e. often expert facilitator or government officials. It is often challenging to be fair to the entire section of community since there will always be losers and gainers in any given combination of resource allocation. Can we claim to have addressed the fairness issue if 90% of vulnerable in a project location have benefited while other 10% are not since community rankings said it is OK to benefit top x number of people ranked as vulnerable? Probably not! Most project implementing agencies would like to see that all vulnerable population in their project are benefited. On the other hand, they would often have to make decisions (or facilitate communities to take decisions in a participatory manner) in order that project finances are 'efficiently' used rather than 'thinned out' (managers may think thinning out is a disaster for project reporting). This brings the discussion to the point that E and F is also closely linked with the 'project resources'. In the current socio-political and economic environment, it appears that bringing E and F to all is not 'just one of the challenges' that development practitioners and researchers would have to deal rather it is the crux of the problem in decision-making. Fairness should be a 'chapter' in the Bible or Quran or Upanishad (or other sacred texts you can think) of every project, whether it deals with development or adaptation, or disaster risk reduction.

18.3.3 Equity and Fairness in CBA: Current Evidence

This section identifies evidence (or lack of it) for achieving E and F through CBA, where and at what level CBA excelled and where it failed. This section also identifies indicators to measure E and F in the context of CBA and enabling conditions for promoting E and F at community level. As for the level, CBA may be able to help improve transparency at the local level but may fail to do so at national and sub-national levels. The reasons for failure of achieving the full potential of CBA in E and F are several: a) participator processes are often implemented as a procedural requirement, wherever they are implemented, with poor consideration of who is included, and how their voices are heard and reflected in the decisions made, b) actions are often not targeted to address the most disadvantaged communities, d) inability of the poor and vulnerable to raise their voice and be heard under some social circumstances that promote power imbalances, and e) actions do not often sufficiently challenge the inequities.

18.3.3.1 In What Aspects CBA Helped in Achieving E and F

Several case studies could be found that vouch for the usefulness of CBA. However, these case studies have not been able to show whether CBA had helped in achieving E and F or do not state if E and F was a motivation to embark upon CBA. Most experiences we have reviewed seem to take CBA as a best practice without any proper justification for it or motivation behind choosing it as a delivery mechanism. Under these circumstances, it has been difficult for us to prove if the concerns of E and F were the drivers in choosing CBA approaches or if CBA approaches have led to E and F.

For instance, a case study of Tach river basin in Thailand highlighted the issues of plans, programs, institutions at the local level (Juntopas 2012). It has provided the evidence that CBA is crucial for successful decision-making. In this case, community based processes have helped in solving the issues such as poor vertical integration, integration across boundaries, sectors, and basin. However, it was not clear whether community engagement has led to any E and F in decision-making or if E and F have contributed to enhanced vertical integration.

In India, it was observed that the disasters could trigger solidarity and altruistic motivations that could help in recovery from disaster. A case study conducted ten months after the 2015 South Indian floods in two affected constituencies (Mylapore and Velachery) of Chennai indicated that the communities tend to help each other with a high sense of participation (Joerin et al. 2018). However, it was not evident whether such an increased sense of solidarity has contributed any better relationship between communities and local government officials, indicating that disasters may not change the inherent institutional structures and related interplay between communities and government officials.

In an evaluation of community based disaster risk reduction programs implemented by UNICEF, it was indicated that the school safety program has led to equitable distribution of benefits from the school safety program or absence of differential treatment of participants due to their caste, economic status or ethnicity (UNICEF 2017). This was possible through formation of village disaster management committees that were ensured to be represented by all sections of a village.

In Papua New Guinea, the issue of public participation has been addressed through two-tier consultation process wherein technical consultations were conducted at the national level and these findings were taken to the local level consultations to contextualize and identify specific adaptation practices. The technical consultations have helped in ironing out the terminology that is being used by various stakeholders that led to proper communication. In addition, the consultation has employed a robust process of climate change vulnerability assessment that has underlined the adaptation practices identified. In this way, various processes being taken up in adaptation decision-making at the national level are able to consider the issue of public participation in adaptation planning (Griffin 2012).

Several such cases could be gleaned from the literatures that attribute success to CBA; however, these cases do not demonstrate how CBA helped in achieving E and F and if E and F played any role in achieving successful outcomes.

18.3.3.2 In What Aspects CBA Failed and Why?

Although the need for E and F in CCA and DRR at the local level is significant, it is not easy to demonstrate the same in practice. Here are some literatures that demonstrated examples of why CBA failed in Bangladesh, Brazil, and Malaysia.

Adaptation decision-making process in Malaysia is centralized (and thus far from CBA) and top down driven (Singh 2012). The federal structure, lack of public consultation, and lack of transparency have posed several bottlenecks in adaptation decision-making and communicating the message of adaptation in an effective manner. It is apparent that in Malaysia, and probably elsewhere too, adaptation should be considered as a rights issue (at par with the right to development).

In Bangladesh, the state is considered to be attributed to insufficient CBA. Huq and Khan (2006) explain state's incremental response to participation as a hindrance to CBA, because policies and guidelines for decentralization of decision-making are vaguely formulated despite regime's support for participation.

Contrary, in the case of water management system in Brazil, despite the state's adoption of a new water resources management system which includes devolving watershed management to local level and ensuring the active participation of a range of stakeholders, the results have not lived up to expectations. Adger and Nelson (2010) show several reasons for such failure. Firstly, power was supposed to be devolved at the river basin management level but neither political nor administrative power is the same as the physical territory of basins. Secondly, although the new water resources management system is cross-sectoral, the approaches of other systems are segmented. Thirdly, each state differently interprets the ideals of civil society, participation, inclusion, and deliberation.

While the above two examples are more related to central/local government, there is a case where relationship within community is attributed to hindrance of CBA. Following a global trend in community-based natural resource management, water resources management in rural drylands is increasingly viewed as local-level responsibility in Africa (Thomas and Twyman 2003). However, the case in Nambia shows marginal and vulnerable groups have the potential to be excluded from access to water unless they are able to offer something in return, or their costs are borne by others within the group (Thomas and Twyman 2003).

18.4 Possible Indicators of E and F in the Context of CBA

Identifying indicators of E and F to climate adaptation should target two main approaches: vulnerabilities and institutionalization. Huyer et al. (2016) argued that climate change vulnerability is determined broadly by geographical, social, economic, ecological, and political factors. In adaptation, especially, it relates the accessibility and domination of resources to adapt to negative shocks and stresses (Ericksen and O'Brien 2007; Sugden et al. 2014). The vulnerability approach that considers the above factors should mainly support communities and locals– those who suffer from direct climate change impacts. Such approach helps them understand how climate change might influence them, what sort of strategies already exist to address climate change impacts, and how their adaptive capacity could be tailored, enhanced, and implemented appropriately to their own local conditions. It will require both qualitative and quantitative measurements as the causes of vulnerability are socioeconomic, environmental, political, ethnic or gendered (World Bank 2013). The below box provides a few examples of indicators for benefits in communities and locals.

Box 18.1: Examples of Indicators for Local Benefits. Source: Authors
Examples of indicators for local benefits

Changes in knowledge of adaptation and available technologies

- Number of local persons who have purchased technologies
- Number of local persons who have access to practical skills
- Number of local persons with improved livelihoods
- Change in local persons' income

Number of local persons who participate in value chains

- Change in local persons' income
- Change in local persons' role

(continued)

Box 18.1 (continued)

- Change in local persons' self-confidence
- Change in local persons' access to and control over resources

On the other hand, institutions control resources, finance, land, technology, and services. Therefore, strengthening the local institutions as well as national ones is an important investment to achieve larger scale and higher-level implementation of climate practices and services beyond individuals' responses (Lipper et al. 2014; Vermeulen 2015). Local institutions could play a role to encourage the actions and interventions of individual local agencies especially those with sufficient local knowledge. However, they tend to be excluded or marginalised from policy making decision processes. When collective action around adaptation management mobilised and multi-stakeholders engaged in local and national planning, which would deliver effective and legitimate policies (Newig and Fritsch 2009), these relevant institutions and processes could be monitored and evaluated with following questions checked.

Box 18.2 Source: Beierle (2002), Beierle and Cayford (2010)
The evaluation criteria for the stakeholder-based decisions:

1. Were decisions more *cost-effective* than alternatives?
2. Did decision increase *joint gains* over alternatives?
3. Was it the *opinion* of participants or authors that decisions were improved?
4. Did other *measures* suggest improvements in quality?
5. Did stakeholders *add information*?
6. Did stakeholders contribute to the *technical analysis* of problems?
7. Did stakeholders generate *innovative ideas*?
8. Did stakeholders introduce a more *holistic perspective*?
9. Did stakeholders bring *technical capacity* to the process?
10. Was there adequate access to information and expertise?

To make the assessment and tracking of indicators of E and F in CBA more comprehensive, the proponent of project cases should include some of the co-benefits– the additional benefits along with the direct benefits of climate change actions. These co-benefits could be in the form of economic, social, and environmental co-benefits. These co-benefits act as hygiene factors in strengthening the main motivation that the participants have to participate in the adaptation interventions. Some of these co-benefits could act as main benefits depending on the actor in question. While individuals participating in adaptation interventions may care more about their own benefits, others may care about how others are benefited. Some of these benefits are accrued from the participation process itself i.e. such benefits couldn't possibly be obtained without participatory processes are put in place or participation do not take place. There are several project interventions where societal adaptation co-benefits are possibly only through participation as these social benefits are channelled only through individual participation.

Box 18.3 Listing Co-Benefits
Economic co-benefits

- What is the total number of jobs created for locals?
- Are locals granted equal access and control of the economic benefits from the project?
- Are locals benefiting from income generated from the project?

Social co-benefits

- Has the project improved educational opportunity for locals?
- Are locals benefiting from improved health and safety?
- Are locals benefiting from increased time savings?
- Are locals benefiting from increased participation in decision-making processes?

Environmental co-benefits

- How are locals benefiting from improved air or water quality?
- How are locals benefiting from disaster risk warning systems?
- How are locals benefiting from enhanced biodiversity conservation?

The question confronting many policymakers is how to implement inclusive climate adaptation reflecting E and F as mentioned above. The easiest answer is to make climate governance more participatory i.e. ensuring that the participatory processes are implemented that satisfies the core principles of participatory development. In many environmental policy areas, including CCA and resilience, participatory forms of governance have been found to improve inclusive decision-making process and results because they incorporate wider-ranging values, accumulate richer information, and generate more equitable and effective outcomes (Fischer 2000; Steele 2001; Beierle 2002; Pellizzoni 2003; Richards et al. 2004; Koontz and Thomas 2006).

The challenge remains as to how to better engage locals into climate policy to achieve more E and F. A possible response to this challenge involves work on multi-level governance. One of the core insights of multi-level governance is that profound transformative change requires stakeholders' engagement at different levels of decision-making (Bache and Fliners 2005; Papadopoulos 2007). It should start at the local community level where the inclusion of stakeholders delivers the implementation on the ground with locally appropriate and tailored solutions towards environmental issues and climate change effects (Macnaghten and Jacobs 1997; Reed 2008). It then extends upward to the institutions that shape policies and the language of E and F in the policies that influences the implementation and potential spread of local actions.

18.5 Ensuring Equity and Fairness

From the foregone discussion, it is evident that participatory processes themselves may not automatically result in E and F. Nevertheless, they do provide conducive environment for them to take place. Hence, it is essential that the projects, programs and policies be implemented with conscious efforts to ensure E and F from the proposal level.

Project proposals for climate finance tend to follow a similar structure i.e. the overview of the project, implementing arrangements, financial information, and tracking performance (Zusman et al. 2016). The template presented in the

Table 18.1 List of questions to be asked during the process of proposal preparation to ensure E and F

Project overview	
Project objective(s)	• Does the objective identify E and F? • Is it clear what the project aim to achieve for E and F specifically? • Is there a discussion of how project activities might change the social roles and/or power relations of locals?
Project activity(ies)	• Are there activities specifically targeting E and F of locals? • Is there a clear procedure and approach to ensure the targeted locals are appropriately invited to join the project activities?
Implementing arrangements	
Implementation team and arrangement	• Does the implementation team have knowledge and understanding of why and how to consider E and F in CBA in their project activities? • Is the implementation team balanced to locals or does it at least include locals? • Is at least one member of the implementation team responsible for social development? • Does the implementing arrangement include host community agencies and organisations with knowledge and capacities to work on local issues?
Institutional partnerships	• Does the lead institution have a track record on CBA or partners with institutions have experience on it?
Financing information	
Budget lines	• Are the budget lines for supporting local activities clear in the budget? • Are the budget lines consistent with the local assessment and the reality of the project?
Evaluating and tracking performance	
Sustainable development potential	• Economic co-benefits • Social co-benefits • Environmental co-benefits (refer full details to the previous box)

Table 18.1 offers guidance questions to highlight and facilitate the integration of E and F considerations into the proposal text. For example, when describing the project overview; it should precisely explain the goal of the project in terms of adaptation and E and F concerns as well as its linkages to relevant national policies and international agreements promoting E and F. For explaining the implementing arrangements and budget information, it is important to identify local agencies and stakeholders' commitment and engagement that would scale up the project impacts. Finally, yet importantly, tracking performance part requires potentials of climate impact and sustainable development. This part would be more comprehensive when the proposal emphasizes co-benefits mentioned earlier related to E and F.

18.6 Conclusions

This paper presented various arguments about E and F and their role in ensuring effective adaptation. It is widely established that ensuring E and F is important for the interventions to sustain and be able to accrue intended benefits to the stakeholders in question. CBA has come to stay as an important approach to design and implement adaptation actions on the ground, both in the urban and rural contexts, because of the wide range of benefits it brings. However, there has not been strong evidence on whether or not the CBA approaches ensures E and F.

In this paper, presented evidences and related discussions indicate that E and F may not necessarily be automatically resulted even in a CBA projects and that explicit efforts are necessary to ensure E and F. Ensuring E and F is about asking specific questions on who are involved in the project or program implementation, who were considered during designing the project or program, and how these members are related to the overall population that is expected to be benefited from the interventions. Interventions that systematically identify the beneficiaries based on vulnerability factors may only end up prioritizing a few among many that could be benefited based on the vulnerability criteria set in an intervention. Hence, revising and broadening the vulnerability assessment methodologies, to consider all circumstances under which exclusion and inequality may take place, is an important first step in ensuring E and F.

From the foregone discussion in the light of limited evidence in the published literature, the following conclusions emerge as relevant for CBA while giving rise to identification of gap for further research. (1) There is a greater demand by researchers and policy makers for promoting E and F at various stages of climate actions that include adaptation and mitigation and at all levels including international, regional, national and sub-national including community level. (2) Differentiating and contextualizing E and F do not appear to be the need of the hour as much as ensuring them in the processes and outcomes. (3) There appears to be very limited or no evidence in the published literature that either proves that CBA has explicitly addressed the E and F issues or if E and F issues were the motivating factors for actors to choose CBA as an approach. (4) However, lack of clear evidence in the literature should not lead to dismissal of need for CBA or that of CBA's ability to promote E and F. Rather, it demands for a well-structured and evidence seeking case study research to show how CBA will influence E and F and the outcome, and what enabling environment will promote E and F in the context of CBA.

18.7 Way Forward

The foregone discussion reveal that the equity and fairness matters are not widely addressed in the DRR and CCA interventions. The primary cause for this appears to be the lack of awareness on the need to ensure equity and fairness. This indicates an important way forward for the CCA and DRR community to invest in capacity building and awareness generation activities across the board, i.e. from the policy makers to the community level, so that the issues related to equity and fairness are clearly addressed at stages of policy conception, and design and implementation of programs and projects. Building the awareness of community members on equity and fairness will help them understand how they are affected by the lack of equity and fairness in decision-making and how to address the problem.

Development practitioners need answers to questions such as how to identify if a particular problem is related to equity and fairness, how to measure (lack or presence of) equity and fairness in a social setup, and how to systematically identify amelioration measures. Hence, the second important way forward is for the development and research community to bring out clear evidence for approaches that work in promoting equity and fairness through systematic case studies, both qualitative and quantitative. Conducting randomized control trials (RCTs) could provide an effective means of bringing out the effectiveness of promoting equity and fairness. However, implementing RCTs could be costly and time consuming.

As seen in the case of India and Japan, equity and fairness can be promoted through bottom up or top-down approaches. Depending the country contexts, the suitability of these approaches may vary. Such a variation in approaches and their effectiveness brings a challenge to practitioners to identify suitable approaches. There is a dearth of practical information on approaches for promoting equity and fairness, suitable conditions in which they work and factors affecting sustainability. The third way forward for promoting equity and fairness is to distil the available knowledge and experiences in the form of a decision support system to help practitioners in identifying simple approaches to address equity and fairness issues.

References

Adger WN, Nelson DR (2010) Fair decision making in a new climate of risk. In: O'Brien B, St. Clair K, Kristoffersen A (eds) Climate change, ethics and human security. Cambridge University Press, Cambridge, pp 83–94

APAN. (2012) Asia-Pacific Climate Change Adaptation Forum. Mainstreaming adaptation into development. Adaptation in action. 12-13th March 2012, Bangkok, Thailand. Available at http://www.asiapacificadapt.net/adaptationforum2012/programme accessed 29-10-2018

Asian Co-benefits Partnership (2017) Achieving social and environmental co-benefits through participatory governance in Seoul, Korea. IGES, Hayama, Japan

Bache I, Fliners M (2005) Multi-level governance: conclusion and implications. In: Bache I, Flinders M (eds) Multi-level governance. Oxford University Press, Oxford

Barrios R, Battle CP (2018) Equity in disaster recovery, mitigation and adaptation. Columbia Law School, Columbia, USA. Retrieved from https://www.law.columbia.edu/sites/default/files/microsites/gender-sexuality/Gender_Justice_Fall_2018/paper_-_colette_pichon_battle.pdf

Beierle TC (2002) The quality of stakeholder-based decisions. Risk Anal 22:739–749

Beierle TC, Cayford J (2010) Democracy in practice: public participation in environmental decisions, Routledge: NY

Ericksen S, O'Brien K (2007) Vulnerability, poverty and the need for sustainable adaptation measures. Clim Pol 7(4):337–352

Fischer F (2000) Citizens, experts and the environment: the politics of local knowledge. Duke University Press, London

Government of Japan (2008a) Act on building unit ownership in Japan. Retrieved from http://www.japaneselawtranslation.go.jp/law/detail/?id=2015andvm=04andre=01

Government of Japan (2008b) City Planning Act, Japan. Retrieved from http://www.japaneselawtranslation.go.jp/law/detail/?ft=1andre=02anddn=1andco=01andia=03andx=0andy=0andky=都市計画法andpage=16

Griffin F (2012) Panel discussion on adaptation decision-making, public participation, and fairness. Adaptation Forum. 13th March 2012, UN Conference Center, Bangkok, Thailand

Huq S, Khan MR (2006) Equity in National Adaptation Programs of Action (NAPAs): the case of Bangladesh. In: Neil Adge MJMW, Paavola J, Huq S (eds) Fairness in adaptation to climate change. The MIT Press, pp 181–200

Huyer S, Campbell B, Hill C, Vermeulen S (2016) CCAFS gender and social inclusion strategy, working paper 171. CGIAR and CCAFS, Frederiksberg

Joerin J, Steinberger F, Krishnamurthy RR, Scolobig A (2018) Disaster recovery processes: analysing the interplay between communities and authorities in Chennai, India. Procedia Eng 212:643–650

Juntopas M (2012) Panel discussion on adaptation decision-making, public participation, and fairness. Adaptation Forum. 13th March 2012, UN Conference Center, Bangkok, Thailand

Koontz TM, Thomas CW (2006) What do we know and need to know about the environmental outcomes of collaborative management? Public Adm Rev 66:111–121

Lievrouw LA, Farb SE (2003) Information and equity. Ann Rev Info Sci Tech 37:499–540. https://doi.org/10.1002/aris.1440370112

Lipper L, Thornton P, Campbell B, Baedeker T, Braimoh (2014) Climate-smart agriculture for food security. Nat Clim Chang 4:1068–1072

Lovell E, Masson V (2014) Equity and inclusion in disaster risk reduction: building resilience for all. Climate and Development Knowledge Network and Overseas Development Institute

Macnaghten P, Jacobs M (1997) Public identification with sustainable development: investigating cultural barriers to participation. Glob Environ Chang 7(1):5–24

Mazepus H, Leeuwen FV (2019) Fairness matters when responding to disasters: an experimental study of government legitimacy. Governance. https://doi.org/10.1111/gove.12440

Merzel C, DAfflitti J (2003) Reconsidering community-based health promotion: promise, performance, and potential. Am J Public Health. Retrieved from http://www.ncbi.nlm.nih.gov/pmc/articles/PMC1447790/

Ministry of Foreign Affairs. (2012) ODA White Paper FY 2011. Retrieved from https://www.mofa.go.jp/mofaj/gaiko/oda/shiryo/hakusyo/11_hakusho_pdf/pdfs/11_hakusho_0202.pdf

Ministry of Health, Labor and Welfare. (2018) Challenges and perspectives of reform of medical care system] Retrieved March 23, 2018, from https://www.mhlw.go.jp/houdou/0103/h0306-1/h0306-1o.html
Newig J, Fritsch O (2009) Environmental governance: participatory, multi-level—and effective? Environ Policy Gov 19:197–214
Nutters H (2012) Addressing social vulnerability and equity in climate change adaptation planning. Adapting to Rising Tides Whitepaper, June 2012
Papadopoulos Y (2007) Problems of democratic accountability in network and multilevel governance. Eur Law J 13(4):469–486
Pauly MV, Willett TD (1972) Two concepts of equity and their implications for public policy. Soc Sci Q 53(1):8–19
Pellizzoni L (2003) Uncertainty and participatory democracy. Environmental Values 12(2): 195–224
Prabhakar SVRK (2013) Resource conservation, tools for screening climate smart practices and role of communities. Presented at World Bank Blended Learning Program on Policies and Practices for Natural Resource Management, 14 March – 31 May 2013, World Bank TDLC, Tokyo, Japan
Prabhakar SVRK, Parwez S (2010) Participatory information management for sustainable disaster risk reduction. In: Natural and anthropogenic disasters: vulnerability, preparedness and mitigation. Secaucus, USA: Springer
Rahman A (2012) Panel discussion on adaptation decision-making, public participation, and fairness. Adaptation Forum. 13th March 2012, UN Conference Center, Bangkok, Thailand
Reed M (2008) Stakeholder participation for environmental management: a literature review. Biol Conserv 141:2417–2431
Richards C, Blackstock KL, Carter CE (2004) 'Practical approaches to participation', SERG policy brief 1. Macauley Land Use Research Institute, Aberdeen
Room H (2018) 3.1 times! Gaps of premium for nursing-care insurance. What is happening?. Retrieved April 24, 2018, from https://hoken-room.jp/nursing/2270
Seoul Metropolitan Government (2014a) We are energy: the people are our energy. Boucher. SMG, Seoul
Seoul Metropolitan Government (2014b) Seoul's exemplary environment policies: pleasant, healthy and sustainable city. SMG, Seoul
Seoul Metropolitan Government (2015) One less nuclear power plant, phase 2: Seoul sustainable energy action plan. SMG, Seoul
Singh G (2012) Panel discussion on adaptation decision-making, public participation, and fairness. Adaptation Forum. 13th March 2012, UN Conference Center, Bangkok, Thailand
Steele J (2001) Participation and deliberation in environmental law: exploring a problem-solving approach. Oxf J Leg Stud 21(3):415–442
Sugden F, Maskey N, Clement F, Ramesh V, Philip A, Rai A (2014) Agrarian stress and climate change in the Eastern Gangetic Plains: gendered vulnerability in a stratified social formation. Glob Environ Chang 29:258–269
Thomas DS, Twyman C (2003) Equity and justice in climate change adaptation amongst natural-resource-dependent societies. Glob Environ Chang:115–124
UNFCCC. (1992) United Nations Framework Convention on Climate Change. Available at http://unfccc.int/key_documents/the_convention/items/2853.php
UNICEF (2017) Evaluation of UNICEF's community based disaster risk reduction and school safety programme, Bihar, India (2011–2016). IPE Global Limited, New Delhi, India
Vermeulen S (2015) Closing the gender gap in climate-smart agriculture. CCAFS Info Note. CCAFS, Copenhagen
World Bank (2006) Module 5: Investment in sustainable natural resource management for agriculture. Available at http://siteresources.worldbank.org/EXTAGISOU/Resources/Module5_Web.pdf
World Bank (2013) Inclusion matters: the foundation for shared prosperity. World Bank, DC

World Commission on Dams (2000) Dams and development, a new framework for decision-making, the report of the world commission on dams. Earthscan Publications Ltd, London. Retrieved from http://www.unep.org/dams/WCD/report/WCD_DAMSreport.pdf
Zusman E, Lee S, Rojas A, Adams L (2016) Mainstreaming gender into climate mitigation activities: guidelines for policy makers and proposal developers. Asian Development Bank (ADB), Manila

S. V. R. K. Prabhakar is currently engaged in policy research in climate change adaptation, disaster risk reduction and natural resource management. He is the Lead Author to the UNEP Geo 6 report, Contributing Author to the IPCC 5th & 6th Assessment Report and Lead Author for the UNEP Global Adaptation Gap Report and ICIMOD HIMAP Report. Reviewer at various peer reviewed international journals and a blogger.

Norkio Shimizu (Ms.) is a climate finance researcher. She joined IGES in 2011 and serving as a Programme Manager of Finance Taskforce in IGES from 2018. Previously, Noriko worked with Friends of the Earth Japan and as a lecturer at Hitotsubashi University in Japan where she worked on environmental integrity especially in the public financial sector. She completed her doctoral program in Tokyo Institute of Technology.

So-Young Lee (Ms., Dr.) is a Research Manager at IGES, Japan where she manages research on sustainability governance and climate co-benefits for the underprivileged. Dr. Lee specialises in Environmental Sociology and has lectured at Korea University, Beijing Normal University, and Waseda University (Japan).

Chapter 19
Gender and Climate Sensitive Disaster Risk Management

Ajinder Walia and Karpoora Sundarapandian

Abstract The Sendai Framework for Disaster Risk Reduction calls for reducing risks to disasters rather than focusing only on the process of management of disasters. The tenet accentuates the philosophy that we need to shift our thinking towards addressing the source of a problem rather than focussing on the outcome of a problem. There is a growing realization that Climate Change is also one of the sources for increase in hydro-meteorological disasters like floods, cyclones, droughts, heat waves, cold waves etc. The impact of climate induced disasters is different for both men and women. Various case studies of climate induced disasters clearly highlight that women tend to be more vulnerable to these disasters due to their pre-existing vulnerabilities. Gender limits both men and women in their abilities to perform various tasks and it leads to the creation of vulnerabilities. Along with this, the biological role of reproduction played by women leads to social discrimination for the rest of their lives. The society needs to be sensitized towards the process in which social norms and stereotypes contribute towards gendered vulnerabilities and overlooking of their capacities in non-disaster as well as disaster times. Research has also shown that given an opportunity to develop skills, women have shown their role as effective contributors to the process of disaster preparedness as well as management. The potential contributions that women have offered to Disaster Risk Reduction around the world need to be highlighted and their leadership in building community resilience to disasters should be acknowledged. Such an empowering process needs to be strengthened in all developmental initiatives to reduce pre-existing vulnerabilities which get augmented in disaster settings. Moreover, there should be a platform for women in disaster management organizations for sharing their needs, experiences, and expertise which would provide impetus to a gender-sensitive approach in the operational aspects of disaster management. A

A. Walia (✉)
Policy, Planning & Cross Cutting Issues Division, National Institute of Disaster Management, New Delhi, India

K. Sundarapandian
Governance and Inclusive Disaster Risk Reduction Division, National Institute of Disaster Management, New Delhi, India

A. K. Gupta et al. (eds.), *Disaster Risk and Management Under Climate Change*,
Disaster Resilience and Green Growth,
https://doi.org/10.1007/978-981-99-4105-6_19

gender-sensitive approach to Disaster Risk Reduction would ensure that the efficacy of our interventions is enhanced and reach every member of the community.

Keywords Gender · Differential impact · Climate sensitive disaster risk management

19.1 Introduction

19.1.1 Climate Change and Disasters

The third United Nations World Conference on Disaster Risk Reduction held in Sendai, Japan from 14th to 18th March 2015 concluded with the endorsement of the 'Sendai Framework in Disaster Risk Reduction (SFDRR) 2015-2030'. India is a signatory to this global action along with representatives of 187 countries. The Sendai Framework for Disaster Risk Reduction calls for reducing risks to disasters by analysing and managing the causal factors of disasters (UNISDR 2015). The tenet accentuates the philosophy that we need to shift our thinking towards addressing the source of a problem rather than focussing on the outcome of a problem. The risk of disasters has been attributed to various human-made factors like increase in urbanization, unplanned development, environmental degradation, and Climate Change.

Climate Change is expected to increase the frequency and intensity of extreme weather events across the globe. Statistics published by the International Disaster Database revealed that in 2017, 318 natural disasters occurred, affecting 122 countries. Impact of the disasters resulted in 9503 deaths, affecting 96 million people, and US$314 billion in economic damages. India has seen a huge burden of natural disasters in 2017, with almost 2300 deaths and 22.5 million people were reported affected, mostly by floods and cyclones (Center for Research on the Epidemiology of Disasters 2018).

Some regions experience more extreme events, such as heat waves and cold waves, high levels of precipitation, extreme floods, droughts, tropical cyclones and storms (Intergovernmental Panel on Climate Change 2001). We need to understand that primary disasters caused by Climate Change can lead to secondary and tertiary disasters as well. In Uttarakhand disaster in 2013, cloudburst and increased rainfall led to floods, which further triggered landslides. Hence, Gender Issues in climate-sensitive disaster risk management cannot be limited only to primary disasters caused by Climate Change, but it has to be understood within the framework that primary disasters can lead to secondary and tertiary disasters as well.

19.1.2 Linkage Between Gender and Climate Sensitive Disaster Risk Management

The impacts of Climate Change and Disasters are different for women and men. Climate Change impacts are far more severe for women, the poor and marginalized groups because of societal inequality such as unequal distribution of power, lack of information, and lack of decision making ability. Women also lack access to productive resources, particularly, land. Women generally perform multiple roles, including food producers and providers as guardians of health, caregivers to the family and community, and as economic actors (Zinnatul-Bassar and Tasnuva 2017). As access to basic assets and natural resources, such as shelter, food, fertile land, water and fuel becomes hampered, women's workload in particular increases manifold. Flood, drought and erratic rainfall, generally results in lack of natural resources, which causes women to work harder to secure livelihoods. Gender inequalities can constrain the influence and control of women and girls over decisions governing their lives as well as their access to resources, thus marginalizing them from planning on how to curb disaster risk.

Literature on 'Gender and Disaster', which has been steadily growing over the past decades, has highlighted the way in which a significant proportion of disaster planning, management and research overlooks gender, despite recognition within the social sciences that there exists a 'gendered dimension' to the responses to any social event. Gender is not a simple quantitatively measured background characteristic; rather it is a fundamental concept in the centre of disaster analysis. Failure to acknowledge this not only runs the risk of overlooking obvious and subtler needs, priorities and capacities that can make all the difference between life and death, but can also diminish the efficiency of our responses to climate induced disasters, and even contribute to creating new categories of victims. Moreover, gender identities and subjectivities are also constructed and reworked through the disaster process to highlight the complexities and contradictions associated with women's responses to a disaster. This focus needs to be understood for gaining a deeper understanding of 'gendered dimensions' of climate induced disasters, which otherwise would complicate attempts to create more gender sensitive frameworks for disasters response.

The pre-existing vulnerabilities in a society frequently have gender dimensions. Disasters increases gender inequality, which can worsen situation for women and vulnerable groups. Biologically speaking, the vulnerability of women is enhanced during disasters as they have to face certain reproductive health problems, which are highly specific to their sex. The private needs of women including specific requirements related to menstruation and pregnancy creates an uncomfortable situation for them. Economically, women are generally considered to be poorer than men on account of their lower incomes, lack of job stability, dependence on the income of men, scant recognition of their productive labour and limited access to resources with potential to produce income (Bradshaw 2004). Lack of credit facilities, knowledge of how to access credit, and marketing and bargaining skills commonly affect women. They have lesser access to resources like transportation, skills and information. They are also the victims of gendered division of labour as majority of them in India are working in underpaid and unrepresented jobs with no union representation. Even if a woman does not hold a paid job, she may have the ability to generate household income from a variety of informal sector activities, whether from the backyard economy or by running small home-based business, which benefits her by combining multiple productive tasks with reproductive ones (ECLAC 2003). Such informal, productive, and reproductive activities are not included in official national accounts and hence do not get accounted in the damage assessment mechanisms, which are operationalized after a disaster.

The economic vulnerability of women makes them vulnerable in the aftermath of a disaster when the government as well as NGOs initiate various economic and livelihood activities. These agencies generally target the income earners of a family, which are generally men, therefore marginalizing women. Sociologically, women have enjoyed an adjunct status to men in the patriarchal Indian society. In many parts of the country, women face a cultural devaluation from the members. Practices of female infanticide and female foeticide uphold this devaluation. In a study conducted in Punjab, which already has a skewed sex ratio against the girl child, about 41.25% of the sample size expressed a favourable attitude towards female foeticide (Walia 2005a). Bihar, Gujarat, Jammu and Kashmir, Rajasthan, Punjab and Uttar Pradesh were the states, which recorded the highest mortality rate for the girl child (Sharma 1995). The patriarchal society coupled with specific gender connotations in India subject women to domestic violence, limiting them to performing household chores, restricting their mobility, and marginalizing them in most of the decisions. During disasters, women still have to cook food and feed the members of the family in certain situations. Her responsibilities are also increased as she has to take care of sick and injured people in her family. They have to apply themselves temporarily to unpaid emergency tasks with an increased amount of unpaid work such as caring of children when schools are shut down because of them being used as shelters for disaster victims, or when the school buildings have been devastated by a disaster. As a family moves to temporary shelters, there is increase in the challenges of women. She experiences discomfort in sharing space with victims in the absence of privacy or safety, lack of separate washrooms and hygiene arrangements. Social norms generally obstruct women to access emergency and post-disaster services. For example, shelters are generally less accessible for women because of their flexibility constraints and childcare responsibilities. Moreover, shelters are also not suited for women because of their hygienic and safety requirements. Women's care for children, the elderly and the sick can become more demanding when Climate Change impacts on health and psychosocial well-being (Pham and Lam 2016).

The incidence of gender-based violence-including rape, human trafficking and domestic abuse also increases during and after disasters (Alston 2013). However, we need to understand that the vulnerabilities created by disasters are not caused by disasters themselves but they are related to the pre-existing vulnerabilities during non-disaster times. These pre-existing vulnerabilities stem out from the root causes of unequal political, social, economic, and cultural norms and practices. Therefore, women are not vulnerable because they are weaker but their restricted access to physical, financial, social, and natural assets along with unequal power relations are pivotal in undermining their capacity to respond (Bradshaw and Fordham 2013; Smyth 2009).

Women experience greater stress due to their multiple responsibilities and generally unequal social status. Stress decreases their resistance to diseases and adversely affects their ability to undertake essential economic and family activities (Kelly 1989). On the other hand, there are very few studies conducted on the vulnerability of men in disasters. However, vulnerability of men in disasters stems from their social and economic roles ascribed to them. Social factors play a role here, where society sees men as 'bread earners of the family'. Men are expected to bear the full financial responsibility for their families. They do not know how to perform household chores and face problems in case they have to do such chores and take care of children. Men are expected to bear the responsibility of protecting their families, which cannot be fulfilled in the event of a disaster. Economically speaking, men generally tend to associate their worthiness with their capacity to earn, which poses a problem for them, as during disasters they lose their sources of livelihood. Psychologically, men are generally socialized not to express their emotions. They tend to suppress their emotions even in times of crisis resulting in stress and trauma.

19.2 Vulnerability and Impacts

19.2.1 Differential Impact of Disasters on Women and Men

Too often, those who are subjected to the impact of disasters are conceptualized as belonging to a homogeneous group called "victims," but this apparent similarity conceals considerable difference: difference in terms of gender, class, race, ethnicity, age, sexual orientation, physical and mental ability, culture, etc. Dealing with difference represents a significant challenge for disaster managers, one that remains largely unrecognized or suppressed. (Fordham 1999).

Weather-related hazards put natural resources used by communities, particularly women, at risk. Women in the developing countries have the responsibility to ensure food, water, and energy for cooking and heating. Drought, irregular rainfall and cyclones result in shortages or lack of resources to fulfil the needs of not only women, but also of their entire families and communities. There are numerous case studies highlighting the differential impact of climate induced disasters on men and women. Gender is an important factor in the dissemination of early warning issued during an emergency. In the 1991 Bangladesh cyclone and flood, emergency warnings were given mainly by loudspeaker and word of mouth. In a highly sex segregated society, warning information was transmitted by the males to their male counterparts in public spaces where the males had congregated on the assumption that this would be communicated to the rest of the family, which by and large did not occur. Those who heard the warning ignored it because cyclones occurring after the 1970 disaster had not caused much devastation. Women in this case had comparatively less knowledge about cyclones and were dependent on male decision-making, perished, many with their children, waiting for their husbands to return home and take them to safety (Baden et al. 1994).

Effective and timely response to early warning warrants an immediate decision on evacuation of a community from the vulnerable area. Evacuation is an important decision, taken in the golden hours of an impending disaster, which can singlehandedly determine morbidity and mortality rate of the vulnerable population. Many women wait for their husbands to return home to take the decision to evacuate, thereby losing precious time that might save their lives and those of their children.

A study in the drought-hit villages of Rajasthan revealed that a drought situation increased the workload on women. Their daily routine in this scenario involved getting up early in the wee hours of the morning to perform household chores, reporting for food for work programme at 6 'O'clock in the morning till 1 p.m. in the afternoon, getting fodder and taking care of the livestock, going to far-off places to arrange drinking water, and performing minor agricultural tasks. Women in these villages faced health problems like malnutrition, anaemia etc. as they tended to eat less food or sacrificed their meals altogether for the males of the household. Moreover, girl children were given lesser opportunities to receive education than their male counterparts in these areas (Walia 2005b).

Sara Ahmed in her research has captured the livid experience of floods in Eastern Uttar Pradesh. She observed that floods in the cultural context added to women's social and economic vulnerability, not only severely and dramatically affecting their lives during actual flood/water-retention periods, but also increasing their marginalisation overtime and testing their endurance. Means of defecation varied depending on where the women lived: typically, a bamboo platform was put just outside the house, or attached to the edge of the roof if that was where they were taking shelter. Sometimes women were simply forced to wade into the flood waters and squat to defecate. It was a bit easier for women sheltering in the jungles as privacy and open spaces were not such a problem. During the devastating floods in Gorakhpur, one of the elderly women in Chittari village reported the problem of defecation by saying "*We can live without food or taking a bath, but defecation is a compulsion. We try to eat and drink less to avoid going to the toilet*" (Ahmed 2004).

A gender analysis of floods in Orissa also highlighted the problem of secluded spaces for defecation for women. Women faced personal discomfort and embarrassment, where in some areas they started reducing the food they ate in order to manage this situation. The lack of space and unavailability of normal local hygienic resources was especially taxing for menstruating women. Women have observed irregularities in their menstruation cycles, emotional trauma including anxiety and depression post-disaster events. Incidents were also reported of pregnant women having false labour pains due to stress of floods and premature deliveries. The victims, needing medical attention could not be transferred to medical care units which resulted in most of the deliveries to took place in the most unhygienic conditions, thereby, risking the well-being of both mother and the new-born. Anxiety, frustration, conflicts and discontent in families rose with food insecurity and loss of property and other belongings. Depression and irritation were found to be common amongst local communities. Incidences of domestic violence also increased as men struggled to cope with the indignity and frustration of enforced inactivity and loss of authority and earning capacity (Behera et al. 2004).

Reports show severe reproductive health problems post-disaster, including early pregnancy loss, premature delivery, stillbirths, delivery related complications and infertility. Pregnant and lactating women need prenatal and postnatal care in the aftermath of a disaster. Social taboos around menstruation and norms about appropriate behaviour for women and girls contribute to health problems in disaster situations.

Relief mechanisms should be practised by understanding the specific needs of women. Participation of women has been observed in disaster management largely related to the role of caring and nurturing. There is limited representation of women on national and local emergency committees and their potential as a resource for organized action at all levels of the managerial process has been seriously overlooked (Noel 1990).There have been instances where certain gender sensitive material has been distributed by men to women passing embarrassing comments. Such gender sensitive material needs to be distributed by women to women. Therefore, an effective relief management strategy has to be provided, by taking care of two parameters namely; addressing the specific needs of women and ensuring women's participation in the relief management process (Walia 2015).

Gender variation in the morbidity and mortality vary according to disaster type and location of the disaster. In 1994, twice as many males as compared to females died in Climate Change related hazards, such as winter storms, thunderstorms and lightning. This happened because men were the ones who participated more in work and leisure related activities and most of these deaths occurred outside home, in vehicles and in the open (U.S. Department of Commerce 1995). Hence, the mortality rate for men and women in any disaster depend upon the location, timing and nature of disaster. It will not be appropriate to say that in every disaster, it is the women and children who have a higher mortality rate. In case of Puttingal fire in Kerala and cyclone Okchi, all the deaths were that of men. Fire broke out in Puttingal temple at about 3:00 a.m. on April 10, 2016 due to bursting of firecrackers. At wee hours of the morning, only men were present in the temple premises and the disaster claimed over 100 lives. Newspapers reported that in Cyclone Ockhi which hit the coasts of Tamil Nadu and Kerala in 2017 killed over 100 fishermen who were in the open sea and had ventured deep leading to their mortality. It was the men who were killed as they had gone into the sea for fishing. Men also experience trauma as disasters survivors but they are generally socialized not to express their emotions. Hence, they tend to externalize their frustrations through alcoholism, gambling, violence and suicides (Peacock et al. 2005). Masculinity norms may encourage risky and heroic action during the search and rescue period, debris removal, and reconstruction, and deter men from approaching agencies for assistance and/ or seeking counselling later. Hence the differential needs vulnerabilities and capacities of men/boys and women/ girls need to be taken into account while designing any intervention in the pre as well as post-disaster scenarios. By taking into account the different impact of disasters on men and women, policies and interventions for sustainable recovery and reconstruction as well as for risk reduction, preparedness and mitigation programs can become more effective and targeted.

19.2.2 Vulnerability to Capacity Building

As stated earlier, the biology of women related to child bearing becomes a source of social and economic discrimination for them in the society. They are socialized accordingly for carrying out specific tasks and tend to get excluded from carrying out all other tasks, which are considered to be the domain of men. However, it needs to be understood that both boys and girls are socialized into acquiring skills related to their roles and responsibilities in the future. This limits them to acquire skills, which are not seen in their respective gendered domains. This limitation creates vulnerabilities for both men and women, especially in disasters when both of them may have to take up roles and responsibilities, which they have not been socialized for. Women are at a more disadvantageous position because of the socialization process. Given adequate opportunity and training, women can be effective contributors to the process of disaster management.

In a study capturing the drought in Orissa where a large number of men migrated to cities in search of jobs, the women respondents stated "*Droughts are strengthening us; we no more despair in its wake and rather have become used to look for coping strategies to overcome the difficulties*". Women felt good about some changes that happened during drought periods. Their access to market areas increased during these years for the purpose of buying and selling. It helped in increasing their know-how about the market as a structure and developed their skills in operating in the market. Women's information about and access to government programmes also increased during the drought years. The knowledge gained regarding wage receipts and procedures in government offices helped women headed households. Women were able to demand their rights from the government; ask for transparency in operations and seek grievances redressal against field level officials. Women gained power for decision making in the absence of migrated men, undertaking work related to community functions, land and asset management, government surveys etc. The new found independence and ability to make decisions on their own remained even after the disaster had abated (Behera et al. 2004).

After the massive damage and destruction caused by Uttarakhand floods in 2013, World Bank and the Global Facility for Disaster Reduction and Recovery (GFDRR) channelized their support to Uttarakhand Disaster Recovery Project (UDRP) to rebuild the homes of disaster survivors after the flood deluge in the state. A key component of the project was to rebuild approximately 2200 resilient houses based on owner-driven housing reconstruction model, which allowed families to rebuild according to their specific needs. Women had joint bank accounts with their spouses, which ensured that women could access formal financial services and promote transparency in aid payments. The houses were also given in the name of joint ownership. Women actively participated in the entire housing reconstruction process from early consultations to rebuilding work. Female social mobilizers in the project engaged with women and included their feedback into the design of new houses. This approach encouraged women to proactively supervise day-to-day masons and construction workers at the site while their husbands were away at work. Active participation of women ensured that they had a voice in Climate and Disaster Risk Management, which boosted their resilience (World Bank and GFDRR 2015). Thus, given adequate opportunity and training, women can be effective contributors to the process of climate sensitive disaster risk management.

Women have not only played an active role after a disaster, but they have also undertaken initiatives to reduce Disaster Risks posed by Climate Change as well. A movement was initiated in Dasholi village of Chamoli district of then undivided state of Uttar Pradesh. The women of Dasholi Gram mobilized women's groups from neighbouring villages. Chamoli district now falls in Uttarakhand. Dasholi Gram Swaraj Mandal, the institution, which covered issues related to conservation of natural resources anchored the endeavour to Reduce Disaster Risk in the region. Local women worked with men, with the aim of preventing deforestation and recharging water resources to improve the ecological profile of the area. Women established that natural resource conservation was a matter for community concern and hence men needed to be involved as partners. The initiative introduced fundamental shifts in gender relations due to women's positioning as community leaders. It reduced the risk of flooding and landslides from extreme weather events, which were likely to be exacerbated by Climate Change. Their actions resulted in increase of forest cover, reduction in the drudgery borne by the village women, and less land damage from floods and landslides (UNISDR 2008). It also highlighted the leadership roles played by women in disaster risk reduction. Women's local knowledge and historical perspective on natural resource-based employment is an essential asset to economic planners working at the community level.

19.3 Conclusion and Way Forward

Gender limits both men and women in their abilities to perform various tasks. This gendered limitation of abilities in the process of socialization leads to creation of vulnerabilities. Along with this, the biological role played of reproducing children leads to social discrimination for the rest of their lives. The health impact of disaster related stress is different on pregnant women, who are living in the affected areas. Postnatal depression in women is the most notable one due to the existing anxiety at the time of disasters and in some cases physical abnormalities are also seen in the prenatal period or at the time of child birth. Along, with this the women living with disabilities are experiencing more problems at the time of disaster because of the discrimination due to gender and disabilities. The society needs to be sensitized towards the process in which social norms and stereotypes contribute towards gendered vulnerabilities in non-disaster as well as disaster times. The impact of climate related disasters is more on women as they have different specific needs as compared to men. The post-disaster environment is a time to question unsustainable practices and the occasion to address the fundamental social issues which create vulnerability. However, conceptualizing women's deprivation basically in terms of her welfare and needs will focus only on the 'victim' aspect of women and we will miss out on the important aspect of women as active agents of change who can transform their own lives, the lives of other women and other members of society (Sen 2000). Our case studies and experiences suggest that disasters create opportunities for women to change their existing roles. They require an opportunity and necessary skills to break the stereotypes. Such empowering process needs to be further strengthened even during developmental initiatives and interventions in non-disaster settings. The pre-existing vulnerabilities can be reduced and they will not get aggravated during disaster times as well. The potential contributions that women have offered to Disaster Risk Reduction around the world needs to be looked into and their leadership in building community resilience to disasters should be highlighted. There should be a platform for women in Disaster Management Organizations for sharing their needs, experiences and expertise so that it can get institutionalized in the operational aspects of disaster management. A gender sensitive approach to Disaster Risk Reduction would ensure that efficacy of our interventions is enhanced and reaches every member of the community.

To address these problems, government has implemented many plans for disaster risk management, and apart from that, the disaster management institutions and departments are also specially working on gender issues. The institutions such as National commission on Women, Department of Women and Child Development accomplish the gender based disaster management strategies at the state and national levels. Also at the state and district levels the Disaster risk Management Program by UNDP provides much educational awareness, capacity building programs related to disaster preparedness and mitigations. And also they provide the gender mainstreaming of Indian government disaster management policy and structures.

In some of the state action plans (SAPCC) in India a 'Gender and Climate Adaptation Policy' are contributed to develop in association with the Climate and Development Knowledge Network (CDKN) along with the NGOs. The Gender Policy Framework for the SAPCC was submitted to the Central Ministry of Environment, Forest and Climate Change (MoEFCC) and is stated to incorporate the gender dimensions to SAPPCCs. These are the few examples of laws and policies implemented by the national and state governments for the purpose to resist gender issues in the societies.

The recommendation for gender related issues would enhance the scope of existing initiatives at the national and state level. As women are the most vulnerable groups during and after disaster situations, a gender sensitive approach is needed for the Disaster Risk Reduction. When if the generation of females put ahead, it could be easier to build resilience for the gender issues at the time of the calamities. For that a basic income-generating projects such as small scale organizations for increasing the employment of women for promoting their empowerment are compulsory to implement. Increasing the number of women participation in all sectors, organizations including disaster management cells and other related sectors, could be one of the ways to promote the participation of local women to reduce the disaster risk at all the phases of a disaster. Therefore, for building resilience, the challenges and barriers of females in the communities should be changed into opportunities by providing community and household capacity building at the national and state levels through providing special attention to the female communities.

References

Ahmed S (2004) The gendered context of vulnerability: Coping/adapting to floods in Eastern India. Working paper for the Adaptive Strategies Project, USAID and ISET

Alston M (2013) Environmental social work: accounting for gender in climate disasters. Aust Soc Work 66:218–233

Baden S, Goetz AM, Green C, Guhathakurta M (1994) Bangladesh cyclone response fails to meet women's needs. BRIDGE Report no. 26: Background Paper on Gender Issues in Bangladesh, Retrieved from http://www.ids.ac.uk/bridge/dgb1.html

Behera A, Mishra P, Mishra S (2004) Voicing silence: experience of women with disasters in Orissa. Graphic Cell, Orissa

Bradshaw S (2004) Socio-economic impacts of natural disasters: a gender analysis, Manual prepared for the Economic Commission for Latin America and the Caribbean (ECLAC) and Women and Development Unit, Santiago Chile. Retrieved from http://www.eclac.cl/mujer/reuniones/conferencia_regional/manual.pdf, pp 20

Bradshaw S, Fordham M (2013) Women, girls and disasters: a review for DFID. Report produced for the Department for International Development, UK

Center for Research on the Epidemiology of Disasters (2018) Natural disasters in 2017: lower mortality, higher cost. Cred Crunch 50:1–2

Economic Commission for Latin America and the Caribbean (ECLAC) (2003) The impact of disasters on women. In: Handbook for estimating the socioeconomic and environmental effects of disasters, Retrieved from http://www.eclac.org/publicaciones/xml/4/12774/lcmexg5i_VOLUME_IVd.pdf

Fordham M (1999) The intersection of gender and social class in disaster: balancing resilience and vulnerability. Int J Mass Emerg Disasters 17(1):15–36

Intergovernmental Panel on Climate Change. (2001) Climate change 2001: IPCC third assessment report. Working Group II: impacts, adaptation and vulnerability. Cambridge, Cambridge University Press

Kelly N (1989) Working with refugee women: a practical guide international consultation on refugee women, Geneva

Noel (1990) The role of women in health related aspects of emergency management, ICN Nursing Review, 36(6)

Peacock W, Morrow B, Gladwin H (2005) Hurricane Andrew: ethnicity, gender and the sociology of disasters. Routledge, London

Pham TDM, Lam TTS (2016) Gender needs and roles in building climate resilience in Hue City, Vietnam. Working paper series 33 for Asian Cities Climate Resilience

Sen A (2000) Women's Agency and Social Change. Development as Freedom, Oxford University Press, India

Sharma S (1995) Socio-cultural practices threatening the girl child. Soc Change 25(3):94–105

Smyth I (2009) Gender in climate change and disaster risk reduction. Dev Pract 19:799–802

U.S. Department of Commerce, National weather Service's office of meteorology (1995) A summary of natural hazard fatalities for 1994 in the United States, Washington, DC

United Nations International Strategy for Disaster Reduction (UNISDR) (2008) Gender perspectives: integrating disaster risk reduction into climate change adaptation: good practices and

lessons learned, Geneva. Retrieved from https://www.unisdr.org/files/3391_GenderPerspectivesIntegratingDRRCCGood20Practices.pdf
United Nations International Strategy for Disaster Reduction (UNISDR) (2015) Sendai framework for DRR Geneva. Retrieved from https://www.unisdr.org/files/43291_sendaiframeworkfordrren.pdf
Walia A (2005a) Female foeticide n Punjab: exploring the socio cultural dimensions. IDEA 10:1. Retrieved from http://www.ideajournal.com/articles.php?id=37
Walia A (2005b) A situational analysis of drought hit villages of Bhilwara, India. Disaster reduction and human security—education for sustainable development: case studies and best practices, a compilation by UNESCO and Kyoto University to the United Nations World Conference on Disaster Reduction (WCDR), Kobe, Hyogo, Japan. pp 82–83
Walia A (2015) ToT module on gender and disaster management. National Institute of Disaster Management, New Delhi
World Bank and Global Facility for Disaster Risk Reduction (GFDRR). (2015). Stories of impact: gender inclusion in post-disaster housing reconstruction in India. Retrieved from https://www.gfdrr.org/sites/default/files/publication/soi-india-gender-housing.pdf
Zinnatul-Bassar ATM, Tasnuva A (2017) Women participation in flood risk management: a case study in Char South Baladoba in Kurigram District. Am J Environ Prot 6(1):7–17

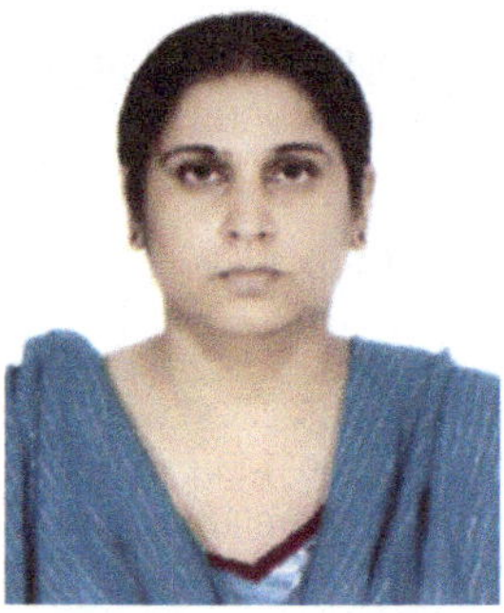

Ajinder Walia is a doctorate in Sociology working as a faculty in National institute of Disaster Management for over fifteen years. She has been working in the area of social issues related to disaster management with a focus on gender and disaster management, community based disaster risk reduction, psychosocial care in disaster management and development of disaster management plans at the village and community level.

Karpoora Sundarapandian is associated with the National Institute of Disaster Management, Govt. of India as a young professional and holds experience in the policy planning and cross cutting issues related to disaster management.

Chapter 20
Climate Change Impact on Landuse and Livelihood in Sundarbans: A Case Study of Sagar Island

Arna Ghosh, Sweta Baidya, and Anil K. Gupta

Abstract Sagar island, situated in the coastal mangrove belt of West Bengal, is one of most climate sensitive regions of the world. Ravaging tropical cyclones and massive erosion due to unprecedented sea level rise has become a sad reality of the island in the last few decades. Loss of livelihood and lack of government interventions has resulted in a horrific deterioration of conditions of the people residing there. This study has identified some of the vulnerable locations, which are in immediate need of attention in terms of infrastructural development and the overall upliftment. The erosion prone locations are identified after a rigorous literature review and with the use of satellite imageries and geoprocessing tools. Land Use Land Cover maps for the years 1990, 2000, 2010, 2020 and 2022 was made to compare the geomorphological changes over the decades and to understand the extent to which climate change has been impacting the ecologically sensitive island. Field survey were conducted to assess the impact of climate change on the lifestyle and livelihood of the people. Finally, a few adaptive strategies have been indicated which can bring forth economic development of the people and to some extent safeguard their life and livelihood.

Keywords Climate resilience · Coastal erosion · Shoreline changes · Adaptation strategies · LULC · Causal loop diagram

---

A. Ghosh
TERI School of Advanced Studies, New Delhi, India

S. Baidya (✉)
National Disaster Management Authority, New Delhi, India

A. K. Gupta
ECDRM Division, International Cooperation, CAP-RES DST Project, National Institute of Disaster Management (Ministry of Home Affairs), New Delhi, India

A. K. Gupta et al. (eds.), *Disaster Risk and Management Under Climate Change*,
Disaster Resilience and Green Growth,
https://doi.org/10.1007/978-981-99-4105-6_20

20.1 Background

Unprecedented temperature rise due to global warming is a topical issue. The average surface temperature is set to rise by 1.5–4 °C by the end of this century. As a result, there would be worldwide changes to the living patterns and habitats. Starting from global sea level rise due to severe melting of polar ice caps and glaciers, to inundation of low-lying plains causing worldwide displacement of people and animal species (Bera and Singh, 2021). Subsidence of the mega deltas like Sundarbans are one of the many problems faced by these delta dwellers (Brown & Nicholls, 2015). Not only this but also the northward migration of animal species and their eventual extinction due to lack of space for movement would become a world crisis. Severe problems of food security will arise in several countries and worst affected will be the developing countries. Tropical cyclones will only become more intense and cause widespread destruction of crops and wreck-havoc on peoples' lives. Coastal region is a delicate ecosystem and it is a home to so many different species of the animal kingdom, plant kingdom and the humans alike. They are the rich hotspots of socio-cultural, historical, and biological diversity. Climate change will greatly alter the ecological balance between all the living forces residing along the coastline across the globe. Therefore, studying the changing patterns of the shoreline has become very important to predict measures to combat and sustain the lives there (Bandyopadhyay, 1997). In the last two decades, studying remotely sensed data to analyse such changing patterns has become of the utmost importance (Bera et al., 2021). By monitoring an element or phenomena at several intervals, the approach of "change detection" can be employed to spot variations in its existence (Singh 1989). Planning balanced and coordinated management plans for the conservation of coastal resources may benefit from the use of GIS (Jayappa et al. 2006).

Sagar Island is the biggest island of the Sundarban Deltaic complex and located in the estuarine ecosystem of the Hooghly River. The island is surrounded by Hooghly River on the west, Muriganga River on the east and Bay of Bengal in the south. The island has an elevation of 6.4 m above sea level and the island is connected to the mainland by a ferry service across the Muriganga River. The island is of great historical importance and houses the Kapil Muni Ashram and witnesses an influx of tourists every year during the great Ganga Sagar Mela. Unfortunately, the current mandir of Kapil Muni is situated at approximately 3–4 km away from its original location. This is the fourth constructed Ashram since, all the previously built ashrams got washed away. Sagar island is a tide dominated deltaic region; hence it is influenced by tidal fluctuations daily and is ravaged by tropical cyclones and is undergoing intense erosion resulting in an alarming receding shoreline (Gopinath and Seralathan, 2005). The island originally comprised of three islands which together came under Sagar Block (Fig. 20.1).

Out of the three, Lohachara island has already seen the face of extinction and the island of Ghoramara is undergoing severe erosion and is likely inundate in the coming decade. Sagar Island tide gauge data from 2002 to 2009 revealed a relative mean sea level rise of 12 millimetres per year, which is significantly more than the 3 mm/year from the prior decade (Hazra et al. 2002). Compared to the global average

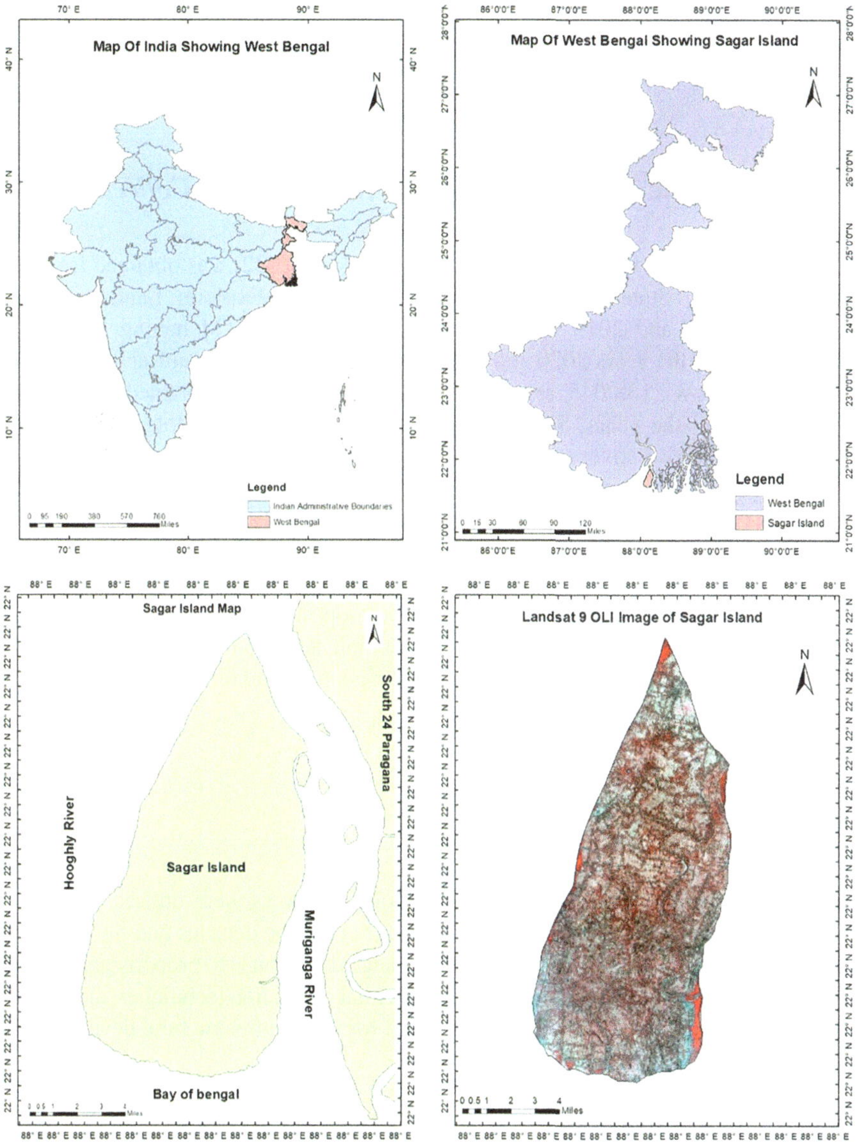

Fig. 20.1 Blocks of Sagar Island

of 3.27 mm/year during the years of 1993 and 2010, this pace of sea level rise is much higher. To analyse the impact of climate change on this island, integated database was generated using both Spatial and non-spatial data.

20.1.1 Satellite Data Collection and Rainfall Data Collection

Multi-dated satellite data (1990, 2000, 2010, 2020 and 2022) was obtained from the website of United States Geological Survey (USGS) earth Explorer. Landsat data for year 1990, 2000 and 2010 was obtained from Landsat 5 TM having Path/Row-138/045 and for the years 2020 and 2022 it was obtained from Landsat 8 and 9 OLI having Path/Row- 138/045 respectively. The images were obtained for second February for all the years. Satellite imagery interpretation was done to determine the loss of mangrove cover over 3 decades 1990–2000, 2000–2010 and 2010–2020. Possible reasons for the clearance of mangroves were identified with the help of literature. Gridded rainfall (0.25 × 0.25) NetCDF data was acquired for the years 2000, 2010 and 2020 from the website Indian Meteorological Department to study the distribution and variation of the total amount of rainfall received by the island. The data was processed on ArcGIS using the appropriate tools to chart which region of the island receives the most amount of rainfall and identify the reason behind it. The snapshots from the field authenticate the geomorphic features identified from the Landsat imageries.

20.1.2 Land Use Land Cover Map

The satellite images were processed and various GIS tools were utilised to prepare LULC maps for the impact analysis. The changes over the decades and the plausible reasons were determined. The goal for such change detection is to properly analyse the most vulnerable location and to bring forth essential adaptation techniques, along with the identification of the location in urgent need for some infrastructural development.

20.1.2.1 Accuracy Assessment of LULC Classes

Accuracy assessment is an integral part of the LULC classification (Rwanga and Ndambuki, 2017). This assessment compares the already classified image to another source, which is taken to be true or accurate ground truth. Ground truth can be collected either in the field or in sources like Google Earth Pro. However, collection of data from the ground is tiring and time consuming, hence the latter option is more viable. There are other methods of acquiring ground truth data, for example high-resolution imagery, existing classified imagery as well as GIS data layers. Google Earth Pro was utilised for the assessment. Accuracy Assessment Formula (Congalton, 1991) used in this study is as follows:

Users Accuracy

$$= \frac{(Number\ of\ Correctly\ Classified\ Pixels\ in\ Each\ Category)}{(Total\ Number\ of\ Classified\ Pixels\ in\ that\ Category)} \times 100$$

Producers Accuracy

$$= \frac{(Number\ of\ Correctly\ Classified\ Pixels\ in\ Each\ Category)}{(The\ Total\ Number\ of\ Reference\ Pixels\ in\ that\ Category)} \times 100$$

$$\text{Overall Accuracy} = \frac{(Total\ Number\ of\ Correctly\ Classified\ Pixels)}{(Total\ Number\ of\ Reference\ Pixels)} \times 100$$

$$\text{Kappa Coefficient} = \frac{((TS \times TCS) - \sum(Column\ Total \times Row\ Total))}{(TS^2 - \sum(Column\ Total \times Row\ Total))} \times 100$$

Here, TS stands Total Sample & TCS stands for Total Corrected Sample.

20.1.2.2 Shoreline Change Detection and Mapping of Water Occurrence Change Intensity

For the purpose of change detection, shoreline changes were mapped. The task was executed using visual interpretation technique. The shoreline was first delineated from the Landsat imageries for the respective years as precisely as possible. The changes in the area of the island were taken as an indicator of shoreline degradation. This exercise enabled to determine the decadal loss of total area. Shoreline configuration of eight selected mouzas were mapped for the period of 1990–2020. The reason behind the humongous changes were identified using water occurrence change intensity data, collected from the website of global sea surface explorer.

20.1.3 Field Data Collection

Field survey was conducted for a period of 5 days from July 31st, 2022 till August fourth, 2022 mainly along the Southern coastal zones of Sagar Island. For the survey a semi structured questionnaire was prepared containing both open ended and close ended questions. Responses from 80 people were recorded about their lifestyle and how climate change is impacting their day-to-day life and occupations alike. Eight mouzas were covered namely Gangasagar, Bankimnagar, Dhablat, Shibpur, Mahismari, Chemaguri, Sumatinagar and Baeguakhali. The responses were limited to 10 people per mouza. Information on their expectations from the government was recorded and this shall be particularly useful for policy makers.

20.1.4 Causal Loop Diagram

A causal loop diagram was prepared on software Vensim PLE x64 for the holistic analysis of climate change impact on the lives of people. The diagram reveals how a single event stirs several other events in momentum, and how that results in in-migration and out-migration of people. The causal loop diagram was prepared based on the responses collected from the field.

This study has three objectives, (i) identifying the impact of climate change on the land-use patterns. (ii) identifying the impact of climate change on the livelihood of people residing in Sagar Island and (iii) to identify the possible adaptation measures.

20.2 The Geographical Change

The striking event that occurred during the first decade under observation is the disappearance of Lohachara island. The island got completely inundated in the year 1996. Further inferences can be drawn from Table 20.1, containing the key observations (Fig. 20.2).

Examining the two satellite imageries of 1990 and 2000 it is also evident that the thick mangrove coverage along the western border of the island, which was along the coastline of the mouzas namely, Mandirtala, Phuldubi, Krishnagar, Naraharipur, Radhakrishnapur, Chandipur, Mahismari and Baeguakhali has reduced considerably. Total mangrove cover in 1990 was 12 sqkm (7 sqkm along the western coastline and 5sqkm along the South-Eastern coastline), which is about 4% of the total land area. A small portion of the land has been accreted along the north-east shoreline of Shikarpur, Ramkrishnapur and Gobindapur mouzas of the island. The total land coverage of mangroves dropped down to 5 sqkm by the year 2000 (2 sqkm along the western coastline of Krishnagar and Naraharipur mouza and 4sqkm along the South-Eastern coastline of Bankimnagar and Chemaguri mouza). About 2% of the land area had mangroves in 2000.

Table 20.1 Showing key observations of satellite imagery interpretation

Decades	Key observations
1990–2000	• Disappearance of Lohachara Island • Mangrove coverage dropped by 58% (12sqkm to 5sqkm) • Possible reason for such clearance could be deforestation for agricultural land and accommodation for the climate refugees of Lohachara island
2000–2010	• An increase of 28% in mangrove coverage was observed (5sqkm to 7sqkm) • An attempt to afforestation and land accretion could be the main reason behind the growth • 3sqkm of land was accreted along the North-Eastern and Northern shoreline
2010–2020	• Decrease of 42% in mangrove was observed (7–4 sqkm) • Intense cyclones, prolonged inundation, increase in salinity and deforestation could be the possible reasons for the clearance

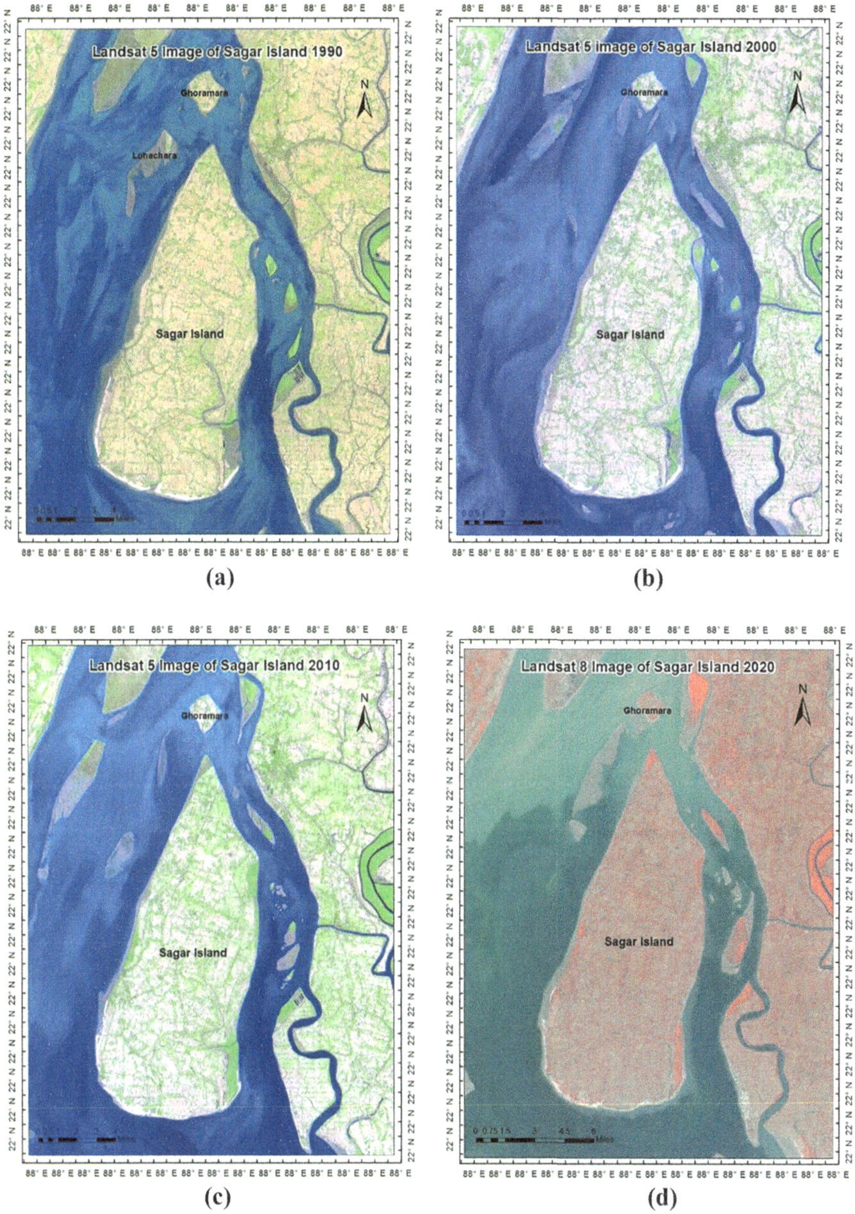

Fig. 20.2 Landsat 5 Image of Sagar Island 1990 (**a**), 2000 (**b**), 2010 (**c**) and 2020 (**d**)

Mangrove coverage area in 2010 increased to 7 sqkm. Total mangrove coverage in 2010 is about 3% of the land area. Satellite imagery of 2010 shows the complete loss of mangroves from the coastline of Mahismari and Baeguakhali Mouza on the

South Western coastline. Accretion has taken place in two places along the North-Eastern coastline of Shikarpur, Ramakrishnapur and Gobindapur (approximately 1sqkm) and in the northern tip of the island, along the coastline of Kashtala mouza (approximately 2 sqkm). In total 3sqkm of land was accreted, which is about 1.2% of the total land area.

The Landsat imagery of 2020 depicts a huge loss of mangroves not only along the western shoreline of the island but also a considerable loss along the South-Eastern coastline. Only 6 sqkm of the total land area is covered with mangroves. Along the western coast, the mangroves are dwindling in fringes along the coastline of Phuldubi, Krishnagar and Naraharipur mouza. The loss of mangroves can be attributed to two reasons. Firstly, a vast portion of the island's mangrove cover was transformed into croplands, settlements, and mangrove swamps. Wave action is another major reason for such changes. Mangroves by nature have the capacity of surviving in saline water and their breathing roots helps them to survive the inundation during high tides. However, prolong inundation due to sea level rise could cause stunted growth with fewer and shorter branches resulting in lower rates of photosynthesis and regeneration and in some cases death. Sea level rise has caused an undue spike in salinity and since some of the species has a limit to its saline water tolerance, needs a daily influx of both fresh water and saline water for their survival. One of the most economically important species of mangroves found on this island *Heritiera fomes,* locally known as *Sundari*, is now on the verge of extinction due to increase in salinity. Along with this, unchecked deforestation to acquire its valuable wood also adds to the reason of its depletion. In 2020, the total area under mangrove is 5 sqkm which is about 1.7% of the land area.

Loss of mangrove cover attributed to either sea level rise or anthropogenic activities also explains the fact why there was a sudden shift in the distribution of total rainfall patterns. Maximum rainfall was initially received by the western coast of the island as shown in figure below. In the second decade under observation, i.e., 2000–2010 mangrove coverage on the western coast dropped from 2 to 1 sqkm, at the same time there was an increase in mangrove coverage on the eastern coastline. It is a fact that water vapour rising from the forest fuels convection and moisture-filled clouds and these factors work together to hasten the creation of rain, which then falls to the ground and is absorbed once more. Wherever there is vegetation, this cycle of absorption, evaporation, and rain occurs. Due to the growth of vegetation on the eastern coastline, the maximum rainfall is now received by the eastern coast as shown in Fig. 20.3 below (Fig. 20.4).

20.2.1 Shoreline Change Detection

Sagar Island has faced tremendous amount of erosion and lack of proper embankments along with a sinical rise in the sea level is gradually leading the island towards its disappearance. The total land area in 1990, 2000, 2010 and 2020 is 252, 248, 239 and 232 sqkm respectively. The percentage of decadal changes for the first decade is 1.7%, for the second decade is 3.6% and for the third decade it is 2.9%.

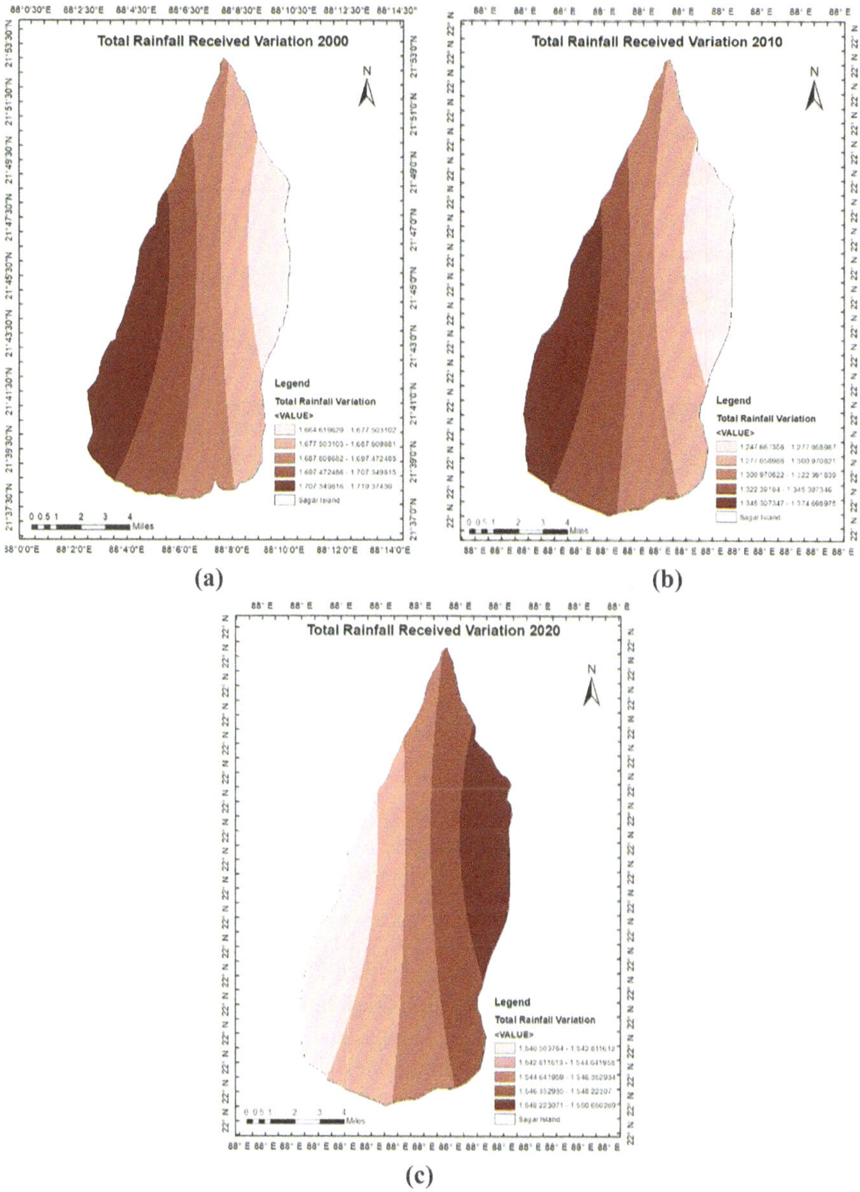

Fig. 20.3 Maximum rainfall received in (**a**) 2000, (**b**) 2010 and (**c**) 2020

Currently, the total area stands at 223 sqkm, which is quite alarming given the fact that in the year 2020 the total area was 232 sqkm. Places of severe coastal erosion, having vegetation especially mangroves are under a very critical condition. As a result, the people residing in the coastal areas are facing severe issues of saline water

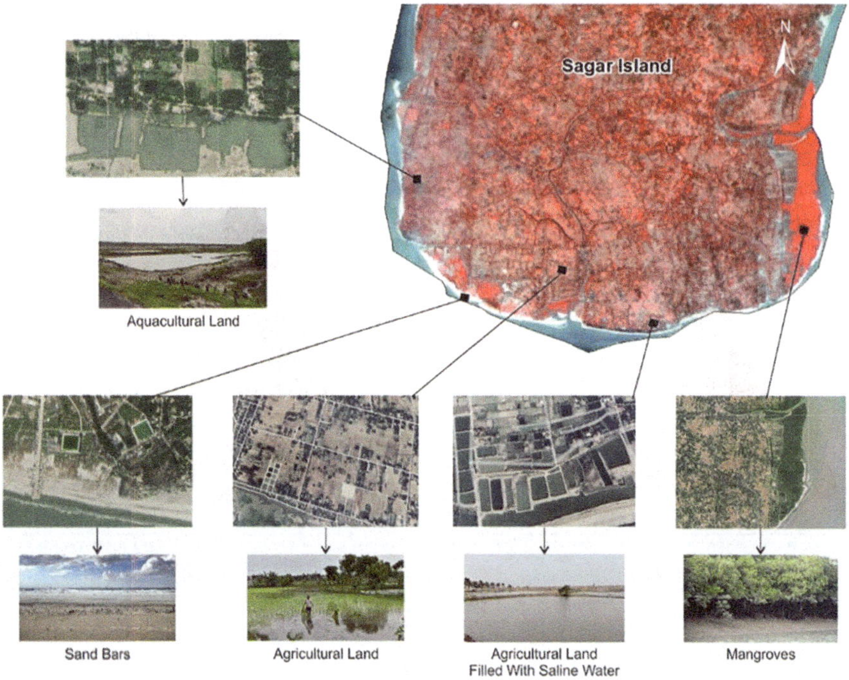

Fig. 20.4 Satellite Image and Google Earth Image with images from the field

intrusion into their agricultural fields and freshwater fisheries (Mukhopadhyay et al., 2018). Mangrove cover needs to be present in a thick stack so that it can provide a padding against the storm surges by dissipating the tidal waves and protect the interiors (Fig. 20.5).

The eastward meandering of the Muriganga River could also contribute to the loss of mangrove cover because they are under the influence of constant wave action along with prolonged inundation during high tides. Severe erosion is observed along the coastline of Baeguakhali, Mahismari, Gangasagar and Dhablat. Constant breach of embankment and flooding during high tides would eventually result in the disappearance of some of the villages. Most particularly vulnerable villages are Dhablat and Shibpur, located along the Southern coastline next to Gangasagar mouza (Table 20.2).

20.2.2 *Water Occurrence Change Intensity and Shoreline Configuration along the Coast of Surveyed Mouzas (1990–2022)*

Water occurrence change intensity map (Fig. 20.6), shows the increase in water occurrence along the entire southern coastline of Sagar Island from 1984–1999 to

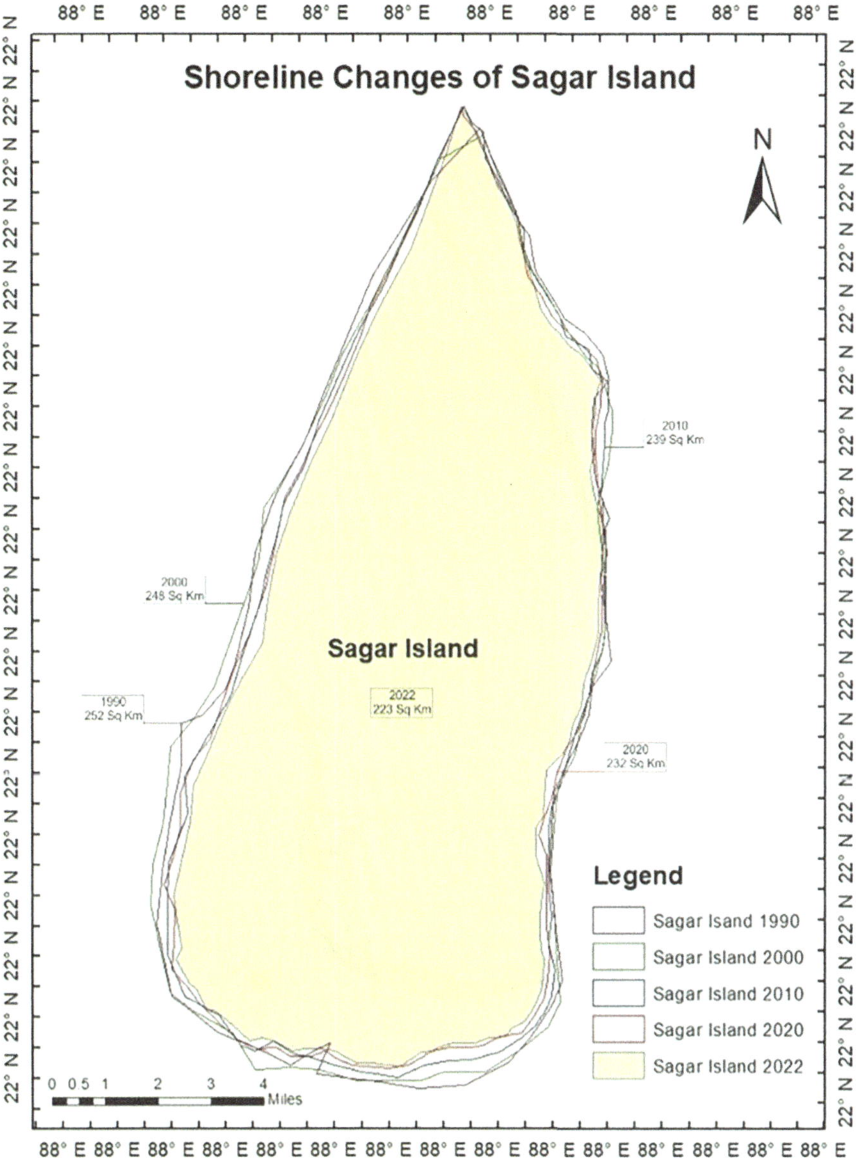

Fig. 20.5 Shoreline changes of Sagar Island

Table 20.2 Showing decadal area loss and percentage of decadal change

Year	1990–2000	2000–2010	2010–2020
Area loss	4 sqkm	9 sqkm	7 sqkm
Percentage of decadal change	−1.7%	−3.6%	−2.9%

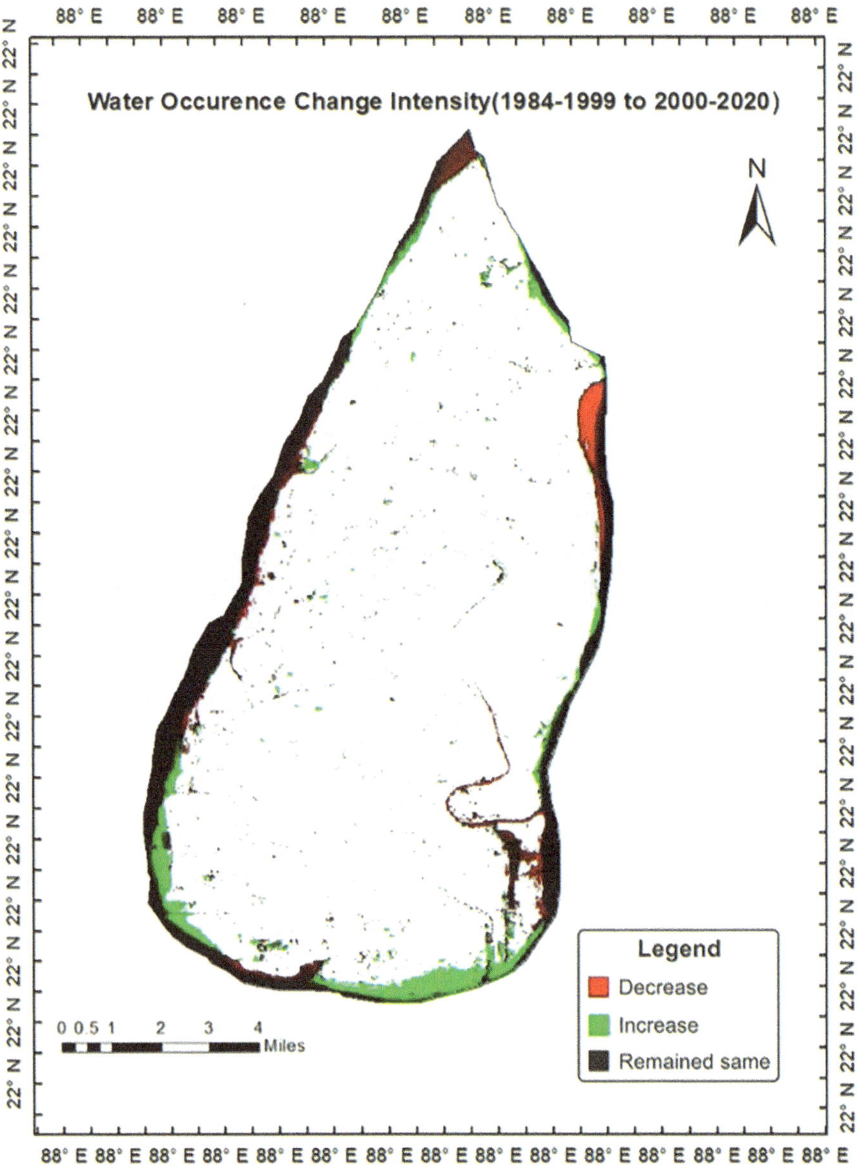

Fig. 20.6 Water occurrence change intensity

2000–2020. Change in water occurrence intensity is unreasonably high along the shoreline of Mahismari, Dhablat and Shibpur (Figs. 20.7, 20.8, 20.9, 20.10, 20.11, 20.12, 20.13 and 20.14).

Shoreline configuration (1990–2022) reveals that among the eight mouzas, Mahismari, Dhablat and Shibpur are the worst affected due to erosion and much

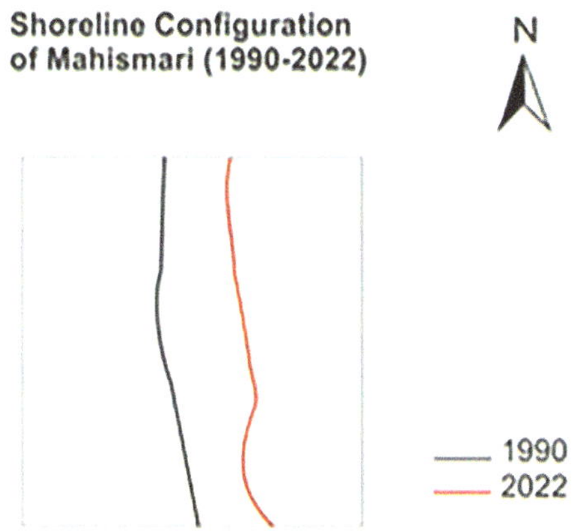

Fig. 20.7 Shoreline configuration of Mahismari

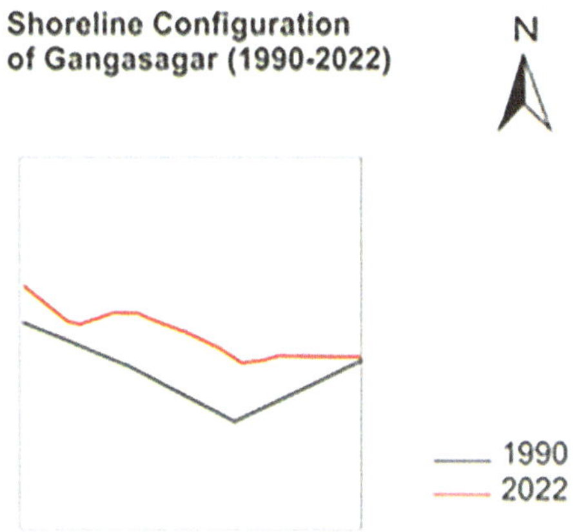

Fig. 20.8 Shoreline configuration of Gangasagar Island

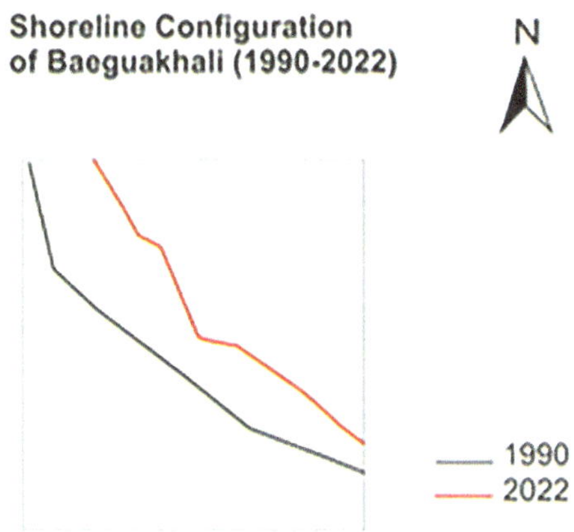

Fig. 20.9 Shoreline configuration of Baeguakhali

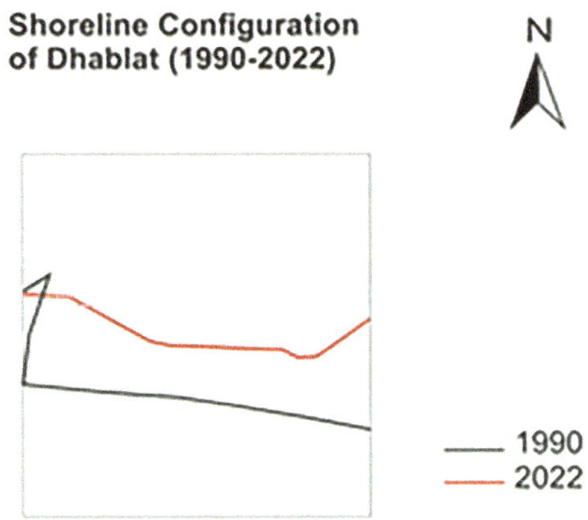

Fig. 20.10 Shoreline configuration of Dhablat

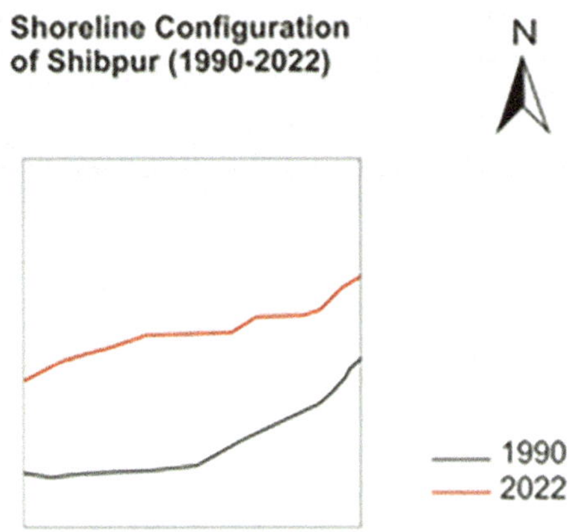

Fig. 20.11 Shoreline configuration of Shibpur

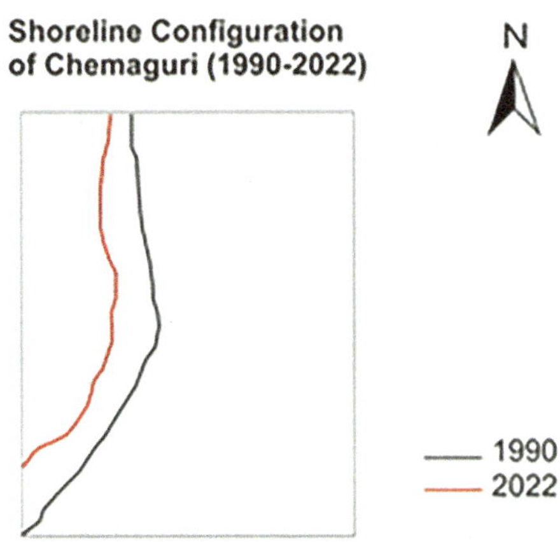

Fig. 20.12 Shoreline configuration of Chemaguri

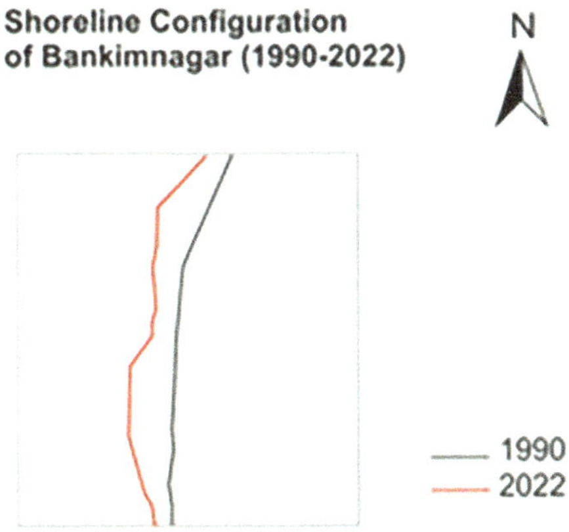

Fig. 20.13 Shoreline configuration of Bankimnagar

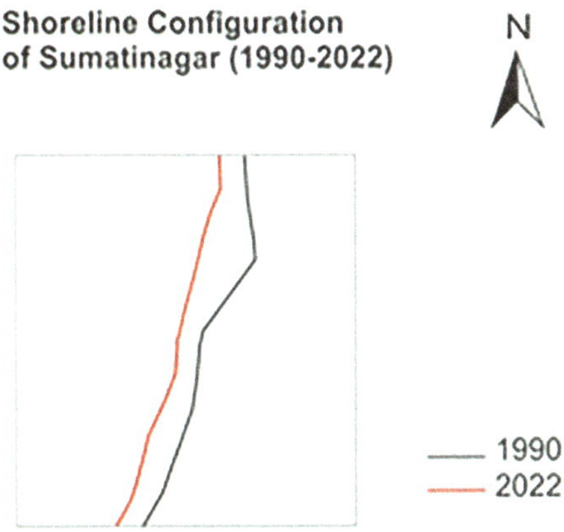

Fig. 20.14 Shoreline configuration of Sumatinagar

of its shoreline has receded. Moreover, there are no vegetation along the shorelines of these three mouzas, which further fuels such an intense erosional activity. Fishermen community of Mahismari explained how there used to be a presence of thick mangrove cover before and the intensity of erosion was not that pronounced then. However, clearance of these natural barriers due to deforestation and unprecedented sea level rise has caused such a tragedy. Locals of Mahismari and Dhablat claimed that, earlier there were several small islands that served as a buffer zone and helped in breaking the tidal currents. However, due to the submergence of those small islands, currently the strong tidal waves directly hit the shore, resulting in massive erosion.

Mangroves act as a protective barrier and they serve as a buffer region to dissipate the wave action, hence, Bankimnagar, Chemaguri and Sumatinagar seem to be less affected amongst the eight mouzas, since there is presence of thick mangrove cover. Prime erosion affected areas requires the much needed acknowledgement and strong initiative must be undertaken for the plantation of mangroves, along with other measures in order to safeguard the island.

20.2.3 Land Use and Land Cover Change

Unsupervised classification yielded the LULC maps for the years 1990, 2000, 2010, 2020 and 2022. This form of classification by utilising GIS tools is widely accepted. Names of the LULC classes were chosen using the visual image interpretation technique and by minutely matching the processed LULC maps to the Landsat imageries (Table 20.3).

Five classes were identified for the LULC map of 1990 (Fig. 20.15) namely, mangroves, tidal creek, settlement & agricultural land, vegetation, and sand bars. LULC map 2000 (Fig. 20.16) has identified 6 classes, open sea, mangroves, tidal creek, vegetation, settlement & agricultural land, and Sand bars. Visibly there is a clear distinction of how settlement area and agricultural land has increased over the decade (1990–2000). Area under vegetation has reduced to 2278.125 hectares (approximately 9% of the total land area) and the reason could be attributed to deforestation for settlement and agricultural purposes. Area under mangroves is approximately 516.3 hectares, which is about 2% of the land area. The increase in settlement could be because of the influx of climate refugees from the surrounding islands, especially from the island of Lohachara. Since, 1990–2000 was the decade during which that island got completely inundated.

Five classes were identified for LULC map of 2010 namely sand bars, mangroves and vegetation, agricultural land, tidal creek, and settlements. LULC map of 2010 shown in (Fig. 20.17) shows a slight recovery in the mangrove coverage. The recovery percentage of mangrove cover over the decade (2000–2010) is about 1%. Reason could be attributed to the fact of land accretion and an attempt to afforestation. The area under settlement is 6683.49 hectares (27%), area under agricultural land is approximately 10,557 hectares (43.62%) and area under sand bars is about 1294.02 hectares (5.34%). Other visible distinction is the formation of sand bars along the coastline of Sumatinagar and Bankimnagar mouza. This region was undergoing accretion during the decade of 2000–2010 (Jayappa et al. 2006, p. 3679).

LULC map of 2020 identified five classes namely, sand bars, mangroves and vegetation, aquacultural land, settlements & agriculture, and tidal creek. This decade (2010–2020) shows an addition of aquacultural land especially along the shorelines since salt water intrusion resulted in the ruin of agricultural land. However, this does not depict the fact that aquacultural lands were missing over the decades of 1990–2010, but the percent increase is quite high to be registered under

Table 20.3 Showing areas of classified LULC classes in hectares

LULC classes	1990	2000	2010	2020	2022
Mangroves	1296 ha	516.3 ha	700 ha	500 ha	400 ha
Vegetation	5991.12 ha	2278.12 ha	5361.23 ha	5819.43	6061.02 ha
Settlement and Agricultural land	11,290 ha	12164.3 ha	17240.49 ha	12,789 ha	7066.08 ha
Sand Bars	384.21 ha	785 ha	1294.02 ha	1107.81 ha	1026.02 ha
Aquacultural Land	–	–	–	3479.41 ha	2757.42 ha

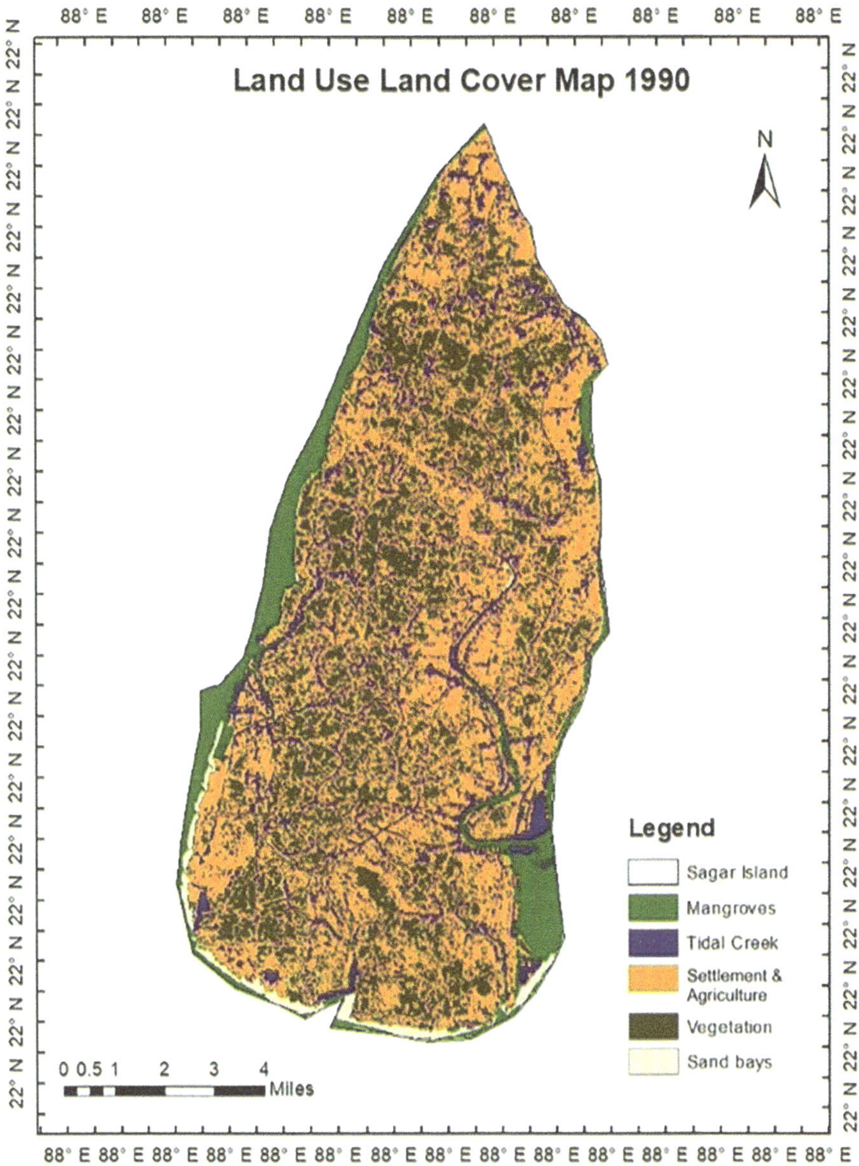

Fig. 20.15 LULC map of 1990

unsupervised classification using geoprocessing tools. Area under sand bars is 1107.81 hectares (4.73%), area under aquacultural land is 3479.41 hectares (14.86%), area under agriculture and settlement is approximately 12,789 hectares (56.65%) and area under mangroves and vegetation is 6419.43 hectares (27.43%). It

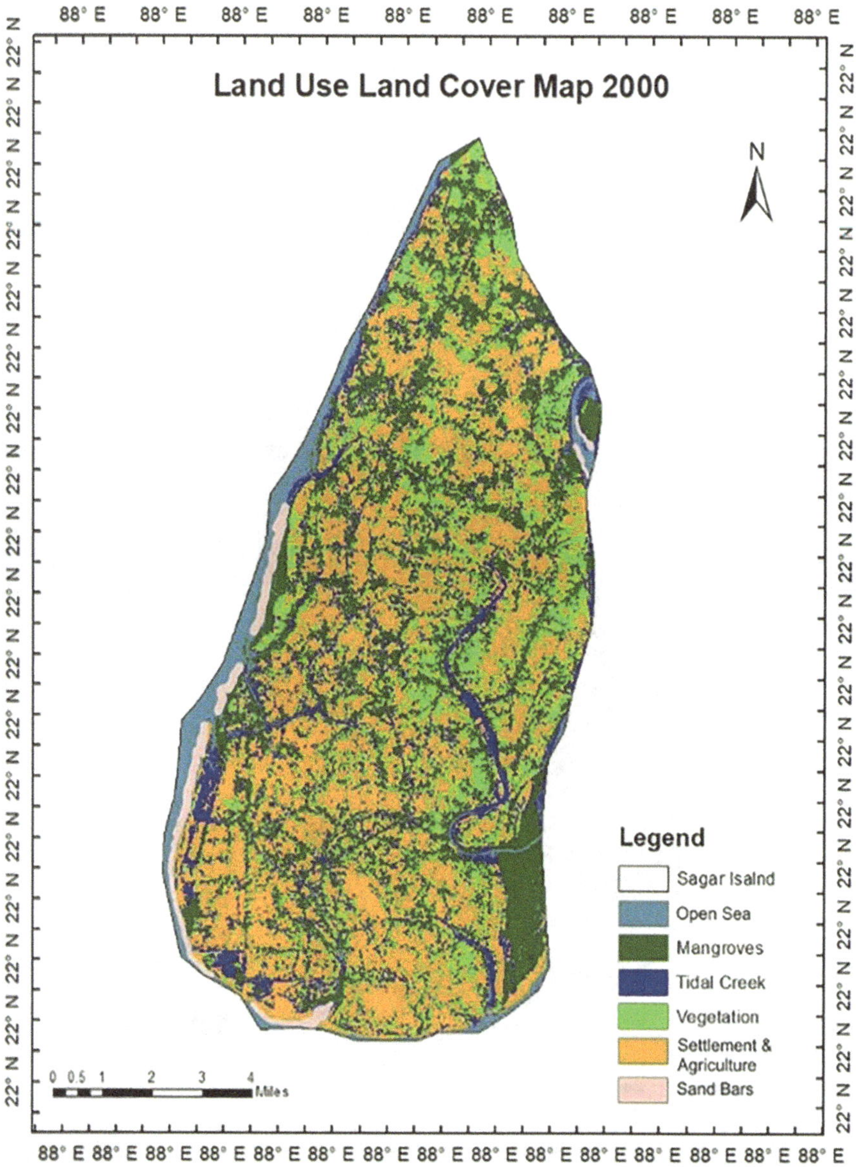

Fig. 20.16 LULC map of 2000

was previously determined that the area under mangroves was about 6sqkm. However, upon comparing the LULC map of 2020 to the LULC map of 2022 (Fig. 20.18), the area under aquaculture has increased. This is because of the tropical cyclone Yaas; water level during this cyclone went up to 25–30 feet resulting in

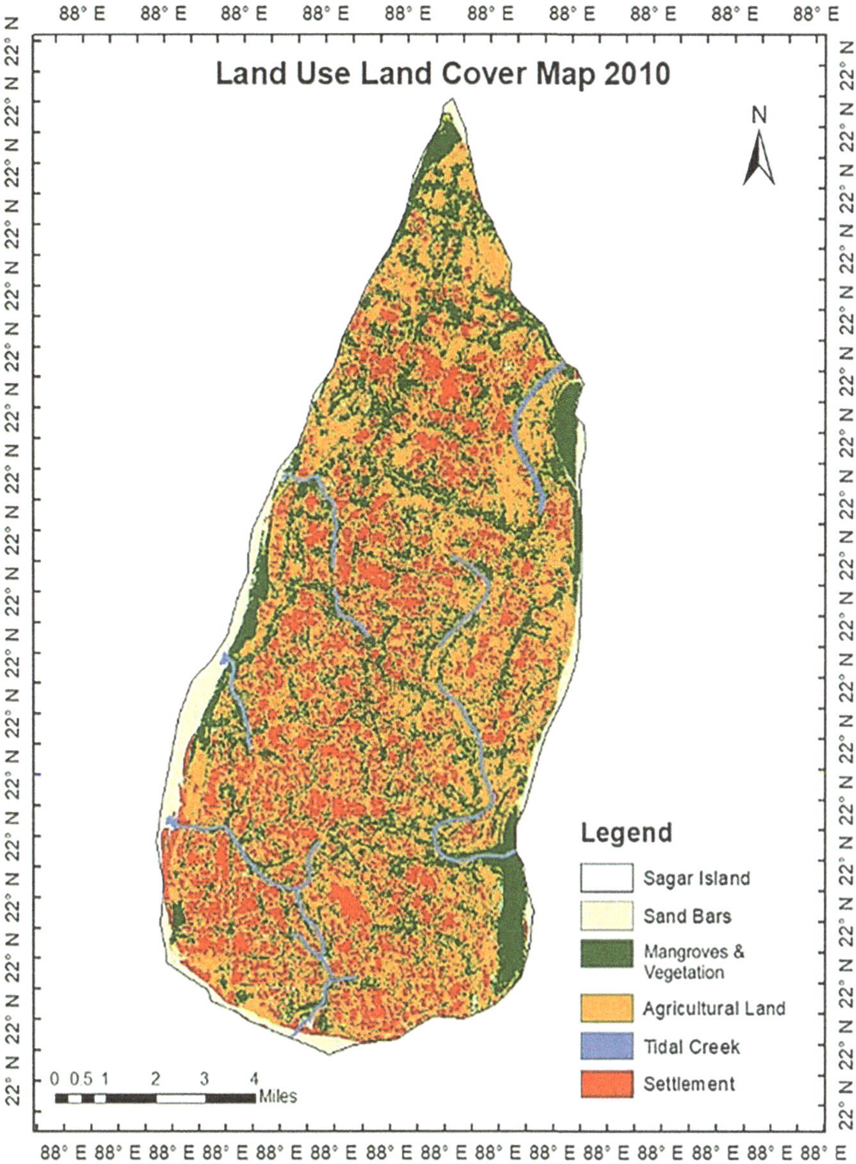

Fig. 20.17 LULC map of 2010

widespread inundation of land. Lack of proper embankment had led to such devastation. Increase in aquacultural land can be spotted along the coastline of Mahismari, Dhablat, Bankimnagar and Sumatinagar mouza. The agricultural land got inundated during the storm and given the poverty-stricken conditions of the locals, most of

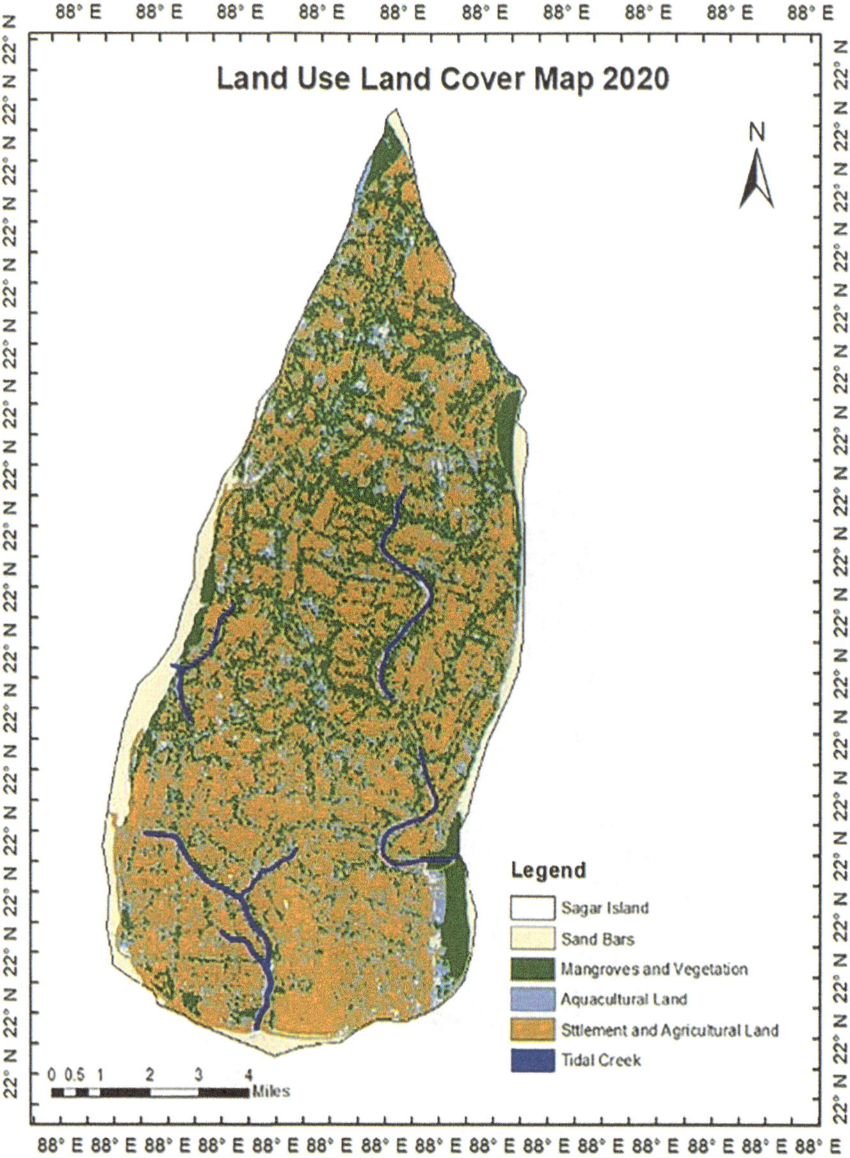

Fig. 20.18 LULC map of 2020

them were unable to recover their land from the saline water. The increase in settlement area was expected over the decade of 2000–2020, but unfortunately, Sagar Island which was once a haven is now a disappearing island and many of the villagers resettled somewhere else and most importantly the young generation is

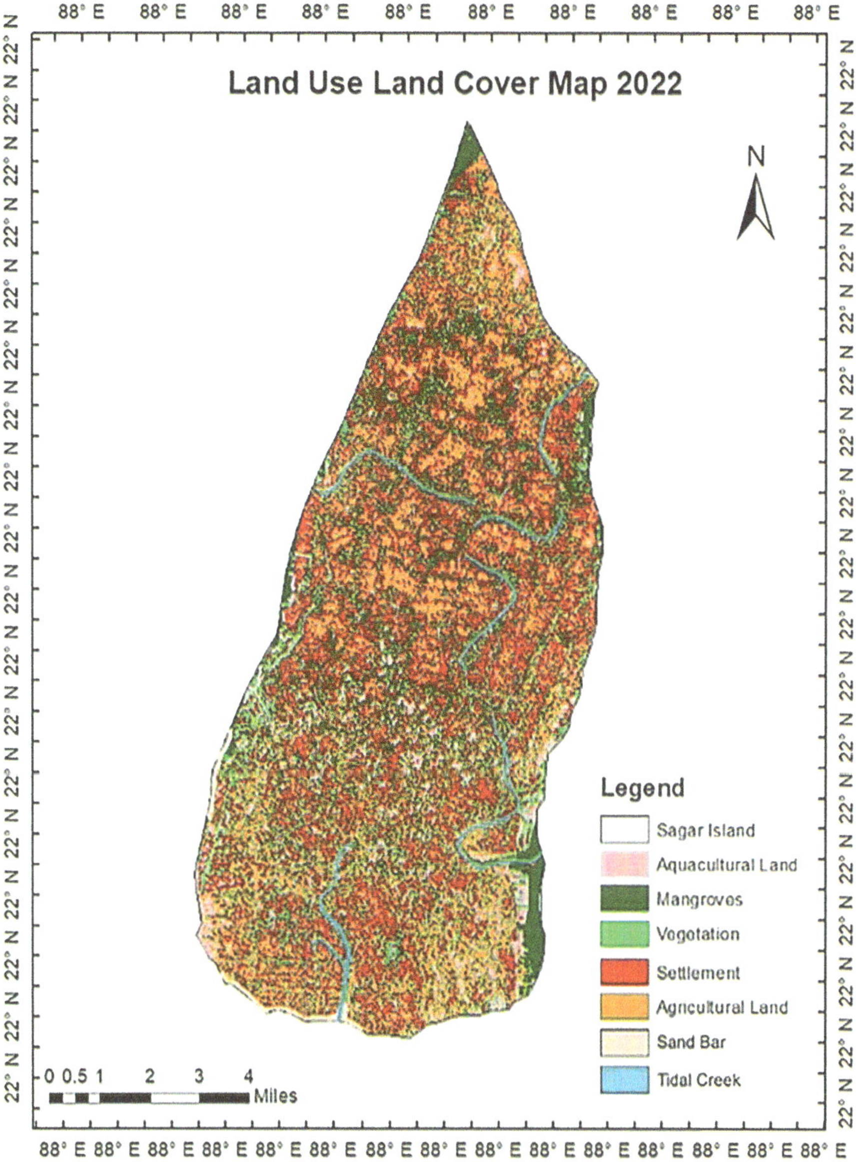

Fig. 20.19 LULC map of 2022

moving out of the island to the mainland in search for jobs. Hence, a drop in settlement cover can be noticed over this decade (Fig. 20.19).

LULC map of 2022 identifies seven distinct classes namely, aquacultural land, mangroves, vegetation, settlement, agricultural land, sand bar, tidal creek.

Aquacultural land covers approximately 2757.42 hectares (12.36%), settlement covers approximately 4017.96 hectares (18%.01), area under vegetation is 6061.02 hectares (27%), area under sand bars is about 1026 hectares (4%) and agricultural land covers 3048.12 hectares (14%). Mangrove coverage is meagre 4sqkm of the land cover. Over the decades, a severe declining trend is noticed in mangrove coverage and an equal rise in aquacultural land along the South-Western, South, and South-Eastern shoreline.

20.2.3.1 Accuracy Assessment of LULC Map of 2022

Accuracy assessment is an imminent component of LULC classification in order to determine the precision. A higher value of accurateness would conform the prominence of the thematic mapping (Mondal et al. 2019, p. 779). An accuracy assessment was conducted for the Landsat 9 OLI (2022) classified image, which was done by drawing an error matrix, calculating overall accuracy and kappa coefficient. Overall classification accuracy of 90% and kappa coefficient of 0.88 was achieved.

20.3 Change in the Livelihood

Responses of 80 villagers were recorded across eight mouzas along the Southern coastline of Sagar Island. 54.8% of the respondents were female and rest 45.2% were male. Female population is more than the male population in these mouzas; since the flow of income from the main occupations, that is mostly agriculture and fishing is dwindling and hence, the male population usually move out to the mainland in search for jobs. Majority of the respondents belonged to the age group of 40–70 and only a handful of the respondents belonged to the age group of 25–35. From the information collected, it was evident that the young generation has moved out in search for jobs to the mainland. Majorly the young men travel to Kolkata, Kerala, Andhra Pradesh, Telangana and Kakdwip. Members of the family who reside in the mainland for job purposes, sends renumerations back to home.

The respondents mostly come from a family of four members, except for one respondent who belonged from a family of eight members. A higher percentage of the respondents live in kutcha houses (66.7%) and this very aspect makes them even more vulnerable. Given the intensity of calamity that strikes the island, even pucca houses fall prey to it. Monthly income of 48% of the respondents lies in the range of 5000–10,000 rupee, which makes it difficult for them to have some savings. Comparatively the fishing community is quite poor. The fisherwomen earn only a mere 100 rupees upon catching 1000 prawn seeds. Catching prawn seeds is something they can do only during the low tide, which is early in the morning. Therefore, their ability to earn more becomes restricted and lack of other sources of income strikes as a great blow to their bare minimum conditions.

Due to breach of embankment during high tides and roaring sea level rise, leading to intensified flooding during tropical cyclones, all the houses that are kutcha is

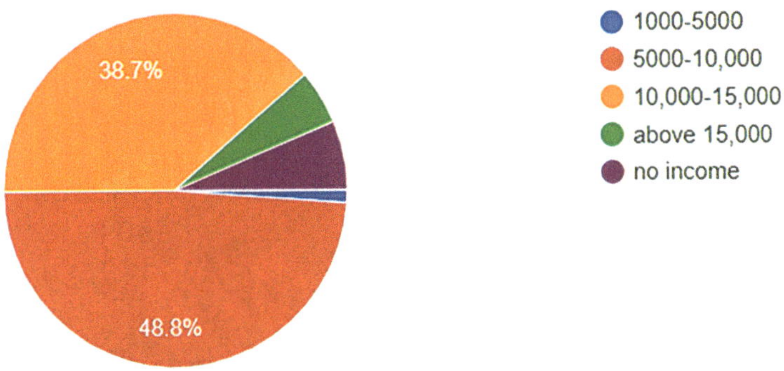

Fig. 20.20 Monthly Income of the Respondents in INR

demolished to the ground. The villagers have no choice but to rebuild their homes at the exact same location, since they are too poor to relocate without external help (Fig. 20.20).

The villagers responded that after every calamity the amount of money that they invest in rebuilding their abode, takes out every penny from their pockets, due to which most of them are in debt. Major occupations of the people are betel leaf plantation (11.3%), agriculture (7.5%), fishing (22.5%), E-rikshaw driver (7.5%), wage labourer (7.5%) and shop owner (18.8%) which includes fruit-sellers, gift shop owner near Kapil Muni Ashram, tea-stall owner, and small diners. Other occupations include pisciculture and horticulture. Respondents who own diners near the Ashram mentioned how their fridges are damaged due to saline water intrusion. Assets such as fridges and stoves get affected and they must invest money not only to repair their homes but also to repair their shops. Villagers involved in agriculture and pisciculture are worst hit because many do not have the capacity of retrieving their land from saline water. Few of the villagers have mentioned that in order to extract the saline water out of their fields, they make use of gypsum or lime, but such techniques are not available to all or more specifically, most of them are not financially well backed to be able to employ such techniques and re-use the field. Respondents who were previously agricultural landowners, has shifted their occupation to being wage labourers and very few of them is using their saline water filled land as a ground to breed salt tolerant species of fish (Figs. 20.21 and 20.22).

One response that was unanimous across the eight mouzas is the fact that there was no agricultural yield across the island during the year 2021 and such a tragedy was caused due to the deadly tropical cyclone Yaas. On the other hand, respondents who were involved in pisciculture, faced similar problems since they used to breed fresh water fish species, which have died due to salt water intrusion. The villagers who are capable, has made attempts to extract the salt from the water but, according to them such attempts are fruitless since, saline water intrusion takes place quite often.

Fig. 20.21 Agricultural land filled with saline water near Manasa Mandir, Dhablat

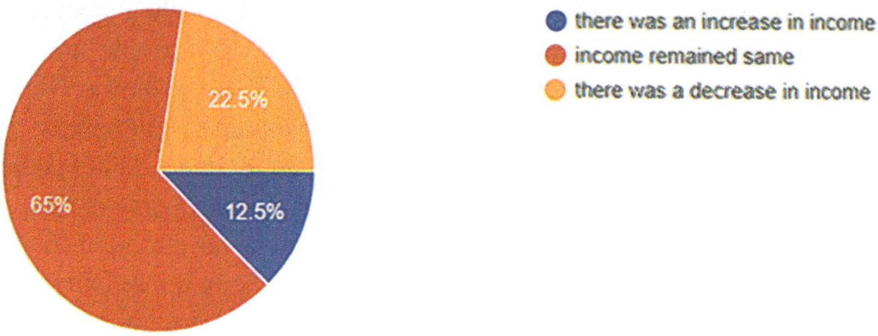

Fig. 20.22 Change in monthly income after shift in occupation

22.5% of the respondents claimed there was a decrease in their income after a shift in occupation. Few of the villagers has shifted to breeding only saline tolerant fish species, but most of them prefer breeding both saline and fresh water species, because the former do not yield much. Betal leaf plantation is one of the major occupations within the island. Betel leaves by nature are very delicate and any minor inconveniences results in crippling of the leaves. The sweet betel leaves are exported out of the island and the non-sweet betel leaves locally known as 'bangla pata' is kept within the island and are distributed across the markets for consumption. Interaction with one of the respondents revealed, that he was an owner of 8-betel leaf plantation in Dhablat, but saline water had taken over the field, during the tropical cyclone Yaas and he was forced to switch his occupation to being a wage labourer. The fact that 65% of the respondent's income has remained the same shows that there is no growth or development-taking place within the island. Moreover, lack of proper governmental interventions has made their condition much more

Fig. 20.23 Breach of embankment

miserable. The condition of the wage labourers is rather tragic during the calamity and the period after, since they do not employed for work. Their lives are solely dependent on daily income and the failure of which leads to grave financial troubles within their households. Not only this but the field owners refuse to employ labourers in the recent years since, their lands do not yield much for their own consumption, let alone accumulating the surplus for the sale in the markets. Eighty percent of the respondents wants to shift to safer locations, but they are unable to do so given their poor conditions. Sixty percent of respondents claimed they have never received any help from the government. One of the respondents had stated very blatantly that they do not wish for any help other than to have better embankments.

The awareness among the villagers were quite high regarding climate change and sea level rise. 73.4% of the villagers were even aware of the fact that their island is slowly disappearing due to massive erosion. The villagers have no problem in securing meals 4 times a day; all the 80 respondents had ration cards through which they receive rice and wheat (Fig. 20.23).

There are three flood shelters namely, Bharat Seva Ashram, Kolkata Vastra Vyavsayi Seva Samity and Ganga Sagar Bhavan, along with one cyclone shelter house in Mahismari. One of the most sinister of all problems was the breach of embankment and lack of proper embankments. The Aila bund that was constructed has been destroyed due to breach and is currently being repaired. Apart from this, lack of mangrove cover along the coast of Mahismari has resulted in rapid erosion of land. The villagers have claimed that previously there used to be a thick coverage of mangroves, which protected their houses from the storm surges. The consolidated mud layers that are at the base of embankment are particularly exposed at Baeguakhali. Riverbanks frequently collapse with joint scars, not only this but sheeting, and gullying of mud banks are a common view in all the eight mouzas. Presence of brick remnants, which were used in the construction of seawalls, found in a linear pattern along the shoreline in Dhablat, proves that erecting seawalls is not a permanent solution (Fig. 20.24).

Fig. 20.24 Destruction caused due to intense wave action in Gangasagar

20.3.1 Causal Loop Diagram for Impact Analysis

Climate change is leading to sea level rise, which is one of the most notorious of all problems faced by Sagar Island. Currently, in many areas, sections of the low-lying shoreline have exposed the clay substrate that usually remains underneath the sand cover and as a result, the sediment mass of the feeder dune system has been scrapped off considerably, with the progressive encroachment of higher water level along the shoreline of Sagar Island (Paul et al. 2016, p. 570). Loop number one, which is a reinforcing loop shows, due to unprecedented sea level rise, massive erosion is taking place and the lack of proper sturdy embankments; land inundation has become a hovering problem across the shoreline. This in turn leads to progressive salinization, which has its own list of cynical problems. Firstly, health issues: increased salinity is affecting drinking water, which is causing stunted growth among children and higher incidence of anaemia among women. Secondly, progressive salinization is damaging agricultural land and fisheries, leading to loss of livelihood. Among coping mechanism, majorly young population are moving out in search for jobs. Even though they send remuneration back to home, it is not

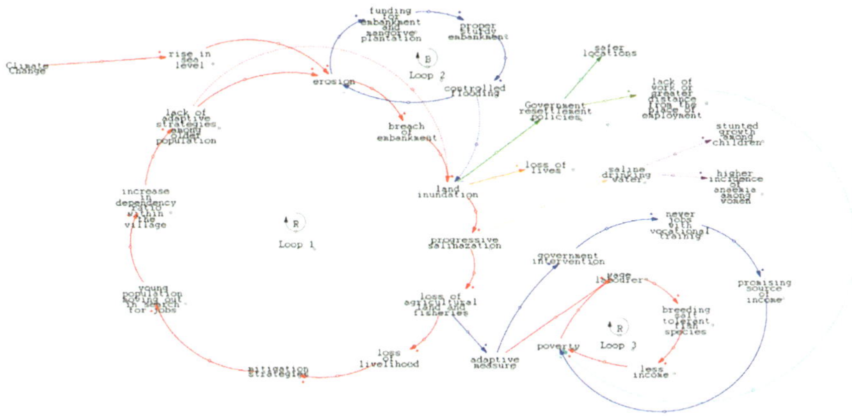

Fig. 20.25 Causal loop diagram

enough since the kind of jobs they are employed to do not pay much. The out migration of younger population is increasing the dependency ratio within the village and lack of coping strategies amongst the older population is making them more vulnerable to erosion and land inundation, ultimately causing them their lives (Fig. 20.25).

Loop number two is a balancing loop, showing if proper funding is sanctioned for embankments, as well as for the plantation of mangroves wherever possible and newer modern technologies are employed to build them, then flooding due to breach of embankment can be controlled and erosion can be kept under check. Loop number 3 is a reinforcing loop, which shows loss of agricultural lands and fisheries, results in unemployment and the villagers move from being owners to part time job employees or wage labourers and since most of the other landowners do not get a stable income from their own land, they refuse to employ wage labourers. Some of the villagers who owned aquacultural land choose to breed saline tolerant fish species, but since, they do not yield much, their financial condition deteriorates. As a result, they are stuck in a constant loop of poverty. However, if one of coping mechanisms include government intervention, such as organising newer opportunities for jobs by giving the villagers' vocational training, then they might get a promising source of income and there would be chance for upliftment of the villagers.

Ultimately, due to land inundation, which is a constant threat to the lives of the villagers, the government intervenes with its several resettlement policies. However, this has two outcomes, firstly they will get safer locations to reside, which means less vulnerable to climate change induced sea level rise. But the second thing that happens, which occurs mainly due to the lack of judgement on the government's side, is that the newer locations they are shifted to, has little or no employment opportunity and their previous location of employment is far from their current residence. For example, the Government started to distribute plots of land in Chandipur to the affected families. The plots of land were reclaimed from an estuary and since this adaptation strategy was not properly calculated, the Government failed to notice, that the place was very poor in terms of road connectivity. Hence, the places of local source of livelihood were not properly connected. Resettlement in this area was rather difficult and since this place was far from the river as well, the migrant families did not find stability in terms of income. They could neither find themselves being involved at some local work to earn a livelihood nor at the embankment, to engage in earthen work (Roy and Guha, 2013). This will ultimately lead to more poverty.

20.4 Adaptive Strategies

Poverty stricken conditions of the villagers do not allow them to adapt to newer better and a promising way of living. They are completely dependent on external help for their survival, which specifically refers to government interventions. The following are the few measures that can be adopted:

- Attempts should be made to generate market for saline tolerant fish species, so that the villagers whose ponds are affected by saline water or even for the farmers who had a shift in occupation to aquaculture, may have a stable income. Saline tolerant fish species like *Liza parsia, Liza tade, Mugil cephalus, Liza macrolepis, Chanas chanos* etc., along with shrimps and crabs can be cultured. Untimely breeding of freshwater fish species before the cyclone periods should be encouraged.
- Vocational training must be provided to the villagers so that they can find employment in other fields. For example, (1) horticulture; the villagers can find an alternative means of livelihood by gardening, preparing tree samplings, rearing mushrooms etc. (2) Breeding rabbits for their meat, fur etc. (3) Preparation of puffed rice by traditional methods or by using a machine can make stable source of income. (4) Honey collection, by preparing beehive boxes next to their houses. (5) There is a great demand for solar lamps, hence, the villagers with the help of the government can purchase all the necessary items, assemble them, and build solar lamps for the sale in the markets. (6) There is ample number of open spaces around the houses, which can be utilised for banana cultivation. (7) Cultivation of Lotus, could secure them a stable income. Lotus seeds, leaves and stem are processed for making several medicines, since they have varied skin and health benefits. Lotus culturing could eventually fetch them a good market. (8) Other job opportunities include food-labelling, preparation of jellies jam and pickles.
- Migrating to safer locations is already one of the coping mechanisms; however, they need government help in order to do so. But, this option comes with a lot of considerations, which are mainly assuring job opportunities in new place of residence. The government should ensure secured livelihood to the villagers, since most of them has been in agricultural or aqua-cultural occupation for way too long and do not possess the ability or the will to shift their occupations.

Therefore, the new place of residence must have proper road connectivity along with the availability of agricultural and aqua-cultural land.

- Safeguarding agriculture by encouraging the villagers to make use of saline tolerant variant of rice like pokkali, Hogla, Gayush, Ghunshi, Kaminibhogh, Morichsal and Talmugur. This can be very useful, since during floods their paddy fields will survive the saline waters. The saline tolerant variant of rice can be used in alternative crop farming technique along with shrimp cultivation. During low saline conditions, rice can be cultivated and during the high saline conditions, shrimp can be cultivated. The waterlogged marshy area no longer requires any additional farming techniques or manures; instead, the leftover waste from the Pokkali harvest is used as a good feed for the shrimps, making the land productive and fertile (Chandramohanan and Mohanan 2011)
- Enhancing eco-tourism. Sagar Island has a great potential towards tourism and if properly manifested, then this might generate a stable source of income for many villagers. The beautiful mangroves along the shoreline of Bankimnagar and Chemaguri could be an attractive tourist spot. The tourists can lodge in Gangasagar, which is the most apt location for lodging and the area near the mangroves can come up with eco huts and a few restaurants along the way. Introducing recreational boating along tidal creeks within the forest belt can also enhance tourism.
- The most important adaptive measure is to ensure proper embankment. For the management of floods, block pitched embankments, mud embankments packed with impervious materials must be used or modern structures with the use of geotextile, riprap boulders and concrete blocks must be erected. For erosion control, long-term measures that must be undertaken are vertical walls, sea walls with a side, which has a gently gradient shoreface and spur cross. For short-term mitigation of erosion, structures like bamboo fencing with clay bags or sand bags, wooden fencing with sand bags as well as bamboo porcupines can be employed (Paul et al. 2016, p. 571).
- Eco-friendly techniques include planting mangroves wherever possible to dissipate the tidal energy. Mangroves must be planted along with the embankments, in order to provide extra padding and safety against storm surges. Planting creepers like thyme, poppies and other kinds of succulent can also be helpful, since their long roots go deep into the soil, binds it firmly, and prevents it from eroding.
- Floods and flash floods salinize the fields. Thus, making using of vermi-compost, heap-compost, gypsum, lime etc., can be used to get rid of the salinity.
- Posts of houses should be made deeper to prevent complete demolition during calamities.
- Keeping all cattle in safe places soon after receiving cyclone warning must be encouraged to prevent loss.

20.5 Conclusion

Sagar Island is one of the most vulnerable locations in the Indian Coastal areas. It is in urgent need for protection against the severe consequences of climate change. Otherwise, given the current rate of sea level rise and the amount of erosion that is taking place, it can be concluded that in the next few decades the island might disappear. The most critical environmental issues faced by the island are: (1) Degraded mangrove cover. (2) Lack of proper embankments/destruction of seawalls. (3) Rising Sea Level. (4) Loss of agricultural land and fresh water fisheries due to saline water intrusion. (5) Ravaging tropical cyclones and (6) Breach of embankments. These issues could eventually give rise to extinction of species and not only this but would also result in bulk climate refugees and would create undue pressure on land resources of the surrounding mainland.

Despite, their critical conditions and heart-wrenching pleads to the government, proper integration of management protocols is still yet to be taken. Hence, the people of the island need immediate government response in terms of monetary aid and job opportunities. This study revealed that out of the eight mouzas, erosion is quite pronounced in Dhablat, Shibpur and Mahismari. Thus, quick action should be taken to relocate the villagers who are residing in that area and proper infrastructural development in these areas is the need of the hour to prevent these villages to get completely wiped off. Advanced geospatial technique must be employed to chart the most vulnerable locations in terms of erosion. Environmental regulations must be carried out with strict administration in order to safeguard the future of the island and its residents. Sagar Island is not only regulated by natural geomorphological processes, but it also undergoes immense anthropogenic activities. In the past three decades, the island became a haven for the climate refugees fleeing from the nearby islands seeking shelter here. However, due to the impacts of climate change the island which was then a haven, is now under the risk of complete inundation like Lohachara Island.

References

Bandyopadhyay S (1997) Natural environmental hazards and their management: a case study of Sagar Island, India. Singap J Trop Geogr 18(1):20–45

Bera A, Singh SK (2021) Comparative assessment of livelihood vulnerability of climate induced migrants: a micro level study on Sagar Island, India. Sustain Agric Food Environ Res 9:2

Bera A, Taloor AK, Meraj G, Kanga S, Singh SK, Đurin B, Anand S (2021) Climate vulnerability and economic determinants: linkages and risk reduction in Sagar Island, India; a geospatial approach. Quat Sci Adv 4:100038

Brown S, Nicholls RJ (2015) Subsidence and human influences in mega deltas: the case of the Ganges–Brahmaputra–Meghna. Sci Total Environ 527:362–374

Chandramohanan KT, Mohanan KV (2011, December) Rice cultivation in the saline wetlands of Kerala—An overview. In *Proceedings of the IInd National Seminar on Genetics, Breeding and Biotechnology (Gregor Mendel Foundation Proceedings 2011), Kerala, India* (pp. 16–17)

Congalton RG (1991) A review of assessing the accuracy of classifications of remotely sensed data. Remote Sens Environ 37(1):35–46

Gopinath G, Seralathan P (2005) Rapid erosion of the coast of Sagar Island, West Bengal-India. Environ Geol 48(8):1058–1067

Hazra S, Ghosh T, DasGupta R, Sen G (2002) Sea level and associated changes in the Sundarbans. Sci Cult 68(9/12):309–321

Jayappa KS, Mitra D, Mishra AK (2006) Coastal geomorphological and land-use and land-cover study of Sagar Island, bay of Bengal (India) using remotely sensed data. Int J Remote Sens 27(17):3671–3682

Mondal I, Thakur S, Ghosh P, De TK, Bandyopadhyay J (2019) Land use/land cover modeling of Sagar Island, India using remote sensing and GIS techniques. In *Emerging technologies in data mining and information security* (pp. 771–785). Springer, Singapore

Mukhopadhyay A, Wheeler D, Dasgupta S, Dey A, Sobhan I (2018) Aquatic salinization and mangrove species in a changing climate: impact in the Indian Sundarbans. World Bank Policy Research Working Paper, (8532)

Paul AK, Subrata J, Amrit K, Abhinanda B, Farhin S, Suman S, Dipankar M (2016) Problems of some erosion affected areas of Sagar Island and proposal for mangrove ecotourism development sites in a sustainable manner. Eastern Geographer, XXII 1:569–582

Roy C, Guha I (2013) Climate change induced migration: A case study from Indian Sundarbanas. ACUMEN-Marian Journal of Social Work 5(1):72-93

Roy C, Guha I (2017) Economics of climate change in the Indian Sundarbans. Glob Bus Rev 18(2): 493–508

Rwanga SS, Ndambuki JM (2017) Accuracy assessment of land use/land cover classification using remote sensing and GIS. Int J Geosci 8(04):611

Singh, A.(1989) Review article digital change detection techniques using remotely-sensed data, International journal of remote sensing 10(6):989-1003

Arna Ghosh holds a masters degree climate science and policy from TERI School of Advanced Science and bachelors in geography. Presently she is working as Senior Analyst at PricewaterhouseCoopers (PwC) Management Consulting ESG Strategy and Transformation Advisory

Sweta Baidya has done her PhD in Oceanography from CSIR-National Institute of Oceanography. She has worked on the Paleoclimate and monsoon shift from Last Glacial Maximum to Holocene. Later she was working on Glaciology in Jawaharlal Nehru University and Ministry of Earth Sciences. She was also teaching Disaster Management in Jamia Millia Islamia University. Later she worked in National Institute of Disaster Management and is currently associated with NDMA (Govt. of India) as a Sr Consultant.

Anil K Gupta Professor of Policy & Strategies, Integrated Resilience. Head of the ECDRM Division, International Cooperation, Programme Director – Centre for Excellence on Climate Resilience, Nodal Officer for Central Ministries at National Institute of Disaster Management, (Govt. of India), New Delhi. He is also a member of the expert Team for WMO Climate Statement and Core Group Member of IUCN-Council of Ecosystem Management.

Part III
Tools and Strategies

Chapter 21
Mainstreaming Disaster Risk Reduction in EIA/SEA for Climate and Disaster Resilient Development

Anil K. Gupta, Sreeja S. Nair, and Shalini Dhyani

Abstract Applying environmental management and ecosystem based approaches in Disaster Risk Reduction (DRR) signals an important paradigm shift in disaster management and risk reduction, as it directly links with people's livelihoods and sustainability of their natural resource base. At the same time, it also integrates adaptations to Climate Change across sectors of development governance and people's actions. This calls for emphasis on better Natural Resource Management (NRM) and sustainable land-use practices within the strategies of disaster prevention, preparedness, post-disaster relief and recovery. Environmental Impact Assessment (EIA) is an environmental regulatory instrument that has a potential role for prudent decision-making and development planning in different phases of Disaster Risk Management (DRM). EIA, of policies and plans, or across regional scales, is known as Strategic Environmental Assessment (SEA) which could facilitate DRR integration into development policy decisions and also spatial planning. This chapter analyzes how DRR may be facilitated through EIA and SEA processes and by integrating EIA with the different stages of Disaster Management (DM). Case examples are presented to discuss the potential role that EIAs can play in facilitating integrated approaches in environmental and Disaster Risk Management. A framework for integrating DRR, environmental management and developmental planning at a local level is suggested.

A. K. Gupta (✉)
ECDRM Division, International Cooperation, CAP-RES DST Project, National Institute of Disaster Management (Ministry of Home Affairs), New Delhi, India

S. S. Nair
National Disaster Management Authority, New Delhi, India

S. Dhyani
Critical Zone Group, Water Technology & Management Division, CSIR-National Environmental Engineering Research Institute, Nagpur, India

IUCN CEM South Asia, Gland, Switzerland

A. K. Gupta et al. (eds.), *Disaster Risk and Management Under Climate Change*, Disaster Resilience and Green Growth,
https://doi.org/10.1007/978-981-99-4105-6_21

Keywords Disaster resilience · Environmental impact assessment · Natural hazards · Risk analysis · Vulnerability assessment · Disaster impact assessment

21.1 Introduction

21.1.1 Paradigm Shift to EcoDRR

The first paradigm shift in DM that took place over the last two decades moved 'response and relief-centric approaches' towards 'mitigation and preparedness-centric approaches', but this resulted in the key emphasis on hard engineering solutions, early warning and response preparedness. Most of these are short-term solutions and are not able to address long term DRR concerns. However, the growing focus on vulnerabilities and exposure has shifted attention towards environment-based approaches. This new emphasis is widely advocated as a second paradigm shift in DRM (Fig. 15.2; Gupta and Nair 2009) which also links with people's livelihoods and sustainability of their natural resource base. Sustainable NRM is recognized as a key element in vulnerability reduction (UNISDR 2011; Inter-American Development Bank 1999) (Figs. 21.1 and 21.2).

Major environmental changes generating or exacerbating hazards and increasing vulnerabilities to disasters are: Climate Change, land-use changes and natural resource degradation (Gupta and Nair 2011). Recognition of the potential role of environmental management in DRR begins with the recognition of 'disasters as human processes interacting with the physical environment' (Box 21.1). Environmental Assessments, therefore, can be utilized to provide scientific and strategic insights on the risks and vulnerabilities that are associated with implementation of policies, programmes and projects and move towards a 'culture of safety and prevention'.

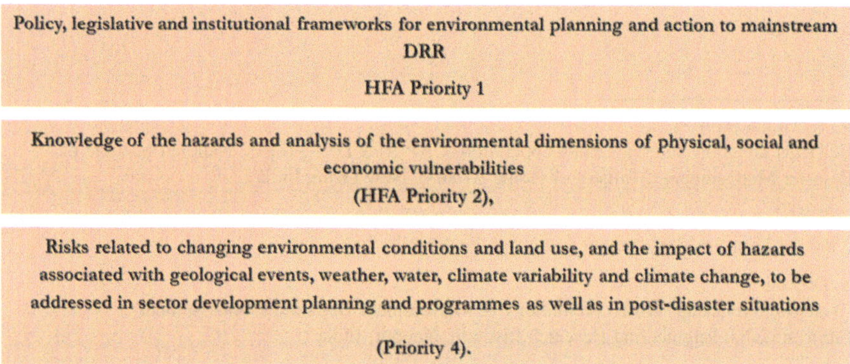

Fig. 21.1 Current HFA priorities (UNISDR 2007) offer environmental opportunities for DRR through the following provisions

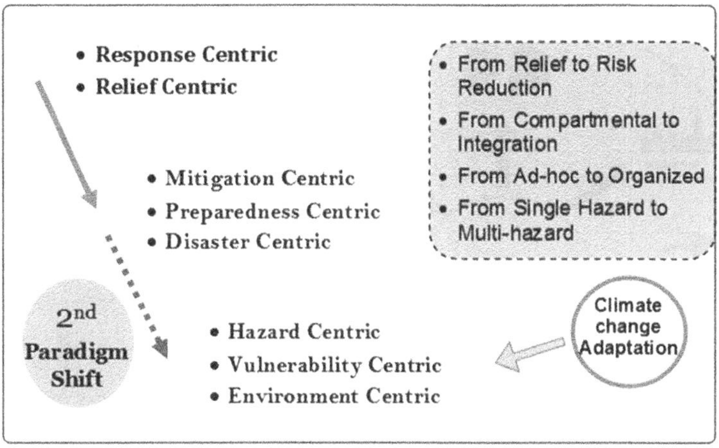

Fig. 21.2 Paradigm shifts in disaster management (Gupta and Nair 2009, Risk to resilience: strategic tools for disaster risk management)

Box 21.1 Classification of Disasters Based on the Origin of Hazards

Natural Hazards

- Geophysical (Earthquake, tsunami, landslides), Hydrological (floods) metrological (Cyclones), climate related (drought), wild fire, biological (Epidemics and Pandemics).

Technological Hazards

- Industrial, chemical, Electrical, mechanical, Nuclear, Radiological, dam break, mining, structural collapse etc.

Civil Disasters and Conflicts

- Civil unrest, strike, war, sabotage, mass poisoning, Bomb blast, Stampede, transport accidents etc.

**Disasters induced by natural hazards could trigger technological hazards and conflicts. On the other hand, a technological or civil disaster can exacerbate the impacts of natural hazards due to increase in vulnerability.

21.1.2 Incorporating Climate and Disaster Resilience into Environmental Appraisal Process

It is a well-recognized fact, that it is not only nature that generates disaster risk but developmental processes that aggravate hazards and shape human vulnerability, which in turn paves the way for disasters. Human activities can alter the magnitude

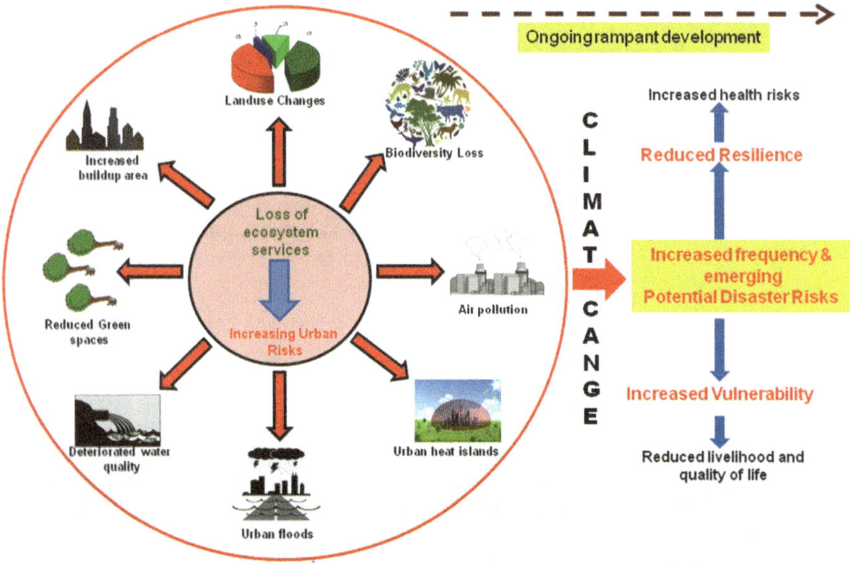

Fig. 21.3 Conventional regional EIA that mostly undermines and ignores the larger and long term disaster risks due to impact of developmental activities. A case of urban expansion (Dhyani et al. 2018)

and frequency of natural hazards, for example flooding, landslide, and desertification, by changing natural hydrological processes and land-use patterns.

Environmental degradation is contributing to increasing disaster losses. For instance, in many countries including India deforestation has disrupted watersheds and resulted in siltation of riverbeds, leading to more severe droughts and floods (Benson and Twigg 2007). The influence of past development on present disaster risk therefore underlines the significance of contemporary decision-making that might impact on disasters experienced by future generations. Environmental Impact Assessments (EIAs) offer to bring together both DRR and development concerns within an environmental management framework. The objective of EIA is to foresee and address potential environmental problems or concerns at an early stage of project planning and, predict the various probable impacts and accordingly assist in designing EMP (Environmental Management Plans) to reduce disaster risks and after effects due to the upcoming developmental project. EIA is a potential tool in identifying key potential environmental impacts and formulating mitigation measures of a project and help planners and government authorities in the decision-making process (Fig. 21.3).

EIAs produce targeted environmental analyses by reporting on current and anticipated future environmental conditions and identifying potential drivers of change. The types and forms of traditional and innovative EIA options are mentioned below:

- EIA of projects (e.g. development projects like water resources, highway, airport, tourism, housing complex, railway, or an industrial project in manufacturing, mining, food, dairy, etc.)
- Strategic Environmental Assessments
- Regional EIA (also known as Country EIA or Cumulative Impact Assessment)
- Carrying Capacity Assessment-based developmental planning process also comes under Cumulative Impact Assessments (Gupta and Yunus 2004).
- Environmental Risk Mapping-Based Developmental Planning (Gupta et al. 2002c)
- Health Impact Assessment (as part of EIA or Risk Analysis) (Gupta et al. 1999).

Early EIAs focused primarily on project impacts on the natural or biophysical environment (such as effects on air and water quality, soil, geology, flora and fauna, noise levels, climate and hydrological systems). Over time, increased consideration has been given to social, occupational health and economic aspects of environmental consequences.

21.2 Rationale and Need

More recently, greater attention has been placed on incorporating hazards and disaster risk analysis into the environmental assessment process itself. ProVention Consortium and the Caribbean Development Bank published a "Guidance Note on Mainstreaming DRR in Environmental Assessments" (Benson and Twigg 2007). This provides guidance in analyzing disaster risk-related consequences of developmental activities as a result of their environmental impacts. Environmental assessments provide opportunities to collate data on natural hazards (*i.e.*, hazard types, magnitudes and probabilities of occurrence) as part of the project appraisal stage, which inform project design and identify risk-sensitive environmental mitigation measures.

Three vital approaches are desired as an important part of the EIA process to make sure that natural pressures and hazard-related factors are properly addressed, evaluated and managed in EMPs of these EIA:

(i) The environmental assessment process is required to include compilation of data on natural hazards and associated risks as a pivotal initial step in broader project scoping and the results are used to determine if disaster risk needs to be addressed and examined in further detail of other aspects of the project appraisal procedure.
(ii) Systematic review and assessment of the potential disaster risks-associated consequences of the project and its impact on the larger surrounding environment in the hazard-prone sensitive areas.
(iii) Environmental issues should be carefully considered and addressed in the design and implementation of post-disaster relief and rehabilitation processes.

The Ministry of Environment, Forests and Climate Change, Government of India has prepared 37 EIA Manuals on major sectors of developmental projects which are listed in the Schedule to the Environmental Impact Assessment (EIA) Notification 2006. These Manuals have been prepared to serve as Technical Guidance Manuals (TGMs) to various stakeholders involved in the environmental clearance (EC) process. The Manual for each sector includes a Model TOR, technological options, processes for cleaner production, waste minimization, monitoring of environmental quality, and related regulations and procedure of obtaining EC.

Numerous examples are there to show how DRR can be mainstreamed in EIAs. For example, in India, environmental concerns of major developmental projects were considered in the fourth Five Year Plan (1969–1978) established in 1980, which marked the beginning of EIA policy and practice in India. Even then, concerns related to catastrophic disaster risks (floods, earthquakes and landslides) and displacement as well as mitigation requirements were already flagged for certain mega-dams and river valley projects (Tehri Dam, Uttarakhand, Teesta Valley project, Sikkim *etc.*). The guidelines required various studies, such as on forests and wildlife impacts in the submergence zone, water logging potential, upstream and downstream impacts on aquatic ecosystems and fisheries, deforestation, e-flow, water-borne diseases, climatic changes, reservoirs carbon sources and seismicity. 'Site clearance' from the State Government, Forest department, National Wildlife Board in case the area is close to a protected or reserve forests including a 'no objection' certificate from the district administration, is the first step towards 'environmental clearance' procedure (Environmental Conservation Team 2005) and is expected to take into account the disaster risks associated with the site and the region.

Data incorporated into the EIA is similar to data requirements for risk and vulnerability assessments related to natural hazards. In the Philippines, the Environment Management Bureau, through a special memorandum dated 11th November 2011, produced "EIA Technical Guidelines incorporating Disaster Risk Reduction (DRR) and Climate Change Adaptation (CCA) concerns"(Environment Management Bureau, 2011) (Fig. 21.4).

As illustrated disaster risk analyses can be incorporated into the EIA process by utilizing the data generated through the EIA process itself. A procedural framework of EIA is given in Figure above that incorporates risk analysis and a DMP as EIA outputs. Environmental clearance of major developmental and industrial projects in India through EIA notification (1994, amendment 2006 and further) under the Environmental Protection Act of 1986, specifically requires that the Environment Management Plan include a Disaster Management Plan and, when necessary, a Rehabilitation Plan. Environmental Impact Assessment Act 2001 of the Federal Republic of Germany envisages for the planning procedure to include environmental assessment as pursuant to the provisions of the applicable Building Codes, which in-turn includes provisions related to disaster resistant housing and infrastructure (Table 21.1).

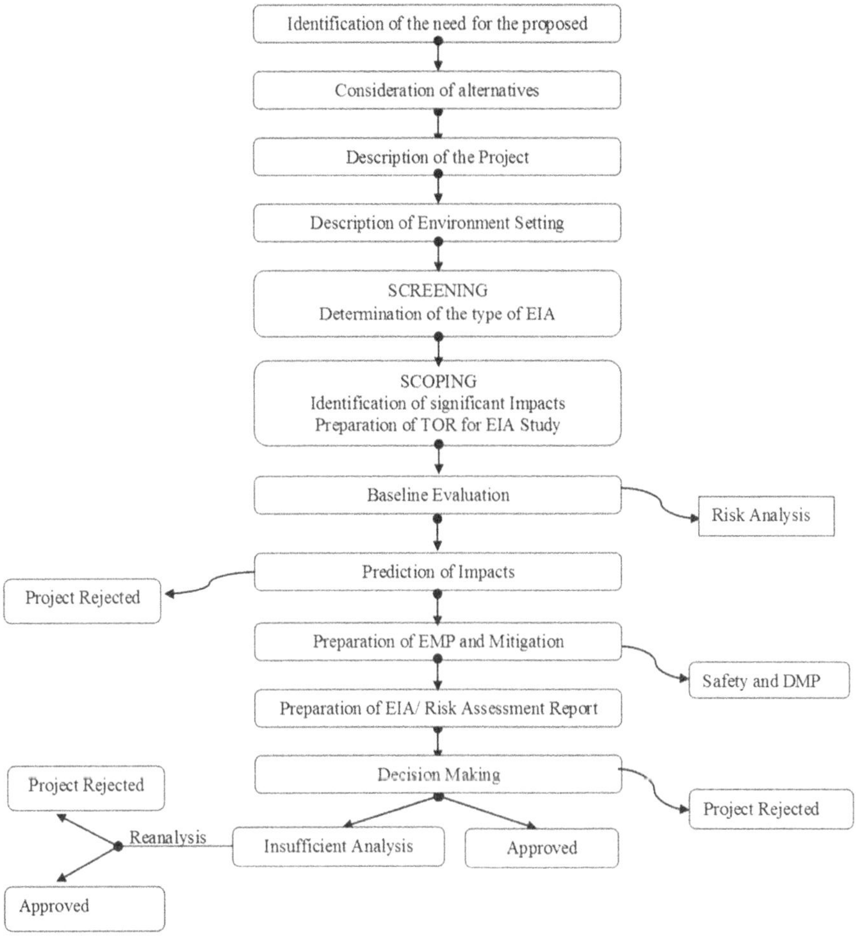

Fig. 21.4 Basic components of EIA study in India (Source: Preliminary Draft of Guidelines on EIA of Water Resources Projects, Bureau of Indian Standards, 2011)

21.3 Incorporating DRR in Strategic Environmental Assessment (SEA)

SEA is the 'formalized, systematic and comprehensive process of identifying and evaluating the environmental consequences of proposed policies, plans or programmes to ensure that they are fully included and appropriately addressed at the earliest possible stage of decision-making on a par with economic and social considerations' (Dalal-Clayton and Sadler 1999, Page 2). SEA is to make sure that broad environmental aspect and considerations are being involved into these higher, strategic levels of decision-making process prior to the identification and design of individual or different projects, ideally the process is based on a participatory

Table 21.1 An overview of various important EIA projects across India to understand disaster risks associated with developmental projects, Source: EIA projects where third author was directly involved as EB (Ecology and Biodiversity) FAE (Functional Area Expert)

Sr. no.	Project	State	Category (A/B)	Risks associated is EMP not implemented appropriately	Associated long term natural disaster risks addressed in EIA reports
1	Proposed Grass-root refinery with Petrochemical Complex, Barmer, Rajasthan	Rajasthan	A	Wetland, Migratory birds	Water Intensive Industry in Desert area of Rajasthan
2	Proposed Solvent Spun Cellulosic fibre Gujarat	Gujarat	B	Effluent discharge	Water pollution during operational phase
3	Comprehensive EIA for proposed LNG Terminal at in Kodinar Gujarat	Gujarat	A	Effect on migratory Birds, Turtle breeding sites, lions	Air and biodiversity loss during construction phase
4	Cumulative Impact of Coal Mining and Thermal Power Project at Singrauli, Madhya Pradesh	Madhya Pradesh	A	Deforestation, Fragmentation, Habitat loss, Fly ash disposal	Soil, Air and Water issues during construction and operational phase
5	EIA studies for the proposed TPP at Kodinar, Gujarat's	Gujarat	A	Effect on migratory Birds, Turtle breeding sites, lions	Air and biodiversity loss during construction phase
6	Carrying Capacity studies for Mining in three districts of Odisha State	Odisha	A	Deforestation, Fragmentation, Habitat loss, Soil, Air, Noise and Water Pollution	Air, water, soil, noise and vibration along with waste disposal issues leading to landslides and floods
7	High Speed Rail Corridor project from Trivandrum to Calicut, Kerala	Kerala	A	Fragmentation, Waste disposal and recycling, Social Environment	Habitat loss, and impacts during construction phase while noise and vibration during operational phase
8	Terrestrial EIA Study for expansion of Thermal Power Project at Udupi, Karnataka	Karnataka	A	Fly ash and air pollution	Area is in vicinity of Western Ghats
9	Environmental Assessment Studies at Zn Mines, Bhilwara, Rajasthan	Rajasthan	–	Overburden dumps, Soil	Revegetation of large area under overburden, tailing waste, leaching

(continued)

Table 21.1 (continued)

Sr. no.	Project	State	Category (A/B)	Risks associated is EMP not implemented appropriately	Associated long term natural disaster risks addressed in EIA reports
10	Assessment of Adequacy of Existing Greenbelt for Thermal Power Plant, Korba, Chhattisgarh	Chattisgarh	B	Fly ash	Recycling and reusing fly ash is an issue for large TPPs
11	EIA study for BS-VI auto fuel quality compliance at a Refinery, Mangalore	Karnataka	A	Air, water, noise and impact on mangroves	Foul smell that leads to poor quality of life for locals and
12	Restoration of the environment and ecology, particularly protection of water bodies due to legal or illegal rat hole coal mining in Meghalaya	Meghalaya	A	Severe impact on water, habitat, fragmentation of forests, biodiversity and downstream communities	Landscape changes from natural forests to grasslands, loss of biodiversity long term damage to ecosystem
13	Illegal encroachment in Udaisagar Lake island, Udaipur	Rajasthan	–	Biodiversity and Habitat loss	Water related issues affecting larger cascade wetland system
14	Assessment of bearing capacity and sustainability around five lakes in Nainital District of Uttarakhand	Uttarakhand	–	Impact on watershed, forests, spring sheds	Water scarcity, landslide and earthquakes
15	EIA of broadening of roads, Srinagar	Jammu and Kashmir	A	Terrestrial ecology, biodiversity, watershed and air	Water scarcity, sub surface flow
16	Proposed Nuclear Power Plant, Madhya Pradesh	Madhya Pradesh	A	Radionuclides	Radionuclides

approach. SEA is also implemented in few of the aspects by many multilateral or bilateral environmental international organisations as well as by a number of governments across the world. At the country programming level, it is also referred as CEA. SEA was developed to address the increasing focus on mainstreaming environmental issues into development policies and planning and DRR is an important aspect of it. SEAs target long term impacts of policies on natural resources and

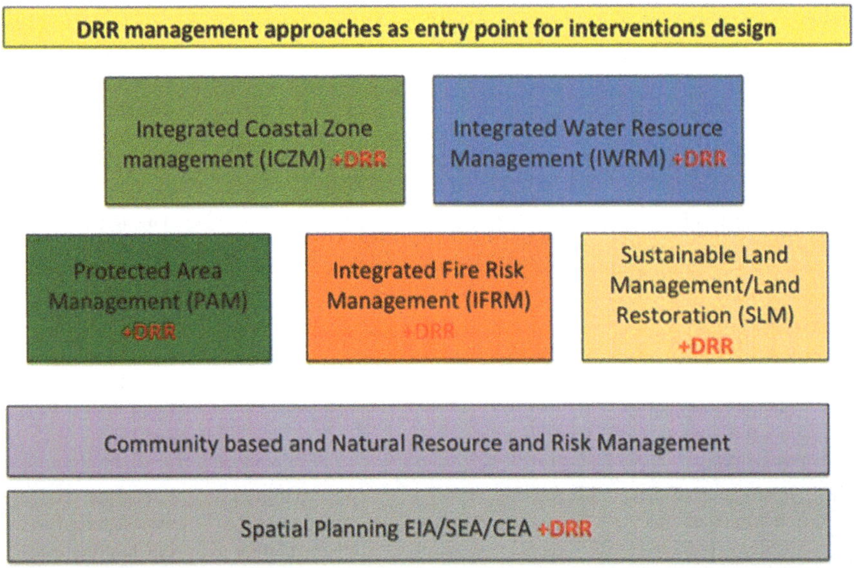

Fig. 21.5 DRR management approaches as entry point for many interventions and policy design

environment and at large how it can affect the society and locals of the country (Fig. 21.5).

SEAs are like EIAs because they also address assessment and mitigation of potential environmental impacts. The main difference is that SEAs may be applied to a national policy for an entire sector (e.g. energy policy), or to programmes covering a geographical area (e.g. a national, state or regional development scheme). SEA aims at environmental sustainability over the long-term and assesses the impacts of developmental activities across wider-spatial extents. It is an important tool for mainstreaming DRR in development policies plans and programmes at different national and sectoral levels (OECD 2008a).

SEA provides systematic analysis of key environmental issues that includes emerging disasters and natural disaster risks that may affect sustainable development of the country and in achievement of the Sustainable Development Goals 2015-30 (SDGs), promises of Paris Agreement 2015-30 and also includes key aspects of Sendai framework of DRR 2015-30. SEA also provides opportunities for overcoming bottlenecks of disaster risks and the environmental implications for key development policies. Figure 17.2 illustrates a conceptual framework for an SEA process, using the example of an SEA in the context of agriculture policy. SEA helps analyze a policy, plan or a decision for its direct and indirect impacts on environmental components, given that decisions made on the ground are often influenced by policies. Impacts and consequences of a policy decision, such as an agriculture policy, on environment and natural resource settings may also alter hazard patterns and modify local vulnerability to the impact of disasters when they occur. SEAs can also help analyze changes in socio-economic conditions that create and influence vulnerability at local levels. Simultaneously, SEAs can assess the potential qualitative and quantitative changes in natural resource systems and the subsequent potential impacts on the level of hazards and social, economic and ecological resilience to absorb the shocks.

The OECD published Advisory Notes (2010) on 'Strategic Environmental Assessment for Disaster Risk Reduction' and identified the following considerations for undertaking SEAs:

(i) Situations that are sensitive and require awareness (*e.g.* post-conflict environments);
(ii) Providing guidance on emerging issues that may need to be more adequately integrated into an SEA (*e.g.* climate risk or status of ecosystem services); and/or
(iii) A key emerging issue or policy area that was not sufficiently addressed within previous SEA guidance.

21.3.1 Concept & Framework

Country Environmental Analysis (CEA) is a variation of SEA and a relatively new analytical tool applied by international agencies as an option for strategic environmental assessment in a regional or country context (Fig. 21.6).

All CEAs are expected to include collation of basic hazard data and background information on past disaster losses to give a preliminary overview of the significance of disaster risk in a country and to provide baseline information that inform future

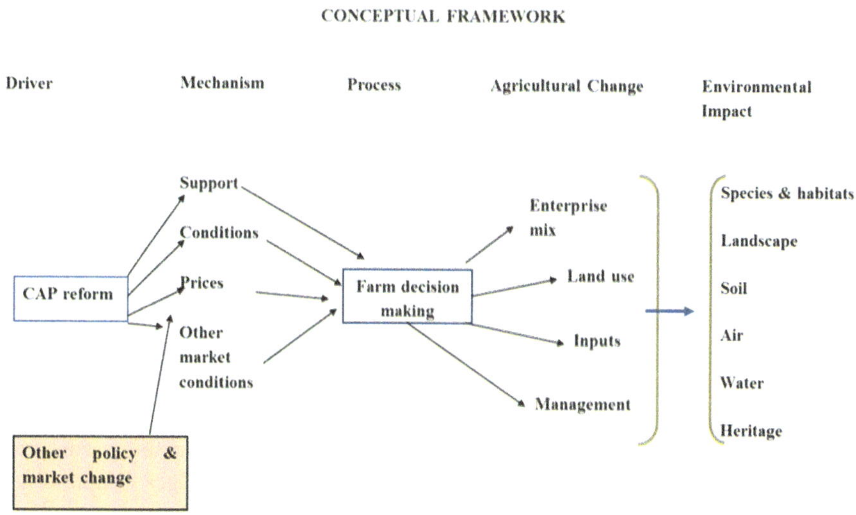

Fig. 21.6 Sample SEA framework for Agriculture Policy (after EEAC, 2008). CAP: Common Agricultural Policy

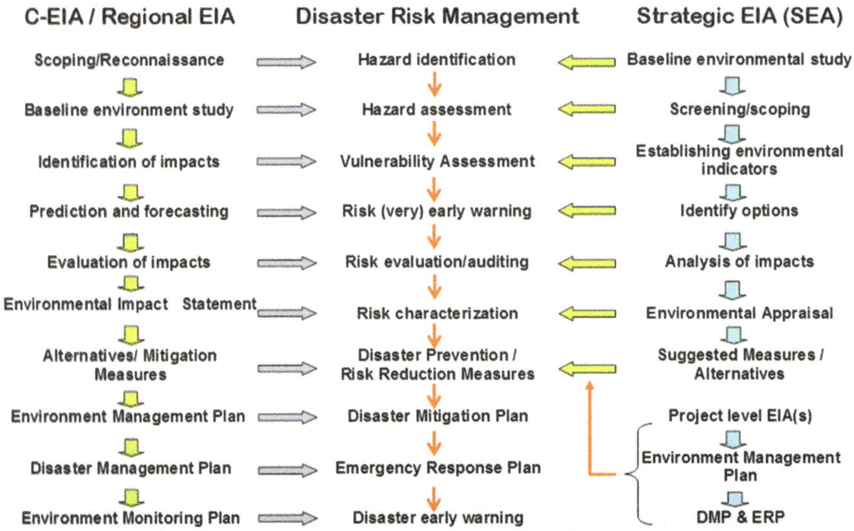

Fig. 21.7 Inputs of EIA and SEA to DRR, (C-EIA—Cumulative EIA; DMP—Disaster Management Plan; ERP—Emergency Response Plan)

environmental assessments of individual projects and country programming (Fig. 21.7).

21.4 Applying EIAs in Dealing with Climate Related Disasters

EIA as a decision-support system and information tool can help in planning throughout all stages of disaster risk management (Figure above). EIA methodology utilizes a range of tools such as matrices, weighted ranking, and computer aided modeling, which helps in determining the relationship and comparing between different actions, environmental changes and their primary and secondary impacts. EIA can be applied to analyze the conditions of hazards, patterns of vulnerability with respect to developmental planning process. EIAs applied in the disaster prevention and mitigation phase can help inform planning for DRR, for instance by providing guidance on choices of mitigation methods, technology investments and site locations for activities. While EIAs have been applied to anticipateBureau of Indian Standards impacts of development in hazard prone areas, in a post-disaster context, conducting rapid EIAs (REA) helps to ensure that sustainability concerns are factored into relief, reconstruction and recovery planning stages (Gupta et al. 2002). The REA does not replace an EIA, but fills a gap in an emergency context until an EIA could be appropriately conducted (Fig. 21.8).

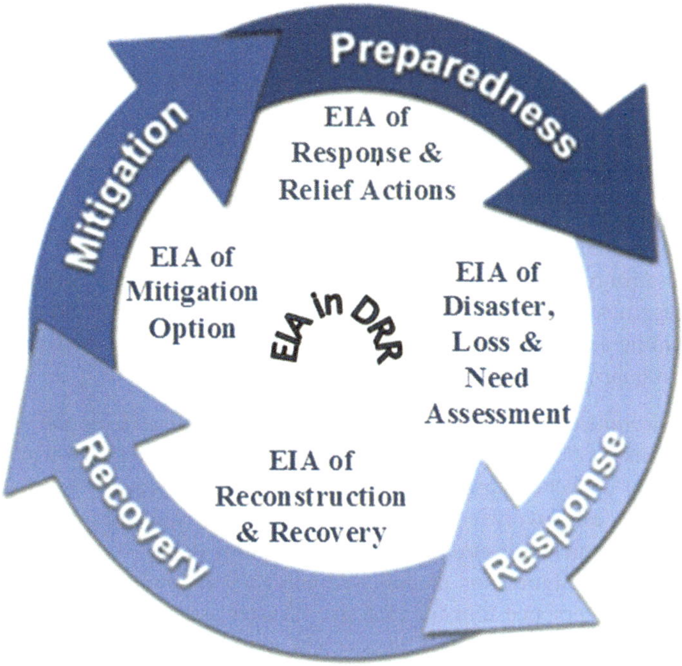

Fig. 21.8 EIA applications in DRR phases (Gupta and Nair 2012)

21.5 EIAs in Post-Disaster Relief and Recovery

Disasters cause significant effects on the ecology, infrastructure, people, livelihoods, and properties. People and their assets are affected either directly in the form of casualties, injuries or damages, or due to indirect disaster impacts on ecosystem productivity, environmental services, and the natural resource base. Disaster impacts may be categorized into the following:

(i) Physical (effects on infrastructure, buildings, physical property, industry, roads, bridges, monuments, etc.)
(ii) Environmental (effects on water, land/soil, land-use, landscape, crops, lake/ rivers /estuaries, aquaculture, forests, animals/livestock, wildlife, atmosphere/ climate, energy, etc.)
(iii) Social (effects on life, health, livelihoods, employment, relations, security, peace, etc.)
(iv) Economic (effects on assets, deposits, reserves, income, commerce, production, guarantee/insurance, etc.)
(v) Ecological (effects on ecosystem integrity and ecosystem-health, structure and functions, productivity, succession, carrying capacity, etc.)

However, environmental impacts are rarely given consideration in damage (and loss) assessments conducted following a disaster, although certain considerations are given to environmental components with direct economic values, *e.g.*, agricultural production. In the aftermath of the Indian Ocean tsunami, Blaikie et al. (2005) suggested that effective recovery and reduction of future vulnerability for local people depended on:

(i) Recognizing that ecosystem services provide the basis for sustainable reconstruction and reduction of future vulnerability;
(ii) Long-term monitoring of both ecological and socioeconomic parameters and a management strategy that encourages adaptation to changing circumstances;
(iii) Providing a clear articulation of the rationale for including biodiversity conservation concerns in reconstruction planning.

Rapid Environmental Impact Assessment (REA) as well as Comprehensive EIA applied in disaster contexts is a tool to identify, define, and prioritize potential environmental impacts in disaster situations. A simple, consensus-based qualitative assessment process, involves narratives, and rating tables that are used to identify and rank environmental issues and follow-up actions during a disaster. The REA (Kelly 2005) is designed for natural, technological or political disasters, and is viewed as a best practice tool for effective disaster assessment and management. REAs can be used shortly before a disaster and up to 120 days after a disaster, or for any major change in an extended crisis. The REA does not provide answers on how to resolve environmental problems. It does provide sufficient information to allow those responding to a disaster to address key issues raised in the REA.

In recent years, there have been innovative applications of EIA in the context of recovery and reconstruction. The Benfield Hazard Research Centre and CARE International have developed more detailed and comprehensive guidelines on rapid environmental assessment (REA) in the context of disaster response. These guidelines focus on the following:

(i) Assessment of a disaster;
(ii) Immediate impacts on the environment;
(iii) Unmet basic needs of disaster survivors (e.g. for fuelwood, building materials) that could lead to adverse impacts on the environment; and
(iv) Potential negative environmental consequences of relief and recovery operations.

The methodology is based on qualitative assessments, drawing heavily on people's perceptions and often based on incomplete data, but provides sufficient information under difficult circumstances to facilitate rapid assessment of needs and priorities.

Post-disaster environmental assessments need to explore whether proposed relief, reconstruction and rehabilitation efforts will have acceptable environmental impacts (e.g. environmentally sound selection of sites for refugee camps and sourcing of reconstruction materials) and whether they will strengthen resilience as well as reduce vulnerabilities to future natural hazards.

UNHCR developed Environmental Guidelines (2005) to incorporate a framework for identifying and addressing environmental issues associated with refugees, returnees and disaster relief activities. Guidelines focus on natural resource deterioration, ecosystem services impairment and their consequences on health and socio-economic well-being of the people. Several donor organizations have established their own guidelines, which include checklists on environmental assessment of disaster relief and humanitarian assistance operations, see for example: Asian Development Bank, 2003 (ADB 2003), Swedish International Development Cooperation Agency, 2002 (SIDA 2002), and the UK Department for International Development, DFID 2003 (DFID 2003).

The Joint UNEP/OCHA Environment Unit is the United Nations mechanism to mobilize and coordinate emergency assistance to countries affected by environmental emergencies and natural disasters with significant environmental impacts. The Joint Environment Unit has developed "Guidelines for Environmental Assessment Following Chemical Emergencies", with the purpose of deploying environmental experts to undertake rapid identification of environment-related problems following an industrial emergency.

21.5.1 Strategies & Gaps

Despite the improved understanding on the potential benefits of applying ecosystem management approaches for disaster risk reduction, the gap between the two sectors is wide especially at the level of policy planning and governance. There is seldom integration in a real sense, except in academic forum. EIAs are often viewed with suspicion, as they are known to simply "rubber stamp" environmental clearances required from the authorities, needed to gain approval of development projects.

Present EIA and SEA practice do not adequately reflect or incorporate disaster risk and disaster mitigation concerns, even though environmental legislation and policy frameworks may already support such integration (e.g. in the case of India). Moreover, despite increasing co-operation between DRR and climate change communities-of-practice, there remains a wide divide across the two communities between practitioners, their approaches, and also the vocabulary used. For example, terms such as vulnerability, mitigation, and risk are understood very differently within each community and terms such as "no regrets measures" and adaptation frequently used by Climate Change experts are less common in DRR parlance. However, this may be changing as there is increased global recognition for integrating DRR and CCA and the new Intergovernmental Panel on Climate Change (IPCC) Special Report on Extreme Events (2012) adopts closer terminology with the international DRR community.

21.6 Opportunity & Approaches

In order to overcome the challenges, several recommendations may be presented, drawing from the context in India:

(i) The presence of environmental policy and EIA/SEA experts is needed at the highest levels of decision-making authority and institutions on disaster management (e.g. the National Disaster Management Authority in India; Ministry of Environment, Forests and Climate Change, Government of India etc.). Conversely, representations of disaster risk reduction expertise are warranted at the highest institutional levels of the environment and natural resource management sectors (e.g. Ministry of Environment, Planning Commission's Environment Division, NITI Ayog, implementing agencies for NAPCC). This would ensure increased integration of environment and DRR concerns in respective policies, programmes and plans.

(ii) Integrating DRR and EMP and practice will require adapting and customizing both the disaster management and environmental management system and governance. DRR and environment agencies need to work together and develop common guidelines for integrating DRR within the EIA process as well as applying EIA in the context of disaster management (e.g. conducting rapid EIA in post-disaster response followed by a comprehensive EIA).

(iii) Mainstreaming Sendai Framework for Disaster Risk Reduction 2015–2030 with its seven clear targets and four priorities in EIA/SEA as well as cumulative impact assessment for action to prevent new and reduce existing disaster risks that include understanding disaster risk; strengthening disaster risk governance to manage disaster risk; investing in disaster reduction for resilience and; enhancing disaster preparedness for effective response, and to "*Build Back Better*" in recovery, rehabilitation and reconstruction.

(iv) There is a need to apply disaster risk-sensitive SEAs and EIAs in the context of local (regional) and national development planning. Requirement of regional EIAs (at district level) should be made a pre-requisite to term planning. For example, five-year planning cycles are common in India, and Regional EIA can facilitate an 'Environmental Action Plan' that is also disaster risk-sensitive at district as well as state level.

21.7 Conclusion

One recent development towards EIA application in disaster management is a Disaster Impact Assessment (DIA) component within the environmental clearance procedure of developmental projects. India's National Institute of Disaster Management (NIDM) and Pakistan's NDMA have recently proposed a collaboration to fill this methodological gap by jointly working on establishing an EIA system that takes into account the project's impact on natural disaster risk.

A number of issues need to be considered when interlinking and integrating disaster and environment aspects in post-disaster planning these include appropriate environment management and quick restoration in the wake of disasters; proper waste management that includes clearing, sorting, recycling/reuse, disposal of disaster debris; assessing water contamination and pollution and also ensuring appropriate handling of hazardous and toxic materials.

21.8 Way Forward

There is a need to update environment-related academic curriculum and training courses to recognize the added value of improved environmental management for disaster risk reduction. Leading Universities in India and the region have already incorporated 'disaster management' as a specialization of environmental sciences, but such efforts still need to be strengthened with case studies of successes and failures, use of interdisciplinary knowledge and modern tools like geo-informatics and space technology, along with modeling impacts of Climate Change and its spatio-temporal impacts at small as well as larger level along with promotion of research in this direction.

References

ADB (2003) Environmental assessment guidelines. Asian Developmental Bank

Benson C, Twigg J (2007) "Guidance note 7: environmental assessment" in tools for mainstreaming disaster risk reduction: guidance notes for development organisations. ProVention Consortium, Geneva, pp 79–89

Blaikie P, Mainka S, McNeely J (2005) The Indian Ocean tsunami: reducing risk and vulnerability to future natural disasters and loss of ecosystems services. International Union for Conservation of Nature Information Paper, February 2005. http://data.iucn.org/dbtw-wpd/edocs/Rep-2005-006.pdf

Bureau of Indian Standards (2011) Preliminary draft of guidelines on EIA of water resources projects

Dalal-Clayton B, Sadler B (1999) Strategic environmental assessment: a rapidly evolving approach. International Institute for Environment and Development. Environmental Planning Issues No. 18, 1999. London

DFID (2003) Environment guide. DFID

Dhyani S, Lahoti S, Khare S, Pujari P, Verma P (2018) Ecosystem based disaster risk reduction approaches (EbDRR) as a prerequisite for inclusive urban transformation of Nagpur City, India. Int J Disaster Risk Reduct. https://doi.org/10.1016/j.ijdrr.2018.01.018

EEAC 2008. SEA framework for Agriculture Policy CAP: Common Agricultural Policy

Environmental Management Bureau. 2011. EIA-DRR/CCA Technical Guidelines, Circular 005/2011, Department of Environment and Natural Resources, Republic of the Philippines. http://www.emb.gov.ph

Environmental Conservation Team (2005) Environment impact assessment process in India and the drawbacks. Bhubaneshwar – 751 007 (on website)

Gupta AK, Kumar A, Misra J, Yunus M (2002) Environmental impact assessment and disaster management: emerging disciplines of higher education and practice. In: Srivastava P, Singh DP (eds) Environmental education. Anmol Publishers, New Delhi, pp 7–23

Gupta AK, Misra J, Yunus M (1999) Environmental-health Assessment of Thermal Power Project within the Scope of EIA and Risk Analysis: Guideview. In: Proceedings of National Seminar on Energy and Environment, Lucknow, July 1999, P 86–95

Gupta AK, Nair SS (2009) Risk to resilience: strategic tools for disaster risk management. NIDM

Gupta AK, Nair SS (eds.) (2011) Environmental knowledge for disaster risk management—concept note. In: Abstract Book of the International Conference 9–10 May 2011, New Delhi. National Institute of Disaster Management, New Delhi and GIZ Germany. P 117

Gupta AK, Nair SS (eds) (2012) Ecosystem approach to disaster risk reduction. National Institute of Disaster Management, New Delhi, p P200

Gupta AK, Suresh IV, Misra J, Yunus M (2002c) Environmental risk mapping approach—risk minimizing tool in developing countries. J Clean Prod 10:271–281

Gupta AK, Yunus M (2004) India and WSSD (Rio+10) Johannesburg: issues of National concern and international strategies. Curr Sci 87(1):37–43

Inter-American Development Bank (1999) Working paper on reducing vulnerability to natural hazards: lessons learned from hurricane mitch a strategy paper on environmental management. Stockholm, Sweden

Kelly C (2005) Guidelines for rapid environmental impact assessment in disasters. Benfield Hazard Research Centre, University College London and CARE International

OECD (2010). Strategic Environmental Assessment for Disaster Risk Reduction. DAC Network on Environment and Development Co-operation (ENVIRONET) at their 8th Meeting on 30 October 2008.

SIDA (2002) Indicators for environmental monitoring in international development cooperation. Swedish International Development Cooperation Agency

Special Report on Managing the Risks of Extreme Events and Disasters to Advance Climate Change Adaptation (2012) Cambridge University Press

UNISDR (2007) Hyogo Framework for Action (HFA) 2005–2015: building the resilience of nations and communities to disasters. UNISDR. Retrieved from UNISDR: https://www.unisdr.org/we/coordinate/hfa

UNISDR (2011) The global assessment report on disaster risk reduction—revealing risk, redefining development. Information Press, Oxford

Anil K. Gupta Professor of Policy & Strategies, Integrated Resilience. Head of the ECDRM Division, International Cooperation, Programme Director - Centre for Excellence on Climate Resilience, Nodal Officer for Central Ministries at National Institute of Disaster Management, (Govt. of India), New Delhi. He is also a member of the expert Team for WMO Climate Statement and Core Group Member of IUCN-Council of Ecosystem Management.

Sreeja S. Nair is presently working as Fellow with the Earth Sciences and Climate Change Division of The Energy and Resources Institute (TERI). She has more than 20 years of experience in the area of Disaster Reduction and Climate Change adaptation and work with National, International and Regional Organisations including National Institute of Disaster Management, UNDP, UNDRR, UN-Habitat, ADPC, RIMES, Aga Khan Agency of Habitat and so on.

Shalini Dhyani is Senior Scientist with Critical Zone Group of Water Technology and Management Division of CSIR-NEERI, India. She is South Asia Chair for IUCN CEM (Commission on Ecosystems Management) (2017–2020). Dr. Shalini is NABET certified Ecology and Biodiversity Functional Area Expert (FAE) and has worked on impact assessment projects across the length and breadth of the country.

Chapter 22
Climate Adaptation and Disaster Risk Reduction Integration through Environmental Legislation in India

Sreeja S. Nair, Swati Singh, and Anil K. Gupta

Abstract Inter-linkages between disasters, environmental degradation, and climatic changes are well recognised as well as linking disaster risk reduction with environmental management. All climate change adaptation measures lead to disaster risk proofing reduction but not vice versa. Integration of DRR with CCA and sustainable development is now becoming prime concern while formulating disaster management strategies worldwide including in India. Assessment and monitoring of environment plays crucial role in generating relevant information that helps in identifying risks, vulnerabilities as well as opportunities to promote resilience of the community. Rules, regulations, and policies for DRR are fundamental to the enhancement of economic, social, and environmental security. It is the first step towards mainstreaming disaster risk reduction into development plans. Law and regulations are needed for implementation of provisioning, planning, operation, and monitoring of the functions suggested as a set of guiding principles, whereas a policy is set of guiding principles and roadmap of actions towards a defined aim or goal. Hence, the goal of sustainable development through disaster reduction needs to be operationalized through a set of policies and an effective legal framework. The laws on environmental protection are becoming more relevant in Natural Disaster Risk Reduction (DRR) in the wake of paradigm shift in disaster management. This paper is an attempt to analyse environmental legislations of India in the context of DRR and CCA and develop a broader strategy for inclusion into developmental plans and policies.

S. S. Nair (✉)
The Energy and Resources Institute, Earth Science and Climate Change Division, New Delhi, India

S. Singh
United Nations Development Programme, New Delhi, India

A. K. Gupta
ECDRM Division, International Cooperation, CAP-RES DST Project, National Institute of Disaster Management (Ministry of Home Affairs), New Delhi, India

A. K. Gupta et al. (eds.), *Disaster Risk and Management Under Climate Change*, Disaster Resilience and Green Growth,
https://doi.org/10.1007/978-981-99-4105-6_22

Keywords Environmental laws · Climate change adaptation · Disaster risk reduction · Sustainable development

22.1 Introduction

Climate Change, land-use changes and degradation of natural resources are major environmental changes that increase the vulnerabilities to disasters (Gupta and Nair 2011). The potential impacts of ecosystem degradation vary and may contribute to increased vulnerable conditions and intensity in occurrence of natural hazards and also reduction in the coping capacity. Some examples include: land degradation, deforestation, desertification, wild land fires, loss of biodiversity, land, water and air pollution, Climate Change, sea level rise, ozone depletion etc. All climate change adaptation measures lead to disaster risk proofing reduction but not vice versa. For example water resource and watershed management can lead to reduced vulnerability to drought and floods. Similarly, coastal zone management, flood zone management, protected area management; forest management, wetland protection etc. are very efficient ecosystem management tools which can help in creating natural buffers for hazard prevention and mitigation as well as for reducing the vulnerability and increasing adapting capacity under changing climatic conditions.

Integrating DRR with CCA and sustainable development is now becoming prime concern while formulating disaster management strategies worldwide including in India. As per UNEP and UNISDR, environmental monitoring and assessment plays an indispensable role in generating information that helps in identifying risks, vulnerabilities, and opportunities to promote community resilience (UNEP, UNISDR-PEDRR 2010). Monitoring and observing environmental factors that signal the onset of a hazard are fundamental to design and implementing of early warning systems. Environmental governance includes policies, legal and regulatory frameworks and institutional structures forms the part of environmental governance and offers unique opportunities for mainstreaming disaster risk reduction into environmental management.

22.2 Rationale and Need for Policies and Laws for Disaster Management

India is prone to multiple natural and technological disasters. High Power Committee on Disaster Management listed 33 disasters categorised under four categories. The classification is based on the origin. There are 11 hazards/ disasters of hydro-meteorological origin. Building Material Technology Promotion Council Released the third edition of the vulnerability Atlas of India for four hazards. One on other part of the country is affected by drought, water scarcity, flood, landslides, lightening, cold wave, heat wave, cyclones etc. throughout the year leading to loss of life, property and irreversible environmental damages. Recent examples are, Kerala Floods 2018, Uttarakhand Floods 2013, Maharashtra drought (2015–2019), Cyclones like Vayu, Titli, Gaza, Okhi, and more recently the extremely severe cyclonic storm Fani and Amphan and 2019 floods in Bihar, Assam, UP are the live example of the countries vulnerability to disasters. Although loss of human life due to disasters is reduced loss of property and human sufferings are still at peak. Serious non-compliance of laws related to land use zonation in coastal, flood plan and mountain areas, dam safety curves and road construction are identified as the key reasons for the disaster losses.

Legislation for protecting and improving environment and DRR are fundamental to the enhancement of social, economic, and environmental security and sustaining developmental gains. It is the first step towards mainstreaming and provides with the framework to build risk reduction strategies into development and reconstruction activities can be empowered. The law can be used to provide penalties and incentives by enforcing standards in construction, land use, tenant's rights and by defining people's rights during relief and reconstruction (Holloway and Pelling 2006).

Law and regulations are needed for implementation of provisioning, planning, operation, and monitoring of the functions suggested as a set of guiding principles, whereas a policy is set of guiding principles and roadmap of actions towards a defined aim or goal. Hence, the goal of sustainable development through disaster reduction needs to be operationalized through a set of policies and an effective legal framework. Legal measures can play important roles in following specific ways:

(a) Legal provision is needed for envisaging disaster prevention through (i) hazard identification and vulnerability reduction, (ii) proactive planning, assessment and audit, (iii) standards methods and codes/standards, (iv) public information, mock-exercise and (v) capacity development and institutional framework (Risk Management).
(b) Legal framework on disaster incident management through (i) incident command/ response management, (ii) resource mobilization and relief coordination, (iii) damage and loss assessment and financing, (iv) services in emergencies – environmental-health (shelter, water and sanitation, food, waste/debris, disaster prevention), power, fuel, roads, timber, disposal, etc. (v) immediate compensation, penalties, responder's safety (Response and Relief Management).
(c) Legal provisions on post-disaster risk reduction: (i) liability for compensation and rehabilitation (ii) enquiries and litigations, (iii) reconstruction (sustainable and safer), (iv) recovery (sustainable livelihood, environment, development) (v) addressing future risk (Reinforcing Risk Reduction).
(d) Legal provisions (policies, acts, rules and guidelines) are required to guide the actions and behaviour of the states, institutions, local bodies, communities, and other actors for planning, provisioning, performance, and monitoring of aforementioned aspects of the disaster management.

Looking to the need and roles of legal provisions in the area of disaster management and risk reduction, it would neither be possible nor feasible to enact new laws to address various aspects of multi-functional, multi-dimensional, multi-disciplinary approach to disaster management. Therefore, constitutional and legal provisions and guidelines related with the following aspects, and their interpretations – functional and judicial, shall be of immense use:

(i) Legal provisions on disaster, safety, and emergency response.
(ii) Legal provisions on environment and natural resources.
(iii) Legal provisions on developmental planning.
(iv) Legal provisions on human rights.

22.2.1 Concepts and Framework of Environmental Law in DRR and CCA

Since the International Decade of DRR in 1990s and following the Hyogo Framework for Action (HFA) in 2005, countries started paying greater attention to DRM as compared to the emergency response and relief centric approach. The HFA 2005–2015 advocates for DRR and mainstreaming in sectoral planning process, and provides impetus for environment based practices. Reid et al. (2005) also identified environmental degradation as a major factor leading to the increasing vulnerability.

India is a party to three Global Agreements pertaining to DRR, CCA and Sustainable Development. The three global agreements are Sendai Framework for Disaster Risk Reduction (SFDRR), The Paris Agreement on Climate Change and Sustainable Development Goals (SDGs). Government of India developed the National Disaster Management Plan (NDMP) in May 2016, reflecting the global priorities as outlined in the Sendai Framework. Prime Minister of India has also set 10-point agenda to address disaster risks during Asian Ministerial Conference on Disaster Risk Reduction (AMCDRR) in November 2016. SENDAI framework focused on DRR as a pre-requisite to achieve SDG, discusses how India has imbibed this essence in its updated national disaster management plan.

To monitor progress and achievement of the global targets of the Sendai Framework, an Open-ended Inter-Governmental Working Group (OEIWG) was set and the group identified 38 indicators for 7 SFDRR targets in December 2016. It is interesting to make note that Monitoring framework of SFDRR clearly depicts that reducing underlying risk factors i.e. environmental and Climate Change related risk are the key outputs of the SFDRR.

Several laws related to protection and improvement of environment and natural resources existed in India before 1950's as well. However, the real momentum for bringing a framework dedicated to environmental protection came after the UN Conference on the Human Environment (Stockholm, June 1972). The National Council for Environmental Policy and Planning was set up in 1972 after the conference, under the Department of Science and Technology in India. The Council later evolved into a full-fledged Ministry of Environment and Forests (MoEF) in 1985. Provisions directly related with natural disaster management, viz.

environmental clearance, EIA and audit, risk analysis, land-use and zoning, emergency preparedness; and environmental services—water, sanitation, waste disposal, preventive-health, including Climate Change Mitigation and Adaptation etc. are the daily business of MoEF (which has now become Ministry of Environment, Forest and Climate Change, MoEFCC). These regulations and policies primarily aiming at environmental quality, resource management and related procedures, have provisions and offer potential interpretations for disaster management and risk reduction (Gupta and Nair 2011).

The laws on environmental protection and quality including natural resource conservation, pollution and waste management, are becoming more relevant in Natural Disaster Risk Reduction (DRR) in the wake of paradigm shift in disaster management. The current emphasis of environment based DRR aims at pre-disaster risk reduction and post-disaster sustainable recovery processes, along with efficient relief management. Whereas the safety and emergency preparedness related laws provide for proper risk assessment, emergency planning and response organization aimed at minimizing the impacts of a disaster event in general and in particular of an accident involving hazardous substance or hazardous process. Related standards and codes ensure preventive environmental-health (food safety and shelter provisions, water and sanitation, waste management and controls of disease outbreak) so as to avoid secondary disasters and/or complex emergencies.

22.2.2 Constitutional Provisions

Indian constitution comprises provisions related to environment and human rights and thus provide for reduction of vulnerability and addressing hazard risk by enhancing capacities. The 42nd Amendment of the Constitution Act of 1976 explicitly incorporated environmental protection as a part of state policy. As per Article 48-A of the Constitution, which is also Directive Principles of State Policy (DPSP) provides that the "*state shall endeavour to protect and improve the environment and safeguard the forests and wildlife of the country*". Article 51-A (g), which is fundamental duty, imposes a similar responsibility on every citizen to "*protect and improve*" the natural environment including forests, lakes, rivers and wildlife, and to have compassion for living creatures. These two Articles directly relates to actions that reduce disasters by addressing hazards and reducing vulnerability of the land and its people.

Article 21 of the Indian Constitution talks about "Right to Life" with dignity and livelihood as a fundamental right and has been employed in a diversified manner in India. The Article states that *"No person shall be deprived of his life or personal liberty except according to procedure established by law"*. Article 14 of the constitution states that *"The State shall not deny to any person equality before the law or the equal protection of the laws within the territory of India*". Article 14 can be used to challenge government sanctions for mining and other activities with high stakes on human rights and environmental impact, where the permissions are arbitrarily granted without adequate consideration of environmental impacts (Gupta and Nair 2011).

22.2.2.1 Common Laws

The Common Law in India continues to be in force under Article 372 of the constitution under which an action might lay for causing pollution of environment, viz., air, water, or noise if it would amount to private or public nuisance. The common law remedies against environmental pollution are available under the law of Torts. Tort is considered as civil wrong other than a breach of trust or contract where nuisance, trespass, negligence and strict liability are four important tort liabilities.

The Indian Penal Code (IPC) of 1860 forms the backbone of criminal law in India. The Code of Criminal Procedure (CrPC), 1973 governs the procedural aspects of the criminal law. Indian Penal Code (IPC), 1860 makes various acts affecting environment as offences which also includes the negligent handling of poisonous substances, combustive, and explosive materials. Criminal Procedure Code, 1973 (CrPC) can also be used to prevent pollution. Section 133 of the CrPC can be used against Municipalities and Government bodies related to pollution and also can be extended in the case post disaster debris and waste management as well.

22.2.2.2 Statutory Laws, Rules and Notifications

The following is list of Acts related to protection and improvement of environment and disaster management, Rules notified and Notifications issued there under.

22.2.2.3 Acts

Indian Forest Act, 1927

The Act was sought to consolidate and reserve the areas having forest cover, or significant wildlife, to regulate movement and transit of forest produce, and impose duty on timber and other forest produces. The Act defined the procedure for declaring forest as reserved, protected and village forest and also levying penalties for not following the provisions of the Act (Section 20, 29 and 28 of The Indian Forest Act, 1927). This act is having provisions for conserving the valuable forest resources although the state monopoly is continued.

The Forest (Conservation) Act, 1980

The main objective of the Forest Act is to provide for the "*conservation of forests and for matters connected therewith or ancillary or incidental thereto*". The Act further restricts the powers of the state in respect of de-reservation of forests and use of forestland for non-forest purposes. By restricting the use of forest land for non-forest purpose, the Act serves as an affective legislation for controlling land-use land cover changes, and unsustainable development in forest areas which leads to degradation of ecosystems floods and landslides. The act highlights that deforestation causes ecological imbalance and leads to environmental deterioration.

Forest Rights Act, 2006 (Ministry of Tribal Affairs)

This Act, by recognising forest dwellers' rights, makes conservation more accountable. Provisions of this Act provides for the sustainable use, conservation of biological diversity and ecosystems and strengthening the conservation regime by strengthening the livelihood and food security of forest dwellers.

Wildlife (Protection) Act, 1972

The Wildlife (Protection) Act, 1972, Amendment 1991 provides for protection of different species of flora and fauna that are endangered or threatened and establishes a network of ecologically-important protected areas. This Act empowers the federal governments to declare any area as wildlife sanctuary, national park or closed area. Wildlife Protection Act is a powerful law, contribute to the maintenance of ecological diversity and balance and also by limiting the human interference in wild life areas and corridors chances of man-animal conflict will be reduced.

The Water (Prevention and Control of Pollution) Act, 1974

The objective of the act as envisaged in the preamble is to prevent water pollution and maintain wholesomeness. The Act does not allow the discharge of pollutants into water bodies beyond a given standard and lays down penalties for non-compliance. *This helps in assuring clean drinking water to every citizen. Besides, by prohibiting discharge of effluents, the act contributes towards protecting and improving the ecosystems and hence reducing vulnerability.*

The Air (Prevention and Control of Pollution) Act, 1981

The Air Act seeks to combat air pollution by prohibiting the use of polluting fuels and substances, as well as by regulating the uses of appliances, which increases air pollution. *By controlling air pollution, the act serves as an important means to mitigate Climate Change. The act reduces the vulnerability to health hazards by prohibiting and controlling industrial activities in control areas.*

Environment (Protection) Act, 1986

EPA is considered as umbrella legislation and cover gaps in the areas of major "*environmental hazards*" which were not covered or mentioned in the previous laws. The Section 3 of this Act empowers the central government to take measures necessary to protect and improve the quality of the environment by setting standards for discharges and emissions; regulation on the setting of the location of industries; management of hazardous wastes, as well as protection of public health and welfare. Section 6 empowers the Central Government to make rules to regulate environmental pollution and section 8 provides that any person handling hazardous substances are required to abide with procedural safeguards where the discharge of any environmental pollution in excess of prescribed standards occurs or is apprehended to occur due to accident or other unforeseen act or event. The Central Government issues notifications under the EPA for the protection of ecologically-sensitive areas or issues guidelines for matters as and when required. *Rules under the EPA Act are of great significance in all phases of disaster management particularly chemical disaster management and protecting and improving vulnerable areas and ecosystems like coastal, mountain, deserts and wetlands.*

Disaster Management Act 2005

DM Act provides for the effective management of disasters and for matters connected therewith or incidental thereto. Section 2 (d), definition of disaster is worthwhile to note. *"Disaster means a catastrophe, mishap, calamity or grave occurrence in any area, arising from natural or man-made causes, or by accident or negligence which results in substantial loss of life or human suffering or damage to, and destruction of, property, or damage to, or degradation of, environment, and is of such a nature or magnitude as to be beyond the coping capacity of the community of the affected area"*. Chapter 2, Section 8, the National Executive Committee comprised of Secretaries of ministries including environment is a key step towards mainstreaming DRR.

Gaurav Kumar Bansal vs Union of India on 8 May, 2017 filed a writ petition under article 32 of the constitution. These two writ petitions were upon the unprecedented flood and landslide disaster that occurred in Uttarakhand in the year 2013. In this petition, the SC observed that, though the disaster led to massive damage to life, limb and property and according to the petitioners, the adverse impact of the disaster would have been mitigated if there had been effective and sound implementation of the DM Act, 2005 and adequate preparedness by the State Government of Uttarakhand. The writ petitions further alleged that many of the other States were also not fully prepared to deal with a disaster and therefore court need to give directions for proper implementation of the Act.

National Green Tribunal Act, 2010

The Tribunal was established on November 18, 2010 under the National Green Tribunal Act 2010 for expeditious and effective disposal of cases relating to environmental protection and conservation of forests and other natural resources including enforcement of any legal right related to environment and hence giving relief and compensation for damages to persons and property and for matters connected therewith or incidental thereto. The Tribunal's dedicated jurisdiction in environmental matters shall provide speedy environmental justice and help reduce the burden of litigation in the higher courts. The Tribunal is mandated to make and endeavour for disposal of applications or appeals finally within 6 months of filing of the same.

The Mines and Minerals (Development and Regulation) Bill, 2011

The act is to consolidate and amend the law relating to the scientific development and regulation of mines and minerals under the control of the Union. Chapter 4 Section 21 (1) (IV) mandates the collection of baseline information on environmental conditions even before the reconnaissance or prospecting operations. Section 21 (1) (v), propose steps to be taken for protection of environment which shall include prevention and control of air and water pollution, progressive reclamation and rehabilitation of the land disturbed by the prospecting operations, a scheme for the plantation of trees, restoration of local flora and water regimes and such other measures for minimizing the adverse effect of prospecting operations on the environment. Such measures are extremely important for reducing the risk of landslides, floods and also for controlling the environmental degradation and damage to ecosystems.

Environment (Sitting for Industrial Projects) Rules, 1999

These rules are precautionary in approach where it prohibits the setting of certain industries in the area within the municipal limits and 25 km belt around the cities having population more than one million, 7 kames around the periphery of wetlands as listed in annex 1, 25 km around the periphery of protected areas including national parks, sanctuaries, biosphere reserves and 0.5 km wide strip of either sides of national highways and railway lines (Rule 2). Rule 2(3) poses restriction on the establishments of new units and expansion or modernization of existing units of industries in the Taj Trapezium zone. Restrictions around wetland under the act will enable protection of those areas and the wetlands protect from floods and droughts.

Wetlands (Conservation and Management) Rules, 2017

Wetlands are conserved and regulated as per the provisions laid down in Ramsar Convention, 1971 to which India is signatory. As declared by UNESCO, World Heritage site, high land or wetlands are identified by the designated authority. Activities like reclamation of wetland, setting up of industries around the vicinity, solid waste dumping, manufacture or handling of hazardous wastes; discharge of untreated wastes and any construction work or any other activity having adverse impact on environment are strictly restricted within wetlands (Rule 4). Rule 5 envisages the constitution of Central Wetlands Regulatory Authority for conservation, preservation and wise use of wetlands. There are total 25 wetland sites as identified by the Ramsar Convention (Rule 3(i)). Wetlands are the most important part of the hydrological cycle and provide wide range of ecosystem services which includes provisional, recreational, regulatory and supporting *services. Wetlands act as natural buffers to control floods and reduce the impact of tides and storm surges and helps in waste assimilation, water purification, erosion control, and groundwater recharge. Further wetlands are ecosystems which support wide range of biological diversity and hence a major source food.*

Plastic Waste (Management and Handling) (Amendment) Rules, 2018

Under this, plastic use is banned in sachets for packing/ selling *gutkha,* tobacco and *panmasala.* This act also prohibits use of recycled plastics/ compostable plastics in packaging of food stuffs. This also mandates recycled carry bags to have specific BIS standards. The recycled plastic bags should have uniform thickness of not less than 20 microns in carry bags etc. Waste pickers under this act are explicitly recognized. *Plastic wastes are harmful to environment since they contaminate surface and ground water, block the water channels (natural and manmade drainage) and increase floods.*

22.2.2.4 Notifications

Coastal Regulation Zone (CRZ) Notification (Revised in 2016)

The main objective of the notification is to ensure livelihood security of the fisher communities and other local communities, residing in the coastal areas and to protect and conserve coastal stretches, its unique environment and to promote development through sustainable manner based on scientific studies and taking into account the dangers of natural hazards in the coastal areas, sea level rise due to global warming etc.". This notification restricts the setting up and expansion of any industry, operations or processes and manufacture or handling or storage or disposal of hazardous substances as specified in the Hazardous Substances (Handling, Management and Tran boundary Movement) Rules, 2009 in the aforesaid CRZ. Section 5 of the notification mandates the preparation of Coastal Zone Management Plans by respective state and UT in consultation with scientific institutions and various stake holders. Guidelines for preparation of Coastal Zone Management Plans are given in Annexure1. *Hazard mapping (Step D) is one of the components as envisaged in the guideline. This notification is one of the key steps towards conserving the marine and coastal resources and improving the livelihood options of the community. Conserving the resources like mangroves, corals and other coastal ecosystems will help in creating natural buffers against tsunami, storm surge and coastal erosion.*

EIA Notification 1994 (Revised 2006)

Section 7 of the notification prescribes the environmental clearance process for new projects, comprised of a maximum of four stages, all of which may not apply to particular cases as set forth below in this notification. These four stages in sequential order includes (i) Stage (1) Screening (Only for Category 'B' projects and activities) (ii) Stage (2) Scoping (iii) Stage (3) Public Consultation and Stage (4) Appraisal. Section 9 of the notifications provides detailed list of factors to be taken into consideration which could lead to adverse environmental effects or the potential for cumulative impacts with other existing or planned activities in the locality. Part III of the Section 9 sets the criteria for "Environmental Sensitivity" taking into consideration of various environmental, cultural, and social factors. Areas occupied by vulnerable groups or sensitive land uses/buildings like hospitals, schools, places of worship; community facilities are included in the sensitive zone. Areas susceptible to natural hazard like earthquakes, subsidence, landslides, erosion, flooding or extreme or adverse climatic conditions are also included under these criteria. Check List of Environmental Impacts (Annexure II of the EIA Notification) is comprehensive and address impact of the project on land environment, water environment, vegetation, fauna, air environment, aesthetics, socio-economic aspects, building materials, and energy conservation. The notification provides for the formulation of the Environment Management Plan (EMP) that consists of all mitigation measures for each of the component mentioned above and activities to be undertaken during the construction, operation and the entire life cycle in order to minimize adverse impacts on environment as a result of the activities of the project. It would also delineate the environmental monitoring plan for compliance of various environmental regulations including the steps to be taken in case of emergency such as accidents at the site including fire. This notification is having provisions for disaster prevention, mitigation, emergency management and integrating environment and disaster management in development planning process.

22.3 Opportunity and Approaches

From the previous sections it is evident that apart from DM Act, other existing environmental laws provide various opportunity for mitigating hazards, increasing coping mechanisms, managing the calamity and post disaster recovery which has been explained in the subsequent sections (Gupta et al. 2013).

22.3.1 Category: Flood

22.3.1.1 The Water Act, 1974

Pre-Disaster

Section 16 (1). The key function of the Central Pollution Control Board (CPCB) is to promote cleanliness and maintain holiness of streams and wells.

Section 24(1) (b).Prohibits on use of stream or well for disposal or polluting matter that will impede the proper flow of the water of the stream and may result into the flood situation.

During Disaster

Section 16(1) and (2)(g). One of the main objectives of the Act is to maintain or restore wholesomeness of water and lay down standards for safe drinking water. This is important for ensuring clean drinking water victims during flood related situations.

Post Disaster

Section 17(1)(a) and (l plans a comprehensive programme for the prevention, control or abatement or pollution of streams and wells. This is of important consideration as during disaster, abundant amount of waste is generated which is generally dumped in the nearby streams. Thus, it becomes important to reduce the impact of this on the environment.

Section 24(2) (a, b, c and d). These clauses of the Act deals with the construction, improvement, and/or maintenance of bridges, weir, dams, sluice or reclamation of land or putting gravel or natural deposits back in the river which are essential maintenance work to overcome the aftermath of the flood.

22.3.1.2 The Forest (Conservation) Act, 1980

Pre-Disaster

Section 2 imposes restrictions on the de-reservation of forests or use of forest land for non-forest purpose and clearing of trees in the forest land which have grown naturally (sub-clause iv). Deforestation is the root cause of ecological imbalance that leads to environmental degradation and increases the occurrence of disasters like flood and drought. Forests act as buffer zones and help in mitigating flood, thus its conservation is important for the protection of human kind.

22.3.1.3 Wetlands (Conservation and Management), Rules), 2010

Pre-Disaster

Section 4 (1) (i), strictly restricts on wetlands reclamation. Conservation of wetlands becomes indispensable as they form an important part of the hydrological cycle and provides with wide range of different ecological services including flood mitigation and erosion control.

Section 4(1)(vi). This section states that construction of boat jetties of permanent nature within 50 m from the high flood level observed in the past 10 years (from the commencement of these rules) is allowed to carry out rescue operations during flood.

During Disaster

Section 4(2)(v)., under this section working of motorized boat, is allowed in the wetlands. As this will speed up the rescue operation during flood.

Section 4(2)(ix). Facilities required for temporary use like pontoon bridges are allowed for construction that can be used during rescue activities.

Post Disaster

Section 4(2)(i): Withdrawal of water or the impoundment, diversion or interruption of water sources within the local catchment area of the wetland system is allowed with prior approval of the State Government.

Section 4(2)(vi): This section allows dredging of wetlands in case of acute siltation. As during floods, heavy amount of silt is carried into the river. Removal of which reduces the impact of flood in the future.

22.3.2 *Category: Cyclone*

22.3.2.1 Coastal Regulation Zone Notification, 2011

Pre-Disaster

Section 3(iv)(b): This section allows for land reclamation activities that involve measures for controlling erosion based on the scientific studies including EIA. The measures to control erosion can include afforestation activities as well as building of structures that can reduce the impact of cyclone in that region.

Section 4.2(i)(d) lay down procedures for clearance of permissible activities in CRZ related to disaster management report, risk assessment report and management plan.

Section 5(iv): Deals with preparation of coastal zone management plans for the purpose of depicting flooding due to tides, waves and sea level rise in the next fifty or hundred years.

Section 8.1(i)(d): Activities like installation of weather radar for monitoring of cyclone movements and prediction by Indian Meteorological Department is permitted in the CRZ-1.

Post Disaster

Section 3(i)(e): This section allows the repair and reconstruction works of dwelling units of local people including fishermen in accordance with local towns and country planning regulations. The cyclone has the potential to destroy the houses, and livelihood of the local communities and this needs to be repaired according to the guidelines of Bureau of Indian Standards (BIS) codes in coastal zone.

22.3.3 *Category: Landslides*

22.3.3.1 Mines and Minerals (Development and Regulation) Act, 2010

Pre-Disaster

Section 32(1) and (8) deals with sustainable development framework for mine closure plan that shall be based on land use plan for the lease area after its closure and include measures to reduce hazards and meets the needs of the host population. Mining is an activity that increases the risk of landslides due to explosions, drilling and other activities. Hence, it becomes the prerequisites for taking up adequate measures to reduce the disaster due to landslides before and after the mining activity.

Section 45(2) deals with the power of central government to issue directions in order to facilitate scientific development and exploration of mineral resources to ensure environment protection and pollution abatement that arises due to mining activity and develop National Sustainable Development Framework in consultation with State Governments.

Section 45(4)(iii)(g) further envisages ensuing minimal ecological disturbance in terms of biodiversity, flora and fauna. Healthy biodiversity reduces the risk of landslides as trees hold the loose debris through their extensive network of root system.

22.3.4 Category: Earthquakes

22.3.4.1 Professional Engineers (Amendment) Act, 2012

Pre-Disaster

Most of the damages in earthquakes are caused due to faulty building design and lack of understanding of seismic knowledge by the engineers, hence, Section 6755.1 (b) lays emphasis on the examination that is to be passed by the applicants. This ensures the selected person has the knowledge of seismic principles and engineering survey principles. Passing of this examination is mandatory for registration as civil engineers.

22.3.5 Category: Drought

22.3.5.1 Biological Diversity Act, 2002

Pre-Disaster

Section 2(f) deals with commercial utilization of biological resources including extracts and genes used for improving crops and livestock through genetic interventions. Drought in our country is always associated with the failure of monsoon which is very erratic. Hence developing drought resistance varieties will help in mitigating the drought situation to a larger extent.

22.3.6 Category: Hazardous Chemicals

22.3.6.1 Water Act, 1974

Pre-Disaster

Section 16(f) deals with adequate measures to be devised by the CPCB for prevention and control of water pollution and prepare codes, manuals for treatment and disposal of domestic and trade effluents properly.

Section 17(f) deals with inspection of sewage and trade effluents and make sure that they comply with the standards as laid down by the boards.

Section 24(1)(a): this prohibits the use of stream or well for disposal of any poisonous, noxious or polluting matter.

During Disaster

Section 32 deals with emergency measures to remove, mitigate any poisonous material that is discharged in any stream or land, to ensure well-being of the people.

Post Disaster

Section 33 gives power to the board to make application to courts for restraining pollution of water. This is required to restricts industry dealing with hazardous chemicals (not properly treating the wastes before discharge) from further polluting the nearby streams and lands where disaster had already occurred. Under extreme case, the board has the authority give directions for the closure, prohibition or regulation of any industry, operation or process (Section 33A).

22.3.6.2 Air Act, 1981

Pre-Disaster

Section 16(1) of the Act deals with the function of the board to prevent, control or abate air pollution as well as carry out research related activities related to air pollution (Section 16(2) (d)).

Section 17(1) (e) deals with the inspection of industrial plant to make sure they comply with standards as laid down by the boards (section 22).

Post Disaster

Section 22A gives power to the board to make application to court for restraining persons from causing air pollution.

22.3.6.3 EPA, 1986

Pre-Disaster

Section 3(2) (vii) has the provision to give power to the central government for laying down procedures and safeguards while handling of hazardous substances.

Section 3(2)(x) provides for the inspection of any premises, plant, equipment, machinery or other processes, materials or substances and take necessary steps for the prevention and control of environmental pollution. Section 6(2)(c), (d), (e) and (f) authorizes the central government to make rules dealing with the management and handling of hazardous substances in different areas. Under this various rules have been formulated including MSICH 1989, EPPR 1996, Hazardous Micro Organisms 1989, MSW 2000, Plastic waste 2011 etc.

Post Disaster

Section 5(a) authorizes the central government for closure, prohibition or regulation of any industry, operations or process that does not comply with the standards or which has caused excessive damage to the environment and human beings by the virtue of its untreated discharge.

22.3.6.4 The Green Tribunal Act 2010

Pre-Disaster

The green tribunal by enforcing the provisions of the other existing environmental regulations contribute to protection of coastal areas, mountain areas, wetlands, protected areas and so on.

Post Disaster

The Act provides for the liability to pay compensation on principle of "No Fault" in certain cases, where an accident, death, injury to any person, or damage to any property or environment has taken place while handling of hazardous substances.

22.3.7 Policies

22.3.7.1 National Level Policies (Fig. 22.1)

International Environment Laws and Implications

During last two decades in India, Environmental law has developed in parallel to the development of global Environmental Law. The first conference on Environment in 1972, popularly known as UN Conference on Human Environment and

National Disaster Management Policy, 2009

Definition of Disaster as per the DM. Act of India recognizes damage to environment as a disaster. Introduction to disaster risks in India (Refer 1.2.1), recognize environmental degradation and climate change as factors increasing people& vulnerability, Objective of the policy (24.1) encourage mitigation measures based on technology, traditional wisdom and environmental sustainability, Section on 'Environmentally Sustainable Development" (5.1.6). 'The policy recognize that 'environmental considerations and developmental efforts need to go hand in hand for ensuring sustainability. Restoration of 'ecological balance in Himalayan regions and raising coastal shelter belt plantations need to be incorporated wherever necessary in disaster management plans. Ecosystems of Forests, islands, coastal areas, rivers; agricultural, urban and industrial environment are also to be considered for restoration of ecological balance and sustainable development. Zonal regulations should ensure preservation of rural habitus, Climate Change Adaptation (5.1.7) with focus on glacial reserves, water balance, agriculture, forestry, coastal ecology, biodiversity and health in order to reduce disaster risks and vulnerability, Section (63.1) Environmental and hazard data for formulation of alternative land-use plans for different geographical and administrative areas with a holistic approach Institutional arrangements (12.2.1) emphasized the face of close interaction with Central Ministries and Departments of Agriculture, Atomic Energy, Earth Sciences, Environment & Forests, Health, Industry, Science & 'Technology and Space; and with academic institutions (NDMA, 2009).

National Forest Policy (Draft), 2018

National Forest Policy is a key policy is having several provisions related to DRR. The policy envisages. The stability through preservation and restoration of the ecological balance that has been adversely disturbed by serious depletion of the forests of the country. The policy recommend checking soil erosion sand denudation in the catchment areas of rivers, maintenance of environmental lakes, reservoirs in the "interest of soil and water conservation, for mitigating floods and droughts and for the retardation of siltation of reservoirs Controlling the extension of sand-dunes in the desert areas of Rajasthan,and along the coastal tracts. This policy by conserving the forests helps in reducing the risks of floods, landslides and enhanced the livelihood options of the people depending on forest. Disaster Risk Reduction measures in the policy are evident from the section 2 i.e. the basic objectives itself (MoEFCC, 2018).

National Environment Policy, 2006

The National Environment Policy 2006 adopts and integrated approach including coastal zone management, management of wetlands and river systems; conservation and development of 'mountain ecosystems; land use planning; watershed management and reducing the impacts natural hazards like, flood, landslides, storm surges and climate change. ELA notification envisages for Hazard Mapping, Vulnerability and Risk Assessment Report as a part of environment management plan of the projects (MoEFCC, 2006).

Urban Sanitation Policy, 2008

Sanitation is a key aspect of disaster management due so its 1nd garbage related health disasters (epidemics) as well as in the post disaster risks in context of hydro-meteorological disasters like flood, cyclone and drought etc. during response and relief phase. While this policy pertains to management of human excreta and associated public health and environmental impacts, it is significance in pre-

Fig. 22.1 Disaster risk reduction related policies

disaster phase in relation to recognized that integral solutions need to take account of other elements of environmental sanitation, i.e. solid waste management, generation of industrial and other specialized/hazardous wastes; drainage; as also the management of drinking water supply. Key Sanitation Policy Issues like managements industrial and other addressing poor and served people, integrated approach, technology choice, occupation and organizational aspects of sanitation, The policy implementation, Policy also emphasized the need of emergency prescribes for a city sanitation plan and its preparedness and response aspects of sanitation management in cities (MoUD,2008).

(iii) National Agricultural Policy, 2000

The National Agricultural Policy of India in 2016 aimed to address the various challenges and opportunities in the agricultural sector of the country. India's agriculture sector accounts for 18 per cent of GDP, and employs around 60 per cent of the workforce, National Agriculture Policy (NAP) aims to attain an annual growth rate of 4 per cent in the agricultural sector over two decades (2000-2020). NAP is another important policy to address the issues pertaining to protection and management of natural resources and ecosystems in rural areas in particular for hydro-meteorological disasters and climate change issues. The new Agriculture Policy of India emphasized on Sustainable agriculture, Food and nutritional security, Technology generation & transfer, risk framework, Agro-ecosystems include mainly the man-made ecological production systems — farms, plantations, ponds, etc. also the natural systems used for bio-productivity purposes, covering the purposes of food, dairy, fisheries and other livestock, management and organizational etc. (MoA,2000).

(ii) National Water Policy, 2012

Objectives of the National Water Policy of India are planning, development and management of water resources. 'The policy by addressing the need of conserving and effectively managing water resources is paving way to mitigate impact of hydro-meteorological hazards and climate change, Water policy recognized water as a key environmental resource for sustaining all life forms, 'Water is part of a larger ecological system' (1.3). Disaster risks like floods and drought are of concerns in the water policy of India. Socio-economic aspects and issues such as environmental sustainability, appropriate resettlement and rehabilitation of project-affected people and livestock, public health concerns of water impoundment, dam safety etc., are envisaged in the policy while planning and implementation of water resources projects. In the policy it is recognized that problems of water logging and soil salinity have emerged in irrigation command areas, leading to the degradation of agricultural land. Besides the physical issues the social problems like equity and social justice in regard to water distribution are required to be addressed. Considering the dectine in quality and quality of ground water due over exploitation in certain parts of the country the concern and need for judicious and scientific resource 'management and conservation is envisaged in the policy. 'The policy recommended water zoning of the country and the development activities including agricultural, industrial and urban development should be guided and regulated in accordance with availability and constraints of water resources. Both in urban and in rural areas adequate safe drinking water facilities should be provided to the entire population, Water environment management with watershed, forestry, soil conservation, and agriculture and river-basin approach is the central philosophy of water policy 3.3 to 3.5). It also prescribes for drainage and irrigation management (6.6), groundwater (7.4), water quality (14), zoning (15), with devoted sections like — floods (17), sea or river erosion (18), drought (19) and use of science-technology and environmental impacts (25). Water Policy is giant leap towards reducing the impact of hazards and vulnerability reduction by assuring drinking and irrigation water supply through structural and non-structural measures. The policy also recommends integration of concerns of water resource 'management in developmental activities (https://nwm.gov.in/sites/default/files/national%20water%20policy%202012_0.pdf) (MoWR, 2012).

Fig. 22.1 (continued)

Development (UNCHED) emphasizes on the healthy and disaster free environment as an extension to the "right to life". The UN conference on Environment and Development in 1992 (popularly known as Rio Conference) was the largest UN conference ever held expands further the concept of sustainable development and reaffirms the importance and centrality of Polluter Pays Principle, Precautionary Principle and Environment Impact assessment (for detail please refer Module-1, Environment Legislation for Disaster Risk Management). The major highlights of the conference were the Rio Declaration on Environment and Development, Agenda 21, Forest principles as well as two legally binding conventions on Climate Change (United Nations Convention on Climate Change) and Biodiversity (Convention on Biological Diversity) that focused at abating global Climate Change and the protection of biologically diverse species. United Nations in 2002, organised the World Summit on Sustainable Development (WSSD) in Johannesburg to re-establish the commitment to build an equitable global society for the protection of environment. The world once again came together at Rio De Janerio in June 2012 (popularly known as Rio +20 Earth Summit) to discuss issues around Sustainable Development. Concern for environment has always been there at global agenda which is evident from various treaties, protocols and conventions from time to time. The table below summarizes status of India vis-à-vis International Environmental Conventions (Table 22.1).

22.4 Integrated Environment DRR Framework

The Disaster Management Act 2005 recognizes damage to or destruction of environment as disaster. The National Disaster Management Authority (NDMA), which is the guiding institution on disaster management in India, has developed guidelines on different aspects of disaster management and also has provisions for various environmental approaches in disaster mitigation and post-disaster management covered under environmental rules and regulations. The 1992 UN Convention on the Protection and Use of Trans-boundary Watercourses and International Lakes calls on each party to define water-quality objectives and to adopt criteria and set guidelines for this purpose. Some bilateral and regional agreements on freshwater and air foresee or mandate water-quality objectives. They address the precursors of the hazards in the river-zones and costal zones known to aggravate the impacts of river or sea erosion, flooding, cyclone or other natural calamities. Such private regulations may constrain behaviour of breaching by exercising a moral or practical (sanctioning) influence and litigants may argue that breach of such codes or standards may be an evidence of malpractice or negligence.

Table 22.1 Handbook on international environment agreements: an indian perspective. Accessed at http://awsassets.wwfindia.org/downloads/mea_handbook_cel.pdf

Convention	Effective From	Year Signed and Enforced
International Convention for the Prevention of Pollution of the Sea by Oil (1954)	1974	1974
The Antarctic Treaty (Washington, 1959)	1998	1983
Convention on Wetlands of International Importance, Especially as Waterfowl Habitat (Ramsar, 1971)	1982	October 1, 1981 (ac)
Convention Concerning the Protection of the World Cultural and Natural Heritage (Paris, 1972)	1978	1977
Convention on International Trade in Endangered Species of Wild Fauna and Flora (Washington, 1973)	1976	1974
Convention on the Conservation of Migratory Species of Wild Animals (Bonn, 1979)	1982	1979
Convention on Early Notification of a Nuclear Accident (1986)	1988	1986
United Nations Convention on the Law of the Sea (Montego Bay, 1982)	1995	1982
Protocol on Substances That Deplete the Ozone Layer (Montreal, 1987)	1992	June 19,1992 (ac)
Convention on the Control of Transboundary Movements of Hazardous Wastes and Their Disposal (Basel, 1989)	June 24, 1992	March 5, 1990
Amendments to the Montreal Protocol on Substances That Deplete the Ozone Layer (London, 1990)	1992	June 19,1992 (ac)
Protocol on Environmental Protection to the Antarctica Treaty (Madrid, 1991)	1998	1992, 1996
United Nations Framework Convention on Climate Change (Rio de Janeiro,1992)	1994	November 1,1993
Convention on Biological Diversity (Rio de Janeiro, 1992)	February 18,1994	June 5, 1992
Convention to Combat Desertification in Those Countries Experiencing Serious drought and/or Desertification, Particularly in Africa (Paris, 1994)	December 17, 1996	October 14, 1994
International Tropical Timber Agreement (Geneva, 1994)	1997	October 17, 1996
Rotterdam convention on the Prior informed Consent Procedure for certain Hazardous Chemicals and Pesticides in International Trade (1998)	2004	2005
Protocol to the United Nations Convention on Climate Change (Kyoto, 1997)	2005	1997
Cartagena Protocol on Biosafety (Nairobi, 2000)	January 23, 2001	January 17, 2003
Stockholm Convention on Persistent Organic Pollutants (2001)	2004	2006
Paris Agreement (2015)	November 4, 2016	October 2, 2016

22.5 Conclusion and Way Forward

Paradigm shift in approach from relief centric to mitigation and mitigation centric approach was witnessed since mid-1990. Hyogo Framework of Action advocates mainstreaming of Disaster Risk Reduction in developmental planning processes. Implication of the HFA is widely reflected in various international and national legislation and policies pertaining to environment, disaster management, land use, forest, water and natural resources. The Sendai Framework makes clear the relationship between the adoption and implementation of DRR strategies and addresses the importance of National and local frameworks of laws, regulations and public policies", with a special focus on implementation of DRR strategies.

Laws and policies related to environment and natural resource management in India covers various aspects of the disaster management and risk reduction by protecting and safe guarding environment. Indian Judiciary interpreted right to pollution free environment and life free of danger as a fundamental right (Article 21). Special emphasis on ecological security is evident from various laws where human beings are also considered as part of larger ecosystems. Forest Rights Act and PESA are the examples. Environmental laws in 70 s and 80 s were focused primarily on the control of pollution as evident from Air Act and Water Act. Indian forest laws are one of the earliest laws to abate the uncontrolled use of timber and other forest products although the laws were not conservation centric till 1980 Forest Conservation Act. During the British period the forest laws were dictated by the colonial interest of reserving forest primary for own vested interests in timber and other forest produces. Ensuring safety measures through law was primarily focused on industrial (Chemical Mining etc.) safety. Oldest laws on safety can be traced back to Explosive Act 1884 to regulate the manufacture, possession, use, sale, transport, import and export of explosives. Insecticide Act of 1968 was enacted to regulate the import, manufacture, sale, transport, distribution and use of insecticides with a view to prevent risk to human beings or animals, and for matters connected therewith. The Factories Act 1948 is having provisions for assuring the safety and health of workers. The Factories (Amendment) Act 1987, (in the aftermath of Bhopal disaster) Chapter 4 added and provisions for setting of industries is added (Section 41(a)). Clearing of causes related to environmental issues were always been a challenge to judiciary and need of involving subject experts was felt. Environmental Appellate Authority and Green Tribunal in 2010 to deal with cases related to environmental damages. Compensation and provisions for relief and environmental rehabilitation fund are emerged under the EPA Act and rules there under. Risk transfer mechanism through insurance and principle of absolute liability were introduced through the Public Liability Insurance Act. India is having a well evolved Techno-legal framework, where codes are prescribed for various hazard prone areas, industries, fire safety and so on (techno-legal is beyond the scope of this paper). Long term developmental issues since the beginning of a project is very well addressed through various provisions of EIA Notification, EMPs, Coastal Zone Regulations, Flood Zone Ordinance etc. Risk Sensitive Land-use Planning is also emerged as a recent trend where hazard profile and state of environment is made as part of the regional and master plans. Provisions for monitoring in the form of Environmental and Safety Audit reports are mandated in EPA and Factories Act. Fortunately, Indian Environmental Laws and Policies have several provisions to address various natural and human induced disasters.

In the post Bhopal Scenario regulations related to the manufacture, storage and handling of chemicals were enacted and Factories Act was amended to consider the offsite issues. After the 1993 Latur Earthquake, BIS codes were revised and new seismic zonation atlas was also developed. Few limitations are (i) cases related to environment come under civil liability and not criminal liability (National Green - Tribunal Act 2010) (ii) Penalties (financial) are not deterrent in many statutory laws and needs amendment (iii) Incentive approach is to be promoted for further encouragement of green development (iv) Liability is more attached to industrial and chemical disasters and natural disasters are yet considered as "Act of God" where no one is held liable (v) Despite the mounting losses due to disasters in the coastal and mountain areas, the regulations are yet to be enforced. Liability on the authority's by-passing the regulations like Flood Zone Regulation, Coastal Zone Regulation etc. to be made more stringent (vi) Relief and disaster management codes needs to evolve further to accommodate environmental damages of disasters. Although Disaster Management Act 2005 consider damages to environment as disaster, there no provision address environmental rehabilitation due to disasters other than the debris and carcasses disposal (vii) Post Disaster Needs Assessment in India yet to cover the environmental damages and losses and needs for the recovery of natural resources. (xiii) Environmentally sustainability mitigation options and the concept of 'greening disaster-response' and 'sustainable-recovery' need to be promoted within the framework of sustainable development by integrating SEA to the developmental planning process. SEA and EIA (already exiting) scope need to necessarily include hazard-risk and vulnerability assessment within the framework. There is an urgent need to revisit the regulations related to development, environment, safety and disaster management and make sure that they are complimenting and not conflicting each other. It is high time to recognise that no disasters are natural and only hazards are natural. Legislations can play a major role in hazard mitigation as well as reducing the vulnerabilities.

References

Gupta AK, Nair SS (2011) Environmental knowledge for disaster risk management—concept note in abstract book of the international conference 9–10 May 2011, New Delhi. National Institute of Disaster Management, New Delhi and GIZ Germany

Gupta AK, Nair SS, Singh S (2013) Environmental legislation for disaster risk management. Environmental knowledge for disaster risk management project, National Institute of Disaster Management and Deutsche Gesellschaft für Internationale Zusammenarbeit (GIZ). GmbH, New Delhi

Holloway A, Pelling M (2006) Legislation for mainstreaming disaster risk reduction. Tearfund, Teddington

Ministry of Law and Justice, GoI (2010) National Green Tribunal Act, 2010

Reid WV, Mooney HA, Cropper A, Capistrano D, Carpenter SR, Chopra K, Dasgupta P, Dietz T, Duraiappah AK, Hassan R, Kasperson R, Leemans R, May RM, McMichael AJ, Pingali P, Samper C, Scholes R, Watson RT, Zakri AH, Shidong Z, Ash NJ, Bennett E, Kumar P, Lee MJ, Raudsepp-Hearne C, Simons H, Thonell J, Zurek MB (2005) Ecosystems and human well-being-synthesis: a report of the millennium ecosystem assessment. Island Press

Sreeja S. Nair is presently working as Fellow with the Energy and Resource Insitute. She has worked with United Nations Development Programme, India, UN Office for Disaster Risk Reduction, Regional Office for Asia Pacific, UNISDR, Bonn, UN Habitat Myanmar, National Institute of Disaster Management, National Disaister Management Authority, Taru Leading Edge, Aga Khan Agency for Habitat etc at variable capacities. She has done her PhD research in the area drought vulnerability and mitigation analysis for Bundelkhand region of central India from discipline of Geography at Indira Gandhi National Open University.

Swati Singh has more than nine years of professional experience in the area of water resource management, climate change adaptation, environmental laws and disaster management with proven contributions to policy planning, research, knowledge building, advocacy and capacity building. She has worked with National Institute of Disaster Management, GIZ, Development Alternatives and Sir Ratan Tata Trust-CInI in various capacities. She is currently working with UNDP as an Adaptation Analyst.

Anil K. Gupta Professor of Policy & Strategies, Integrated Resilience. Head of the ECDRM Division, International Cooperation, Programme Director - Centre for Excellence on Climate Resilience, Nodal Officer for Central Ministries at National Institute of Disaster Management, (Govt. of India), New Delhi. He is also a member of the expert Team for WMO Climate Statement and Core Group Member of IUCN-Council of Ecosystem Management.

Chapter 23
Pathways for Integrating Climate-Disaster Resilience into Planning: Scaling Sub-National Studies to National Policy Paradigms

Shashikant Chopde, Anil K. Gupta, Dilip Singh, and S. A. Wajih

Abstract Climate-induced disasters have caused huge damage worldwide. It has profound implications for development characterized by large-scale investments in infrastructure and poverty reduction programmes. This calls for urgently integrating climate-induced disaster- risks in development. While the integration has happened at national level in India through policies and plans, it is far from satisfactory at the subnational levels (state, district and lower levels). The findings from three case studies in diverse socio-economic and hazard contexts across three states in India have helped understand the challenges and barriers to integrate climate- and climate-induced disaster- risks in development planning at the subnational levels. Further, the implications of research findings are drawn for national and international policies and treaties. The research indicates key institutional and capacity gaps of departments at district level and respective urban local bodies. These gaps exacerbate climate-induced disaster- risks to achieve desired level of sustained productivity and functionality of infrastructure in varied sectors/sub-sectors as well as hampering the effectiveness of poverty reduction programmes.

Keywords Climate and disasters · Disaster management · Resilience

---

S. Chopde (✉)
DRR Expert, Institute of Social and Environmental Transition (ISET), Boulder, CO, USA

A. K. Gupta
ECDRM Division, International Cooperation, CAP-RES DST Project, National Institute of Disaster Management (Ministry of Home Affairs), New Delhi, India

D. Singh
Climate Change Resilience and Energy Unit, UNDP, New Delhi, India

S. A. Wajih
Gorakhpur Environmental Action Group, Gorakhpur, India

A. K. Gupta et al. (eds.), *Disaster Risk and Management Under Climate Change*, Disaster Resilience and Green Growth,
https://doi.org/10.1007/978-981-99-4105-6_23

23.1 Introduction

Globally, damages from disasters are increasing and likely to significantly increase with Climate Change. A large share of these damages is from climate-induced disasters such as floods, droughts and cyclones. Climate-induced hazards affect poor people disproportionately, by impacting livelihoods, reduced crop yields or destruction of homes (IPCC 2012). The impact impoverishes them further. A whopping 26 million people are pushed into poverty every year due to extreme weather events, 90% of which can be attributed to Climate Change (Hallegatte et al. 2017).

High dependence of India's rural population on climate-sensitive agriculture sector for livelihoods is considered as one of the major reason for increased indebtedness among farmers especially the small and marginal farmers, who do not have necessary (access to) resources to counter challenges faced by climate-extremes such as floods and droughts. Primarily the increasing gap between agriculture income and household expenses on issues such as health have increased indebtedness of farmers, with most of the small and marginal farmers not having access to institutional credit forcing them to borrow from local moneylenders at exorbitantly high rates (Financial Express 2017, July 28). As per National Sample Survey Organization (NSSO) 70th Round (year 2013) about half of the agricultural households in the country were indebted.

Apart from strengthening agriculture systems (production and marketing) poverty can also be alleviated by investments in infrastructure. Infrastructure fosters better growth and also leads to increased employment and poverty reduction. Enhancing poor households' access to health, education, water and sanitation can significantly promote inclusive development (NSSO 2013). However, with increasing climate-induced disasters, investments in infrastructure without due consideration to climate-risks can lock the finances for decades into not so productive infrastructure.

Another important dimension to address poverty is by strengthening rural-urban linkages. It will need providing adequate infrastructure and services in the rural areas and small urban centres and effective linkages between the rural areas and the urban centres, as the small towns and municipalities are known to drive the economy of rural areas around them.

Hence, investments in development specifically need to integrate climate-induced disaster- risks for India to meet the SDGs by 2030.

23.2 Rationale and Need

However, the extent of integration of CCA-DRR into developmental plans and activities reduces significantly as we go down to the subnational levels—state, district and lower levels. Interestingly, the recent National Disaster Management Plan of India (version 2019) emphasizes integration of Disaster Risk Reduction

(DRR) into development. It specifically defines roles of various central and state agencies and promotes cooperation of the centre and state governments, and coordination among various ministries and departments. The Plan follows the Disaster Management Act (2005) and National Policy on Disaster Management (2009) and substantiates the paradigm shift from reactive and relief-centric to a proactive disaster risk reduction approach across various phases of disaster management. In addition, NDMA has come out with several guidelines on holistic management of various disasters including covering dimensions of prevention and mitigation, and robust recovery. At the State level, the State Action Plan on Climate Change (SAPCCs) and State Disaster Management Plan (SDMPs) advocate for an integrated approach to development by considering climate- and disaster- risks. Here, although vehicles for integration such as State Disaster Management Authority (SDMAs) exist, there is lack of effective instruments to translate such plans into practice. Going further down to district levels, the potential agencies for this integration is the District Disaster Management Authority (DDMAs) and District Planning Committees (DPCs)—both of which have neither the needed resources nor the technical know-how. In fact, the DPCs are almost dysfunctional in most of the states even where they were constituted. Even the DDMAs are still to delve into effective prevention, mitigation and robust recovery—the core realm of mainstreaming integration.

The research was conducted in three districts across as many States in India--Gorakhpur (Uttar Pradesh), Almora (Uttarakhand) and Puri (Odisha). The case study areas taken together significantly capture the diversity of types of major disasters in the country. As part of the research, barriers to integration of climate-induced disaster- risks in development planning were assessed at the district level with the objective of drawing lessons for actions at state and national levels.

The research involved using established tools and strategies. Climate Resilience Framework (CRF) and Shared Learning Dialogues (SLDs), developed through a decade of implementation experience, have been used.

23.3 Concept and Framework

23.3.1 Overall Approach

Climate Resilience Framework (CRF) was used as a guidance tool through a process of Shared Learning Dialogues (SLDs) for understanding various components of vulnerability (the left-hand loop) and identifying actions for building resilience (right-hand loop).[1] The resilience building interventions can be categorized as:

(i) Systems (flexibility, redundancy and safe failure): the characteristics of physical infrastructure and ecosystems such as water supply, irrigation, roads,

[1] For more details please refer to https://www.i-s-e-t.org/resource-climateresilience-training.

groundwater and surface water natural systems, that allows maintaining the functionality of systems even during disaster events;

(ii) Institutions (decision-making process, codes, financing norms etc.): that foster learning and change for adaptive response capacities of formal and informal structures, for example any department's internal procedures and norms for decision-making, designing, financing, implementation and monitoring. These procedures are norms are stipulated and dependent on policies, programmes and plans laid down by higher level institutions—at meso (state) and macro (national) levels; and,

(iii) Change agents (responsiveness, resourcefulness and capacity to learn): capacity to organise and re-organise as needed to identify, anticipate and plan for disaster events. It also relates to their ability to mobilise resources and assets from within their own departments or if needed from other departments through collaboration. The change agents considered both departments and communities.

The SLDs are used to unpack complexities of systems (physical, environmental, institutional and organizational) by engaging diverse set of stakeholders in a structured process. These engagements can take a variety of forms, from large-scale workshops, to homogenous focus group sessions, to medium-sized multi-stakeholder gatherings, to one-on-one meetings. They can take place at various levels—national, state and local level or exchanges between different localities. Structuring this process requires a facilitator – an actor or group of actors seen as credible and legitimate by all stakeholders, who can plan strategically and take advantage of opportunities to engage different actors and broker relationships .

23.3.2 Methodology

The District/ State Disaster Management Authority at the District/ State level (DDMAs/ SDMAs) are provided for, by the DM Act 2005 as institutions for promoting integrated and holistic response on disaster management across various departments/ bodies at sub-national levels in India. Hence, we considered SDMAs and DDMAs as the focal points to assess issues of mainstreaming DRR-CCA in Development Planning in the three selected locations—Almora, Puri and Gorakhpur.

SLDs were conducted in each of the three locations, wherein departmental knowledge (on linkages between various climatic extremes and disasters, existing infrastructure and environmental system, development planning mechanism etc.) and external experience (Climate Change, concepts of resilience, mainstreaming theories and practices etc.) to understand vulnerabilities and identify resilience actions, were brought on the same platform. The SLDs were conducted over two rounds in all the three locations following a similar practical approach. An example of range of issues covered over the two rounds in case of Almora is presented below:

Table 23.1 First round SLD

1. State Level Consultation with DMMC and departments	2. Districts level consultation with DDMA and departments	3. Consultations with each prioritized district level department
• Introduction to programme • Nature and extent of disasters and impacts in the state • Initiatives by state in different phases of disaster management- preparedness and mitigation, response and recovery • Major Developmental programmes (Flagship programmes) • Financing mechanisms	• Introduction to programme • Nature and extent of disasters and impacts in the district • Issues of inter- departmental coordination in different phases of disaster management • Priority sectors/departments to meet	• Hazard ranking • Nature and extent of damage to assets and infrastructure related to department • Reasons for impacts (organizational, institutional, funds, infrastructure attributes) • Challenges in different phases of disaster management- preparedness and mitigation, response and recovery

23.3.3 First Round SLD

It involved conducting consultations in sequence (left to right) as shown in table below (Table 23.1):

23.3.4 Second Round SLD

Sequence of consultations from left to right as shown in following table (Table 23.2):

23.4 Opportunity and Approaches

23.4.1 Review of Data/Literature

- Data archives from each department on various issues covered in SLDs;
- State Disaster Management Plan (SDMP) and State Action Plan on Climate Change (SAPCC);
- Reports and publications such as State's annual budget and expenditure, census data and economic survey; and,
- Relevant national level mission documents, programmes, plans and guidelines.

Table 23.2 Second round SLD

1. Districts level Consultation with SDMA and Departments	2. Consultations with each district level department	3. State level consultation with DMMC, Department and DMMAs of other districts prone to disasters
• Key findings from the First Round • Feedback	• Programme/schemes and budget • Opportunities to integrate disaster preparedness, mitigation and recovery in developmental programmes/ schemes • Suggestions changes in programmes/ schemes, financing norms, technical and administrative approval systems, standards and codes, targeting for mitigating impacts • Organization procedures • Monitoring systems	• Findings from two rounds of consultations on mitigation actions- reducing vulnerability and exposure and feedback • Relevant issues from other districts • Capacity needs and financing channels

23.4.2 Key Findings from the Three Cases

23.4.2.1 Description of Cases: Context and Hazards

The three cases capture the diversity of major weather-related disasters in India. These locations have experienced extreme rainfall events, cyclones, extreme hot and cold days for long durations and hailstorms. These have manifested in flash floods, riverine and deltaic flooding and waterlogging, droughts, landslides and heat and cold waves impacting all the sectors of development.

Almora (Uttrakhand)

Almora district is centrally located in multi-hazard prone districts of northern Himalayan State of Uttrakhand, spread over 3139 sq. km. The population of Almora is 622,506 (Census 2011), of which about 90% is rural and agriculture is the major occupation of people in the state. Almora is an important town of the district. The climate of Almora district varies in places depending upon its elevation. In summer, near the river valleys the temperatures can go as high as 40 C, the winter temperatures drop to below 0 °C at higher altitudes. The average temperature though ranges from 31.2 °C in summers to 0.1 °C in winters. This district receives an average annual rainfall of about 1027 mm.

Hazard Profile

Cloudbursts, landslides, flash floods and earthquake are major hazards in the districts. The district has observed some very severe devastation in year 2010 due to flash floods and landslide caused by cloudbursts. Almora is susceptible to earthquake and falls under zone IV and V. The district faced a severe flash-flood event in 2010 caused by cloudburst. Subsequently heavy rainfall and landslides episodes in 2013 and snowfall in 2014 are some major recent disasters making district vulnerable to multiple hazards.

Most of the GCMs project increase in total rainfall in July, August and September though a few indicate reduction too. In contrast, the winter rainfall is likely to increase.

The maximum and minimum temperatures in summer and winter are likely to increase as evident from almost all the GCMs.

Puri (Odisha)

Puri is located in the eastern State of Odisha with geographical area of about 3479 Sq.km. The population of Puri district is 16,98,730 (Census 2011). Puri district comprises 11 blocks and four Urban Local Bodies (ULBs) including Puri Municipality and agriculture remains the main source of livelihood. Puri district has tropical climate and receives an average rainfall of 1424 mm.

Hazard Profile

Multiple natural hazards and new types of vulnerabilities are emerging in the district, floods and cyclones being the most prevalent. Based on the discussions with various departments and local community, the types of hazards and months of their occurrence are shown in the graphic.

The total number of cyclones is likely to reduce. However, the high intensity cyclones are observed to rise from pre-warming (before 1950's) to post-warming era.

The cumulative two- and three-day extreme rainfall that significantly contributes to flooding/ waterlogging in Mahanadi deltaic areas of Puri district is likely to occur once in 2 years.

Gorakhpur (Uttar Pradesh)

Gorakhpur district is located in the eastern part of Uttar Pradesh and covers an area of 3484 Sq. km. Gorakhpur city itself is an important religious, cultural, commercial, educational and medical center – serving the hinterland of eastern Uttar Pradesh in the Gangetic plains. Gorakhpur is one of the most populated districts of UP with population of 44,36,275 (Census 2011) of which about 80% is rural and agriculture remains the main source of livelihood. The district comprises 7 tehsils, 19 blocks and 8 ULBs.

The climate of eastern UP varies from semi-arid to sub- humid and experience four seasons. The summers are very hot, while the winters are cool and dry. Currently, the summer average maximum temperature soars to as high as 37.5 °C and winter average minimum is 10.2 °C. During the later part of summer season and during monsoon, the humidity levels increase significantly. The mean annual rainfall varies from 80 cm in the south to 140 cm in the northern parts.

Hazard Profile

Gorakhpur district is majorly affected by flooding from Rohini, Rapti, Aami, Kuano and Ghaghra rivers. It is almost an annual phenomenon. In addition to this, this district also experiences dry spells during the monsoon season which can cause drought-like conditions.

Vulnerabilities, Risks and Capacities across Three Districts

In this section we present analyses of findings of SLDs, archived data of departments, and various state government reports related to development planning, disasters and Climate Change. The results are presented by sectorally—social, economic and environmental (adapted from Chopde et al. (2016). Though environment is not a separate sector per se but it needs special attention so it has been presented distinctively (Tables 23.3, 23.4 and 23.5).

23.5 Strategy and Gaps

From the above tables, the vulnerabilities, risks, and capacities can be categorized into the three core elements of Climate Resilience Framework (CRF)- fragile physical and environmental systems, institutional gaps, and lack of capacity of communities and departments. Addressing these can go a long-way in promoting resilient infrastructure and development:

23.5.1 Fragile Physical Systems

- The infrastructure like roads, schools, health, irrigation and electricity are either situated in disaster-prone areas and do not follow resilient design standards or are very old or underdeveloped (as against needed scale). At some locations/ sectors it is combination of all the three. The thrust is on constructing new infrastructure as per existing technical norms and merely replacing the damaged infrastructure, as against adopting technical standards of resilient designs to build-back-better.

Table 23.3 Social sector

Sector/ dimensions	Almora (flash floods, landslides, water shortages, cold waves, forest fires)	Puri (floods/cyclones/ drought)	Gorakhpur (flooding/ waterlogging/ water scarcity in summers
Poverty	Almost 50% population below poverty line (BPL); significant proportion of scheduled caste (SC)/ scheduled tribe (ST) population	Significant proportion of SC/ST population	High population density and low literacy rates; 30% population BPL and significant SC/ST population
Employment	76% farmers are small and marginal; High dependence on agriculture for livelihoods; less diversified livelihoods; High unemployment levels	High dependence on agriculture for livelihoods; less diversified livelihood; 92% farmers are small and marginal A third of total workers employed less than 6 months in a year	High dependency on Agriculture and allied sectors; about 78% of land holdings are small and marginal
Housing	High proportion of semi-permanent and temporary houses with more concentration in just two blocks	More than 50% of houses semi-permanent and temporary	About a third of houses are semi-permanent and temporary
Water supply	High turbidity in local streams supplying domestic water supply causes breakdown of pumps and allied equipment no backup power for rural water supply systems so power disruption impacts water supply Inadequate field staff	Contamination of drinking water sources	High dependency on groundwater for water supply; India Mark II Hand pumps get submerged; inadequate funds for elevating pumps
Health	Some healthcare centres are sited on landslide prone areas	Some health care infrastructure situated in disaster prone locations even inundating the approach road and the infrastructure do not follow resilient design norms; inadequate and DM trained personnel at field level and boats;	Most healthcare centres are sited on waterlogging prone areas The approach roads to most centres get damaged due to flooding/ waterlogging Power disruptions impede functioning of centres
School/ Education	Inadequate funds for periodic maintenance and timely repairs of schools including for retrofitting	Approach road to primary and secondary schools get inundated	Most schools are sited on waterlogged prone areas
Agriculture	High Proportion of cultivable area unirrigated;	More than 50% Rabi season cultivated area	Waterlogging and poor damage making land

(continued)

Table 23.3 (continued)

Sector/ dimensions	Almora (flash floods, landslides, water shortages, cold waves, forest fires)	Puri (floods/cyclones/ drought)	Gorakhpur (flooding/ waterlogging/ water scarcity in summers
		unirrigated Inadequate trained person-power at field level of agriculture department	unfit for cultivation Food storage godowns sited on waterlogging prone locations
Animal Husbandry and Fisheries	Unsafe cattle keeping locations Nutritional feed inadequate against required to cattle during cold wave' Inadequate and trained person-power at field level of Animal Husbandry department	Unsafe cattle keeping locations; inadequate green fodder storage facilities for supply during flooding Inadequate and trained person- power at field level of Animal Husbandry department	Unsafe cattle keeping locations; Inadequate water availability for cattle during summers Inadequate fodder availability during waterlogging Increase in disease incidence in cattle in floods
Fisheries	No consultation	Inadequate and trained person- power at field level of Fisheries department Lack of scientific methodology for assessing fishery production and loss Conventional fishing boats not of adequate strength	No consultation
Urban	Almora Municipality Few slum pockets; haphazard developments; water supply and drainage infrastructure not resilient; poor financial resources and high dependency for finances from State and Central government	Puri Municipality High slum population; hazard development; road and electricity infrastructure not resilient; weak coordination between the Municipality and various parastatal agencies on infrastructure development; poor financial resources and high dependency for finances from states and central government	Gorakhpur Municipal Corporation High slum population; haphazard development; road and electricity infrastructure not resilient; very poor drainage infrastructure poor financial resources and high dependency for finances from State and central government.

This approach increases risks in most cases to future disasters. Though some anecdotal evidences from Almora indicate that reconstruction of infrastructure has been undertaken adopting higher technical and financing norms but it is yet to be mainstreamed.

Table 23.4 Economic sector

Sector/ dimension	Almora (flash floods, landslides, water shortages, cold wave, forest fires)	Puri (floods/cyclones/ drought)	Gorakhpur (flooding/ waterlogging/water scarcity in summers)
Roads	Roads constructed in cutting instead of balancing the cutting and filling. It makes the slopes unstable At most locations cross-drainage works do not follow resilient norms due to higher cost Unavailability of required width of land for roads from forests Inadequate budget for maintenance/ repairs/ renovation Inadequate and trained field staff	Inadequate funds for maintenance and periodic repairs of roads At most locations cross-drainage works do not follow resilient norms due to much higher cost	No consultation
Irrigation/ Flood Control	Inadequate knowledge of local geology causing subsidence of canals. Need higher norms of financing in hilly areas but they are same as for plains	Lack of fortnightly/seasonal forescasts on sudden high inflows into (majorly silted up) Hirakud dam during mid- sept-oct causes sudden release of water downstream into Mahanadi delta; Reduced discharge capacity of main rivers and its secondary and tertiary tributaries in delta due to siltation	High temperatures crack embankments.; situation of cracked embankments worsens when exposed to intense rain; Inadequate field staff
Electricity	Ageing electricity infrastructure	Ageing electricity infrastructure; most cables over the ground; corrosion of poles and conductors due to salinity Lack of funds for maintenance	Ageing electricity infrastructure; rapidly growing demand

- Cross-drainage works of roads have inadequate capacities to channelize runoff generated from even moderately high rainfall events due to budget constraints for constructing them with adequate cross-section and appropriate type (e.g. box culverts).
- In most cases alternative power backup system especially at the sub-district/ block levels are practically absent; hence damage to the main infrastructure causes breakdown of services/ functions to be provided by that system.

Table 23.5 Environment sector

Sub-sector/ dimension	Almora (flash floods, landslides, water shortages, cold wave, forest fires)	Puri (floods/cyclones/ drought)	Gorakhpur (flooding/ waterlogging/water scarcity in summers)
Forest/ Land/Rivers/ Streams	Hilly terrain-difficult to access; Rapid loss in vegetation making the slopes unstable; Large forest areas of species that catch fire easily; Capacity of local streams inadequate for increasingly intense rainfall events; and High turbidity in local streams supplying domestic water supply	Reduction in forest and open land areas for flooding buffering due to land conversation High groundwater salinity	Reduced discharged capacity due to siltation of riverbeds of major streams Combined with increasing high intensity precipitation events cause floods High dependency on groundwater for water supply Landuse improper; drainage congestion due to obstructions by embankments and roads.

- The Government considers damages of only infrastructure while flow-losses (such as reduction in employment, erosion of livelihood base of the communities etc.) are often not accounted for. The ground reality is that there is enough evidence of increased erosion of livelihood assets and financial base of communities due to disasters.

23.5.2 *Fragile Environmental Systems*

- High environmental degradation viz. groundwater salinity, soil erosion, and unstable hilly slopes, low land productivity;
- Improper land use that reduces flood buffering/ absorption capacity of natural system;
- Reduction in capacities of natural drainage/ rivers/ streams due to infrastructure without considering long-term (negative) consequences. For example, jacketing of the rivers by embankments has led to increased siltation in the river raising its bed, thereby reducing its discharging capacity.

23.5.3 *Institutional Gaps*

The institutional gaps relate to technical standards, regulations, and departmental or organisational systems such as Office Circulars, Government Resolutions and norms for financing, that govern the way an organisation plans, designs, grant approvals, executes and monitors its infrastructure and development programmes. It is a critical element that enables or constrains resilient-development.

- Appropriate technical standards for various components of infrastructure (for example, retaining walls of roads in hilly areas) need to be developed and supported through enabling norms of (increased) financing, (lower) technical norms (for example, reducing dependability requirement for design of reservoirs in drought-prone area from 75% to 50 percent), (lower) Benefit-Cost-Ratio especially in hilly terrains, and adequate budget. Evidences from cases indicate that one or more of above have constrained development or reconstruction/ repairs of infrastructure in a resilient manner.
- Need to strengthen systems for timely technical and administrative approvals and disbursement of funds across various levels of departments. Delay has seen to affect cost overrun and delay in realizing benefits from infrastructure.

23.5.4 Lack of Capacities of Communities and Departments

23.5.4.1 Communities

- High poverty levels and significant population of socio-economically weaker sections of the society; high dependence of population on climate-sensitive agriculture sector combined with inadequate irrigation development, and limited options for livelihoods in (secondary and tertiary) sectors. The situation worsens further with recurrent disasters eroding the valuable livelihood assets of such community, in some cases leading to increased indebtedness;
- Severely constrained capacity of communities to finance reconstruction of damaged assets/ houses in a resilient way due to high poverty; and, poor knowledge of disaster prevention, mitigation, preparedness and management.
- Technological know-how on resilient measures (Agriculture, Housing, Drainage, Sanitation, etc.)

23.5.4.2 Departments

- Very limited capacity of various departments at the field level for assessing loss, implementation and monitoring infrastructure development and restoration, and disaster response; inadequate maintenance and repair budgets; lack of resilient design technical standards, guidelines, procedures and unclear financing channels for developing resilient infrastructure especially at and below the state level; lack of coordination between departments with mandates for planning, constructing, maintenance and repairs, and restoration of sectoral infrastructure. For example, between the Irrigation Drainage Division and Puri Municipality, with the former involved in planning and development of sewerage system/drains, the responsibility of maintaining the same is entrusted to the later. In addition, there is lack of effective coordination between departments for response, temporary restoration and reconstruction of infrastructure.

- Weak finances of Urban Local Bodies due to lower collection/ recovery of taxes/ fees, constrains their ability to operate and maintain various municipal services effectively; more importantly for reconstruction/ repairs of infrastructure damaged due to disasters. Their capacities are further constrained by limited technical manpower. Hence, appropriate assessments and systems need to be put in place to increase their capacities.
- There is no systematic damage database at the district level that assimilates damages incurred by asset and infrastructure of departments. The scale of data being collected by departments varies across departments with reporting being at either revenue tehsil level or block or division/sub-division levels. In absence of systematic data compilation system, it becomes practically difficult to assess the damages at district and sub-district levels.
- The main coordinating agency at district level, the DDMA, has limited staff and inadequate technical capacities to promote resilient practices and approaches across activities of various departments. Hence, there is dire need to strengthen DDMAs with adequate staffing and providing them access to pool of experts through funding mechanism like State Disaster Response/ Mitigation Funds.

23.6 Conclusion and Way Forward

23.6.1 Zooming Out: Larger and Strategic Implications for State and National Actions

The SLD process fostered shared-learning horizontally across departments at the district level and vertically between the district and state level departments. For example, the SLD process provided platform at district level for departments to understand and appreciate challenges of inter-departmental coordination that exacerbate vulnerability. In addition, the process provided an opportunity for state level departments/ authorities to understand issues from their field level district and lower level offices. The findings led to revising the District Disaster Management Plans (DDMPs) of the three districts and making recommendations to State Level Departments including the State Disaster Management Authority (SDMA, Odisha; Disaster Management and Mitigation Centre–DMMC, Uttrakhand; and, SDMA, Uttar Pradesh). Apart from the development issues that have been flagged for resilient development in the foregoing, we present an indicative list of strategic actions required to address challenges of sustainable development 2030 in face of projected increasing intensities and frequencies of climate-induced disasters:

(i) The *Climate Resilience Framework and Shared Learning Dialogues are useful tools* for mainstreaming DRR-CCA in Development. There is need to scale-up such sub-national experiments to national level;

(ii) *Increasing investments in resilient-agriculture*: Given the fact that agriculture is still the backbone of majority of rural population despite its declining contribution to GDP, strengthening systems for sustaining and increasing resilience of agriculture to disasters need to be devised and implemented through concerted efforts. For this, three specific actions could be promoted: promoting sustainable integrated agriculture-horticulture-aquaculture-animal husbandry model of agriculture; increasing vigour of implementation of District Agriculture Contingency Plans by creating systems of decentralised storages of inputs (such as needed varieties of seeds), building capacity and reach of extension system, and technologies/ techniques to reduce production losses due to untimely hailstorms/ rainfall; the MNREGA programme could be used to enhance supply of agriculture labour especially when there is shortage during peak agriculture season. That way, MNREGA can also meet the challenges of providing 100 days of employment, while making the agriculture system climate-resilient.

(iii) *Strengthening support mechanism for agriculture:* Such as improvement of drainage and waterbodies conservation (as demonstrated in Gorakhpur) or plantation in hilly terrain (Almora) through MNREGA can help increasing effectiveness of drainage/run off and hence saving crop damages.

(iv) *Diversifying livelihoods through enhancing effectiveness of social programmes* using area-based approach: Tackling poverty for achieving SDGs will need increased and sustained level of investments in resilient infrastructure, job creation in secondary and tertiary sectors, and enhancing effectiveness of social programmes. On social programmes, states and districts will need to be prioritised not just on poverty levels but also by considering dimensions of vulnerability in social, economic and environment sectors, as presented earlier. For this, an area-based approach to development could help to ensure convergence of all relevant programmes. Relevant advances in Science and Technology, such as remote sensing and GIS, for monitoring, evaluation and dissemination could be of great use to foster convergence and effectiveness;

(v) Urban-rural linkages are important to understand, for poverty alleviation. The urban areas in India are increasingly facing problems of flooding/ waterlogging. The problems not only impact the economy of urban areas but also the economy of rural areas around/ connected to it. Hence, adequate emphasis for building urban resilience to increasing climate-induced disaster-risks is very much needed. This will need investments in building resilient infrastructure not just in urban and rural areas but also between them for smooth flow of goods and services.

(vi) Coherence and Convergence of programmes/ schemes for increasing their effectiveness:

- Coherence in planning and implementation of components within and across schemes is much needed to realise their full productivity and potential, leading to enhancing effectiveness to address challenges posed by climate-induced disasters;
- For the needed convergence and effectiveness of development programmes/schemes coordination mechanisms at the state and district level will need to be strengthened by forming empowered inter-sectoral coordination committees. To start with, the capacities of offices of District Collector/Divisional Commissioner/ sub-national Development Boards/ Urban Local Bodies needs to be developed by creating and providing tools (such as dashboards) and models for integration of risk reduction, effective coordination across departments and convergence of schemes.
- Spatial screening for risks and fund allocation for (higher) disaster-prone areas:
- All infrastructure projects (such as roads, irrigation and drainage) need to be screened for potential impacts on environment and climate risks. Specific tools such as project appraisal templates need to be developed for granting approvals to foster such screening;
- Areas specifically prone to climate-induced disasters have been spatially identified in the country but it needs to be updated periodically given the changing scenario of vulnerabilities every 3–5 years by incorporating further details regarding severity/recurrence. In addition, Government Orders or Circulars are needed for using such information for planning and development of projects backed with appropriate/increased level of budgetary allocations across various development sectors and sub-sectors;
- Within a specific geographical area (state/district/block/urban area), higher financing norms for investment in new infrastructure are needed in specific administrative divisions that are disaster-prone.

(vii) *Enhancing O and M* funds *for building resilience against small magnitude and frequently occurring disasters*: annual or biennial climate-induced disasters, though they are of smaller magnitudes (for example, floods occurring every year or in 2 years), cause significant cumulative damages in longer term to assets of community as well as public assets and infrastructure. Hence, provision for allocating adequate funds for O and M or special repairs -as the case may need, is imperative to ensure that the assets and infrastructure remain useful and productive during their entire life;

(viii) *Incremental risk* management *approach:* Given the uncertainty in Climate Change projections and related changes in intensities and frequencies of disasters, an incremental risk management approach is essential—revising design norms of infrastructure and/ or implementation approach. It will need a robust monitoring and evaluation system of all major programmes/schemes that provide feedback for revision in design and/or implementation approach;

(ix) *Prioritizing the vulnerable:* Women, children and the aged in rural areas, and the slum population in urban areas are generally are more vulnerable. Thus, concerns of these groups need to be integrated and articulated well in relevant policies, plans and programmes of various ministries/ departments and disaster management.

(x) Strengthening links between Sustainable Development Goals (SDGs), Sendai Framework for Disaster Risk Reduction (SFDRR) and Paris Agreement: Integrating DRR and CCA into development furthers achievements on SFDRR, SDG and the Paris Agreement (Gupta et al. 2016).

References

Chopde S, Singh D, Gupta AKA, & Wajih S (2016) Unpacking mainstreaming DRR-CCA in the sub-national level development planning: insights from three states in India. GEAG, Gorakhpur (UP, India) and ISET, Colorado (US)

Financial Express, July 28, 2017. Family debts main cause of Indian farm suicides. https://www.financialexpress.com/india-news/family-debts-main-cause-of-indian-farm-suicides/739125/

Gupta AK, Chopde S, Singh S, Wajih S, Katyal S (2016) Prime Minister's Agenda 10: India's disaster. *Risk management roadmap to climate resilient and sustainable development*. GEAG, Gorakhpur (UP, India) and ISET, Colorado (US)

Hallegatte S, Vogt-Schilb A, Bangalore M, Rozenberg J (2017) Unbreakable : building the resilience of the poor in the face of natural disasters in climate change and development. World Bank, Washington, DC. © World Bank. https://openknowledge.worldbank.org/handle/10986/2535 License: CC BY 3.0 IGO

IPCC (2012) Managing the risks of extreme events and disasters to advance climate change adaptation. In: Field CB, Barros V, Stocker TF, Qin D, Dokken DJ, Ebi KL, Mastrandrea MD, Mach KJ, Plattner G-K, Allen SK, Tignor M, Midgley PM (eds) A special report of working groups I and II of the intergovernmental panel on climate change. Cambridge University Press, Cambridge, UK, and New York, NY, USA, 582

National Disaster Managemen Plan (2019). National disaster management plan. A publication of the National Disaster Management Authority, Government of India. May 2016, New Delhi, India

NSSO (2013). Proceedings of National Seminar on Results of NSS 66th Round, Bangalore, 20–12 June 2013

Shashikant Chopde has 25 years of experience in policy- and action- research and training in water management and rural livelihoods including 15 years on CCA, DRR, Urban Resilience and Mainstreaming CCA and DRR in development planning. He has been providing trainings on the themes as resource person to Central Government Ministries, State and District level departments, and national and international practitioners.

Anil K Gupta Professor of Policy & Strategies, Integrated Resilience. Head of the ECDRM Division, International Cooperation, Programme Director—Centre for Excellence on Climate Resilience, Nodal Officer for Central Ministries at National Institute of Disaster Management, (Govt. of India), New Delhi. He is also a member of the expert Team for WMO Climate Statement and Core Group Member of IUCN-Council of Ecosystem Management.

Dilip Singh has over two decades of experience in the development sector and has been working on climate change adaptation, resilience building and mainstreaming CCA-DRR in development planning for more than a decade. His experience covers the functional areas of decentralized participatory planning, project management, institutional development, monitoring and evaluation, qualitative and quantitative analysis, socio-economic impact assessment, and capacity building.

S. A. Wajih is a development professional working as researcher, directing projects, trainer and facilitator for last 40 years. He is the president of Gorakhpur Environmental Action Group since its inception and is currently President of the organization. Having published more than 60 papers Dr Wajih holds Ph. D. in Ecology and represented on board of various organizations like Oxfam, ISET, SPWD, IYF, and VHAI.

Chapter 24
Insight for Climate Resilience and District Level Developmental Planning for Disaster Risk Reduction in Himalayas: A Case of Uttarakhand

Ashish Panda, Taniya Gupta, and Anil K. Gupta

Abstracts The Himalayan ecosystem is the key factor to ecological security of the Indian landmass providing forest cover, feeding perennial rivers which are the source of drinking water, irrigation and hydropower, inhabiting biodiversity, providing a rich base for agriculture system and spectacular landscapes for sustainable tourism. For centuries, snow (ice) has supported human survival in hilly regions. Fallen snow on the land provides the moisture for farming and pastures. Glacial snow melts and feeds the streams that are the lifeline of the hilly settlements. Due to Climate Change Himalayan glaciers have lost its volume endangering the flow of perennial rivers, especially in lean seasons, which (Northern rivers) are largely fed by melting glaciers. For past few decades the pattern has been observed that snow falls for shorter period and does not turn into hard ice making glaciers prone to melt early during summer. On top of it inconsistent and intense rainfalls has accelerated the melting of glaciers causing flash floods, washing away the homes, fields, trees and livestock. This chapter is focused on the Climate Change impacts on Indian Himalayas and suitable policy planning for reducing the impacts and increasing the overall well-being of the Himalayan people.

Keywords Climate resilient · Disaster risk reduction · Developmental planning

A. Panda (✉)
National Institute of Disaster Management, New Delhi, India

T. Gupta
CCDRM Project, National Institute of Disaster Management, New Delhi, India

A. K. Gupta
ECDRM Division, International Cooperation, CAP-RES DST Project, National Institute of Disaster Management (Ministry of Home Affairs), New Delhi, India

A. K. Gupta et al. (eds.), *Disaster Risk and Management Under Climate Change*,
Disaster Resilience and Green Growth,
https://doi.org/10.1007/978-981-99-4105-6_24

24.1 Introduction

The Himalayas, being the third largest body of snow on Earth after the Antarctic and Arctic poles, are known as the Third Pole (Huettmann and Roy 2013). Presently 10% of the Earth's landmass is covered with snow, out of which nearly 0.77% is in the Himalayas (Shiva n.d.). It has the maximum concentration of glaciers, outside the polar region. Nearly, 9.04% of the Himalayas are covered with glaciers, which feed the Lifeline Rivers of Asia and support nearly half of the human population. The Indian Himalayan Region (IHR) extends over 2500 kilometres in length between the Indus and the Brahmaputra river systems with geographical coverage of over 5.3 lakh km sq. Himalayas (including Indian Himalayas) is the source of eight largest rivers of Asia and is known as "water tower of Asia"(Cuo et al. 2014). Nearly 900 million people of Indo-gangetic regions, whose lives are sustained through the perennial rivers flowing all the way back from Himalayan glaciers, are under the trap of Climate change and the disastrous impact of degrading Himalaya (Mani 2009). Thus, Climate Change in this region is a matter of global concern.

India homes 5243 glaciers with the area of 37,579 km^2 and containing 142.88 km^2 of ice (Shiva). The receding of glaciers has accelerated with global warming, which also causes glacial lakes overflow and the phenomena of Glacial Lake Outburst Floods (GLOFs). Climate Change adversely affects the Himalayan ecosystem through altered precipitation, increased temperature and sporadic droughts. It may initially bring in widespread flooding, then drought in summer. Glacial runoff in the Himalayas is the largest source of fresh water for northern India and provides most of the water to the Ganges, Indus, Brahmaputra, Irrawady, Yellow and Yantze rivers. In the Ganga, the loss of glacier meltdown would reduce July–September flows by two thirds, causing water shortages for 500 million people and 37 percent of India's irrigated land (Shiva n.d.).

The economy in the Himalayas ranges from the subsistence level economy of the migratory pastoralists (sheep/cattle farmers) and small farmers to the prosperity of the fruit and tea estate owners. Local crafts and skills, casual labour, employment in the fruit processing industries, tea gardens, tourism and pilgrimage etc. too play role in the economy of Himalayan region. Over-exploitation of the natural resources has led to loss of dense forest cover, soil erosion, increased silting of water bodies, delayed springs, and biodiversity loss of flora and fauna. Economic changes, increase in population, urbanization, tourism, the building of communication lines into the remotest areas etc. along with Climate change impact have further accumulated stressed the Indian Himalayas. In coming years, Climate change impact is predicted to intensify and be direr with devastating effects on food-energy-water nexus.

The Indian part of Himalayas extends over 12 states covering over 16% of the country's geographical area (ENVIS Center on Himalayan Ecology 2018). The pristine ecological balance of the region is withering away and is under tremendous stress of Climate Change. Uttarakhand.

24.2 Rationale and Need

A 4 × 4 assessment, Climate Change and India, 2010 (Kumar et al. 2010) has been carried out by the Ministry of Environment & Forests, Government of India which indicates the projected change in climate in the year 2030 in comparison to the year 1970 with increase in Extreme Temperature and Precipitation. Intensity in Extreme Precipitation is also projected to increase with more number of rainy days. According to this report, the following projections were established:

- Annual mean temperature to increase from 0.9 ± 0.6 °C to 2.6 ± 0.7 °C in the 2030s
- Net increase in temperature from 1.7 °C to 2.2 °C with respect to the 1970s
- With respect to 1970s, winter temperatures (October, November and December) to decrease by 2.6 °C.
- Thermal discomfort is likely to increase in the 2030s, with the THI (Thermal Humidity Index) > 80
- Annual Rainfall may vary between 1268 ± 225.2 and 1604 ± 175.2 mm.
- The precipitation shows a net increase (5% to 13%) in the 2030s with respect to the simulated rainfall of the 1970s in the Himalayan region by 60–206 mm. Some areas of Uttarakhand may show increase in precipitation by even 50%.
- Increase in the sediment yield in the Himalayan region up to 25.
- Western Himalayan states have shown significant decreasing trend of 2.1 and 2.7 days per decade respectively. The report claims that the pre-monsoon maximum temperatures have increased over the Himalayan region.

The Himalayas is one of the world's most sensitive hotspot to Climate Change manifestations (Singh et al. 2011). Climate Change is real and globally it is pacing at an unprecedented rate threatening the reservoir of natural resources and the environment associated with rapid urbanization, exploding population and economic development. Among the many threats imposed by Climate Change in the Himalayan region undoubtedly glacial retreat and its consequences on surface water flow is predominant but many other threats like drought, flash flood, extreme rainfall, forest fire, low value agriculture and loss of biodiversity are not to be disregarded.

Ecosystem is the network of biotic element (animal, plants, and microorganisms) and abiotic element (environment and physio-chemical factors) which functions due to energy gradient. Imbalance or disruption of any element affects the whole system, which is interlinked. Unfortunately this climatic change is extreme and adverse thus detrimental to ecosystem and its components.

Ecosystem for decades is being subjected to human-induced pressures, such as excessive use of natural resources, fragmentation and degradation of natural habitats, etc. Change in plant phenology is one of the earliest observed responses to global Climate Change and can substantially damage plants and animals life cycle that depend on seasonally available resources (Kumar et al. 2010). It has been well established that species in some ecosystems are very strongly adapted to the long-prevailing climatic pattern that they become vulnerable even to slightest changes in ecosystem. Increased level of carbon dioxide in the atmosphere and consequent global warming also have a potential effect on the flowering period of plants (Kumar et al. 2010).

Deviation in the arrival of monsoon, long winter dry spells, increased frequency of forest fires during winter, early flowering/fruiting of native trees, such as *Rhododendron spp*. and *Myrica esculenta*, etc. are another consequences of climate change (Kumar et al. 2010).

24.3 Impact of Climate Change

24.3.1 Glacial Retreat

Glaciologists expect glaciers all around the world to continue to retreat in coming years due to continuous warming of the planet. Recent surveys have recorded a 10% reduction in the area of glaciers in the last three decades, from 48,860 square kilometres (18,865 sq. miles) in the 1970s to 44,438 sq. km (17,158 sq. miles) today (Greenpeace 2007). However, at which rate the Himalayan glaciers are retreating is still not known accurately due to insufficient research and knowledge of glacial dynamics of Indian Himalayan glaciers. The melting glaciers would mean floods (fast snow melting coupled with intensive rainfall) and fast run-offs in the rivers in the short term and droughts and water scarcity in the long term. Glacial run-offs will cause area's erosion, river discharge and sediment patterns, which will directly impact the hydropower reservoirs, new and planned constructions, agricultural lands, irrigation canals, streams as well as the cropping patterns in the Gangetic Plains and Himalayan region (Table 24.1).

Table 24.1 Retreat of some of the Himalayan Glaciers monitored by GSI (Source: Geological Survey of India 2011)

S no.	Name of the glacier	Period of monitoring	Average retreat (m/y)
1	Bandarpunch	1960–1999	25.5
2	JaundarBamak	1960–1999	37.3
3	JhajjuBamak	1960–1999	27.6
4	Tilku	1960–1999	21.9
5	Gangotri	1935–1996	18.8
6.	Pindari	1906–2001	15.21
7.	Chipa	1961–2000	26.9
8.	Ramganga	1962–2002	48.8
9.	Bara Shigari	1906–1995	29.8
10.	Sara Umga	1989–2004	43.3

24.3.2 Agriculture and Vegetation Shift

Tropical and moist deciduous forests, which once were luscious green, covering entire Sub-Himalayan area, has mostly been deforested for commercial lumber and agricultural land purposes. The Himachal region, once housed species of pine, oak, rhododendron, poplar, walnut and larch, has also been almost completely deforested (Kumar et al. 2010). Remains of forest cover are now available only in inaccessible areas and on steep slopes. Unfortunately, the projected impacts of Climate Change (continued rise in temperature, glacial retreat etc.) are most likely to influence the hill agriculture of Himalayan regions. If the warming of temperature continues to spike up and upward shifts in agriculture maintains loss of permanent pastures and grassland, which are already in very low area. It will cause less availability of fodder and ultimately will adversely affect livestock sector and agriculture.

Unjustified and illogical land use has brought cultivable land at the brink of extinction. Deficit in food production is growing in recent times in Jammu and Kashmir (Kumar et al. 2010). Horticultural crops such as Apple is showing decline in production and cultivable land for apple, particularly due to decreasing trend in snowfall. This trend of decline in area under apple and other fruit is comparatively higher in marginal and small farms. Apple productivity recorded a drop of about 2–3% over the past years, in some cases with marginal farmers, the decrement dropped up to 4%.

24.3.3 Forest Density

More than 65% of Himalaya's geographical area is under forests, representing one-third of the total forest cover and nearly half (46%) of the very good forest cover of the country (Kumar et al. 2010). Forests provide life supporting, provisioning, regulating, and cultural 'eco-system' services to millions of people. Over 9000 Himalayan glaciers and high altitude lakes form a unique reservoir storing about 12,000 km^3 of fresh water (Arora et al. 2010). A study on impacts of Climate Change on forests in years 2050's and 2080's claims *"shifts in forest boundary, changes in species-assemblage or forest types, changes in net primary productivity, possible forest die-back in the transient phase, and potential loss or change in biodiversity"* (Kumar et al. 2010). It is projected by 2050's majority of the forest biomes in India will be extremely vulnerable to the anticipated changes in climate and almost 70% of the vegetation will be not adaptable enough to then existing eco conditions.

India is the sixth most vulnerable country in the world in terms of facing extreme weather events, as per the 2016 Global Climate Risk Index Report released by Germanwatch (2018). The report claims that India lost over 21 Billion US Dollars' worth of properties in 2016.

24.3.4 *Health*

Climate Change is believed to have negative effects on ecologically vulnerable population groups, tribes and the poorest. These could include direct health impacts such as heatstroke, and indirect impacts such as increased diarrhoea risk from water contamination via flooding, or higher risk of mortality from the impact of large-scale loss of livelihoods. Qualitative description of the impacts of Climate Change on health issues in India such as nutrition availability, communicable diseases, heat stress, vector-borne diseases vis-a-vis heat waves, extreme weather events, increase in mean annual temperature, air quality, aeroallergens, droughts, and increase in UV radiation loss in forest litter and wood used for heating purposes in the cold season.

24.4 Causes

24.4.1 *Dam Constructions*

In the last few years, Pakistan, India, Bhutan and Nepal have prepared plans for several hundred dams building in the Himalayas, with possibility of capacity additions of over 150,000 MW in the next 20 years. Most of these dams are proposed in the high-seismic zones in the Himalayan, which is prone to landslides, flash floods, and earthquakes. The melting of glaciers due to Climate Change will phenomenally increase the inflows to these dams challenging its safety and increase the problem of sedimentation in reservoirs etc. The accumulation of sediment behind these dams also deprives downstream plains of nutrients and silt deposits, the source of their fertility (Fig. 24.1).

24.4.2 *Forest Fire*

Forest fire in western Himalayan region has become a new normal causing large scale dent to the forest ecosystems. More than 50% in Himalayan forests in Uttarakhand are prone to high incidence of fire, mostly due to human activities. In India, majority of fires are due to humans and natural causes of fires such as lightning are very less (Bahuguna and Singh 2002) and causing adverse ecological, economic, health and social impacts in the region. Causes of forest fires in the western Himalayas of India are due to (Bahuguna and Singh 2002) burning of grass by villagers for a good growth of grass in the next season; use of fire for extracting and collecting honey, gum etc. Accidental fires are also the cause of carelessness of humans, such as throwing of burning.

Frequent drought is also a known cause of forest fire. Drought reduces the barrier to fire (natural and man- made) by reducing fuel greenness, snow pack and moisture

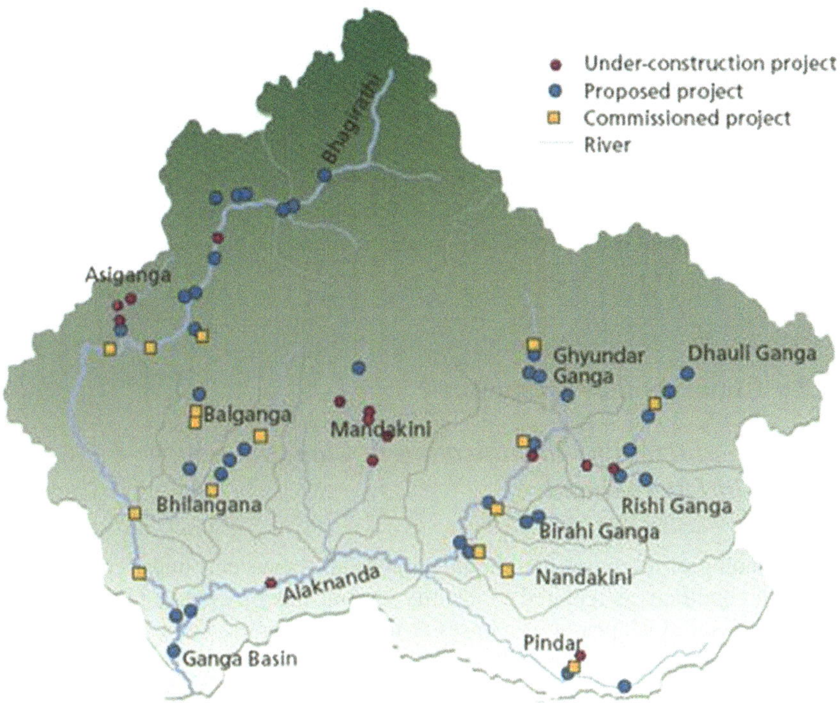

Fig. 24.1 Dams in Uttarakhand; Source: Wildlife Institute of India

Table 24.2 Forest fire in Indian Himalayan, Source—Ministry of Environment, Forest and Climate Change

Sr. no.	Year	Loss in hectares	Revenue loss (Rs. in lakhs)
1	2010	1655	2
2	2011	238	0.3
3	2012	3025	43
4	2013	388	4
5	2014	942	23.5
6	2015	710	8
7	2016	4538	46
8	2017	1251	18
9	2018	4538	86

present in the vicinity of forest. Drought sets back tree growth for years. Loss of foliage and carbohydrate reserves during drought period impairs growth in coming years of drought (Table 24.2).

24.4.3 Outmigration

About 86% of the out migrants are males with particularly 51.4% comprising of age group 30–49 years and 35.5% of age group 15–29 years (Mamgain and Reddy 2017). The increasing trend of this migration has affected the gender responsibilities particularly increased the hardship of women. In addition, the ecosystem services provided by the forests, biodiversity and the natural landscapes is not being able to link to the livelihood sources of rural population due to the huge outmigration (Joshi 2018).

24.5 Concepts and Framework

24.5.1 Vulnerability Identification to Climate Change in Uttarakhand

Understanding of responses and the functioning of ecosystems under continuously changing conditions is required to identify the impacts of Climate Change and to develop advanced adaptation responses at the crucial times. Local species extinctions need to be key issues to address, as this change is irreversible. This is crucial, especially because of a strong link between biodiversity and ecosystem functioning in the maintenance and existence of ecosystem services. When it comes to having a planned process to tackle the climate change issues to prepare a climate resilience system and society, it is even more necessary to learn about the sectoral vulnerabilities in the region. Such an effort was undertaken in one of the prominent hilly town of Almora in Uttarakhand. Shared learning dialogues with the state government officials as well as focused group discussions with stakeholders and local communities of Himalayan region is proved to be key sectors in identifying vulnerabilities of community and local ecosystem in fighting Climate Change issues. Important sectors with their vulnerabilities can be understood as under:

24.5.1.1 Agriculture

Abundance of small, marginal and fragmented land holdings are predominantly dependent on rain-fed agriculture. The land are mostly owned by men and tilled by women. The sector constantly dwindles with the problem of soil erosion and high production cost due to inadequacy in infrastructure and farm machinery. Other issues faced by this sector are reduction in crop genetic diversity and low/unavailability of inputs and access to last mile connectivity. Migration of local community to economically viable towns/cities is also a challenge to this sector.

24.5.1.2 Disaster Management

Densely populated mountain eco-system, slight disturbance in the ecology of the region is sure to jeopardize the life support strategy of the region as also in the downstream areas. Drift in monsoon pattern by ±20–30% in a season can cause agricultural droughts or severe floods and can significantly affect the crop productivity and even disturb the hydrological cycle. Uttarakhand is increasingly becoming more susceptible to ecosystem damages like soil erosion, landslides, prolonged dry spells, glacier recession, erratic precipitation, extreme climate events and rapid loss of habitats and biodiversity.

24.5.1.3 Energy

Reduction in snow cover in the Himalayan glacial systems means decreased discharge in the rivers thereby lower generation in Hydro Power Plants, which would widen the deficit of power supply demand in Uttarakhand. Extreme weather events also pose threats to physical infrastructure assets.

24.5.1.4 Forestry and Biodiversity

Gradual shift of plant species towards higher altitudes, forcing local communities to shift to new agricultural and fodder species is the emerging challenge due to the rise in average temperature in the region. The biodiversity in the higher altitudes will have less place to occupy and will be at increased risk of extinction. Increase in global temperatures will cause an increased occurrence of GLOFs threatening the existence of alpine meadows. Anticipated occurrence of heavy rainfall followed by long period of droughts will put strain on availability of water resources in the hills. The variation in rainfall and temperature may lead the variations in phenological events of many species, particularly the dominant oaks. The regeneration of many species is dependent on the monsoon rainfall i.e. Sal, Banj oak and Telonj oak. In addition, increased tourism related activities are causing severe stress on the fragile ecosystems.

24.5.1.5 Human Health

Extreme weather events and the resurgence of disease organisms and vectors is posing a severe impact on health in this region. Morbidity and mortality due to exposure to thermal extremes such as cardiovascular and respiratory diseases are emerging. Climate Change has impacts on activity and life cycle of infective vectors and parasites, leading to upsurge in incidences of vector-borne diseases, e.g. malaria, dengue fever and several types of encephalitis. Altered local ecology of water-borne and food-borne infective agents because of higher temperatures, leading to changed incidences of diarrhoea and other infectious diseases such as cholera. Altered food (especially crop) productivity due to changes in climate, weather events and associated pests and diseases often lead to malnutrition, hunger and consequent impairment of child growth and development.

24.5.1.6 Livestock and Animal Husbandry

Changes in the climate modify the reproduction, maturation, survival rate and dispersal of vector species and consequently bring about changes in the time, duration, severity, morbidity and mortality of the diseases borne by these vectors. Decrease in the availability of fodder and water is leading to discouragement of animal rearing in the light of unavailability of food for domestic animals.

Two important viral diseases (foot and mouth and peste des etits ruminants diseases), used to occur during monsoon have now started occurring during the peak of winter and the months immediately following the winter.

24.5.1.7 Roads

Blasting, excavation and chipping of mountain slopes for the construction of roads have resulted in frequent phenomenon like soil erosion, which leads to uprooting of large trees, destruction of lower plants and siltation of rivers. It also creates geological disturbances causing the movements of slip zones, cracks, fissures and weak planes. Road-cutting operations, downhill movement of landslide material and disposal of excavated mass from road construction degrades the natural landscapes as well as natural inclinations of hills. It also disturbs natural drainage thereby inviting occurrence of flash floods.

24.5.1.8 Urban Development

Unplanned growth of towns causes immense pressures on the urban infrastructure and services, resulting in degradation of the urban environment and of natural resources. Depletion of forest area, loss of biodiversity, potential urban pollution in the form of air, water shortages and water quality, noise pollution, issues of solid and liquid waste management and landslides. Unplanned industrial activities and the virtual absence of adequate and appropriate wastewater treatment facilities result in untreated sewage, wastewater and solid wastes entering drains, lakes, water bodies and rivers.

24.5.1.9 Water Resources

Pressures on natural resources and environment, lack of water availability in the rivers, decrease in crop yields, jeopardizing food security, increased health concerns due to extreme events such as floods and droughts. Despite immense availability of water in the state, water is scarce for the local people, both for drinking and domestic use and for irrigation, which is due to the lack of proper enforcement of water management policy, lack of regulation on groundwater, poor water supply systems in urban and rural areas.

The discharge of the water sources has shown a considerable depleting trend during the last decade. Rural drinking water supply schemes (Uttarakhand Jal Sansthan) have been badly affected.

The natural rate of groundwater recharge is reckoned to be 31% of the total annual rainfall. Therefore, the soil characteristics and land use play a major role in recharging sloppy aquifers. Deforestation, grazing and trampling by livestock, erosion of fertile top soil, forest fires and development activities (e.g. road widening, mining and building construction) leads to the reduction in the infiltration rate and sponge action of the land and thus failure of the watershed. This results in an unchecked flow of water during the monsoon, causing a sudden swelling of streams and rivers, resulting in floods in the foothills and plains and there are droughts in the villages located on the slopes of the mountains.

24.6 Approaches of Climate Change Adaptation

Adaptation is the process of change to suit a changing situation or environment. It is a never-ending and continuous process. It means dealing with both sudden events and slow ongoing changes. It may include climate or weather, social and economic stresses, changes in ecosystem, policies, regulations, infrastructure, etc.

Four 'pillars'—science, action research, piloting and communication—are essential in finding solutions to climate change issues, if integrated in the developmental plannings in the Himalayan region. Science helps in filling knowledge gaps and answers key questions in the context of change in the Himalayan region and the scope of adaptation options. Scientific findings need to be proven on the ground and effectively communicated to the policy makers. We gain deeper understanding through action research on key areas of adaptation and systematically test practices and approaches on the ground to document their benefits and challenges. To ensure impact, the findings of action research need to be further developed through piloting and documentation. Pilot projects must be founded on a solid base of scientific evidence. Documenting the results of pilot approaches through scientific analysis is important to ensure their credibility at a larger scale. Finally, communication and dissemination of knowledge is essential for taking forward the achievements of the other pillars of action, and ensures that findings and results reach the intended audiences.

24.6.1 Medium to Long-Term Adaptation

- Generate knowledge base on impacts, driving forces of vulnerability and adaptation mechanisms, and disseminate among stakeholders at different levels

- Test adaptation options (e.g. drought resistant crops, improved irrigation systems)
- Elaborate an enabling policy framework allowing communities to cope and adapt
- Set up hydro-meteorological monitoring stations
- Conduct impact and sensitivity studies (e.g., crop models or sensitivity of water resources)

24.6.2 Climate Change Response Mechanisms

After the monitoring of the anticipated and observed impacts of Climate Change, strategic and inclusive initiatives, adaptive practices and mitigation practices are indispensable to tackle the Climate Change. The following measures need to be placed at ground level to augment sustainable development in the state:

- Diversified farming system, Farming on leased land common property resource development, Short duration, high value crops on flood prone areas
- Plough bullocks in support of small farmers, Backyard poultry, Chick rearing centers, Flood resistant animals like ducks and geese
- Village level seed banks, fodder banks and Tree nurseries
- Index based insurance for crops, Crop insurance schemes for production lines that have proven to be flood resistant.
- ICT-enabled climate information and services to farmers
- System for groundwater management
- System of rice intensification (SRI)
- Introduction of millet into the Government Public Distribution system (PDS) to encourage its production
- Inland fisheries cooperatives, Common Interest Groups (CIG) for goat rearers and Goat crèche (Goat Kids Rearing Centre)
- Non-farm service sector livelihoods
- Planned migration for alternative income opportunity
- Community managed livestock vaccination service and Livestock insurance system
- Drainage system with open ditches, culverts, De-silt drainage canals and strengthen bunds
- Socially inclusive, gender-sensitive early warning system
- Growing of flood resistant vegetation and earthling near the bank of rivers

24.6.3 Climate Change Adaptation in Uttarakhand

Uttarakhand is considered to be the Devbhumi of India, a hub of pilgrimages and Hindu temples in the country. This state is situated in the northern part of the country in the lap of Himalayan mountain ranges. It is the most vulnerable Himalayan region to climate related hazards and changing climate. Nearly 3/4th of the state's population is rural and depends on agriculture and allied practices for livelihoods. Climate Change will have the worst impacts on the agricultural based livelihoods (economy) of the state.

There is a need to involve communities in Uttarakhand so that resilience could be developed in various livelihood sectors. One such successful attempt has been made by the CHEA (Central Himalayan Environment Association) in the farming sector, which implemented a project for building climate resilience farming through community approach. The clear objectives of the project were:

- Capacity building of communities and local institutions i.e. Van panchayats for forest management
- Implementation of solutions apt for sustainable agriculture at community level and friendly to local environmental, social and cultural context.
- For this particular project, CHEA focused on forest management through capacity building of Van Panchayats. The key stakeholders in the project were Van panchayats of 15 selected villages of Almora District, market association of Haldwani and Nainital and farming communities (men and women).

CHEA adopted a participatory process for forest management and strengthening agriculture. Villages were divided into areas coined as 'Toks'. CHEA helped in training of community members for teaching forest management. Their imitative was not of teaching only, CHEA team also tracked down the natural resources availability, forest area, community practices, agricultural practices, flora and fauna diversity and animal pool and learnt thoroughly about the way of living of communities for many weeks. CHEA held meetings with farmers and Van Panchayat members regarding training, action taken, record maintenance and addressing community grievances.

Research and understanding of issues with scientific experts, forming focused group discussions and forming farmer clusters at Van Panchayat level has been conducted. This method assisted farmers at every level of agriculture and forestry practice by disseminating community based scientifically advanced techniques and practices. Cultivation of ginger, turmeric, garlic and herbs were suggested as new practices looking into the local ecosystem and its availability.

The first phase of the project (2006–2008) worked with 7 Van Panchayats of Lamgarh block benefitting approximately 400 families by strengthening their capacities to fight Climate Change (Bhamra et al. 2014). In its second phase (2008–2013), the project was extended to 15 Van Panchayats of Almora district benefitting 8117 people of the local villages and 1124 ha land was covered in this project (Bhamra et al. 2014). Major achievements and outcomes of the projects were:

- Adoption and encouragement of new agricultural practices looking into livestock and human needs need of communities and sustainability of ecology.
- Conversion of wasteland into growing green fodder for farmers.
- New mechanisms adoption for Van Panchayats for forest management.
- Providing market linkages to farmers for selling their agricultural and forestry produces.

24.7 Conclusion and Way Forward

The Himalayan mountain system is extremely vulnerable to global warming. Adaptation and mitigation measures intended to cope with Climate Change can create opportunities as well as offset the dangers of a warming planet. There is lack of studies and basic data in this area, which is one of the biggest limitations in understanding the accurate impacts of Climate Change on Indian Himalayas. Uncertainty about the rate and shift of the Climate Change make conditions of Himalayan region more gruesome. To reduce uncertainty, there is a need for well-equipped baseline stations, long-term monitoring, networking, open data exchange, and cooperation between all Himalayan states and countries. Community involvement must be emphasized in the developmental planning process. For achieving sustainable livelihoods, we need to build the capacity to adapt and strengthen the socio-ecological system in the face of Climate Change. Adaptive policies and climate resilient plans need to be practiced, to reverse the human drivers of Climate Change, into sectors viz. land use, water management, disaster management, energy consumption, and human health etc. Hazard mapping would help both decision-makers and local communities to understand the current situation and then it would be possible to anticipate or assess the flexibility to adapt to future changes through proper planning and technical design. There is a need to train our farmers about weather forecasting and early warning systems to minimize losses in climate-related risks through awareness generation camps for climate adaption strategy.

Traditional knowledge can be used in planning adaptation in response to Climate Change at household level. In the absence of additional input from policy makers and the scientific community, well-known and locally employed coping strategies, such as rain-water harvesting system, use of alternate source of energy, using traditional practices in the agriculture field and shifting of cropping pattern, are mostly used. We need to strengthen community-level seed banks for contingency and alternative cropping strategies, to stock, seeds of main and contingency crops along with that of fodder and green manure plants etc. The Temple Committees and religious groups may play a vital role in the revival of the traditional crop diversity of the region by persuading local communities.

At local government level, there is a need for a planned approach to tackle climate related issues that should be in synchronization with nationally recognized and scientifically advanced techniques. The local plans should deal with the local issues as per the need and requirement of the local community. Participatory management approach needs to be followed while preparing developmental plans. This approach should be inclusive of knowledge repository that local communities carry drawing upon the changing environment. The local population, especially youth needs to be lured to stay back in their locales and not migrate to other parts of the country by way of providing them various alternate means of employment in their own villages/towns. This responsibility should well be executed in the State Disaster Management Plans, District Disaster Management Plans and other developmental plans (Table 24.3).

Table 24.3 Erratic rainfall pattern of Almora, Uttarakhand (Source—Pandey 2018)

Year	Jan		Feb		March		April		May		June		July		August		Sept		Oct		Nov		Dec	
	R/F	% Dep	R/F	% Dep	R/F	% Dep	R/F	% Dep	R/F	% Dep	R/F	% Dep	R/F	% Dep	R/F	% Dep	R/F	% Dep	R/F	% Dep	R/F	% Dep	R/F	% DEP
2013	50.6	10	166.0	255	13.9	−71	11.9	−53	5.5	−88	310.5	135	258.1	−14	244.6	−12	55.4	−63	28.2	−51	9.6	19	9.3	−56
2014	68.5	48	85.6	83	49.7	5	29.3	15	42.1	−10	52.3	−60	421.9	41	91.2	−67	67.3	−55	37.9	−34	0.0	−100	56.9	171
2015	39.5	−15	32.1	−32	129.4	173	55.3	117	11.5	−75	129.2	−2	290.3	−3	177.1	−36	33.4	−78	14.0	−76	1.1	−86	3.6	−83
2016	3.8	−92	14.2	−70	62.2	31	7.6	−70	56.4	21	126.4	−4	425.5	42	231.6	−16	77.9	−48	0.0	−100	0.0	−100	0.4	−98
2017	26.0	−44	5.1	−89	15.6	−67	24.1	−5	109.9	135	133.6	1	451.6	51	196.7	−29	215.6	44	0.1	−99	0.0	−100	6.0	−72

Annexure

References

Arora BR, Gosain AK, Juyal N, Kumar KK, Kulkarni AV, Sarwade RN et al (2010) Report of the study group on Himalayan glaciers. IIT Bombay

Bahuguna VK, Singh S (2002) Fire situation in India. Int Forest Fire News 26:23–27

Bhamra A, Bisht H, Mathew R, Reen R, Kulshreshta S, Prajapati S (2014) Climate adaptive practices: grassroots initiatives. development alternatives. UNICEF

Cuo L, Zhang Y, Zhu F, Liang L (2014) Characteristics and changes of streamflow on the Tibetan plateau: a review. J Hydrol Reg Stud 2:49–68

ENVIS Center on Himalayan Ecology. (2018). Indian Himalayan States. Retrieved from: http://gbpihedenvis.nic.in/him_states.htm

Geological Survey of India (2011) Report of the study group on Himalayan glaciers. Geological Survey of India

Germanwatch (2018) Global climate risk index 2018

Greenpeace. (2007) Expedition documents melting Himalayan glaciers. Retrieved from Greenpeace (Greenpeace East Asia): http://www.greenpeace.org/eastasia/news/stories/climate-energy/2007/himalayan-glacial-melt/

Huettmann F, Roy AK (2013) Three poles: the Arctic, Antarctic and Himalayas all connect. Retrieved from Daily News-Miner: http://www.newsminer.com/opinion/community_perspectives/three-poles-the-arctic-antarctic-and-himalayas-all-connect/article_8a954e96-ca8d-11e2-97b5-0019bb30f31a.html

Joshi B (2018) Recent trends of rural out-migration and its socio-economic and environmental impacts in Uttarakhand Himalaya. J Urban Reg Stud Contemp India 4(2):1–14. Available at: https://home.hiroshima-u.ac.jp/hindas/PDF/jurci/4_2/01_joshi.pdf

Kumar KK, Unnikrishnan AS, Kumar SN, Aggarwal PK, Vivekanandan E, Sharma S (2010) Climate change and India: a sectoral and regional analysis for 2030s. Ministry of Environment, Forest and Climate Change

Mamgain RP, Reddy DN (2017) Out-migration from the hill region of Uttarakhand: magnitude, challenges, and policy options. In *Rural labour mobility in times of structural transformation* (pp. 209–235). Palgrave Macmillan, Singapore

Mani R (2009) Impact of climate change on major river basins in India: The Indo-Gangetic-Plains. Retrieved from India Water Portal: http://www.indiawaterportal.org/articles/impact-climate-change-major-river-basins-india-indo-gangetic-plains

Pandey RE (2018) Climate change adaptation in the Western Himalayas: household level perspective on impacts and barriers. Ecol Indic 84:27–37

Shiva V (n.d.) Climate change in the Himalayas. Retrieved from Navdanya: http://www.navdanya.org/climate-change/in-the-himalayas

Singh SP, Basignana-Khadka I, Karky BS, Sharma E (2011) Climate change in the Hindu-Kush Himalayas. The state of currnet knowledge. International Centre for Integrated Mountain Development, Kathmandu, Nepal

Ashish Panda is a Science Graduate and an MBA (Disaster Management). He is presently a researcher on Climate Change issues. He has 25 years of experience of working in various ministries & departments of Government of India. At present, he is Deputy Secretary, NITI Aayog, Government of India.

Taniya Gupta is a Research Assistant at National Institute of Disaster Management. She is a graduate in Agriculture from Banaras Hindu University and postgraduate in Development Studies. She also has a diploma degree in Journalism. Ms. Taniya is very young professional and has one international research paper to her credit. She has worked with Centre for Science and Environment for a project on Economic, Social and Health issues of Chemical farming in the villages of Rajasthan.

Anil K. Gupta Professor of Policy & Strategies, Integrated Resilience, Head of the ECDRM Division, International Cooperation, Programme Director - Centre for Excellence on Climate Resilience, Nodal Officer for Central Ministries at National Institute of Disaster Management, (Govt. of India), New Delhi. He is also a member of the expert Team for WMO Climate Statement and Core Group Member of IUCN-Council of Ecosystem Management.

Chapter 25
Risk Management for Averting, Addressing and Minimizing Climate Related Loss and Damages

Kirtiman Awasthi and Somya Bhatt

Abstract Several efforts are being undertaken to address Climate Change through mitigation and Climate Change Adaptation actions through international commitments, national strategies and voluntary actions. However, it has also been well established that adaptation and mitigation efforts will not be able to completely prevent the negative impact of Climate Change and some amount of residual risk remains because of both extreme weather events like floods and slow-onset events like salinization. Recently, Loss and Damage (L&D) has become an issue of high interest in academia, politics, and development cooperation and accordingly discussion and concept papers are increasingly entering and fuelling the debate, with Paris Agreement (2015), reaffirming the Warsaw International Mechanism for Loss and Damage as the main vehicle and setting the stage for more focused deliberations under the UNFCCC process. This study discusses climate risk assessment in the context of L&D, discusses the Climate Risk Management approach and its framework and lists the possible options to reduce L&D at all stages of climate risks.

Keywords Loss and damage · Climate change · Climate risk management

25.1 Introduction

Ever since the establishment of anthropogenic factors as the main driving force behind rapidly changing climate and its impact on important sectors of the economy like agriculture, water, health, infrastructure, efforts are being undertaken to address Climate Change through mitigation and Climate Change Adaptation actions through international commitments, national strategies and voluntary actions. However, it has also been well established that adaptation and mitigation efforts will not be able to completely prevent the negative impact of Climate Change and some amount of residual risk remains because of both extreme weather events like floods and

K. Awasthi (✉) · S. Bhatt
Deutsche Gesellschaft für Internationale Zusammenarbeit-GIZ, New Delhi, India
e-mail: kirtiman.awasthi@giz.de

A. K. Gupta et al. (eds.), *Disaster Risk and Management Under Climate Change*, Disaster Resilience and Green Growth,
https://doi.org/10.1007/978-981-99-4105-6_25

slow-onset events like salinization (IPCC 2014). This phenomenon is called loss and damage (L&D) that represents actual and/or potential manifestation of impacts associated with the current climate and future Climate Change that negatively affect human and natural systems. L&D is usually the result of climate-related sudden-onset events and/or gradual changes in interaction with a specific development path that either reduces or exacerbates the risk of L&D (UNFCCC 2012a, b, c). In fact, weather and climate-related loss and damage coupled with slow-onset events, such as rising sea levels and soil degradation have increased dramatically over the past few decades. Subsequently, the L&D debate has become an important additional pillar of the international climate negotiations under the UNFCCC (UNFCCC 2018a).

Early attempts (1991, even before the UNFCCC) of the Alliance of Small Island States (AOSIS) representing amongst others developing countries highly prone to climate-related disasters to take up the issue of L & D in the international climate debate (establishment of an insurance pool) had not been successful. In the UNFCCC process, the topic gained importance since the Conference of the Parties in Warsaw (COP 19 in 2013). Referring to the Bali Action Plan (decided during COP 13, in 2007) that stated the need to consider risk sharing and transfer (e.g. insurance), parties decided in Cancun (COP 16 in 2010) to establish a Work Programme on Loss and Damage. After incremental progress was made over several conferences prior to COP 19, the Warsaw International Mechanism (WIM) to address L&D associated with the adverse impacts of Climate Change was established in 2013 (UNFCCC 2012a, b, c; UNFCCC 2018a). At COP 21 in Paris, L&D was incorporated under Article 8 of the Paris Agreement, which includes, among other things, the decision to continue with the WIM, the establishment of a clearinghouse for risk transfer in order to deal with L&D and the intention to set up a task force to develop recommendations for integrated approaches to avert, minimise and address displacement related to negative Climate Change Impacts. Article 8 also suggests several areas of intervention for comprehensive climate risk management (UNFCCC 2018a) (Fig. 25.1).

25.2 Rationale and Need

25.2.1 Key Discussion Areas for L&D

The Warsaw Mechanism for L&D stimulated a lot of debates around Loss and Damage bringing together a range of actors and stakeholders into it. The current debate on Loss and Damage is more focused on the gaps and challenges of climate disaster risk reduction and post-disaster relief and Climate Change adaptation, where there is a considerable lack of feasible and practicable concepts on risk assessment, risk transfer mechanisms, and assessment of loss and damage in disastrous events.

Potential L&D includes a range of negative effects and is not limited to tradable and, in monetary terms, quantifiable items. **Non-economic L&D** negatively affects

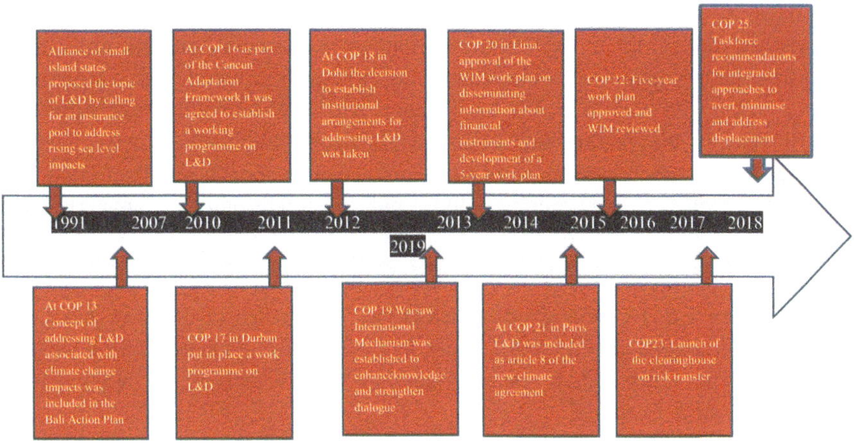

Fig. 25.1 Loss and damage chronology at UNFCCC (UNFCCC 2018b)

human welfare. Indeed, in developing countries, this may be even more substantial than economic losses. Currently, there is no commonly accepted definition on Loss and Damage in place, nor commonly acknowledged concepts or instruments that can assess and measure **economic** (measurable by market prices) and **non-economic** values (UNFCCC 2012a, b, c). The existing definitions do not differ between L&D that could be avoided by mitigation or adaptation and L&D that is considered unavoidable no matter how ambitious are mitigation and adaptation efforts (Verheyen 2015). This categorization is, however, highly important as it highlights the fact that enhanced global efforts to reduce greenhouse gas emissions and put in place effective adaptation and risk reduction measures can reduce future L&D to a certain extent (Verheyen 2015). Those impacts that are **unavoidable/residual L&D** and are beyond risk tolerance levels will need to be addressed using a range of other approaches such as risk transfer instruments (e.g. insurance in the case of extreme events), retention measures, and transformative measures such as planned relocation. In general, L&D can result from a range of negative Climate Change impacts occurring along a continuum of climatic hazards (and interactions) that runs from extreme weather events like cyclones to gradual changes like sea-level rises, melting glaciers, etc. Furthermore, L&D consider **direct impacts** (e.g. housing damage) as well as **indirect impacts** (e.g. migration due to housing loss) (UNFCCC 2012a, b, c). The actual manifestation of direct and indirect and of economic and non-economic L&D depends not only on mitigation and adaptation efforts but also on the specific development path chosen by a country.

Besides the discussions on the weak points of risk management of climate-related disasters, the subject of (adverse) slow-onset impacts is being discoveredand this field is to a much lesser extent developed. Concepts of assessing and addressing the slow-onset risks as part of L&D are in a initial stage and have not yet fully entered the development and adaptation strategies and measures in the scope of development cooperation and financial mechanisms. What is not measured is not managed and measuring slow-onset and indirect impacts from multiple stress factors with Climate Change as one of them are a great challenge. Uncertainty on the other hand (in particular, exogenous shocks cannot be predicted), paired with complexity, plays a major role here and calls for concepts and instruments that explicitly take uncertainty as a core feature into account next to other relevant parameters, such as the direct and indirect dimensions of slow-onset climate impacts and their impact on global / regional / local public goods and collective goods – material and non-material (Surminski 2012).

Current discussions highlight the following dimensions of Loss and damage:

(i) **Sudden L&D versus gradual L&D** (a consequence of disasters and slow-onset impacts): The first are losses that occur through one extreme weather event, the latter occurs gradually and can be best recognized by comparing longer time intervals.
(ii) **Economic vs. non-economic L&D**: The first is related to hardware (such as houses, land or vehicles), the second to non-material values, such as loss of a relative (in some concepts a human life loss includes a material and a non-material loss; in fact, it has both aspects) or the loss of a place that is felt at home
(iii) **L&D of materials vs benefit**: This refers to an economic view of the cost and benefit concept: Is it the material that determines our life quality or the benefit that we receive from the material, and what benefit is it (can we receive the same benefit from other sources)? If we apply the term of benefit, we can include the benefit of material and non-material sources. This may be a good starting point when discussing relief / coping schemes.
(iv) **Direct vs indirect L&D:** The first is the L&D that occurs as a direct consequence of a weather event or Climate Change impact. The second is a consequence of this direct loss or damage that may lead to adverse impacts in other fields that are linked to it.
(v) **L&D that can be coped with and those that lead to a significant decrease in life quality and adaptability** of the affected people (or threaten their entire existence like in the case of small island states). Life quality here means more than the livelihood; it includes all benefits that contribute to a person's / community's well-being. Some losses create a remaining "pain" which results in a net loss of life quality after all relief and coping measures have been applied. Such a loss could be the loss of a critical amount of arable land that is needed to feed a community as a result of a slow-onset climate impact.
(vi) **Pre and Post-L&D**: The first regards the ex-ante situation before L&D has happened and mainly deals with risk assessment, disaster preparedness, avoidable risk, and risk transfer. The second regards the situation ex-post situation when L&D has occurred and deals with L&D assessment, coping and compensation.
(vii) Losses of **private and public goods**: This encompasses individual losses or losses of the common property of a group ("club") as private goods, and

public goods on the local, regional or global level. Benefit from and interests in protecting the goods/values from the loss will have to be considered.

(viii) **L&D that decrease (short-term or long-term) adaptive capacity and those that don't**: Adaptive capacity is crucial for all living species to survive. And it is crucial for other species to benefit from the large genetic pool. For human beings, adaptive capacity requires flexibility (incl. ability to cope with losses) and innovation capacity (incl. learning by trial and error). If the ones that are the drivers for adaptation in a community are lost (e.g. through migration due to slow-onset impacts), the remaining community may lose adaptation capacity. Next to this, (clean) life space, nutrition, and water are a basic condition to survive. A loss of living space may lead to reduced adaptation capacity. However, this would need an in-depth investigation of the single case.

(ix) **Insurable vs non-insurable L&D**: There are risks that can be insured via a market-based approach and that compensate for L&D. This may include public subsidies for the insurance system if it is a merit good. Not all risks can be insured, but remain L&D. This largely applied to non-material risks. Here, further elaboration is necessary to become clear about the range of risks and the strategies and measures to assess and address the risks and the loss cases appropriately.

(x) **Positive impacts of Climate Change as negative L&D**: Widely neglected are the positive impacts of Climate Change that provide new opportunities in some areas. In how far these can be balanced with losses and damages needs to be discussed. We think that balancing is only applicable where the same people who are affected by the threat of L&D also have new opportunities form Climate Change or when on the national level a transfer mechanism from those who have a net benefit to those who have a net loss is installed/discussed.

(xi) **L&D at international vs. at the national level** (negotiating vs. managing): This refers to the level of managing L&D and the links between these levels.

(xii) **Differentiated L&D:** A loss, even though it may cover the same material or non-material loss, may not be the same for everyone. E.g., one person/community may suffer a lot from losing a traditional place, while others cope with it much easier. This may be due to weaker adaptive capacities and therefore require special support.

(xiii) Another challenging factor in assessing L&D is attributing Losses and Damages to anthropogenic (man-made) Climate Change in settings with multiple stress factors (There are multiple interacting systems and stresses in the countries and regions, where Climate Change is one of the parameters; hence loss and damage cannot be attributed to one single cause and needs to be discussed in the larger context of environmental degradation and climate change, where Climate Change is linked to impacts on natural and human systems; IPCC 2014); e.g. understand the role of Climate Change in migration patterns.

25.2.2 Need for a Comprehensive CRM Framework

Climate Change impacts are increasingly being framed in the context of risk. The IPCC has built, on key concepts from the disaster risk management discourse, and introduced the concept of climate risk in its Fifth Assessment Report (Working Group II AR5[1]). Emphasizing the link of Climate Change mitigation, adaptation and sustainable development the IPCC AR5 risk concept serves as a valuable complement to the previously used concept of vulnerability to Climate Change. It broadens the perspective to climate-related impacts triggered by extreme events and slow-onset changes. Any efforts to address climate risk and reduce vulnerability can influence the extent of Loss and Damage. This includes a range of options such as interventions to enhance adaptive capacities of communities, disaster risk reduction, poverty reduction, investments in climate resilient infrastructure. In this context, a comprehensive climate risk management approach that combines climate risk assessment and such options under one umbrella provides a way forward. Thus, CRM has become the overarching methodological framework for assessing Climate Change impacts and subsequent adaptation requirements Accordingly, both climate science and international climate negotiations stress the urgent need to develop and implement effective climate risk assessment and management approaches to avert, minimize and address L&D.

The Special Report on Extreme Events (SREX) by IPCC and its fifth Assessment Report (IPCC 2014) identified the development of a CRM framework for comprehensively reducing, preparing for, and financing climate-related risk, while tackling the underlying risk drivers, including climate-related and socio-economic factors. These risks associated with the hazard, vulnerability, and exposure have been illustrated in figure below (Fig. 25.2).

IPCC has attributed slow-onset events and some sudden-onset extreme climatic events to anthropogenic greenhouse gas emissions based on event-based trend analysis. This approach also identifies the difference between frequent and rare events, which are a key feature of climate-related risks. Risk assessment can be applied to sudden-onset and slow-onset climate-related processes unfolding over timescales from hours to days to months and years (Fig. 25.3). Climate-related impacts and the subsequent L&D is part of the overall CRM framework and needs to assessed and addressed as part of the overall CRM framework.

The six-step customizable Climate Risk Management (CRM) Framework presented here is designed to support public sector institutions, governments (when applicable) and other intended users to help them respond to climate-related risks at national, sub-national and local levels. The framework involves a comprehensive assessment of the climate and disaster profiles of an identified region—also known as the system of interest - for which a context specific climate risk assessment

[1]WGII: Working Group II assesses the vulnerability of socio-economic and natural systems to climate change, negative and positive consequences of climate change and options for adapting to it.

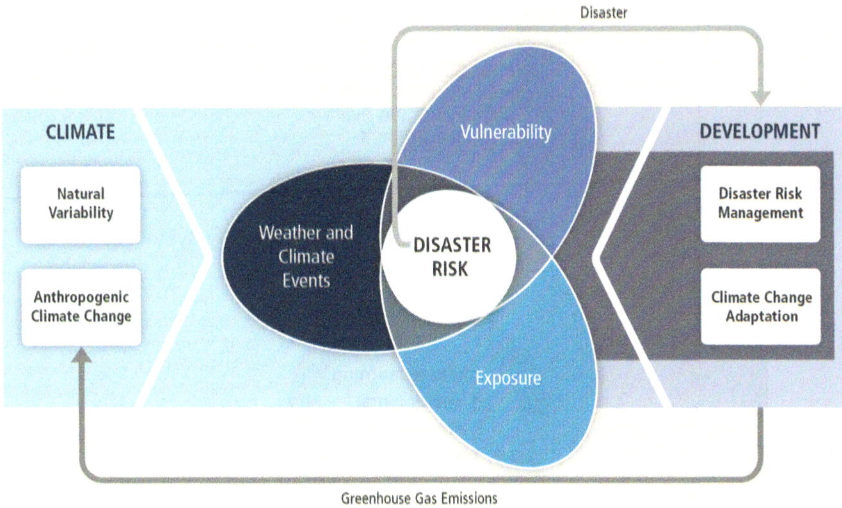

Fig. 25.2 Risk as a function of hazard, exposure, and vulnerability. IPCC 2012 and IPCC 2014

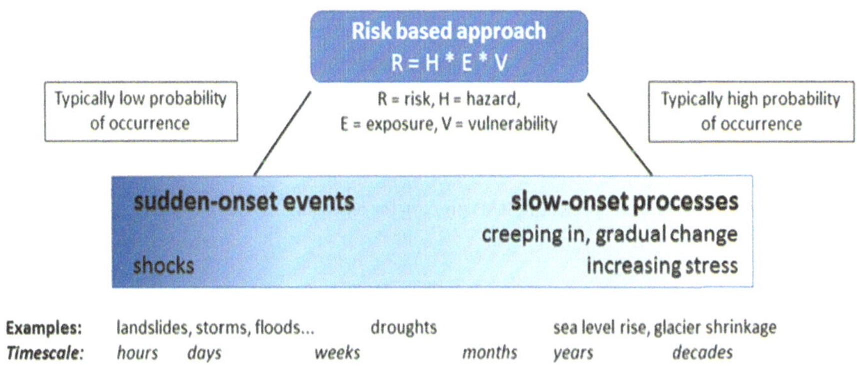

Fig. 25.3 The risk concept as applied to sudden-onset and slow-onset processes

methodology is then deployed. This methodology helps identify the potential risks faced by the region while evaluating the risk tolerance capacity of the region's affected stakeholders. Based on these assessments, appropriate risk management options are designed, through the use of decision-making techniques. These techniques help minimize vulnerability to hazards and increase the capacity of the system of interest to cope with potential climate-related risks. This dynamic framework can adapt to modification and revision of decisions based on any new evidence or data on local conditions over time. In comparison to traditional DRR and CCA frameworks, this customized framework approaches risk management at different geographical, institutional and periodic levels. It is designed to not only help leverage the existing

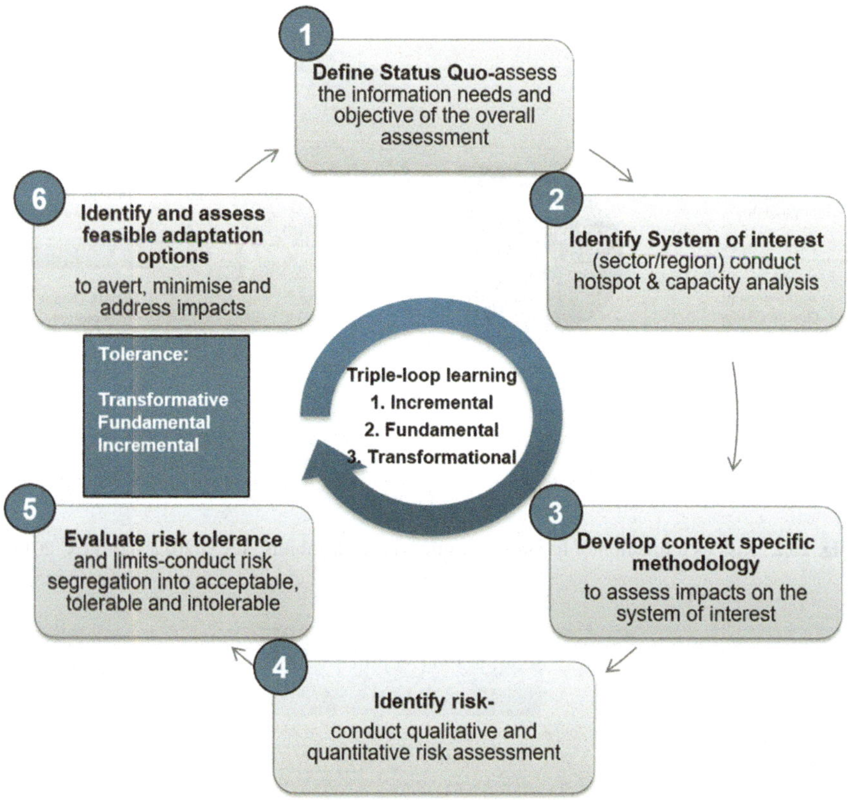

Fig. 25.4 Climate risk management (CRM) process for assessing Loss and Damage (GIZ, IIASA, KPMG, & IIT D, 2019)

expertise of various stakeholder groups, but also to help assess the gap areas in data availability, institutional linkage mechanisms, and technical expertise essential to minimizing loss and damage in the medium- and longterm.

Six step climate risk management (CRM) process thatoperationalizes climate risk management at scale is (see Fig. 25.3) (Fig. 25.4).

25.2.3 Operationalising the Framework

The six steps can be broken down as described in the following discussion.

25.2.3.1 Step 1: Define Status Quo

Step 1 assesses the status quo, screens the information and data requirements and frames the objective of the overall CRM framework application. This step comprises highlighting data needs and potential gaps in data availability as well as quality. The first step defines the overall objective of the CRM framework while showcasing the need for developing interlinkages between Climate Change and disaster risk management in terms of assessment of risks and associated data as well as institutional analysis. This step is key to the whole process as it helps in setting the purpose/ objective of the CRM. A clear definition of the purpose of CRM is a prerequisite to its planning. The purpose should clearly identify which questions the assessment will answer. A CRM is always directed towards a particular user or audience who will be using the results of the assessment, e.g. a state or national government. Ultimately, the purpose of the assessment, its level of complexity and the approach to communicating the results will depend on the specific audience of the assessment and the objectives defined at the beginning of the assessment. Some of the uses of the doing a CRM can be identified adaptation options/risk management strategies, identify particularly vulnerable people, regions or sectors, allocate adaptation funds to high-risk regions, sectors or groups of people or conduct scientific research. It is also important to define the purpose of close collaboration with stakeholders.

25.2.3.2 Step 2: Identify System of Interest

Step 2 takes the debate from a more general level to more specific levels. This step identifies the concrete system of interest by conducting climate-related risk hot spot and capacity analysis. Step 2 is used to clearly define and delineate the boundaries of the system of interest on which the CRM framework will be applied. For example, for the current project based on stakeholder consultations, a decision was made to utilize the administrative boundaries of states as the system of interest for the framework being developed. While Step 1 defines the objective of the framework highlighting the interlinkages between Climate Change and disaster risks, Step 2 involves detailed climate-related risk and capacity analysis which will result in the identification of suitable illustrative systems of interest also called hot-spots. The system of interest depends on the purpose of the assessment which may be carried out at different scales, i.e. at the national, subnational or local level. The system of interest can be delimited by either socio-economic boundary (e.g. country, state, district, community, groups within a community) or natural/ecological boundaries (e.g. river basin, sub-basin, watershed, agro-climatic zone). Assessments of the latter types of systems can, however, pose some serious challenges when collaboration between different political administrations becomes necessary. Data availability is the most important deciding factor when selecting methods, tools, and levels of detail for the assessment. Limitations in available data can significantly reduce the range of potential methods and tools to be used in the assessment.

The assessment in this step can be carried out using a mix of tools, primarily desk research (secondary data collection) of literature available on Climate Change in the region such as studies on Climate Change in the area, regional climate policies like State Action Plans on Climate Change (SAPCC), etc. This research has to be bolstered by information obtained from stakeholder interactions (primary data collection) through interviews, surveys, etc. Additionally, on-ground data collection and analysis including climate modelling can also be used to support the decision-making process, if time and economics permit. For example, Geographic Information Systems (GIS) information can be utilized for tracking geographic patterns and trends, through the lens of CCA and DRR.

25.2.3.3 Step 3: Develop Context Specific Methodology

Once the system of interest is identified, a context-specific methodology will be developed to implement the framework based on the information gathered in the previous two steps. This methodology will be customized based on local conditions and affected stakeholders of the system of interest. Under this framework, three types of studies have been identified, which draw inputs from both DRR and CCA. These studies will feed into the framework and support the development of a context-specific methodology.

Step 3 develops a context-specific multi-method approach customized to the region(s) of interest to assess potential climate-related impacts. Products may comprise general informational studies (building on what is available on hazards and impacts, backward-looking climate-risk analysis (broad risk assessment and scenarios using available data and information on risk) and forward-looking scenario and risk-based model analysis (detailed risk assessment and scenario generation, attribution assessment) may be selected depending on the available data, resources, and expertise as well as expected output to be generated in the specific context. Next, to the traditional assessment of market-based impacts, the context of L&D requires shedding light on non-monetary impacts as well as effects on informal economic activities.

Since this step involves the development of a customized methodology to be implemented in the system of interest, there is a need to consider sources of information within its boundaries as well as others who might be outside the system, especially entities/organizations that bring the relevant scientific and technological expertise required for the methodologies above.

A sample set of information providers to consider has been listed below:

- National and local policy actors such as MPs, MLAs (Elected Representatives), District Collectors, Department of revenue, agriculture, forests, science and technology, statistics, public works, and urban and rural development.
- Local communities and community self-help groups
- Non-governmental / social organizations

- Research, academic institutions, and think tanks for inputs on empirical skills and statistics

The study selection would be conducted based on available information from the public domain and those provided by the relevant stakeholders. It is only in Step 4 of the framework that these methodologies are applied on the system of interest for risk identification.

Further, as part of this step, it is proposed to conduct an analytical impact chain assessment to provide potential Climate Change-related impacts. These impacts can be categorized as Direct/Indirect and Economic/Non-Economic based on its quantification exposure, sensitivities. A major limitation the framework can encounter at this step is the unavailability of information and data sources.

The information collected in Steps 1 and 2 is then fed into the three studies identified in this step to develop a customized methodology that can be further implemented in the system of interest through field trials, stakeholder interactions with all relevant stakeholders. The results of this step will be further fed into the next step, through the formulation of impact chains to scope out the different direct and indirect impacts and risks (economic and non-economic).

25.2.3.4 Step 4: Identify Risks

Step 4 identifies risk, which is determined by the risk drivers—hazard, exposure, and vulnerability. Climate risk assessments go through a structured process for calculating risks and the benefits of relevant adaptation measures. Thereby, direct as well as indirect, economic as well as non-economic effects have to be assessed, e.g. by employing impact chain logic. A comprehensive approach needs to align top-down insight from expert-based methods and tools with bottom-up information on households' and communities' risks gathered through participatory processes.

This impact/risk chain approach (as a function of hazard, vulnerability, and exposure) can be one of the approaches to identify risk. It summarises relevant and observable impacts for each system of interest, induced by climate stimuli, taking into consideration socio-economic and environmental sensitivity factors. This analysis provides inputs on indirect socio-economic and human development impacts, which is derived from direct biophysical impacts via direct socio-economic impacts (Fig. 25.5).

An impact chain is an analytical tool that helps the user better understand, systemize and prioritize the factors that drive risk in the system of concern. The structure of the impact chain developed according to the IPCC AR5 approach is based on the understanding of risk and its components. In accordance with the IPCC AR5 definitions, we understand 'impacts' as the basic building blocks of cause-effect chains from hazard to risk. One of the major limitations of this step is that there are a large number of impacts that can be determined on the application of the methodology. However, practically only a limited number of impacts and risks can

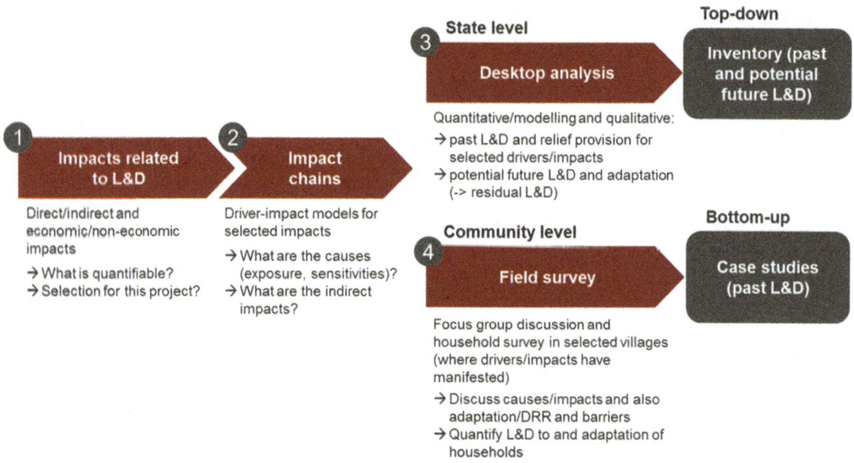

Fig. 25.5 Example of the impact chain approach (GIZ and Adelphi 2015)

be (and are usually) assessed due to data gaps, resource, time and expertise limitations.

25.2.3.5 Step 5: Evaluate Risk Tolerance and Limits

Step 5 sets out to understand what risk means to those potentially affected. Once an understanding of the risks affecting the identified region/area is acquired, it is critical to understand the ability of the affected governments, communities and/or private sector in the region to reduce and adapt to the climate-related risks, also known as the risk coping capacity. While risk identification involves an assessment of monetary and non-monetary assets, risk evaluation focuses on socio-economic assessment to understand the strengths and weaknesses of communities in a region. This step will help the users in formulating appropriate risk management actions at different scales (Fig. 25.6).

To evaluate risk tolerance levels, the key elements that need to be considered are as follows:

(i) The affected stakeholders in the region based on the risks identified in Step 4.
(ii) Social and cultural assets such as available with the affected stakeholders such as access to infrastructure, education and health facilities.
(iii) Economic assets are available with the affected stakeholders such as land, access to jobs and access to loans.
(iv) Resource expertise for judgment to categorize the qualitative impacts into risk levels.
(v) Thresholds for categorizing communities and/or regions into acceptable, tolerable and intolerable risk levels.

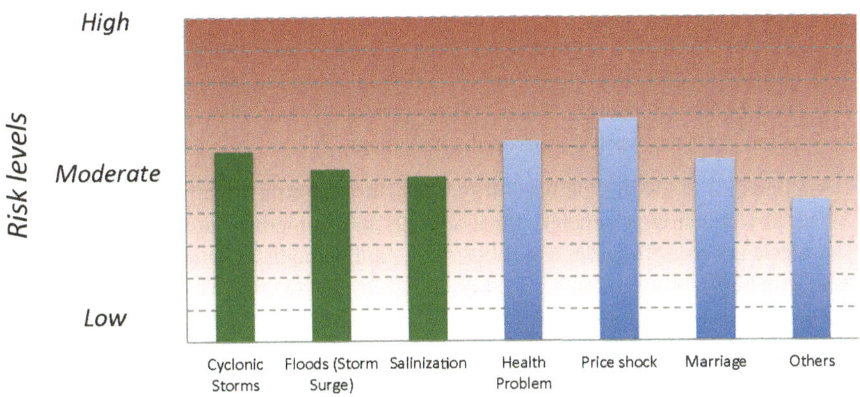

Fig. 25.6 The risk space in a coastal state as evaluated from household responses (mean) (GIZ,, and Adelphi 2015)

This involves evaluating risks by establishing risk tolerance thresholds. Two basic approaches can be: (i) semi-quantitative surveys or focus group-based assessments, which gauge risk tolerance from reported risk perceptions and risk responses; (ii) risk-based modelling formalizing risk-based decision-making building on modelled risk perceptions, e.g. to understand government actors risk tolerance for dealing with climate-related risks.

25.2.3.6 Step 6: Identify Feasible Options

Step 6 assesses potential options to reduce climate risks and identifies risk management options for residual risk, which potentially cause loss and damage. There is a wide spectrum of potential risk reduction, preparedness and risk financing measures that can be taken to reduce or finance risk. While risk reduction and prevention are at the center of attention, it is important to identify options to deal with residual risks that could potentially lead to loss and damage. The following elements need to be considered while implementing this step to efficiently identify feasible options:

- Identified risks and their intensity in the region (outcome from Step 4)
- Risk tolerance levels of affected stakeholders (outcome from Step 5)
- Available and implementable interventions based on identified risk levels and risk tolerance (Kates 2012)
  - Incremental Interventions: an extension of actions and behaviours that already reduce the losses, or enhance the benefits, of natural variations in climate and extreme events. Standard disaster risk reduction and Climate Change adaptation interventions addressing specific risks, e.g. raising dykes
  - Fundamental Interventions: Non-standard interventions in the system of interest, e.g., opening floodplains instead of a dyke

  - Transformative Interventions: adopted at a much larger scale or intensity and are truly new to a region or resource system, while transforming places and shifting locations .g., migration from floodplains to cities to provide alternative livelihood opportunities via access to new labour and other markets; or options taken by force, such as forced displacement.

- Methods for assessing the feasibility and prioritizing the identified interventions. The key decision-parameters for assessing and implementing risk management projects are risk-effectiveness, robustness, equity and distributional concerns, and acceptability.
- Affected stakeholders and experts to be consulted for prioritizing interventions
- There are various decision-supporting techniques to prioritize options which strongly building on participatory decision methods. Cost-Benefit Analysis (CBA) is one such tool that can be used to assess the economic efficiency of the identified interventions subsequently prioritizing risk management investments. There are other decision-making techniques such as Cost-Effectiveness Analysis (CEA), Multi-Criteria Assessment (MCA) and other robust decision-making approaches.

Adaptation Pathways is another important tool that addresses various risks and uncertainties while defining feasible actions and tipping points. The methodology builds on a planning framework that identifies various types of actions that help to understand where, when and if adaptation is needed. The methodology also supports in assessing if the adaptation options are feasible and possible.

Adaptation tipping points, being the central element of adaptation pathways methodology, define the conditions under which actions cannot anymore satisfy the pre-specified objectives, thus constituting hard or soft adaptation limits. Beyond the tipping points, additional new actions are required for leading to a pathway, to operate within the risk space.

The key opportunities, characteristics, and applicability of different decision-support tools for assessing risk management options have been summarised below (Mechler 2016) (Table 25.1):

While these different decision-support tools are applicable for different objectives and circumstances, multiple approaches can also be combined for the assessment.

25.3 Approaches for CRM

A lot of approaches already exist in the field of short-term risk assessment and management, mainly in the field of extreme events; existing approaches do often not sufficiently address long-term, slow-onset changes due to Climate Change. Further, risk and vulnerability assessments often do not meet the information needs of policy-makers and local governments in order to manage the risks of Climate Change and associated losses and damages effectively. The approaches and methods used to date for risk assessment and management do not take adequate account of loss and

Table 25.1 Characteristics and applicability of different decision-support tools for assessing risk management options Source: CBA-Cost Benefit Analysis; CEA-Cost-Effectiveness Analysis; MCA-Multi-Criteria Analysis. Adapted from Mechler 2016

Decision tool	Opportunities	Challenges	Application
CBA	A rigorous framework based on comparing costs with benefits	Need for monetizing all benefits, difficulty in representing plural values	Well-specified *hard-resilience* projects with economic benefits
CEA	Ambition level fixed, and only costs to be compared. Intangible benefits part. Loss of life does not need to be monetized	Ambition level needs to be fixed and agreed upon	Well-specified interventions with important intangible impacts, which should not be exceeded (loss of life etc.)
MCA	Consideration of multiple objectives and plural values	Subjective judgments required, which hinder replication	Multiple and systemic interventions involving plural values
Robust approaches	Addressing uncertainty and robustness	Technical and computing skills required	Projects with large uncertainties and long timeframes
Adaptation pathways	Scenario-based decision-making at decision points depending on future system changes	Considerable investment into scenarios and stakeholder interaction	Flood risk management

damage, the associated (non-) economic costs or slow-onset events brought about by Climate Change (UNFCCC 2012a). This underscores the challenges involved in developing and rolling out comprehensive methods and measures for assessing and managing Climate Change Risks.

In order to effectively assess and address L&D, it is important to manage it as part of a broader climate risk management cycle. To conceptually localize the subject of Loss and Damage, it has to be included as part of the concepts of Climate Change Adaptation, Disaster Risk Management, and Sustainable Development. CRM brings an opportunity to bring these lines of work together under one umbrella for a broader sustainable and climate resilient goal. In addition, it crucial to consider the Climate Change Loss and Damage Mechanism emerging into an institutional arrangement with the state-of-the-art expertise of assessment and addressing methods and clear rules on compensation schemes on the national level, which may include a compensation fund (e.g. to subsidize insurances that are not marketable due to the risk structure). Next, to this, climate finance mechanisms (possibly together with a share of official development assistance, ODA that is officially earmarked for Climate Change adaptation support) in vulnerable countries, serves as a complementary mechanism to avoid loss and damage and to utilize opportunities from Climate Change to reduce the net losses. Prior to the issue of the Technical Paper, the Work Group on Loss and Damage had called for inputs for all the thematic areas.

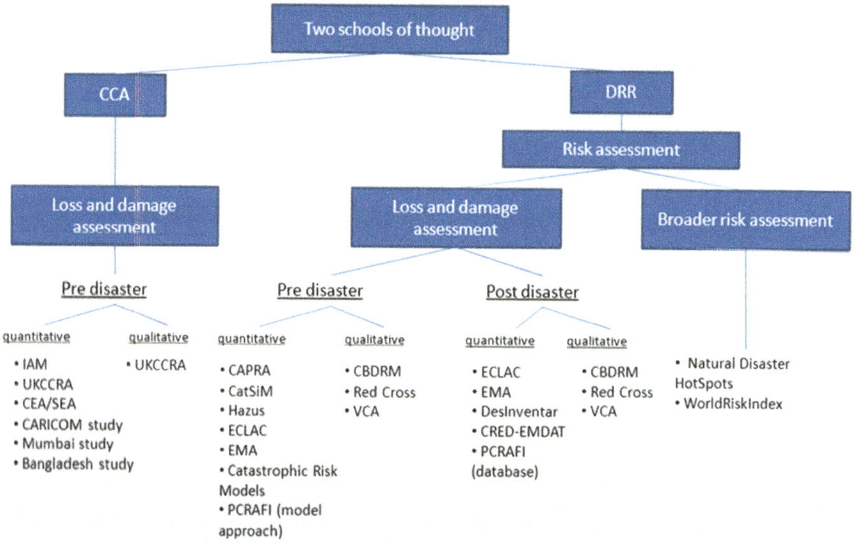

Abbreviations: CAPRA - Comprehensive Approach for Probabilistic Risk Assessment; CARICOM - Caribbean Community; CATSIM - Catastrophe Simulation model of the International Institute for Applied Systems Analysis (IIASA); CEA - Country environmental analysis; CBDRM - Community based disaster risk management; CCA - Climate Change Adaptation; CRED - Centre for Research on the Epidemiology of Disasters; DRR - Disaster Risk Reduction; Hazus - Hazards U.S.; ECLAC - Economic Commission for Latin America and the Caribbean; EMA - Emergency Management Australia; IAM - Integrated Assessment Model; PCRAFI - Pacific Risk Assessment and Modelling; SEA - Strategic environmental assessment; UKCCRA - Climate Change Risk Assessment of the Department for Environment, Food and Rural Affairs of the United Kingdom of Great Britain and Northern Ireland; VCA - Vulnerability and capacity assessment.

Fig. 25.7 Inventory of concepts and instruments on Loss and Damage (UNFCCC 2012a)

From the various contributions, the authors structured existing approaches as shown in the following scheme (Fig. 25.7):

As shown in the scheme above, the contributions came from two thematic fields, CCA, and DRR. Both have in common that Loss and Damage occur as the part that cannot (pre-L&D) or could not (post-L&D) be avoided by the strategies and measures that the two approaches offer. The scheme also shows that far more quantitative assessment methods and concepts have been developed than qualitative approaches. This may be because insurance schemes apply specific mathematical methods that work with quantitative approaches.

The inventory of the approaches shows that there are assessment methods in place that can more or less be applied to the L&D case. However, it is not a toolbox of complementary measures and some important tools are missing. An inventory is the first step and helps to become aware of what is there and what is missing. After the first inventory

25.4 Mainstreaming CRM as Part Climate Change and Disaster Management in India

There is a growing evidence that Loss and Damage from slow-onset changes is occurring. For example, according to a Swiss Re study the severe flash floods in Chennai in November 2015 were the largest disaster, causing estimated economic losses of USD 2.2 billion. It further estimated insured losses at around USD 755 million, making these floods the second costliest insurance event in India (Businessline, H n.d.; Swiss Re 2016). The economic survey of India, in its report of 2017, estimated that currently, India incurs losses of about $ 9–10 billion annually only due to extreme weather events. Of these, nearly 80% losses remain uninsured (India 2017). The losses due to slow-onset events like salinization and rainfall variability on various sectors are going to be much higher. For instance, the agriculture ministry in a recent report said productivity decrease of major crops would be marginal in the next few years but could rise to as much as 10–40% by 2100 unless farming adapts to Climate Change-induced changes in weather. On the other hand, the cost of adapting to the current and projected Climate Change impacts is estimated to be one trillion USD until 2030 for India. Within India, the existing framework for addressing L&D can be seen about its existing activities on CCA, disaster management, and sustainable development, as well as to the numerous state and central sector-development programmes and schemes that are in place. Loss and Damage due to Climate Change is as much a social issue in India as it is an environmental or developmental issue. Hence it cannot be seen in isolation to all other challenges that the country is currently is facing.

As mentioned earlier, there is an intrinsic relationship between Climate Change vulnerability and disaster risk management when dealing with the development of a robust L&D framework. It thus becomes imperative to explore plausible interlinkages between DRR and CCA institutional setups within the country, which can then be used to establish a distinct institutional framework for L&D in India.

Mainstreaming is not new for most policy makers. Experiences exist with cross-cutting issues like environment, poverty reduction, gender and also climate change adaptation and disaster risk reduction. Mainstreaming climate risk management can thus build on a series of lessons learnt, which help to overcome typical constraints and barriers (e.g. mainstreaming fatigue, concentration on short-term strategies) and imply, amongst others. Identify suitable entry points for L&D, i.e. ongoing processes where L&D consideration is beneficiary (e.g. SAPCC, SDMP process, sectoral or spatial planning). The most relevant sectors/ministries for mainstreaming L&D/CRM are environment, disaster risk management, planning and finance; depending on the national L&D risk further important sectors include agriculture, water, urban & housing, health/social welfare etc.

The DRR framework in India is well established and has strong connections with the local government departments. Thereby, the country's DRR policies, plans, and institutions have also been established much before CCA policies and institutions.

The Ministry of Home Affairs (MHA) is responsible for the overall coordination activities of Disaster Management (DM) across India. MHA is nationally supported by the Cabinet Committee on Security (CCS) and the National Crisis Management Committee (NCMC). A National Disaster Management Authority (NDMA) has been established as the national level agency responsible for the preparation and implementation of DM plans and functions (National Disaster Management Plan, NDMA 2016).

Primarily all disaster-related work is undertaken by the state government with support (on request) from the central government. Each state has its disaster management institutional framework set up. The guidelines developed by NDMA will assist state governments to develop their respective state DM plans. NDMA has been empowered to authorize rescue and relief provisions and procurement at the time of disasters. Additionally, a National Disaster Response Force (NDRF) has been set up to assist relevant State Government/District Administration in the event of an imminent hazard event or its aftermath.

On the other hand, CCA institutions and policies are relatively new and some policies are still under review. It was in 2008 that a National Action Plan on Climate Change (NAPCC) was developed, which comprises of eight national missions targeting different sectors such as Solar, Energy Efficiency, Sustaining the Himalayan Ecosystem and building Strategic Knowledge on Climate Change among others. Simultaneously the Ministry of Environment and Forests (MoEF) was handed the additional responsibility of Climate Change and renamed to Ministry of Environment, Forests and Climate Change (MoEFCC). This assessment of India's climate and disaster risk profile coupled with the existing institutions and their linkages will help prepare the foundation for an L&D framework.

State Action Plan on Climate Change (SAPCC) and State Disaster Management Plan (SDMP) are the formal institutional platforms at state-level to mainstream climate risk and disaster risk into development planning (Fig. 25.7). Few recent studies (Bahadur 2016) (Dubash and Jogesh 2014) of SAPCC and SDMP, respectively of different states provide useful insights about the process. The plans prepared either in the Climate Change context or the disaster management context should not run parallel to the development plans that the state governments make. For effectively addressing the climate risk and the disaster risk it is important to synergize different plans into development planning. Further, it must be noted that, as the planning process, the SAPCC, as well as SDMP, should be seen as dynamic processes that are periodically revised. For this purpose, as well as effective implementation of the plans, an appropriate institutional structure must be conceived and nurtured. While climate policy framework and implementation mechanism has institutional has evolved in India at national and state level, mainstreaming climate risk management for averting, minimizing and addressing L&D into national and state level schemes and plans requires developing a shared policy vision. For the purpose, a joint or common vision for CCA and DRM has been suggested as an important first step (Michael Howes et al. 2014) and is in-line with the precepts of joined-up Government and Network Governance. As with the growing science base, systematic integration and meaningful participation of sectors and stakeholders

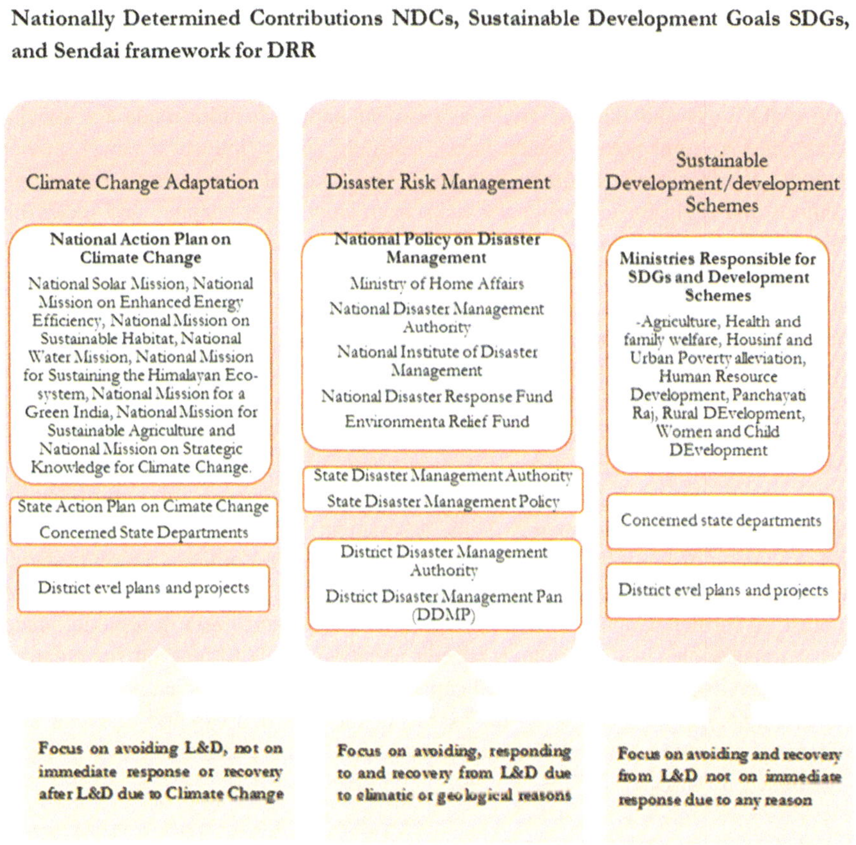

Fig. 25.8 Institutional interlinkages between Climate Change adaptation, disaster risk management and sustainable development for L&D in India

should become a common feature to ensure buy-in. Information and knowledge management systems and adequate communication systems can make a significant contribution to meaningful participation and informed decision making (Fig. 25.8).

25.5 Conclusion and Way Forward

Recently, Loss and Damage has become an issue of high interest in academia, politics, and development cooperation and accordingly discussion and concept papers are increasingly entering and fuelling the debate. At the current state, there are more open questions than answers. While traditional DRR and CCA policy typically acts within an incremental adjustment learning loop, climate-related risks discussed in the context of L&D may not only require new innovative response measures, but particular attention has to be paid to locally-applicable techniques for understanding risks and risk management interventions, such as through Vulnerability Capacity Assessments (VCAs) and community-led focus groups. This will help make Loss and damage (risk) assessment approach tailor-made for special areas, such as rural areas, coastal areas, sensitive zones (industrial zones, power stations, water supply infrastructure etc.), and natural ecosystems. Enhancing methods and tools for assessing the risks from slow-onset changes and receiving more information and data, such as melting of glaciers, sea level rise, salinization or the degradation of ecosystems and ecosystem services, but also heat as a threat to human and animal health will be critical in understanding climate risk. The evidence thus generated needs to be considered and utilized in adaptation and DRR strategies to reduce net L&D. Further addressing L&D requires innovative risk transfer mechanisms that could evolve from inventory of insurance models (pros and cons in different settings), gap analysis and development and test of practicable (innovative) insurance models; paradigm shift in risk reduction approaches (e.g. no settlements at riverbanks with disastrous floods) and setting up of compensation mechanism for loss and damage. Thus, assessment of climate risk, identification and evaluation of feasible options (interventions, measures, policies, among others) will result in better implementation of the six-step CRM framework in the system of interest, to further scaleup for addressing climate risks and L&D in regional, state and national scale. In this context, special attention needs to be paid for creating enabling environments (e.g. technical capacity, skills, fiscal tools, etc.) for utilizing the available methods and tools for assessing the risk of Loss and Damage and effective Loss and Damage; enhancing cooperation for DRR and CCA in local stakeholder networks.

Acknowledgements The chapter is based on the outcomes of two studies undertaken by GIZ on Loss and Damage and Climate Risk Management as part of Indo-German bilateral cooperation project- Climate Change Adaptation in Rural Areas of India. The authors acknowledge contributions from Dr. Reinhard Mechler and Dr. Thomas Schinko IIASA, Dr. K S Kavi Kumar Madras School of Economics and Climate Change teams from Adelphi, KMPG India and IIT Delhi for conducting the studies.

References

Bahadur AE (2016) Strengthening disaster risk management in India: a review of five state disaster management plans. Overseas Development Institute

Businessline, H (n.d.) Chennai floods largest natural catastrophe in India in 2015: Swiss Re, Chennai, Tamil Nadu

Dubash N, Jogesh A (2014) From margins to mainstream? Climate change planning in India as a 'Door Opener' to a sustainable future. Centre for Policy Research (CPR), Climate Initiative, Research Report

Garg A et al (2015) Climate Change and India: Adaptation Gap (2015) - A preliminary assessment. Working Paper, Indian Institute of Management Ahmedabad (W.P. No. 2015-11-01)

GIZ, Adelphi (2015) *Climate-related loss and damage in India. Results of a rapid assessment in Tamil Nadu and Odisha. Internal report*

GIZ, NIDM, IIASA, KPMG, IIT D. (2019) *Climate risk management (CRM) framework for India.* https://www.weadapt.org/knowledge-base/climate-services/climate-risk-management-framework-for-india

India TO (2017, August) Climate Change costs India $10 billion every year: Government. New Delhi

IPCC (2012) Managing the Risks of Extreme Events and Disasters to Advance Climate Change Adaptation, IPCC 2014. Accessed at https://www.ipcc.ch/report/managing-the-risks-of-extreme-events-and-disasters-to-advanceclimate-change-adaptation/

IPCC (2014) Fifth Assessment Report (AR5), IPCC, 2014, Accessed at: https://www.ipcc.ch/report/ar5/

Kates WR (2012) Transformational adaptation when incremental adaptations to climate change are insufficient. Proc Natl Acad Sci USA 109:7156–7161

Mechler RA (2016) Identifying the policy space for climate loss and damage. Science 354:1–3

Michael Howes et al (2014) Towards networked governance: improving interagency communication and collaboration for disaster risk management and climate change adaptation in Australia. J Environ Plan Manag 58:775–776

National Disaster Management Plan, NDMA. (2016). Retrieved from http://ndma.gov.in/images/policyplan/dmplan/National%20Disaster%20Management%20Plan%20May%202016.pdf

Surminski SL (2012) *Current knowledge on relevant methodologies and data requirements as well as lessons learned and gaps identified at different levels, in assessing the risk of loss and damage associated with the ad.* Retrieved from https://unfccc.int/sites/default/files/background_paper_full.pdf

Swiss Re (2016) Natural catastrophes and man-made disasters in 2015: Asia suffers substantial losses https://www.swissre.com/dam/jcr:b4a158fc-2669-4b70-b817-96db6a8705e9/sigma1_2016_en.pdf

UNFCCC (2012a). Retrieved from https://unfccc.int/sites/default/files/non_econ_losses_synopsis.pdf

UNFCCC (2012b) Loss & damage: evidence from the front lines, UNFCCC. Retrieved from Accessed at: https://unfccc.int/files/press/media_outreach/application/pdf/cop18_mw_loss_and_damage.pdf

UNFCCC (2012c) *Current knowledge on relevant methodologies and data requirements as well as lessons learned and gaps identified at different levels, in assessing the risk of loss and damage associated with the adverse effects of Loss and Damage associated with adverse*

UNFCCC (2018a) *Loss and damage online guide.* Retrieved from https://unfccc.int/sites/default/files/resource/Online_guide_on_loss_and_damage-May_2018.pdf

UNFCCC (2018b) *Milestones, UNFCCC.* Retrieved from : http://unfccc.int/files/adaptation/cancun_adaptation_framework/loss_and_damage/image/jpeg/milestones.jpg

Verheyen R (2015) Loss and damage due to climate change: attribution. Int J Global Warming 8:2. https://doi.org/10.1504/IJGW.2015.071968

Kirtiman Awasthi works at GIZ India as Senior Policy Advisor-GIZ. He has extensive experience of working on project planning, management, research, knowledge management and community support in the field of Climate Change, Environment & Natural Resource Management. Kirtiman was involved in the development of State Action Plans on Climate Change (SAPCC) since 2010.

Somya Bhatt is working as Policy Advisor with GIZ on Climate Change Adaptation. Her work is focused on vulnerability & risk assessments, community-based adaptation measures, state action plans on climate change, information & amp; knowledge management and capacity development for adaptation. She is a trained trainer on integrating climate change adaptation into development planning.

Chapter 26
Integrating Climatic Disaster Risk Reduction and Climate Change Adaptation into Project Management Cycle

P. G. Dhar Chakrabarti

Abstract Disasters are severe disruptions in the normal functioning of a community or a society due to exposures of vulnerable conditions to hazardous events, leading to damage and losses to human, material, economic, or environmental assets. To reduce the vulnerability to extreme climatic events, long term, medium, short and immediate measures are needed to be identified and integrated into the planning processes. Disaster Risk Reduction (DRR) and Disaster Management (DM) together form part of Disaster Risk Management (DRM), while the combined processes of Climate Change Mitigation (CCM), Climate Change Adaptation (CCA), DRR, and DM contribute to the overall process of sustainable development. The present study deals with the similarity in the goals of DRR and CCA, discuss about their area of convergence and divergence, and processes a way by which it can be integrated into Project Management Cycle.

Keywords Disaster risk reduction · Project management cycle · Climate change adaptation

26.1 Introduction

In the context of climate change the hydro-climatic 'hazardous events' (extreme temperature, too much or too little of rains, sea storms, high runoff, etc.) cause a wide range of disasters such as heat and cold wave, drought, flood, flash floods, glacial lake outburst floods, cloudbursts, lightning, landslides, forest fires, cyclones, hurricanes, storm surges etc., commonly described as hydro-meteorological disasters. To reduce the impacts of these hazards, long-term, medium, short and immediate measures are needed to be identified and integrated into the planning processes. Long-term measures are required for mitigating anthropogenic climate change,

P. G. D. Chakrabarti (✉)
Environment and Disaster Management at Ramakrishna Mission University, Narendrapur, Kolkata, India

A. K. Gupta et al. (eds.), *Disaster Risk and Management Under Climate Change*, Disaster Resilience and Green Growth,
https://doi.org/10.1007/978-981-99-4105-6_26

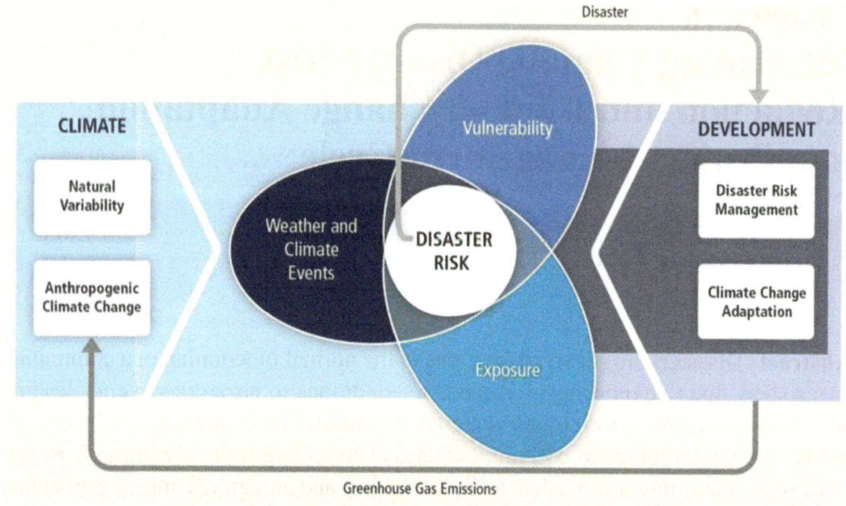

Fig. 26.1 Conceptual framework on climate change, CCA and DRM (IPCC 2012)

medium and short-term measures for reducing vulnerabilities and exposures and immediate measures for disaster response, relief, and rehabilitation.

The long-term measures generally fall in the category of Climate Change Mitigation (CCM), medium and short-term measures are usually the concerns of both Climate Change Adaptation (CCA) and Disaster Risk Reduction (DRR), while immediate measures are the tasks of Disaster Management (DM). The combined processes of CCM, CCA, DRR, and DM contribute to the overall process of sustainable development. Figure below explains these inter-related processes (Fig. 26.1).

The Sendai Framework for DRR(successor of Hyogo Framework)acknowledges that addressing Climate Change as one of the drivers of disaster risk' provides 'an opportunity to reduce disaster risk in a meaningful and coherent manner' (UNISDR 2015). UN Sustainable Development Goals adopted 'urgent action to combat climate change and its impacts' as one of the 17 goals for transforming the world and incorporated disaster risk reduction in the targets of eight of these goals (UN 2015, Transforming our world: the 2030 agenda for sustainable development). Paris Climate Agreement which is legally binding called for enhanced understanding, action and support on eight specific issues of disaster risk reduction. These include: (a) early warning systems; (b) emergency preparedness; (c) slow onset events; (d) events that may involve irreversible and permanent loss and damage; (e) comprehensive risk assessment and management; (f) risk insurance facilities, climate risk pooling and insurance solutions; (g) non-economic losses; and (h) resilience of communities, livelihoods and ecosystems. The trinity of global development frameworks—Sendai Framework, Sustainable Development Goals and Paris Climate Agreement—all adopted in 2015 have created synergies and opportunities for integrating DRR and CCA, as never before (United Nations Climate Change Secretariat, 2017).

26.2 Rationale & Need

DRR and CCA share the common goal of reducing vulnerabilities and exposures to extreme climatic events that lead to disasters. They also share common tools to access, analyse, monitor and address risks. Various risk reduction measures, like drought proofing, flood protection, tree plantation, green belt development, alternative livelihood development etc. have similarities with adaptation programmes.

Therefore, integration of DRR and CCA would be necessary to avoid duplicity, to derive maximum benefits from scarce resources and also adds value to the projects through lessons learned from the respective perspectives. The inclusion of Climate Change issues in disaster risk reduction projects would enrich the projects and make them more relevant to the emerging concerns. For example, risk management tools to assess and reduce climatic risks by various structural and non-structural measures of prevention and preparedness can facilitate the tasks of Climate Change Adaptation. The global and national discourses on CCA and DRR originated from the same concern for sustainable development in the 1987 report of the United Nations (UN 1987). However, these branched off separately in post-1990 initiatives. DRR took the route of International Decade for Natural Disaster Reduction (IDNDR), Yokohama Strategy and Hyogo Framework of Action, while CCA took the pathway of UN Framework Convention on Climate Change (UNFCCC), Kyoto Protocol and Conference of Parties (COP). The two agendas intersected when the Fourth Assessment Report of Intergovernmental Panel on Climate Change, (AR4, IPCC) came out with its findings that rising temperature of the globe would increase the frequency and intensity of climate-related disasters (Fig. 26.2).

26.3 Concept & Framework

More and more countries are developing legislations, policies and strategies for integrating DRR and CCA in a common framework. Almost, all national legislation on disaster management, enacted post-IPPC-AR4 makes explicit provision on integrating DRR with CCA. Government of the Maldives adopted the world's first Strategic National Action Plan (SNAP) that integrates DRR and CCA in the common development planning. In 2009, Philippines passed the *Climate Change Act*, which emphasizes the need to integrate disaster risk reduction into climate change programs and initiatives. The following year the country passed the Philippine *Disaster Risk Reduction and Management Act*. It mandates the State to "develop, promote, and implement" a comprehensive National Disaster Risk Reduction and Management Plan that would inter alia institutionalize arrangements for reducing

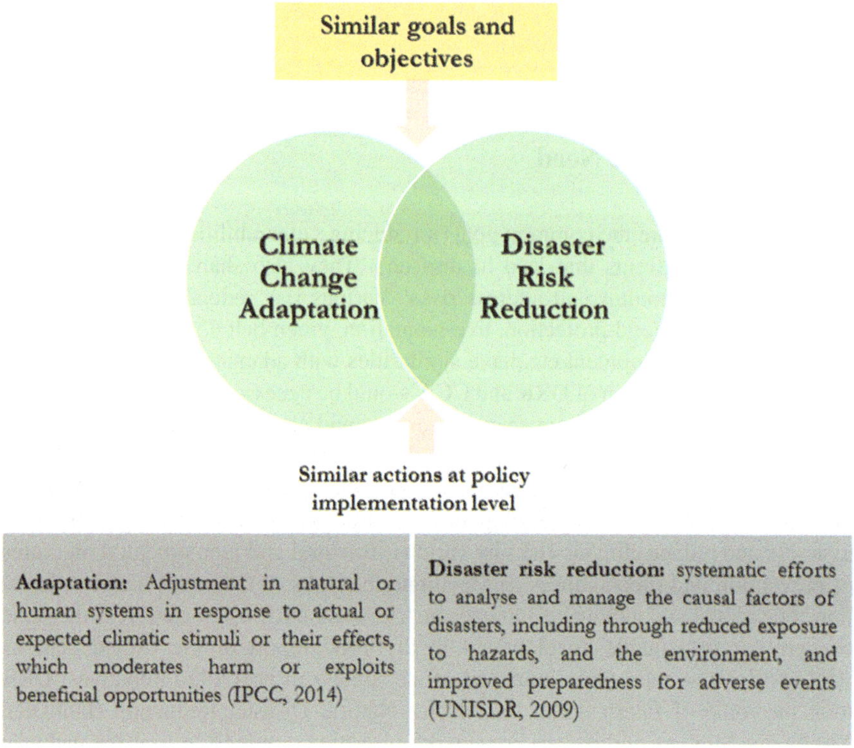

Fig. 26.2 Similarities between CCA and DRR, (Chakrabarti, 2015)

disaster risks, including projected climate risks, and enhancing disaster preparedness and response capabilities at all levels. Many countries in Africa and Latin America are also adopting similar approaches. National Policy on Disaster Management of India announced, "Synergies in approach and strategies for CCA and DRR shall be encouraged and promoted" (NDMA 2009).

Japan enacted its Climate Change Adaptation Act in June 2018, which is first of its kind in the world. The Act sets up clear roles of national and local governments, private sectors, and citizens to promote Climate Change Adaptation efforts in various fields, including integration with disaster risk reduction, through reliable scientific information (A-PLAT n.d., kein Datum) (Fig. 26.3).

DRR and CCA can be integrated by identifying the areas that create divergence as well as those that create convergence between the two. Also, by developing institutional, strategic and technological linkages that promote better integration. While at the national and provincial levels coordination mechanisms can be established for better institutional linkages, at the local levels DRR and CCA must converge with clear plan of action, funding arrangements and guidelines for implementation and monitoring.

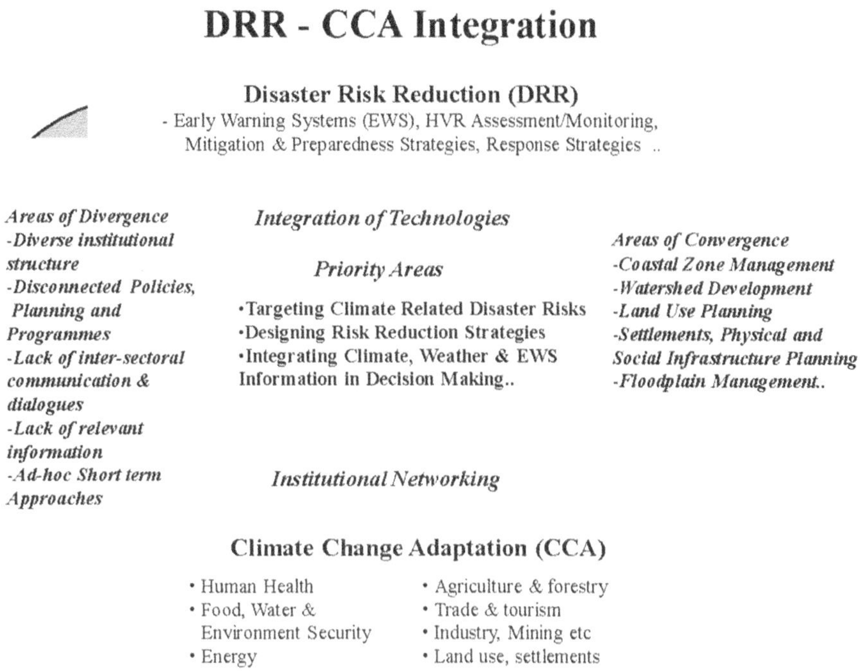

Fig. 26.3 Integration of CCA and DRR (Chakrabarti 2010)

26.4 Strategies & Gaps

Implementation of global and national agendas of integrating CCA and DRR in plans, programmes and projects would require sizeable investments. Several national governments have started allocating separate budget for CCA and DRR, but these are woefully inadequate to meet the adaptation needs of the developing countries. Globally, Adaptation Fund established under Kyoto Protocol could mobilize hardly US$ 512 million since 2010. Green Climate Fund, with promises of US$ 100 billion annually from developed countries have gathered pledges of only US$ 10.3 billion as on August 2018.

Since it is unlikely that the national governments and international organisations would be able to meet the requirement of funds for CCA and DRR entirely, and given that separate funds would not be available for integrating CCA and DRR, it is highly imperative that tools and techniques should be developed for integration and mainstreaming CCA and DRR under the existing Plans, Programmes and Projects (PPP) through proper designing, implementation and monitoring of the PPPs. Countries are investing large sums of money on development projects across sectors, which can contribute to the development process for both CCA and DRR, provided that the projects include a holistic understanding of Climate Change risks and disasters of past and future.

Three separate studies on tracking public investments on DRR in Philippines (Chakrabarti 2012a, b), Indonesia (Darwanto 2012) and India (Chakrabarti 2012a, b) demonstrated that although direct public investments on DRR are less than one percent of the total budgetary allocations of federal governments, indirect investments with elements of DRR embedded in plans, programmes and projects were as high as 20–30%. These provide enormous opportunities for mainstreaming DRR in development.

Development projects, particularly large development projects, provide good opportunities for integrating and mainstreaming CCA and DRR. Conversely, without effective mainstreaming, development projects may exacerbate existing climate and other risks of disasters or unwittingly create new risks of disasters.

The best way to ensure that CCA and DRR are mainstreamed into the projects is to integrate this into the Project Cycle Management (PCM). PCM is 'the process of planning, organizing, coordinating, and controlling of a project effectively and efficiently throughout its phases, from planning through execution, completion and review to achieve the pre-defined objectives or satisfying the project stakeholder by producing the right deliverable at the right time, cost and quality' (European Commission 2004).

There are six phases in PCM - programming, identification, appraisal, financing, implementation and evaluation - as shown in the diagram below. Disaster risk reduction can be mainstreamed into each of these six phases of the project cycle (Fig. 26.4).

Programming No project is taken up in isolation. The objectives of national and sectoral plans are to give the rationale for the project, while strategic considerations of resources and economy, and justify its size, scope and location. Using this as backdrop the project is formulated defining its goals and objectives and outlining the specific tasks to be taken up in the project. CCA and DRR can be factored as a proxy objective of every project irrespective of its primary aim, when the project formulation goes through the analysis of threats and opportunities of the project since Climate Change risks and disasters may create potential threats for smooth functioning of any project and may hamper successful project delivery.. Once the strategic goals of CCA and DRR are factored into the objectives of the project subsequent tasks of mainstreaming becomes easier to be accomplished. Therefore, measures for climate adaptation and risk reduction would be useful for protecting the gains of the project even though it may involve marginally higher investments.

Identification Once the project objectives are clearly defined and the tasks outlined, the next phase in the project cycle is to identify the elements of the tasks that need to be performed. This includes designing various structural and non-structural elements of the project, and analysing the needs and interests of all possible stakeholders particularly the vulnerable sections of population. This would provide an opportunity for detailed risk analysis of the project and identify the hazards, vulnerabilities, exposures and risks of Climate Change and disasters that may affect the project. This would further provide an opportunity to assess whether the project would have any adverse impact on the communities or on the

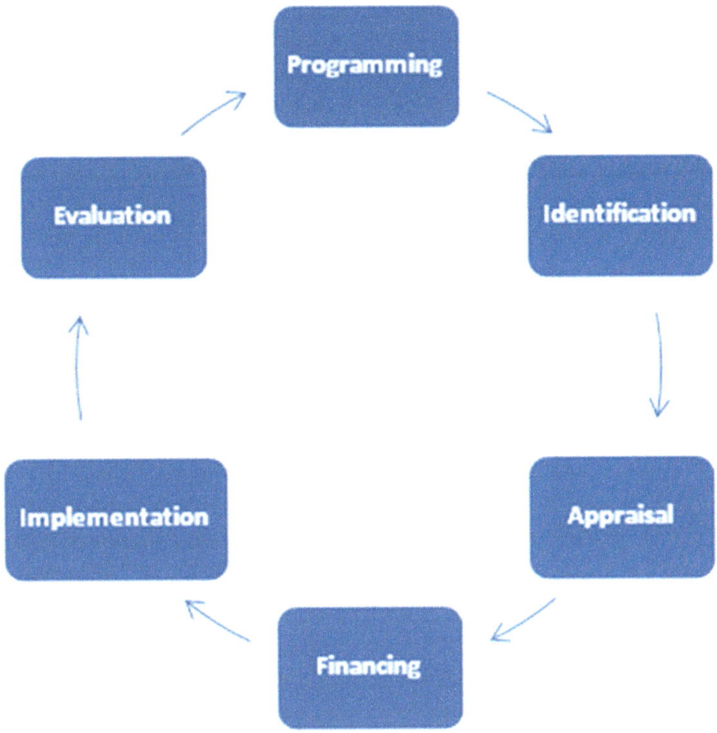

Fig. 26.4 Project cycle (European Commission 2004)

environment and accordingly to design the specific elements that should be built into the project to minimize such impacts.

This by no means is a simple task and would require comprehensive understanding of the risks of Climate Change and disasters at the project level. Therefore special studies have to be commissioned to downscale Climate Change projections and make detailed study of hazards, vulnerabilities, exposures and risks at the local level and get the same validated by independent experts. The project has to be designed taking into account the risks of Climate Change and disasters so as to contribute towards the process of adaptation and risk reduction, and also makes sure that it does not contribute in creation of any new risks of disasters.

Appraisal Once the project objectives are defined, tasks identified and structural and non-structural elements designed, the next phase of the project cycle is the appraisal of the project. In this phase, all relevant aspects of the project are studied, taking into account costs and benefits of the project, views of the stakeholders, feasibility and other issues. Logical or results-based management frameworks, and activity and implementation schedules, are developed and the required inputs are calculated. The outcome assist in taking a decision of taking the project forward, or not. In this phase cost-benefit analysis (CBA) is a good entry point for incorporating

CCA and DRR elements into the project that may enhance project costs but provide longer term benefits to the society. In the conventional CBA indirect and usually longer term benefits of risk reduction are not always appreciated as recommended methodologies for estimating such benefits since these are difficult to quantify and hence considerations of short term economic benefits prevail over other considerations, unless there are strong political commitments for risk reduction. As methodologies for assessing indirect costs and benefits improve there would be greater acceptance there would be greater acceptance of this tool in project appraisal.

Financing Once a decision has been taken to go ahead with the project various options for financing the project, such as budgetary support, equity, market borrowing, bonds, external assistance etc. are explored and decisions as deemed appropriate are taken. Financing is not always a separate stage in project cycle as financial decisions may be taken at different points in the cycle—e.g., at the end of the identification or appraisal phases—depending on the particular procedures being followed. This provides an entry point for mainstreaming CCA and DRR into the project cycle management in as much as it opens up considerations of various options of risk financing and risk transfer that may reduce the immediate burden of investment while incorporating elements of risk reduction into the project.

Implementation The implementation phase is important as in this phase the project managers have to ensure that all the designed standards and specifications are complying and that there is no let up or compromise in the agreed parameters and processes. This is also the phase to closely monitor the progress, to adjust to changing circumstances and to do mid-course corrections necessary for achieving the objectives of the project.

Evaluation The project cycle ends with auditing and evaluation of the results achieved and the lessons learnt from the project, which are useful for deciding extension or replication of the project. Every project has its logical framework with built in input, process, output and outcome indicators which should measure the progress achieved at every stage. Concurrent evaluations are conducted during the course of project implementation, while performance auditing and result framework analysis are conducted after the conclusion of the project.

Assessing the impacts of the project is more difficult as the impacts are felt over a longer time frame. There are well defined *theories of change* that can assess long term impacts of the projects. Tools of anticipatory impact analysis such as Environmental Impact Analysis (EIA), Climate Impact Analysis (CIA) and Disaster Impact Analysis (DIA) are also available for analysing the possible impacts of the projects. These can also be combined in one single impact analysis covering all three aspects in an integrated manner.

26.5 Opportunity & Approaches

Among the various toolkits available for mainstreaming CCA and DRR in project cycle management the following have been applied with varying degrees of success:

(a) Marginal Investment Analysis
(b) Cost Benefit Analysis
(c) Multi-purpose Development Projects
(d) Cost Effectiveness Analysis
(e) Multi Criteria Analysis
(f) DIA (on lines of EIA/SEA)
(g) Check Lists for DRR

26.5.1 Marginal Investment Analysis

Mainstreaming CCA and DRR does not necessarily mean that huge and fresh investments have to be made for DRR in every project—existing investments can be so designed and calibrated that these do not exacerbate the latent risks or create new risks of disasters. Incorporation of elements of risk resilience in the concept, design, management and evaluation of existing and new programmes, activities and projects may necessitate additional investments. The tools of marginal investment analysis are useful to determine the effectiveness of such additional investments for CCA and DRR.

When marginal costs of projects are higher than marginal benefits there are strong economic justifications for going ahead with projects with elements of CCA or DRR incorporated, as appropriate in given contexts.

Constructions of school, hospitals, roads, bridges and buildings are investments that are routinely made in public or private sectors in every country in different hazard zones, but these infrastructures are not always resistant to the risks of these hazards. Making these infrastructures hazard resistant is an additional investment of a marginal nature which most countries fail to do, thereby making these vulnerable to earthquake, landslides, flood or cyclone. Similarly, there are many non-structural measures, which have the potential to increase the resilience of the structures. The marginally higher costs in earthquake resistant buildings is 2.5% for structural elements and 0.8% for non-structural elements (Pereira 1995), but the benefits are higher than the replacement costs of these structures if these collapse in earthquakes.

Infrastructure, agriculture, health, forestry and water sector projects can be so designed as to factor projected climate change. These may entail additional investments, but marginal benefits of such investments may far exceed the marginal costs (Fig. 26.5).

There are hundreds of development plans, programmes and projects across different sectors that can be calibrated to incorporate elements of CCA and DRR for sustainable development. The marginal costs involved in such exercises are

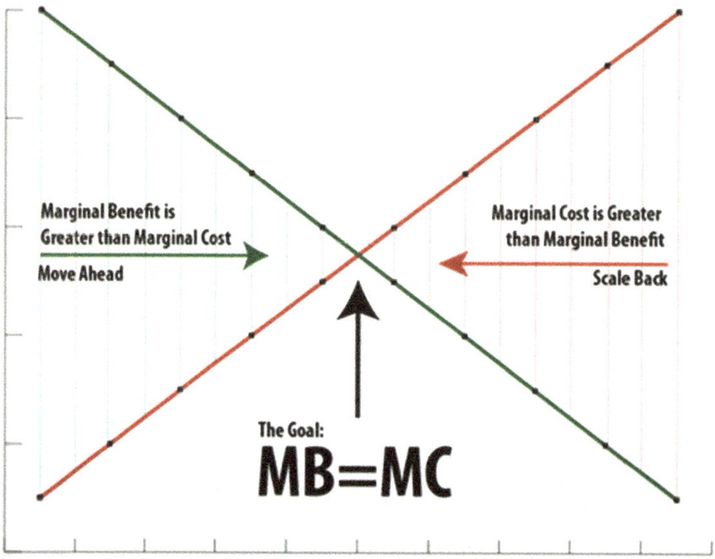

Fig. 26.5 Marginal investment analysis

generally cost effective, even though the processes involved may not always be so easy. It may also require reassessment and redesigning of the programmes and projects and development of new standards, codes, guidelines and capacities, and buy-in from stakeholders who are used to business-as-usual approaches and practices, even in hazard prone countries where repeated climate-related disasters have caused recurrent damage and losses to lives and assets.

26.5.2 Cost Benefit Analysis

Elements of cost-benefit analysis are inherent in marginal benefit analysis, but full scale Cost Benefit Analysis (CBA) for CCA and DRR is usually applied for new and exclusive risk mitigation and climate adaptation projects. The tasks involved are not simple and straightforward. The costs of projects can be estimated in terms of projected investments, but the benefits cannot be calculated so easily, as most of the benefits are in the nature of preventing or reducing direct and indirect damage and losses that did not take place. The future risk patterns are not known clearly due to the dynamic nature of hazards, vulnerabilities and exposures. Secondly, location specific nature of risk patterns coupled with divergent perspectives on the effectiveness of risk reduction strategies complicates evaluation of costs and benefits. Further, the way, costs and benefits of the projects are distributed among various vulnerable groups cannot be calculated so easily. Data on vulnerabilities and impacts are also not available in many contexts and, more importantly, on many values. The techniques applied for quantifying avoided losses and valuing non-market benefits or costs also vary and are more often controversial. The choice of discount rates affects results of analysis and, despite extensive research and agreement among economists, often remains controversial among policy makers other stakeholders. As Benson and Twigg (2004) had put it:

Despite these limitations, CBA tools are finding increasing acceptance in analysis of CCA and DRR projects. CBA goes beyond the conventional direct costs (land, labour, materials, capital, management, maintenance etc.) and direct benefits (such as power or irrigation, agricultural production, rise in income etc.) to factor indirect costs (such as environmental and ecological costs, costs of new vulnerabilities, exposures and risks created or existing risks exacerbated) and indirect benefits (environmental and ecological benefits, benefits of damage and loss prevented etc.).

Reinhard Mechler, one of the pioneers of CBA on DRR projects, has catalogued evidences of useful application of CBA (Mechler 2005). The People's Republic of China invested USD 3.15 billion in the 1960s and 1970s on flood control measures that averted damages over USD 12 billion (Benson 1998). The mangrove plantation project of IFRC in Vietnam at a cost of USD 7.2 million significantly reduced costs of maintenance of dykes besides saving lives and property with a Cost Benefit Ratio (CBR) as high as 52 during 1994–2001 (IFRC 2002, World Disasters Report). The integrated water management and flood protection scheme for Semarang, Indonesia had an Internal Rate of Return (IRR) of 23% and CBR of 2.5 (Mechler 2004).

The combined disaster mitigation and preparedness program in Bihar and Andhra Pradesh had a cost-benefit ratio of 3.76 (Venton and Venton 2004). In a community based disaster preparedness programme in Bangladesh implemented over a period of fifteen years the benefits exceeded costs between 3.05 and 4.90, even though many supposed benefits had to be excluded from the calculation due to difficulties in collecting data on the benefits (IFRC 2012, The Long Road to Resilience: Impact and Cost-Benefit Analysis of Community Based Disaster Risk Reduction in Bangladesh). CBR of 2.04 was recorded in a livelihood-centred CCA and DRR programme in Nepal (Wilenbockel 2011). The Nepal earthquake destroyed houses and infrastructure, but many buildings and infrastructure constructed as per earthquake resistant technology survived even in most affected areas. In Kathmandu Valley, 160 school building retrofitted under an ADB supported school safety programme withstood the shock of 7.8 magnitude earthquake (ADB 2015).

Japan invested heavily on various structural and non-structural measures for disaster risk reduction during the 1960s (around 7–8% of national budget) and is reaping tremendous benefits of reduced risks of disasters even though the country was struck by some of the worst natural and manmade disasters in the subsequent years. If Japan had not made these investments, the extent of devastations of Great Hanshin Earthquake in 1995 and Great East Japan earthquake in 2011 could have been much worse (GFDRR 2012).

Japan developed significant policy and legal frameworks and guidelines for disaster impact assessments of development projects. These include Policy Evaluation Act 2001 and Technical Guidelines for Cost Benefit Analysis of Public Work Projects 2004. Regulatory Impact Analysis (RIA) is legally mandatory since 2007 to

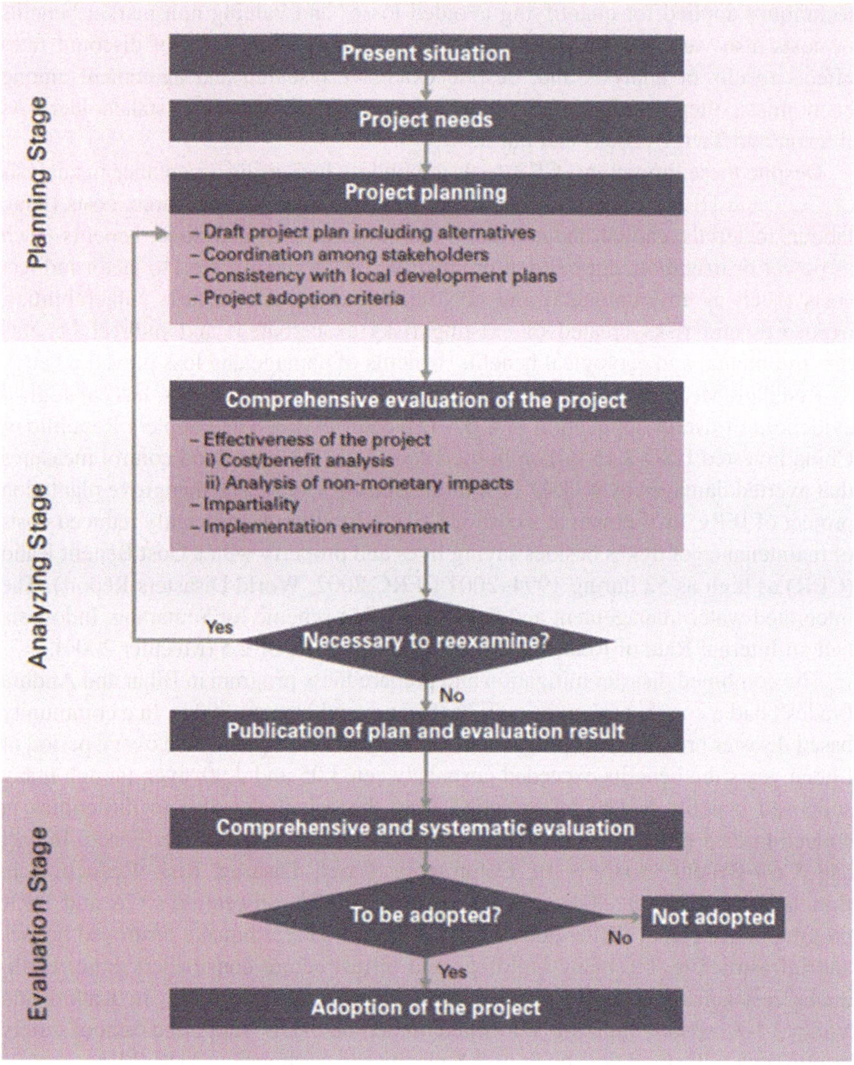

Fig. 26.6 MLIT process of project evaluation (Source: www.mlit.gov.jp)

improve effectiveness and transparency in the process of regulatory establishment. Ministry of Land Infrastructure Transport and Tourism (MLIT) has established a comprehensive process (Fig. 26.6) which ensures that all construction projects are designed in such a manner that no new risks of disasters are created while existing risks are reduced through co-benefits.

26.5.3 Multi- Purpose Development Projects

In developing countries disaster risk mitigation projects do not receive high priority due to competing demands of scarce resources from other priority sectors where benefits of investments are more direct and immediate. One way out of this impasse is to design the projects in such a way that long term benefits of CCA and DRR are combined with more short term benefits of priority sectors. Such dual or multi-purpose development projects can reduce the risks of extreme climatic events and at the same time provide direct economic benefits that would enhance both cost-benefit ratio and internal rate of return thereby justifying the costs of investments.

One of the most common examples of such multiple purpose development projects are large hydroelectric dams that generate electricity, provide irrigation and at the same time protect downstream locations from the risks of floods. The greatest example is Three Gorges Dam of China that combines multiple co-benefits far outweighing the costs. It supports the largest power station of the world with installed capacity of 22,500 MW; avoids emission of greenhouse gas equivalent to 100 million tonnes of CO_2; releases 12 cubic kilometres of water for agriculture and industry; enhances shipping capacity of Yangtze River by ten times, and most importantly, from the perspectives of disaster risk reduction, reduces the potential risks of major downstream flooding from once every 10 years to once in 100 years (Chinese National Committee on Large Dams, Three Georges Project 2012). The Yellow River of China which was once known as the river of sorrow for the recurring floods is now the river of prosperity for mainland China.

There are many examples of innovative multi-purpose development projects that resulted co-benefits of CCA and DRR. One such example is the Metropolitan Area Outer Underground Discharge Channel in Greater Tokyo Area built at a total cost US$ 13 billion to mitigate overflowing of the city's major rivers and waterways (Fig. 26.8). It is world's largest underground floodwater diversion facility, which helped the metropolis to adapt to the growing incidence of rains and typhoons and reduce risks of flood.

The city of Kuala Lumpur, Malaysia designed and developed the 9.7 km long Storm water Management and Road Tunnel (SMART), which is the longest storm water tunnel in South East Asia. The tunnel has three levels, the lowest for drainage and the upper two layers for road traffic (Fig. 26.7). This simultaneously solves the problem of flash floods and reduces traffic jams during rush hours. This has another advantage—it ensures regular maintenance of the drain that otherwise would be used only sporadically (World-Bank 2010) (Figs. 26.8 and 26.9).

Most of the countries in developing world have huge deficits in critical infrastructure. Countries are introducing significant policy reforms to create enabling environment for private and foreign direct investments in infrastructure, manufacturing, tourism and other sectors. This offers ways for innovative designing of infrastructure helping in DRR in a cost effective manner.

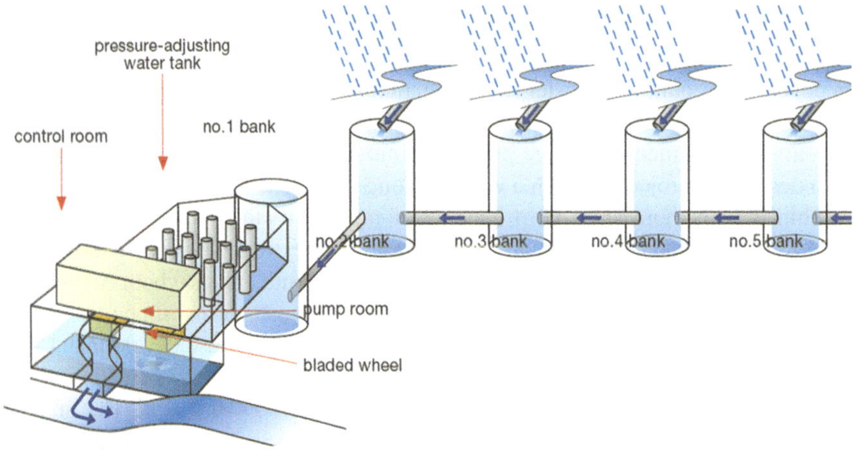

Fig. 26.7 Greater Tokyo underground discharge channel (Source: www.mlit.gov.jp)

Fig. 26.8 Greater Tokyo underground discharge channel (Source: www.mlit.gov.jp)

26.5.4 Cost Effectiveness Analysis

Cost-effectiveness analysis (CEA) is used to find the least cost option for meeting selected physical targets of adaptation or risk reduction. CEA is applied in assessing adaptation options in areas where adaptation benefits are difficult to express in monetary terms, but costs can be quantified, in terms of human health, freshwater systems, extreme weather events, and biodiversity and ecosystem services. The steps followed in CEA approach is the following (UNFCCC 2011):

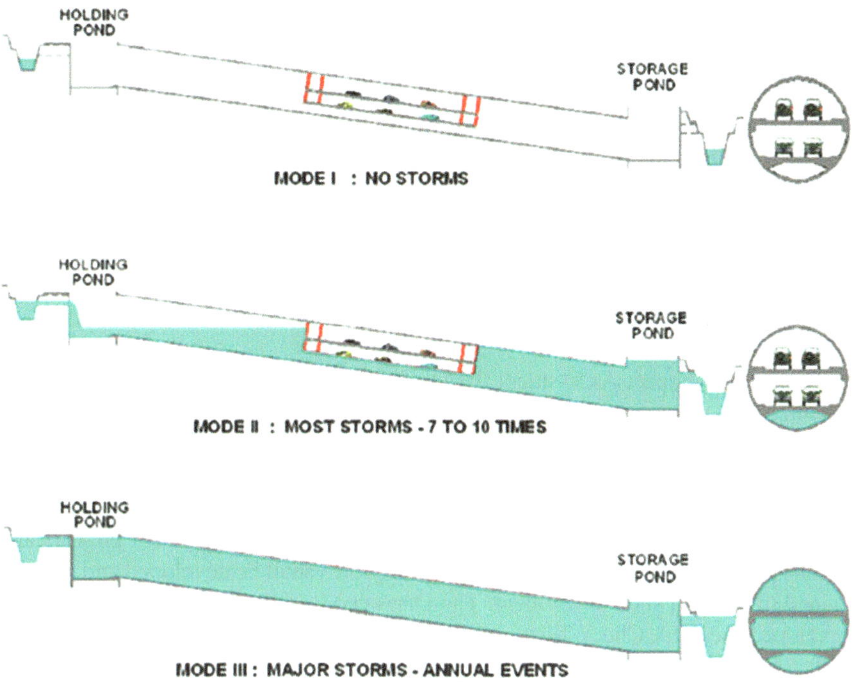

Fig. 26.9 Three modes of SMART Tunnel (Source: www.smarttunnel.com.my)

(a) Agree on the objective and identify potential options.
(b) Establish a baseline to agree on a set of indicators for evaluating and tracking benefits
(c) Quantify and aggregate the various costs in various options
(d) Compare the cost effectiveness of the different options
(e) Determine the most cost-effective option

26.5.5 Multi Criteria Analysis

In Multi Criteria Analysis (MCA) different options are assessed against a number of criteria. An overall score for each option is obtained by providing weights to each criterion. The highest score is then selected for adaptation. MCA was used by the Least Developed Countries (LDCs) in preparation of their National Adaptation Programmes of Action (NAPAs). Steps followed in MCA are the following:

(a) Agree on the adaptation objective and identify potential adaptation options
(b) Agree on the decision criteria.
(c) Score the performance of each adaptation option against each of the criteria

(d) Assign a weight to criteria to reflect priorities
(e) Rank the options

The main result of an MCA is a rank order of adaptation options and an appreciation of the weaknesses and strengths of the attributes of each of the options. An MCA can also be conducted in conjunction with other assessment approaches (CBA and CEA) to provide a more solid foundation for informed decision-making.

26.5.6 Disaster Impact Analysis

Environmental Impact Assessment (EIA) of development projects has been institutionalized in many countries with established norms, procedures and tool kits. Drawing on the analogy of EIA some countries have initiated Disaster Impact Assessment of selected development projects. Sri Lanka adopted its Roadmap for Disaster Risk Management (2005–2015) stipulated that Disaster Impact Assessment (DIA) shall be integrated into the approval process of all development projects and for this purpose necessary guidelines and procedures shall be developed and persons trained to conduct such assessment. The Disaster Management Centre of Sri Lanka developed the DIA Check Lists in four parts—assessing risks, incorporating risk reduction measures into designs, monitoring during construction and maintenance, and analysing post-disaster impact assessment. DIA checklist was introduced and field tested in road sector and is proposed to be introduced in other sectors.

Under the Comprehensive Disaster Management Project (CDMP) Bangladesh had introduced Disaster Impact and Risk Assessment (DIRA) for analysis of all development projects, but detailed sector specific guidelines and tools for DIRA have not been developed, which has made the programme weak.

26.5.7 Checklists for Disaster Risk Reduction

Government of India issued notification in 2009, which stipulates that any new project costing more than INR1000 million must necessarily have a *Check List for Natural Disaster Impact Assessment*, which would provide complete information on the hazards, risks and vulnerabilities of the project. This would include not only the probable effects of natural disasters on the project but also the possible impacts of the project in creating new risks of disasters. The costs involved in the prevention and mitigation of both types of impacts shall be built into the project costs and accordingly its economics and viability of the project shall be worked out. This checklist was meant to be a kind of self-assessment by the project implementing agencies rather than independent evaluation by a body of experts.

26.6 Conclusion and Way Forward

The benefits of integrating CCA and DRR in development plans, programmes and projects are well established. These are mandated under the global and national frameworks on Climate Change Adaptation as well as disaster risk reduction. Existing or projected public and private investment on adaptation and risk reduction would always fall way short of actual requirements. Therefore, there is no other alternative but to integrate CCA and DRR to avoid duplication of efforts and expenses, enhance efficiencies and effectiveness and promote synergies and cohesiveness in policies, plans and projects. The best way to integrate is to mainstream CCA and DRR in all existing and future plans, programmes and projects, wherever these are relevant and feasible. Tested tools and techniques are available and these can be further adapted according to the specific requirements of countries, sectors and projects.

References

ADB (2015, May 12) Schools with earthquake-proof technology survive Nepali disaster. Retrieved from ADB: https://www.adb.org/news/features/schools-earthquake-proof-technology-survive-nepali-disaster

A-PLAT (n.d.) Climate change adaptation information platform. Retrieved November 1, 2018, from A-PLAT: http://www.adaptation-platform.nies.go.jp/en/index.html

Benson C (1998) The cost of disasters: development at risk? Natural Disasters and the Third World, Oxford

Benson C, Twigg J (2004) Measuring mitigation: methodologies for assessing natural hazard risks and the net benefits of mitigation—a scoping study

Chakrabarti DGP (2012b) Understanding existing methodologies for allocating and tracking DRR resources in India

Chakrabarti PGD (2010) Integrating disaster risk reduction with climate change adaptation: new initiatives in South Asia. In: Shaw R, Pulhin JM, Pereira JJ (eds) Climate change adaptation and disaster risk reduction: issues and challenges. Emarald

Chakrabarti PGD (ed) (2015) Climate change and sustainable development. Oxford University Press

Chakrabarti PGDhar (2012a) Tracking public investments on disaster risk management in India

Chinese National Committee on Large Dams, Three Georges Project (2012)

Darwanto H (2012) Disaster risk reduction investment tracking: case study Indonesia

European Commission (2004) Project Cycle Management Guidelines, Vol-1,

GFDRR (2012) Measuring the cost-effectiveness of various DRM measures. Knowledge Note 6:1

Government of India (2009) National action plan on climate change

IFRC (2002) World disasters report. IFRC, Geneva

IFRC (2012) The long road to resilience: impact and cost-benefit analysis of community based disaster risk reduction in Bangladesh

IPCC (2007) Fourth assessment report, Working Group-II, Cambridge

IPCC (2012) Special report on managing the risks of extreme events and disasters to advance climate change adaptation. Cambridge University Press

IPCC (2014) Fifth assessment report, synthesis report

Mechler R (2004) Semarang case study. GTZ

Mechler R (2005) Manual on cost benefit analysis of natural disaster risk management projects in developing countries. UNISDR

NAPCC (2009) National action plan on climate change. Government of India. Retrieved from http://www.moef.nic.in/downloads/home/Pg01-52.pdf

NDMA (2009) National policy on disaster management 2009. NDMA

Pereira J (1995) Costs and benefits of disaster mitigation in construction industry. USAID

Secretariat, U. N. (2017) Opportunities and options for integrating climate change Adaptation with the Sustainable Development Goals and the Sendai Framework for Disaster Risk Reduction 2015–2030

UN (1987) Report of the world commission on environment and development: our common future. Oxford University Press

UN (2015) Transforming our world: the 2030 agenda for sustainable development

UNESCAP (2017) Mainstreaming disaster risk reduction for sustainable development: a guidebook for the Asia-Pacific

UNFCCC (2007) Report of the conference of the parties on its thirteenth session, held in Bali from 3 to 15 December 2007. UNFCCC

UNFCCC (2010) Report of the Conference of the Parties on its sixteenth session, held in Cancun from 29 November to 10 December 2010. UNFCCC

UNFCCC (2011) Assessing the cost and benefits of adaptation options: an overview of approaches. Bonn

UNFCCC (2015) Paris Climate Agreement

UNFCCC (2017) Opportunities and options for integrating climate change adaptation with the Sustainable Development Goals and the Sendai Framework for Disaster Risk Reduction 2015–2030

UNISDR 2009 Terminology on disaster risk reduction

UNISDR (2015) Sendai framework for disaster risk reduction 2015–2030. UNISDR

UNISDR, Sendai Framework for Disaster Risk Reduction 2015–2030

Venton CC, Venton P (2004) Disaster preparedness programmes in India a cost benefit analysis. HPN

Wilenbockel D (2011) A cost-benefit analysis of practical action's livelihood based disaster reduction project in Nepal. Practical Action

World-Bank (2010) Natural hazards and unnatural disasters: the economics of effective prevention. The World Bank

P G Dhar Chakrabarti held many administrative, policy making and academic assignments. He was Chief Executive Officer, Emergency Relief Organization of J&K Government, Executive Director of National Institute of Disaster Management, Secretary of National Disaster Management Authority and founder Director of SAARC Disaster Management Centre. He was Member of UN Secretary General's Advisory Group on Central Emergency Response Fund for two consecutive terms and served in expert groups of UNISDR, UNOCHA, UNFCCC, UNESCAP, the World Bank and other agencies.

Chapter 27
Nature Based Solutions for Disaster Risk Reduction: Concepts and Overview

Shalini Dhyani, Muralee Thummarukudy, and Anil K. Gupta

Abstract Increased frequencies of extreme climate events have been a matter of grave concern for the South Asia region. Disasters and environment have been well observed for being interlinked in multiple ways. Deforestation and forest degradation, rapid land-use change, and overexploitation of natural resources are compounding natural vulnerabilities. For reducing disaster risk and human well-being there has been increased national as well as international attention on the role of healthy ecosystems. Scientists have explored and identified profound linkages between faulty environmental management practices and increasing disaster risks. This understanding so far has been crucial to address and manage disaster intensity and frequency that augments environmental damage affecting quality of life. Nature based Solutions (NbS) can be considered as a broad overarching concept and approach that substantially addresses disaster risk reduction and climate change adaptation. NbS has full potential as a tool to help India in achieving the targets of Paris Climate Agreement, 2015 and other relevant goals of international agreements. It is pertinent to understand the environment-disaster interface so that NbS can be integrated in ongoing developmental planning. There has been substantial progress made across the world related to NbS as a concept, policies, and also its implementation. Present paper takes stock of knowledge available about the growing importance and application of NbS across India through various examples and ongoing efforts. Study attempts to provide the comprehensive international and national

S. Dhyani (✉)
Critical Zone Group, Water Technology & Management Division, CSIR-National Environmental Engineering Research Institute, Nagpur, India

IUCN CEM South Asia, Gland, Switzerland
e-mail: s_dhyani@neeri.res.in

M. Thummarukudy
G20 Global Initiative Coordination Office, UNCCD, New York, NY, USA

A. K. Gupta
ECDRM Division, International Cooperation, CAP-RES DST Project, National Institute of Disaster Management (Ministry of Home Affairs), New Delhi, India

A. K. Gupta et al. (eds.), *Disaster Risk and Management Under Climate Change*, Disaster Resilience and Green Growth,
https://doi.org/10.1007/978-981-99-4105-6_27

overview on NbS for EcoDRR. Paper presents a synthesis of robust understanding about NbS for EcoDRR and targets to attract policy makers as well as important stakeholder groups in protection and management of environment to include NbS as an important approach to meet targets of an array of international agreements by using a common approach.

Keywords Nature-based solutions · Disasters · India · Climate change adaptation · Ecosystems

27.1 Introduction

A disaster as a sudden, calamitous incident resulting in huge losses many times exceed the community's ability to cope using their own resources. Disasters have been intricately linked to progress and expansion, and to a magnitude they also depend on development concerns about the environment. Degradation of environment characterized by deforestation, biodiversity loss, impromptu development, etc., so far has been few of the foremost considerations for worsening their causative vulnerabilities. Ecological degradation magnifies the intensity of natural disasters and often transforms a natural hazard or climatic extreme into a catastrophe.

South Asia region is world's most densely populated area and home to more than 500 million poor people of the world (UN Habitat 2010). Region is also one of the mega-diversity centers and biodiversity hotspots and has been historically vulnerable to several natural hazards. Exposure linked to geo-climatic and socio-economic circumstances makes India one of the most disaster prone countries on the planet (Fig. 27.1). In last three decades (1990 onwards) the country has undergone rapid temperature rise and alteration in rainfall patterns (Rathore et al. 2013). The intensification is in terms of number of incidents, magnitude and intricacies of these disasters that has triggered and heightened responsiveness and necessity for pre-disaster preparedness and mitigation efforts as a pertinent obligation (Fig. 27.1; Table 27.1).

The depreciation of environmental sustainability has been a crucial trigger for upsurge in physical and socioeconomic exposures throughout the country resulting in colossal loss of human lives and property (Table 27.1 below).

Disasters are no more centered in remote and isolated natural pockets of the country. Urban areas are new emerging centers of natural as well as man-made disasters. As per, vulnerability atlas of India, 38 cities are in seismic vulnerability areas adding to their exposure to environmental and human induced calamities. This is distinctly marked by amplified consistency and deliberations of urban catastrophes observed by the country in last few decades (table above). Indian cities are gradually emerging as exposed centers to climate-meteorological calamities (Revi 2008).

Fig. 27.1 Showing Impact of climate-meteorological associated disasters in different parts of India (Clockwise showing Delhi Smog, 2016; Nagpur Flash Flood, 2018; Chennai Flood, 2015 and Himalayan Tsunami in Uttarakhand, 2013)

27.2 Rationale and Need

Increasingly, planners, scientists and engineers are finding that using green approaches or a mix of green and grey solutions can reduce disaster intensity in long term. Enhancing resilience against uncertainties using NbS is less of a choice but a compulsion (Dhyani et al. 2020). Despite the growing international support for NbS, the reality is that they have not yet been mainstreamed in the existing disaster risk management policy and practice (Dhyani and Thummarukudy 2016). Present overview provides development in DRR arena post 2015 after the three major global agreements (Sendai Framework for DRR 2015, Paris Agreement 2015 and SDGs (Sustainable Development Goals) 2015 have been negotiated and agreed upon. Present chapter intends providing guidance that includes concepts and lessons on the benefits of NbS and how NbS can be integrated in DRR strategies at the local and national levels. Chapter is largely an outcome of the literature review, supported by case studies from India on NbS and approaches. Local and indigenous communities in India for millennia have been informally using NbS to address DRR and climate change adaptation. In modern times NbS must be considered as a *work-in-progress*, as its' solicitations are still being established and verified.

Table 27.1 Major Indian disasters in last few years that have resulted in huge economic losses; loss of human lives and long term impacts on ecosystems and human well-being

Disaster	Causes	Economic losses	Loss of human lives	Future impact
Kerala Flood and Landslide (2018)	37.5% excess rainfall in 2.5 months and degradation of forests with no preparedness	≈ Rs 15, 000–20,000crore s (till Aug 20, 2018)	≈370 killed (till Aug 20, 2018)	≈10,000 km of roads washed away, 1 lakh houses damaged. Considerable wash off in topsoil resulting poor land productivity in both the tea and rubber markets as well as paddy cultivation. (Kerala floods 2018: Businesses stare at a loss of over Rs 20,000 crore)
Chennai Flood (2015)	Excessive rainfall exacerbated by very strong El Nino in the Pacific Ocean, the record-warm Indian Ocean, unplanned development choking drains and filling up wetlands due to lack of preparedness	≈Rs 10,00,000 crore	≈400 killed	Chennai floods are having worldwide economic impact, particularly in IT (Information Technology) sector
Jammu and Kashmir flood (2014)	Unprecedented rainfall, unplanned urbanization due lack of preparedness	≈Rs 5400–5700 crore	280–300 killed	2.53 lakh damaged houses
Himalayan Tsunami (2013)	More than 400% of average rainfall, deforestation, and unplanned development on the banks of rivers	≈Rs 12,000 crore	4000 killed	A total of 2119 houses were fully damaged. 2070 km roads and 145 bridges damaged and washed off

27.3 Concepts and Framework

Evolution in the paradigms of reducing disaster risks and adapting to the impacts of climate change in a framework that is shifting from 'response and relief' to 'mitigation and preparedness' has invited lot of focus to NbS. NbS is a fairly new concept and can be well defined as an umbrella term that covers an array of environmental protection and management approaches introduced by IUCN (Jones et al. 2012; Cohen-Shacham et al., 2016) to address increasing global disaster and climate risk

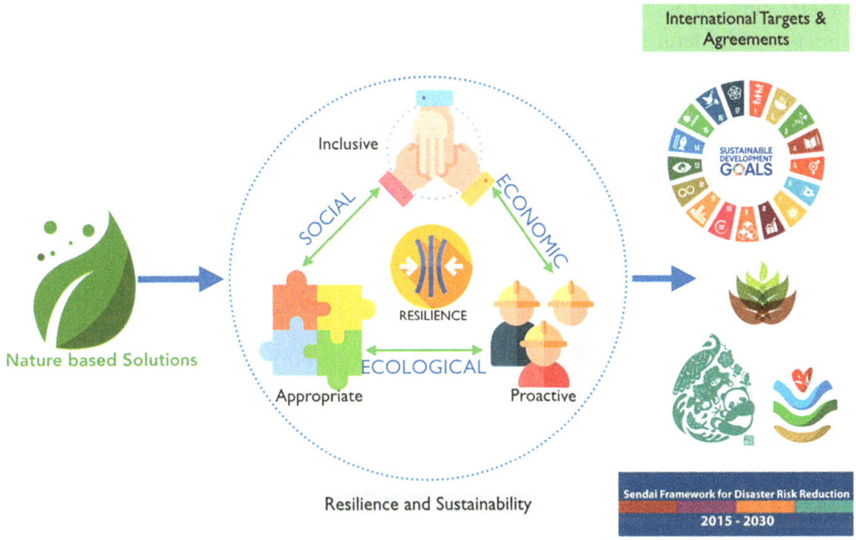

Fig. 27.2 NbS for addressing societal challenges and long term biodiversity benefits that ensures human well-being

challenges (Fig. 27.2). NbS was proposed by IUCN to respond and help achieving promises of climate change negotiations "*as a way to mitigate and adapt to climate change, secure water, food and energy supplies, reduce poverty and drive economic growth*" (Balian et al. 2014; Cohen-Shacham et al., 2016). Late 2000s was the year when the World Bank first introduced the 'Nature-based solutions' approach (MacKinnon and Hickey 2009) for promoting nature as a vital solution to address risks and disasters linked with climate variability and vulnerability.

Globally, IUCN pioneered NbS in UN climate dialogues and negotiations, and biodiversity conservation as a pertinent approach that is well addressed in NbS (IUCN 2009; Balian et al. 2014; Cohen-Shacham et al., 2016). IUCN recognized the vital requirement to link NbS with global policy and related implementation approaches on ground (MacKinnon et al. 2011). NbS is not only restricted to dialogues and discussions on ecosystem services and natural capital build up and the term has been regularly used to provide information related to soft engineering practices that increase resilience from disasters and climate change (Marton-Lefèvre 2012; Van Wesenbeeck et al. 2014). The European Commission has adopted and included NbS as an important area in its Horizon 2020 research programme though it has its explicit focus on urban sprawls.

The scope and definition for NbS is broad, and the core emphasis lies on enhancing the resilience of landscapes as well as communities by building back the natural stock against the traditional engineered approaches (Maes and Jacobs 2015; Eggermont et al. 2015). NbS have been well-defined as engagements stimulated by or sustained by impersonating nature to support people in ecological, societal and financial tasks in sustainable ways (European Commission and Directorate-General for Research and Innovation, 2015) (Fig. 27.2).

European Commission has expressed NbS as "actions which are inspired by, supported by or copied from nature" (European Commission and Directorate-General for Research and Innovation, 2015). IUCN definition for NbS reflects, "*Actions to protect, sustainably manage and restore natural or modified ecosystems that address societal challenges (e.g. climate change, food and water security or natural disasters) effectively and adaptively, while simultaneously providing human well-being and biodiversity benefits*". Maes and Jacobs (2015) define NbS as "*Transition to use ecosystem services with reduced inputs of non-renewable natural capital and enhanced investment in renewable natural processes*". Eggermont et al. (2015) has deliberated significant characteristics of NbS (No or nominal external interferences in ecosystems, interferences in managed ecosystems that involve sustainability and multi-functionality, and establishment or expansion of ecosystems to foster green or blue set-ups) to recover the flow of variety of ecosystem services.

27.3.1 Understanding Linkages Disasters, Ecosystems and Society

Terrestrial and marine ecosystems function as a global carbon sinks, absorbing approximately 50% of total annual emissions from fossil fuels (IPCC 2014). Managed healthy ecosystems deliver natural protection for many natural hazards. Landslides, floods, desertification, tsunamis, flash floods, forest fire and drought are just a few examples of such natural disasters. Impact of forest degradation in Himalayas results in massive landslide incidents every year (Fig. 27.3). Deforestation affects flow of provisioning (fodder, fuel wood *etc.*) and regulating (soil erosion control) ecosystem services (for both upstream and downstream people). Reduced flow of ecosystem services leads to enhanced drudgery of women and weaker sections of the society in Central Himalayas (Dhyani et al. 2011; Dhyani et al. 2013; Dhyani and Dhyani 2016). Disaster shows socio-ecological linkages and may aggravate long term societal risks (Dhyani and Dhyani 2016; Dhyani et al., 2020; Dhyani et al. 2021b).

Biodiversity conservation and ensuring the flow of ecosystem services are basis for planning solutions to catastrophes, related to climate change, reducing disaster risks, alleviation of poverty and promotion of green economy. NbS addresses several policy objectives and is more action-oriented. NbS bring multiple co-benefits. Some of these benefits are reducing costs, generating more livelihood and alternative livelihood opportunities that develop into green economy *etc*. However, NbS require a lot of understanding during implementation to address knowledge gaps, managing the trade-offs, implementing successful on ground applications, including natural aspects and finances of NbS. NbS are here and they make a perfect short term as well as long term sustainability and financial sense, provide co-benefits and multiple

Fig. 27.3 Disaster have socio-ecological linkages. e.g. Forest Degradation in Central Himalayas affecting women drudgery, malnutrition and health

benefits beyond climate change mitigation and adaptation (Fig. 27.3). NbS approaches can include activities that reduce emissions from deforestation and forest degradation (REDD, REDD+), promoting climate resilience through adaptive agriculture, developing carbon sinks along coasts to harness the power of ecosystems for climate change adaptation. Ecosystems are natural safeguards and are more effective and inexpensive options than to maintain conventional engineering approaches, *viz.* concrete protection walls on river banks to reduce flood risks (Figs. 27.4).

Similarly, catchment area treatment by plantation measure to improve water infiltration for replenishing ground water sources and aquifers is cheaper and more sustainable than constructing a new water supply system.

27.3.2 Nature-Based Solutions: Addressing Socio-Environmental Challenges

NbS have broad portfolio and provide diverse solutions to address important global challenges. Community centric, participatory initiatives to co-design, co-create and co-manage ('*co-co-co*') are promoted in NbS hence, NbS stands different from more conventional, conservative, and top-down initiative. Integrated governance-based approach for developing and managing NbS are preferred (Van Ham and Klimmek 2017). The goal to strengthen economic growth and sustainability using NbS has found special place in European Commission and IUCN documents (Cohen-Shacham et al., 2016; Maes and Jacobs 2015). NbS supports replicability options and opportunities of its solutions (Van Ham and Klimmek 2017). EU has identified NbS approaches to cover urban regeneration and human well-being, watershed management, coastal resilience, enhancing the insurance value of ecosystems, sustainable use of matter and energy, and increased carbon sink potential. Report by the

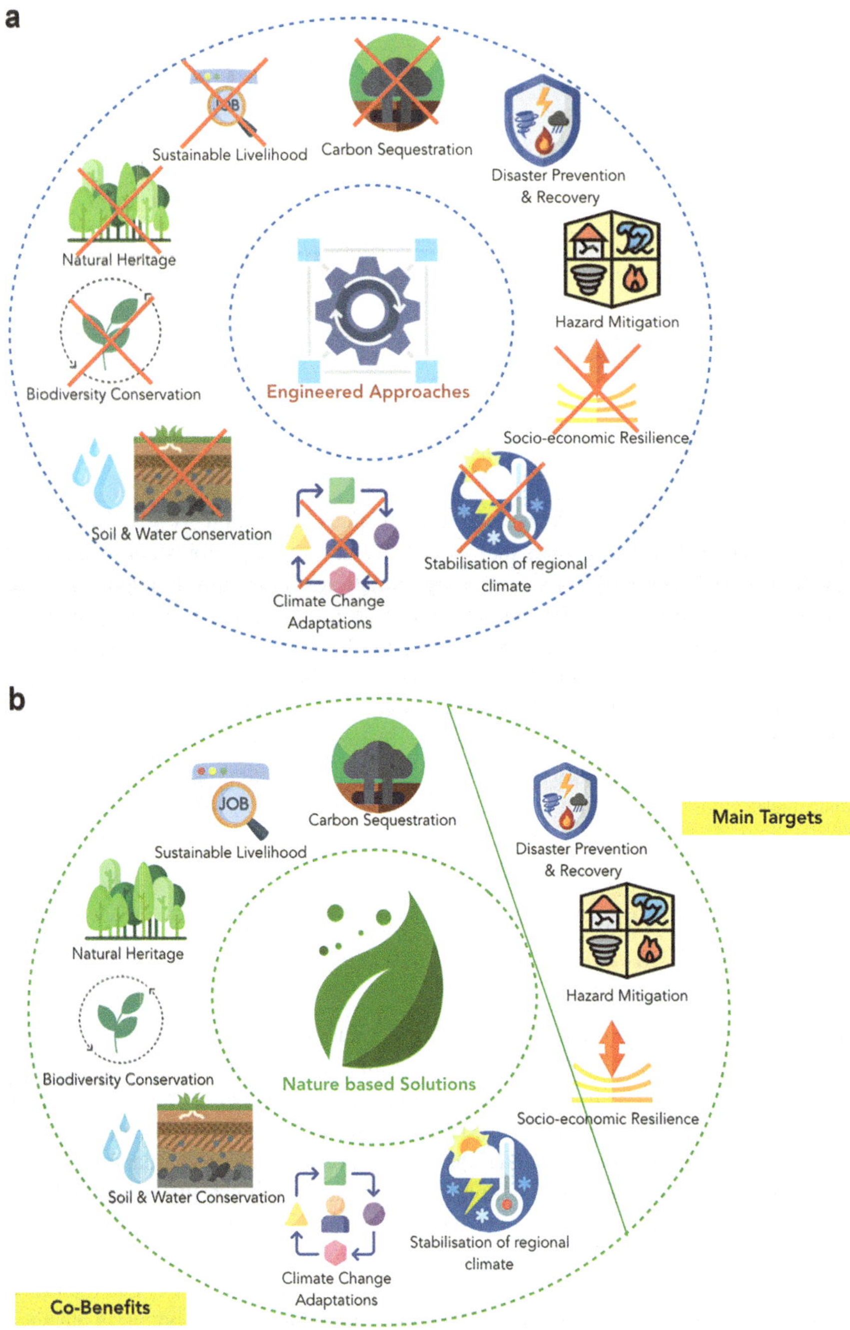

Fig. 27.4 (a, b) Comparing conventional engineering approaches with nature based solutions (NbS) that provide co-benefits along with targeting hazard mitigation and climate change adaptation

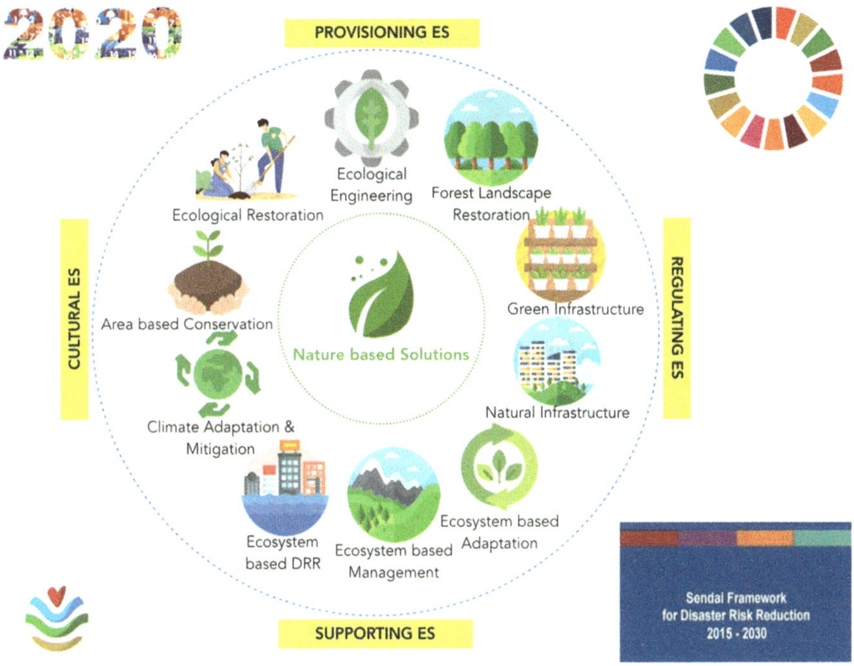

Fig. 27.5 NbS concepts involving ecosystem based approaches that provide constant flow of ecosystem services and fulfilling promises of international agreements and national targets

European Commission's expert group (Sutherland et al. 2014) lists 310 actions (*e.g.* Protection and expansion of forests to absorb gaseous pollutants, plantation of wind breaks for soil conservation, conservation of urban green spaces and planning green roofs for various co-benefits *viz.* promotion of biodiversity, carbon storage and flash flood management) as NbS examples.

NbS is a proven strategy to not only to reduce disaster risks but also provide variety of ecosystem services and nature's contribution to people for human well-being (Fig. 27.5) . NbS implementation is cost effective and also involves people and communities hence, has potential to realize Aichi Targets for nature conservation and localizing and achieving SDGs, 2015.

27.3.3 Nature-Based Solutions (NbS): From Theory to Practice

Well-managed, healthy and diverse ecosystems are crucial for existence and supporting human well-being, resulting in resilient and prosperous society (Diaz et al. 2015). NbS facilitates judicious utilization of natural resources for human

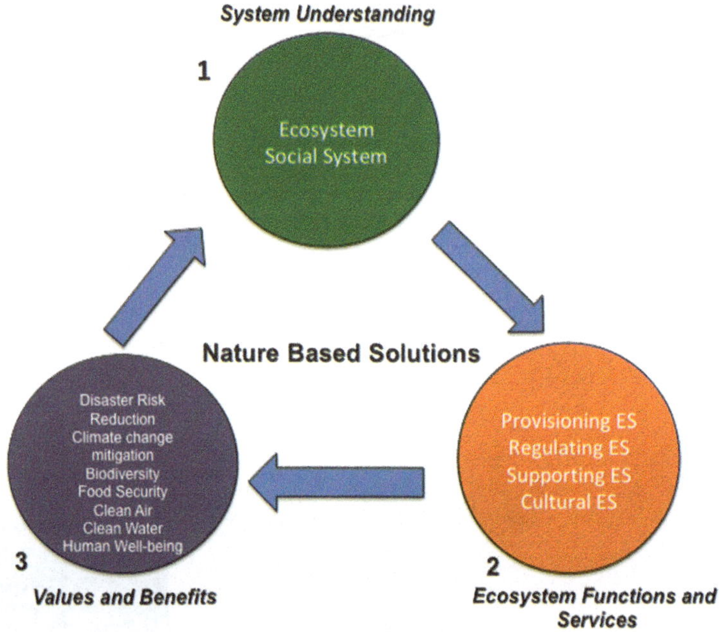

Fig. 27.6 Nature-based Solutions (NbS) involves system understanding for planning effective approaches that ensure restoring ecosystem functions and services for human wellbeing though nature's benefits to people

wellbeing with minimum impact on surrounding environment (Yigitcanlar and Teriman 2015). NbS are "*no-regret*" approaches as they combine major global challenges and also help securing livelihood opportunities for locals. '*Quest for Sustainability*' is a significant apprehension in disaster management because of the enhanced understanding of '*disasters' as environmental progressions* rather than '*environmental extremes*'. Well-managed, healthy ecosystems function as a critical natural arrangement that safeguards societies (Sudmeier-Rieux and Ash 2009; Sudmeier-Rieux et al. 2011). NbS puts categorical prominence on mainstreaming biodiversity conservation with targets for sustainable and climate resilient progress (Balian et al. 2014; Eggermont et al. 2015), and suggests advanced, implementable 'solutions'. NbS planning and design promotes identification of the issue and better and holistic understanding of the ecosystem (Fig. 27.6). By integrating societal needs along with natural processes, we can address grass root issues and root causes that can help to achieve additional values and benefits for human well-being, resilient society as well as ecosystems. Designing NbS is an iterative process and several cycles of planning and design need to be followed (Fig. 27.6).

27.3.4 *Nature Based Solutions (NbS) for Ecosystem Resilience and Disaster Risk Reduction (DRR)*

Insignificant land management, unsustainable manipulation of natural resources and degraded ecosystems are major emphasized features and considerably debated fundamental drivers of disaster risk (Hughes 2017; Diaz et al. 2015). As environmental impressions of disasters are being acknowledged nations are unambiguously demanding to reinforce the safeguard and management of ecosystems for building resilience to disasters and climate change (PEDRR 2010). Ecosystem resilience plays a fundamental dual role, both as a strategy for climate resilience building against climate mitigation and adaptation. Increasingly, hybrid grey-green approaches that optimize ecosystem functions by inclusive engineering solutions are being implemented to serve as protective barriers to local livelihoods, biodiversity conservation, and carbon sequestration. Some of the green resolutions are now applied in urban provision for water conservation as effective, flexible, economical, enduring and versatile than conservative 'grey infrastructure' preferences (*e.g.* Gill et al. 2007; Pugh et al. 2012; Raje et al. 2013; Dhyani et al. 2018).

27.4 Opportunity and Approaches

2015 was a critical year in terms of foremost international treaties and expanding worldwide acknowledgment of NbS. The year presented a historic and unprecedented opportunity to embark on taking new options and opportunities for improving human wellbeing and also the present ecosystems' state. All these historical agreements and frameworks recognize the role of ecosystem as a crosscutting concern in disaster risk reduction. NbS approaches not only address human needs but also make positive contributions for meeting the Sustainable Development Goals (SDGs). In one fifth of the earth's population, South Asia has a critical role as well as a lot of opportunities in the global achievement of the promises of Paris climate agreement and SDGs through NbS (Fig. 27.7 below).

SFDRR (Sendai Framework of Disaster Risk Reduction) prioritizes risk governance (Priority Action 2) in developmental planning. It also prioritizes investment in resilience integration and resilience inclusive planning (Priority Action 3) (Kelman 2015). SFDRR agenda comprises of various primacies and obligations, not only for reducing disaster risks, adaptation to climate vulnerability but also in sustainable development. The pace of development coupled with climate change is steering us to adopt different approaches and requires innovative institutional, business, governance, and finance models, approaches and frameworks. It also involves working with a multidisciplinary and multi sectorial approach. India has also agreed on the Intended Nationally Determined Contributions (INDCs) and target of reducing the emission of Green House Gases (GHG) to 35% by 2030 (India's Intended Nationally Determined Contribution: Working Towards Climate Justice). Indian policies

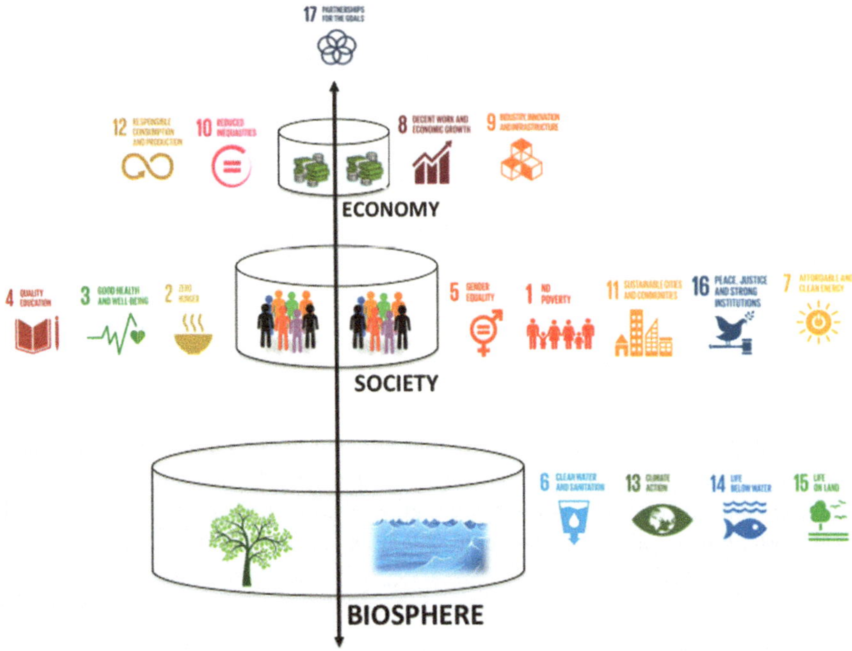

Fig. 27.7 Importance of Nature based Solutions (NbS) to develop healthy Biosphere against natural and man-made disasters that can help developing resilient societies for hassle-free economic development and achieving SDG (2015–2030)

documents have inclusion of many NbS approaches in core. Indian Government has been promoting NbS through National Action Plan for Climate Change (NAPCC, 2008) that includes National Mission on Sustaining Himalayan Ecosystem, Sustainable Agriculture, Green India Mission, Water Mission, Clean Energy, Sustainable habitat *etc.*

27.4.1 Environmental Engineering, Environmental Restoration, and Forest Landscape Restoration

The National Mission for Green India (GIM) is important among climate action missions under the NAPCC, 2008. GIM is an integrated approach for improving the green cover in the country and is one of the priority areas for NAPCC that enhances carbon sink potential as a co-benefit. GIM has adopted an inclusive cross-sectorial strategy that is intended to be applied for public as well private lands by inclusive community involvement in all its key phases of application and implementation. The strategy focuses on contributions from different partners with mutual agenda for aiming to essential need-based interventions in a rapid pace.

Box 27.1 Bonn Challenge a Nature Based Solution (NbS) for Achieving the Goals of Paris Agreement, 2015

The goal of Paris Agreement of living below 2 °C equals approximately 800 gig tonnes $C0_2$ equivalent ($GtC0_2e$) . Since, current Nationally Determined Contributions (NDCs) would only reduce the 14 $GtC0_2$ e/year emissions gap by roughly 5 $GtC0_2$ e/year, and land use accounts for 25% of global emissions, countries are considering NbS as part of their climate change mitigation and adaptation efforts. Conservation International's (CI) work on climate policy targets helping countries to make Paris agreement a success by considering full advantage of the NbS towards achieving the goals. The Paris Agreement clearly recognizes that these goals are tough to be achieved \without using nature's carbon sink potential to reduce the climate vulnerability by enhancing resilience of people. India has promised to revegetate approximately 18 million ha of degraded forest areas during COP 21 under Bonn Challenge. Reforestation and restoration of degraded and deforested lands stimulates stabilizing the microclimate and supports economic upliftment of local and regional communities. The Bonn Challenge is a pertinent global initiative to rehabilitate 150 million hectares of the earth's degraded areas by 2020 and an additional of 350 million hectares of such areas by the year 2030. Bonn Challenge is supervised by the Global Partnership on Forest Landscape Restoration, along with IUCN Secretariat. The Bonn Challenge is not an innovative global commitment to restore degraded forests, but rather an approach that is practical and realizes many on- going international commitments for conservation and protection of nature and biodiversity. The challenge is inclusive and realizes the CBD (Convention on Biological Diversity) Aichi Target 15, the UNFCCC REDD+ goal, and also Rio + 20 neutral goal of land degradation.

(Dhyani et al., 2020)

Box 27.2 National Green Highways Mission, 2016: NbS Approach to Increase Carbon Sinks Potential

India looks at National Highways Greening projects as a tool for ecological and economic development of the country, and also as an initiative towards India's commitment at CoP 21 (Conference of the Parties) Summit for developing additional carbon sink of 2.5 billion tons annually. In 2015, Ministry of Road Transport and Highways (MoRTH), Government of India, realized the immediate need and importance of developing green corridor along the national highways. The Mission got underway in July 2016 with starting the initial plantation drive along 1500 kms of national highways.

National Green Higways: An NbS approach to increase Carbon sink potential

(continued)

Box 27.1 (continued)

The task of planning, implementation and monitoring roadside plantations along one lakh kms network has been entrusted to the National Green Highways Mission under the National Highways Authority of India (NHAI). Besides providing a green canopy and a safeguard against rising pollution, the NGHM aims to create jobs for one million youth and generate alternate income source for local communities. The tree plantation drive is also an intrinsic part of the government's commitment to the world community at the Paris Climate Change meet (CoP 21) to do more to protect the environment.

(Green Highways Division (National Highways Authority of India)

27.4.2 Issue Specific Approaches for Ecosystem Conservation

Issue specific EbA have emerged from a long prevailing history of using environmental approaches of nature conservation and management to adapt to climatic variability and reduce risks from emerging and ongoing natural hazards. Two important approaches to adaptation that have gained currency over the past few years are Ecosystem-based Adaptation (EbA) and Ecosystem-based Disaster Risk reduction (EcoDRR) (Dhyani and Thummarukudy 2016; Gupta and Nair 2012). EbA and EcoDRR are kept under the umbrella of NbS. Remunerations of the latest era of the second archetype change in disaster administration acknowledges '*ecosystem approach to disaster risks reduction (EcoDRR)*'. Methods grounded in ecosystem correspondingly deal with the remunerations of a community-based method for its stress on livelihood, health and food safety within the framework of vulnerability reduction. It also targets cutting down externalities in dependency, refining on self-assurance and native assets for disaster mitigation and attentiveness.

Box 27.3 Fodder Banks to Ease on Women Drudgery and Pressure from Forests for DRR: Success Story from Central Himalayas

Animal husbandry is vital segment of mountain community structure. Inaccessibility of green fodder throughout winters in higher Himalayan region has constantly remained a grave concern. Women in hills are customarily involved in gathering fodder from forests and other areas. Fodder bank is an important NbS that was applied to address degradation of forests, women drudgery and slope stabilization in Kedarnath Valley in Central Himalayas. Fodder bank approach was developed for exploiting fast growing and excessive biomass yielding nutritive species. Pilot was developed to understand functioning and magnitude to which Fodder Bank, can help by reducing short and long-term pressures on forests, and reducing disaster risks. Nurturing fast developing and

(continued)

Box 27.3 (continued)
high biomass yielding nutritious fodder species in agro-pastoral set ups eases the drudgery of women and also reduces deforestation. Communities based on their requirement preferred the native tree and shrub species, and their native awareness about species, with respect to improved lactation, enhanced nourishment etc. Women were instructed in cultivating high biomass yielding forage species in cropland bunds and kitchen gardens. Livestock holders and cultivators were also skilled to create animal sheds on systematic technical lines offered with economical nourishing and watering schemes along with accurate aeration using locally existing supplies. Apart from contributing to fodder bank prototypical site improvement, women have also initiated producing high biomass yielding fodder grasses and shrubs on their insufficient cropland and kitchen garden bunds. This practice helped reducing pressure on forests in addition to reduced landslide incidences and soil erosion risks in the village. At the same time a lot of nutrient rich fodder in village vicinity helped increasing milk yield from mulching animals that gave better economic benefits to locals.

(Dhyani et al. 2013; Dhyani and Dhyani 2016; Dhyani et al. 2021a)

National Mission for Sustainable Agriculture (NMSA), 2008 helps enhancing agricultural productivity especially from rain fed areas dependent to monsoon rains. NMSA takes its main mandate from Sustainable Agriculture Mission under NAPCC, 2008 by endorsing and facilitating integrated farming to enhance water use efficiency and also to support soil health management. This results in synergizing resource optimization, conservation and management. NMSA aims to accommodate the crucial proportions of 'water use effectiveness', 'nutrient supervision' and 'livelihood modification' by implementation of sustainable progress methodologies. These approaches that are also crucial NbS methodologies include environmental responsive skills, implementation of energy proficient gears, natural assets conservation, assimilated farming, *etc.*

Box 27.4 Sequestering Carbon through Indigenous Adaptive and Climate Resilient Organic Agriculture Practices: A Case of Garhwal, Central Himalayas
Conventional agricultural approaches supported on salvaging of accessible natural resources are going to be the significant sustainable production procedures and leads to adaptive agriculture.

***Baranaaja* a diverse mixed cropping system:** *Baranaaja* is a conventional diversified cropping practice of the region. *Baranaaja* technique is defined, as a 'synchronized scheme' in which crop seeds are cultivated in combinations and the interactions spatio-temporal. It has been explained as a

(continued)

Box 27.4 (continued)
cropping arrangement that comprises of 12 and many more crops (cereals, millets, legumes, oil yielding seeds) cultivated in 'synergistic' mixtures.

Back to native crops: Cereals, pseudo-millets, millets, and legumes have excessive nutritive and environmental significance. Customary diversities that can endure climatic variations and instabilities for delivering sustenance yields are preferred.

Native soil managing practices: Legume yields fix atmospheric nitrogen and have massive prospects to realize the nitrogen supplies for soil. Legume crops are sustainable source of nitrogen for the region, as the landholdings are insignificant.

(Misra et al. 2008)

27.4.3 Natural and Green Infrastructure

Ecosystem services notion has been extensively acknowledged for conservation of natural landscapes but now has also been receiving reputation for human controlled landscapes to advance attributes of human well-being in urban populations (De Groot et al. 2010; Lafortezza et al. 2013; Dhyani et al. 2021b). Though, assimilation of ES in urban provisions still undergoes countless challenges, yet the perception has assisted to appreciate the coupled interactions to support in taking decisions (Gret-regamey et al. 2017). Combination of nature approach in urban expanses has not been a new thought as apparent in viewpoints of established landscape urban planners (McHarg 1967, 1969). Ebenezer Howard's "Garden city concept" signifies the combination of urban, nature and their reliance on each other (McHarg 1969; Norman 1973). Approaches that endorse integration of ecosystem approaches in urban planning are predominantly green infrastructure (GI), ecosystem based adaptation (EbA), and ecosystem based disaster risk reduction (EcoDRR) under the overarching concept of NbS. GI is a reasonably old approach that was, largely established to regulate urban sprawl by augmenting the natural setups and assimilating green areas (Bowler et al. 2010; Haifeng et al. 2015). The fundamental of GI idea is associated to connectivity and multi-functionality (Benedict and McMahon 2006; Hansen and Pauleit 2014; Dhyani et al. 2022). Being deprived of the awareness on NI and GI sensible provision of resilient cities is not conceivable. National Mission on Sustainable Habitat (NMSH) under NAPCC, 2008 addresses urban resilience to climate vulnerability, by involving mitigation and adaptation into urban planning and management practices (Kadaverugu et al. 2022). NMSH emphasizes on bio-chemical transformation, wastewater utilization, sewage exploitation and recycling choices wherever imaginable (National Mission on sustainable habitat 2011).

Box 27.5 Constructed Wetlands a NbS for Managing Waste Water in Urban Areas

For growing urban sprawls in India regular water shortage during summers has become common Decentralized wastewater treatment systems are now been considered for peri-urban and urban areas, where sewage network of centralized systems is under developed. Treated water of Nagpur, Maharashtra is supplied to thermal power plant at Koradi (Dhyani et al. 2018). Water requirement of thermal power plant (about 121 MLD) is now met by treating sewage of 265MLD. This reduces fresh water usage for the purpose and the company is able to reduce its expenses on water. Plant based technology (NbS) has potential to treat urban wastewater while providing alternative livelihood to locals. Treated water quality meets the specified water quality norms of state as well as Central Pollution Control Board (CPCB) (VI fresh water category). Constructed wetlands use specific plants, such as elephant grass (*Pennisetum purpureum*), cattails (*Typha sp.),* reeds (*Phragmites sp.),* canna and yellow flag iris (*Iris pseudacorus*) that are normally found around natural wetlands and have effective filtration and treatment capability. Treated water can be reused for gardening, agriculture etc. Phytorid technology carried out by onsite treatment and reuse of grey water up to 95%, attracts a total of 5 credits in Indian Green Building Certification (IGBC) (Mhaske et al. 2014). Constructed wetland technology in combination with rain-water harvesting can bring added benefit. Adding to the monetary benefits rainwater harvesting structure waives off 5% on property tax for new constructions, as it is now a mandatory requirement for all buildings across many states of India.

27.4.4 Integrated Management of Coastal Zone and Water Resources

Most of the disasters are water-related, with an impact on water supply and its provision for multi-sectorial water use. Integrated Water Resource Management (IWRM) can be strengthened through greater recognition of ecosystems that structure efficient and effective water management approaches (Dalton and Murti 2013). IWRM as a strategic coordination process is designed to maximize the returns from good water management for human well-being. NAPCC, 2008 pronounces the descriptions of National Water Mission (NWM) to safeguard integrated water resource management to protect water reduce wastage and guarantee more reasonable distribution.

27.4.5 Area-Based Conservation Including Protected Area Management Approaches

Though Protected Areas (PAs) are very western approach to conserve biodiversity and ecosystems and addressing societal challenges was not a primary objective or concern for developing them however over the time they have helped in providing co-benefits and DRR and EbA are few of them. Protected areas are increasingly being recognized as important approaches for their role in improving DRR (Dudley et al. 2010). Protected areas in India have limited human occupation or at least the exploitation of resources as per Wildlife Protection Act, 1972. There are varieties of protected areas, which are protected under different level of protection enabling laws of each country or by the regulations of the international organizations like IUCN. Protected areas of India under the Protected Area Network (PAN) have been distributed categories *viz.* Biosphere Reserves, National Parks, Wildlife Sanctuaries, Conservation Reserves, Community Reserves and Marine Protected Areas. The role of PAs can be well acknowledged in reducing DRR and climate vulnerabilities as a co-benefit and not as immediate NbS. Some of the relevant ones are giving space to floodwaters by absorbing present and after effect of floods using natural vegetation; retaining natural vegetation that helps to stabilize soil; limiting pressures (especially grazing pressures) on land and reducing desert formation and extensions by managing population of drought resistant and arid and semi-arid plants to withstand extreme conditions; limiting human interferences in the most fire affected regions; maintaining traditional and customary cultural protection systems that help wildfire protections; protecting inclusive natural systems that are associated with natural fire regimes to ensure short as well long term ecosystem and ecological stability; reduce or mitigate landslides and climate induced hazards and other extreme events.

27.5 Strategies and Gaps

Mainstreaming and scaling up NbS has been less than what was expected. There are many success stories, that have not been fully mainstreamed. In-sufficient implementation efforts are also due to less interest of engineering community who are less proficient in ecological approaches and have more focus on providing quick but short term solutions using grey instead of green approaches. This lacuna in application reduces chances of further replication and further implementation. There is vital need of demonstrating the societal benefits by using NbS applications, a lot of case studies are already available locally we need to showcase them and build up on them for up scaling. Introducing different capacity building approaches for NbS and increasing the reach of existing courses by involving universities and private sector. Facilitate dialogue and networking on increasing awareness through standalone courses, online courses and linking up with ongoing efforts. Dialogue facilitation on increasing co-operation for NbS by discussing the need of the key policy stimulus, potential partners' availability of international support and partnerships and also using national missions for mainstreaming NbS. Mainstreaming NbS in relevant national policies and plans as a compulsory approach can help to reduce disaster risks in long run. Using new generation tools for integrated watershed management, REDD, REDD+ that include community for plantation and conservation to increase sink potential, Ecosystem Protecting Infrastructures and Communities (EPIC) and Payment for Ecosystem Services (PES) (Monty et al. 2017) can help to achieve long-term benefits.

27.6 Conclusion and Way Forward

It is already recognized that there are noteworthy rolling leads on EbA for CCA and EcoDRR in the country available with communities, corporate systems including the extensive network and presence of various academic and R&D organizations. NbS has a huge potential and future for India to not only in addressing DRR and CCA but also localizing and achieving SDG, 2015, Aichi Targets, Bonn Challenge targets, land degradation neutrality (LDN), Nationally Determined Contributions (NDC) and also Glasgow, 2021 and Paris Promises, 2015. There are variety of solutions and ongoing practices that are available for different climatic and non-climatic challenges with local communities residing in different agro-climatic zones of India that are efficient and important NbS solutions. NbS needs to customize as per area, local/regional issues of disaster intensity and frequency. India specific standards, criteria and indicators need to developed for different expectations and agro-climatic zones. These desire technical and scientific assessment for appropriate NbS for specific contexts by certified or accredited NbS community. A well-coordinated and mainstreamed effort to integrate NbS in different policy framework and practices can happen by focusing on convergence of policies that stretches across different government line departments, NGOs *etc*. Required, adequate and sufficient finances and funds using PPP (public private partnerships) model and moreover, constant determination to engage disaster hit communities in the project planning, execution and decision making process will help in NbS mainstreaming. Potential benefits of NbS based partnership with corporate, industry and private sector can help identifying potential risks, opportunities, sharing new knowledge, practice for wider application and endorsement. PPP and CSR (corporate social responsibility) inclusive models of NbS application can help improving relationships among Govt., NGOs, Pvt. Sector and academia working in rural as well as urban ecosystems.

Acknowledgements First author thanks Knowledge Resource Center of CSIR-NEERI for plagiarism check using i-thenticate software under the number CSIR-NEERI/KRC/2018/OCT/WTMD/1.

References

Balian E, Eggermont H, Le Roux X (2014) Outputs of the Strategic Foresight workshop "Nature-Based Solutions in a BiodivERsA context", Brussels, June 11–12

Benedict MA, McMahon ET (2006) Green infrastructure: linking landscapes and communities. Island Press, Washington, DC, p 2006

Bowler DE, Buyung-Ali L, Knight TM, Pullin AS (2010) Urban greening to cool towns and cities: a systematic review of the empirical evidence. Landsc Urban Plan 97(3):147–155. https://doi.org/10.1016/j.landurbplan.2010.05.006

Cohen-Shacham E, Walters G, Janzen C, Maginnis S (eds) (2016) Nature-based solutions to address global societal challenges. IUCN, Gland, Switzerland, p xiii + 97

Dalton J, Murti R (2013) Utilizing integrated water resource management approaches to support disaster risk reduction. In: Sudmeier-Rieux K, Estrella M (eds) The role of ecosystems in disaster risk reduction. United Nations University Press, Bonn, Germany

De Groot RS, Alkemade R, Braat L, Hein L, Willemen L (2010) Challenges in integrating the concept of ecosystem services and values in landscape planning, management and decision making. Ecol Complex 7:260–272. https://doi.org/10.1016/j.ecocom.2009.10.006

Dhyani S, Dhyani D, 2016. Strategies for reducing deforestation and disaster risk: lessons from Garhwal Himalaya, India. Ecosystem-based disaster risk reduction and adaptation in practice by Fabrice G. Renaud, Karen Sudmeier-Rieux, Marisol Estrella, Udo Nehren (eds.) 507–528. Springer International Publishing. ISBN 978-3-319-43631-9

Dhyani S, Karki M, Gupta AK (2020) Opportunities and advances to mainstream nature-based solutions in disaster risk management and climate strategy. In: Dhyani S, Gupta A, Karki M (eds) Nature-based solutions for resilient ecosystems and societies. Disaster resilience and green growth. Springer, Singapore. https://doi.org/10.1007/978-981-15-4712-6_1

Dhyani S, Lahoti S, Khare S, Paras P, Verma P (2018) Ecosystem based disaster risk reduction approaches (EbDRR) as a prerequisite for inclusive urban transformation of Nagpur City, India. Int J Disaster Risk Reduction. https://doi.org/10.1016/j.ijdrr.2018.01.018

Dhyani S, Maikhuri RK, Dhyani D (2011) Energy budget of fodder harvesting pattern along the altitudinal gradient in Garhwal Himalaya. India J Biomass Bioenergy 35(5):1823–1832

Dhyani S, Maikhuri RK, Dhyani D (2013) Utility of fodder banks for reducing women drudgery and anthropogenic pressure from forests of Western Himalaya. Natl Acad Sci Lett (Springer Link) 36(4):453–460

Dhyani S, Majumdar R, Santhanam H (2021b) Scaling-up nature-based solutions for mainstreaming resilience in Indian cities. In: Mukherjee M, Shaw R (eds) Ecosystem-based disaster and climate resilience. Disaster and risk research: GADRI book series. Springer, Singapore. https://doi.org/10.1007/978-981-16-4815-1_12

Dhyani S, Murthy IK, Kadaverugu R, Dasgupta R, Kumar M, Adesh GK (2021a) Agroforestry to achieve global climate adaptation and mitigation targets: are south Asian countries sufficiently prepared? Forests 12(3):303. https://doi.org/10.3390/f12030303

Dhyani S, Singh S, Basu M, Dasgupta R, Santhanam H (2022) Blue-green infrastructure for addressing urban resilience and sustainability in the warming world. In: Dhyani S, Basu M, Santhanam H, Dasgupta R (eds) Blue-green infrastructure across Asian countries. Springer, Singapore. https://doi.org/10.1007/978-981-16-7128-9_1

Dhyani S, Thummarukudy M (2016) Ecological engineering for disaster risk reduction. Environ Sci Pollut Res 23(19):20049–20052

Diaz S et al (2015) The IPBES conceptual framework—connecting nature and people. Curr Opin Environ Sustain 14:1–16

Dudley N, Stolton S, Belokurov A, Krueger L, Lopoukhine N, MacKinnon K Sandwith T, Sekhran N (eds.). (2010). Natural solutions: protected areas helping people cope with climate change, IUCNWCPA, TNC, UNDP, WCS, The World Bank and WWF, Gland, Switzerland, Washington DC and New York, USA

Eggermont H, Balian E, Manuel J, Azevedo N, Beumer V, Brodin T, Claudet J, Fady B, Grube M, Keune H, Lamarque P, Reuter K, Smith M, van Ham C, Weisser WW, Le Roux X (2015) Nature-based solutions: new influence for environmental management and research in Europe. Gaia 24(4):243–248

European Commission and Directorate-General for Research and Innovation (2015) Towards and EU research and innovation policy agenda for nature-based solutions & re-naturing cities. European Commission, Brussels, Belgium. https://doi.org/10.2777/765301

Gill S, Handley J, Ennos R, Pauleit S (2007) Adapting cities for climate change: the role of the green infrastructure. J Built Environ 33(1):15–133

Green Highways Division (National Highways Authority of INdia). (n.d.). Retrieved from http://www.nationalgreenhighway.org/

Gret-regamey A, Altwegg J, Siren EA et al (2017) Integrating ecosystems in spatial planning—a spatial decision support tool. Landsc Urban Plan 165:296–219. https://doi.org/10.1016/j-landurbplan.2016.05.003

Gupta AK, Nair SS (2012) Ecosystem approach to disaster risk reduction. National Institute of Disaster Management, New Delhi. www.nidm.gov.in

Haifeng L, Wenbo C, Wei H (2015) Planning of green space ecological network in urban areas: an example of Nanchang, China. Int J Environ Res Public Health 12:12889–12904. ⟨https://www.ncbi.nlm.nih.gov/pmc/articles/PMC4627006/pdf/ijerph-12-12889.pdf⟩ (Accessed 15 August, 2018)

Hansen R, Pauleit S (2014) From multi-functionality to multiple ecosystem services? A conceptual framework for multi-functionality in green infrastructure planning for urban areas. Ambio 43(4): 516–529

Hughes AC (2017) Understanding the drivers of Southeast Asian biodiversity loss. Ecosphere 8(1): 1624

IPCC (2014) Climate change, 2014: synthesis report. Contribution of Working Groups I, II and III to the Fifth Assessment Report of the Intergovernmental Panel on Climate Change [Core Writing Team, R.K. Pachauri and L.A. Meyer (eds.)]. Geneva, Switzerland: IPCC

IUCN (2009) Position paper for UNFCCC COP15, Copenhagen. IUCN, Gland, Switzerland

Jones HP, Hole DG, Zavaleta ES (2012) Harnessing nature to help people adapt to climate change. Nat Clim Chang 2(7):504–509. ⟨http://citeseerx.ist.psu.edu/viewdoc/download;jsessionid=2E767B63FBF5F3805136C3120614A9A8?doi=10.1.1.701.4069&rep=rep1&type=pdf⟩ (Accessed 15 January 2018)

Kadaverugu R, Dhyani S, Dasgupta R, Kumar P, Matli C (2022) Urban sustainability and resilience building: blue-green infrastructure for air pollution abatement and realizing multiple co-benefits. In: Dhyani S, Basu M, Santhanam H, Dasgupta R (eds) Blue-green infrastructure across Asian countries. Springer, Singapore. https://doi.org/10.1007/978-981-16-7128-9_18

Kelman I (2015) Int J Disaster Risk Sci 6:117. https://doi.org/10.1007/s13753-015-0046-5

Lafortezza R, Davies C, Sanesi G, Konijnendijk CC (2013) Green infrastructure as a tool to support spatial planning in European urban regions. iForest 6:102–108

MacKinnon K, Dudley N, Sandwith T (2011) Natural solutions: protected areas helping people to cope with climate change. Oryx 45(4):461–462

MacKinnon K, Hickey V (2009) Nature-based solutions to climate change. Oryx 43(1):13–16

Maes J, Jacobs S (2015) Nature-based solutions for Europe's sustainable development. Conserv Lett. [online journal]

Marton-Lefèvre J (2012) Nature at the heart of urban design for resilience. In: Otto Zimemrmann K (eds): Resilient Cities 2 (K. Otto Zimemrmann, Ed.). Cities and Adaptation to Climate Change Proceedings of the Global Forum, 2011. 113–118 pp

McHarg IL (1967) An ecological method for landscape architecture. Landsc Archit 57(2):105–107

McHarg IL (1969) Design with nature. Natural History, New York

Mhaske AR, Taley SM, Biniwale RB (2014) Removal of turbidity from sewage water by phytorid sewage treatment plant: a study using the response surface methodology. Intern J Agric Eng 7(2):365–372

Misra S, Dhyani D, Maikhuri RK (2008) Sequestering carbon through indigenous agriculture practices. LEISA INDIA 10(4):21–22
Monty F, Murti R, Miththapala S, Buyck C (eds) (2017) Ecosystems protecting infrastructure and communities: lessons learned and guidelines for implementation. IUCN, Gland, Switzerland, p x +108
National Mission for Sustainable Agriculture (NMSA) (2008) Retrieved from https://nmsa.dac.gov.in
National Mission on Sustainable Habitat (2011) January 27. Retrieved from Centre for Science and Environment: https://www.cseindia.org/national-mission-on-sustainable-habitat-1977
Norman L (1973) The effect of sir Ebenezer Howard and the Garden City movement on twentieth century town planning. Rickmansworth, Hertfordshire, United Kingdom
NAPCC (2008) India's National Action Plan on Climate Change. Government of India. Available at http://www.moef.nic.in/sites/default/files/Pg01-52_2.pdf. Accessed 19 Jan 2017
PEDRR (2010) Demonstrating the role of ecosystem-based management for disaster risk reduction. Partnership for Environment and Disaster Risk Reduction. www.pedrr.net
Pugh TAM, MacKenzie AR, Whyatt JD, Hewitt CN (2012) Effectiveness of green infrastructure for improvement of air quality in urban street canyons. Environ Sci Technol 46(14):7692–7699
Raje S, Kertesz R, Maccarone K, Seltzer K, Siminari M, Simms P, Wood B, Sansalone J (2013) Green infrastructure design for pavement systems subject to rainfall-runoff loadings. Transp Res Rec 2358:79–87
Rathore LS, Attri SD and Jaswal AK (2013). State level climate change trends in India. Met Monograph. Environment Meteorology-02/2013. Pp. 147. http://www.imd.gov.in/section/climate/StateLevelClimateChangeMonoFinal.pdf
Revi A (2008) Climate change risks: an adaptation and mitigation agenda for Indian cities. Environ Urban 20(1):207–229. ⟨http://journals.sagepub.com/doi/pdf/10. 1177/0956247808089157⟩ (accessed 19 August 2018)
Sudmeier-Rieux K, Ash N (2009) Environmental guidance note for disaster risk reduction: healthy ecosystems for human security, Revised edition, IUCN, Gland, 2009
Sudmeier-Rieux K, Ash N, Murti R (2011) Environmental guidance note for disaster risk reduction, revised edition. IUCN, Gland, Switzerland
Sutherland WJ, Gardner T, Bogich TL, Bradbury RB, Clothier B, Jonsson M, Kapos V, Lane SN, Möller I, Schroeder M, Spalding M, Spencer T, White PCL, Dicks LV (2014) Solution scanning as a key policy tool: identifying management interventions to help maintain and enhance regulating ecosystem services. Ecol Soc 19(2):3
UN Habitat (2010) State of Asian Cities 2010/11. State of Cities- Regional Reports. 162/10E 279. ⟨http://www.indiaenvironmentportal.org.in/files/The%20Stat%20of%20Asian%20Cities%20201011.pdf⟩ (Accessed 12 August 2018). ISBN9789211322743
Van Ham C, Klimmek H (2017) Partnerships for nature-based solutions in urban areas—showcasing successful examples. In: Kabisch N, Korn H, Stadler J, Bonn A (eds) Nature-based solutions to climate change adaptation in urban areas. Theory and practice of urban sustainability transitions. Springer, Cham
Van Wesenbeeck BK et al (2014) Damming deltas: a practice of the past? Towards nature-based flood defenses. Estuar Coast Shelf Sci 140:1–6
Yigitcanlar T, Teriman S (2015) Rethinking sustainable urban development: towards an integrated planning and development process. Int J Environ Sci Technol 12(1):341–352

Web References

Bonn Challenge. Retrieved from http://www.bonnchallenge.org/content/india

India's Intended Nationally Determined Contribution: Working Towards Climate Justice. (n.d.). Retrieved from http://www4.unfccc.int/ndcregistry/PublishedDocuments/India%20First/INDIA%20INDC%20TO%20UNFCCC.pdf

Kerala floods: Businesses stare at a loss of over Rs 20,000 crore. (2018, April 20). Retrieved from rediff.com: http://www.rediff.com/business/report/kerala-floods-businesses-stare-at-a-loss-of-over-rs-20000-cr/20180820.htm

Shalini Dhyani is Principal Scientist with Critical Zone Group of Water Technology and Management Division of CSIR-NEERI, India. She is Asia Vice Chair and Steering Committee member in IUCN CEM (Commission on Ecosystems Management) (2022-2025). Shalini is Visiting Scientist CIFOR, Indonesia, UNESCO-TWAS Associate and lead author IPBES.

Muralee Thummarukudy is currently the Director of G20 Global Initiative Coordination Office on Land, UNCCD, Germany. Muralee holds a PhD in Environmental Engineering from Indian Institute of Technology, Kanpur. He is an Alumnus of the International Leadership Academy (United Nations University) and also attended University of California, Berkley for the Beahers Environmental Leadership Programme.

Anil K Gupta Professor of Policy & Strategies, Integrated Resilience. Head of the ECDRM Division, International Cooperation, Programme Director, Centre for Excellence on Climate Resilience, Nodal Officer for Central Ministries at National Institute of Disaster Management, (Govt. of India), New Delhi. He is also a member of the expert Team for WMO Climate Statement.

Chapter 28
Multi-Hazard Risk and Integrated Approach to Resilience

Fatima Amin, Kopal Verma, and Pritha Acharya

Abstract Disasters occur as a result of a combination of risks and vulnerabilities. Every year, natural and anthropogenic hazards cause damage around the world, thousand of fatalities or injuries and other societal impacts as the frequency of destructive natural events is expected to rise as a result of climate change. Risk management is concerned with reducing vulnerability in order to protect the community. The multi-hazard disaster risk reduction approach involves assessing a composite risk from all hazards in order to conduct integrated planning. When hazards are viewed in isolation, mitigation measures proposed as a solution for one risk may create vulnerability for another. The interdependencies between different types of hazards must be considered in a multi-hazard approach. Poverty causes vulnerability, but it is not the same as vulnerability. Poor people are forced to settle in vulnerable areas as a result of rapid urbanisation and scarcity of land, and can be avoided through a multi-pronged approach that considers the structural, social and economic part. Community knowledge and education are critical components of disaster risk reduction.

Keywords Multi hazard approach · Risk · Resilience · Risk financing

F. Amin (✉)
Division, Environment Climate and Disaster Risk Management, National Institute of Disaster Management, New Delhi, India

K. Verma
G20, Ministry of External Affairs, New Delhi, India

P. Acharya
ECDRM Division, CAP-RES DST Project, Centre for Excellence on Climate Resilience, National Institute of Disaster Management (Ministry of Home Affairs), New Delhi, India

A. K. Gupta et al. (eds.), *Disaster Risk and Management Under Climate Change*, Disaster Resilience and Green Growth,
https://doi.org/10.1007/978-981-99-4105-6_28

28.1 Introduction

Disasters are the result of a combination of hazards and vulnerability. Risk management is about reducing that vulnerability to protect people and property; it is about saving lives and livelihoods. Risk must be taken in to consideration in every aspect of development in order to be managed effectively: reducing the extent and severity of natural, technological and environmental hazards contributes to poverty reduction, improved health and resilience, This need for an integrated approach to risk management is reflected in the internationally agreed goals and targets defined in the Sendai Framework for Disaster Reduction, but also in the Paris Agreement on Climate Change and in the 2030 Agenda for Sustainable Development (Fig. 28.1).

UNESCO has made a strong commitment to the UN Plan of Action on Disaster Risk Reduction for Resilience: Towards a Risk-informed and Integrated Approach to Sustainable Development, which was revised in April 2016. "By operating at the

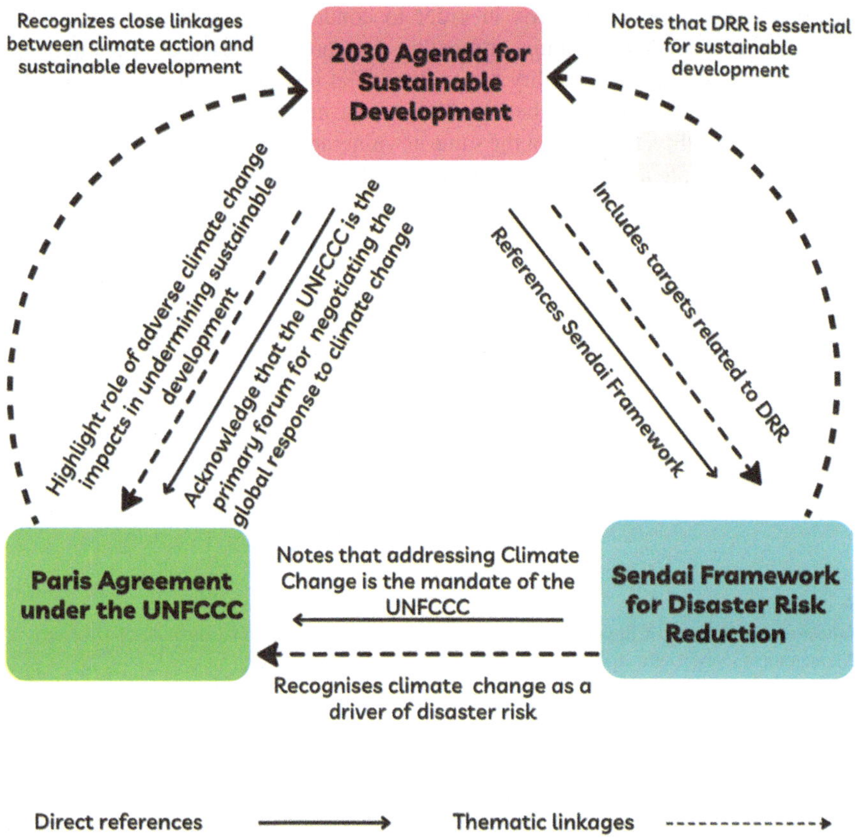

Fig. 28.1 Sendai framework for SDG's

interface between the natural and social sciences, education, culture and communication.

Promote community resilience through multi-hazard early warning systems, capacity building and policy recommendations on issues that range from water management to address risks of floods and droughts, to the VISUS methodology to ensure the structural safety of educational facilities, to preparedness to specific hazards such as earthquakes and tsunamis. Through the inclusion of business and policy, it facilitates the implementation of practical and innovative solutions to reduce global risks.

The Sendai Framework has played a significant role in shifting the focus from a reactive "disaster management" to a more proactive "risk management" approach. It has also, importantly, widened its scope by placing the spotlight on disaster risks originating from man-made/technological hazards.

In a **multi-hazard approach,** the interdependencies between different types of hazards need to be considered, e.g. natural hazards which could trigger technological accidents, so-called "NATECH" events. The widely known accident at the Fukushima Daiichi nuclear power plant in 2011, triggered by a tsunami following an earthquake, shocked the international community as it demonstrated the vulnerability of technology and human beings in the face of strong natural forces. After Chernobyl, it was the second nuclear accident classified the highest on the International Nuclear Event Scale. I believe that much more work is needed to duly take account of natural hazards – which are on the rise due to climate change and increasing extreme weather events - and their potential to trigger such technological accidents.

In a **multi-stakeholder approach,** all relevant stakeholders—"risk owners", as man-made/technological hazards result from human activity and risk creation—need to be aware of their shared responsibilities for reducing such hazards and risks.

Integrated approach implies full integration across **all relevant sectors and institutions**—industrial safety, environmental protection, extractives/mining, water resources management, land-use planning, environmental assessment, emergency response, to name a few. Cross-sectoral cooperation is an essential prerequisite to comprehensively understand the risks emerging from technological hazards and to develop coherent prevention and preparedness policies, strategies and measures. In the spirit of the Industrial Accidents Convention, inter-institutional cooperation needs to happen horizontally, between the respective national-level authorities, across the national, regional and local levels. Only then will disaster risk governance be effective. Cooperation is also crucial among countries sharing borders, rivers or watersheds. An integration of approaches here will occur through **trans-boundary cooperation** for joint disaster prevention and preparedness, e.g. when harmonizing off-site contingency plans, or developing bi- or multilateral agreements.

Box 28.1 Conceptual Framework of the Holistic Approach to Disaster Risk. Adapted from Cardona et al (2012)

What is the multi-hazard disaster risk reduction approach? How is it different from other approaches used in disaster risk reduction?

Multi-hazard disaster risk reduction approach is to assess a composite risk from all hazards so that integrated planning can be undertaken. Looking at hazards in isolation may result in a situation where mitigation measures proposed as a solution for one risk may create vulnerability for another hazard.

The national vision as enshrined in the National Policy on Disaster Management, 2009 is to build a safe and disaster resilient India by developing a holistic, proactive, multi-disaster oriented and technology-driven strategy through a culture of prevention, mitigation, preparedness and response. Thus, since the beginning, we have focused our approach towards multi-hazard disaster risk reduction.

National Disaster Management Plan, 2016 also highlights the multi-hazard approach at all stages of disaster risk reduction. It emphasizes upon risk assessment to be carried out with a multi-hazard concept leading to fool-proof land use planning, promoting skill development for multi-hazard resistant construction, strengthening ability of communities to manage and cope with disasters based on a multi-hazard approach and also on ensuring that multi-hazard resistant features are incorporated in planning and execution of social housing schemes.

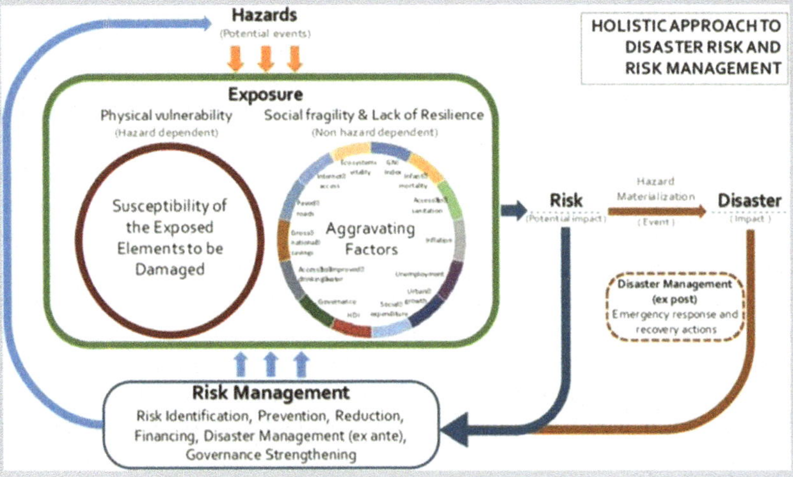

28.1.1 Present Scenario

28.1.1.1 Current Trends and Challenges

Multi-hazard risk assessment (MHRA) is a way to understand the hazards, vulnerability (both intrinsic and extrinsic) and risks arising from the geographic location and socio-economic backdrop. Complexities will arise in understanding and comprehending these multi-layered, location-based information of different themes. However, resolving this information for different stakeholders is also carried out along with the assessment itself. Risk communication and dissemination is an activity undertaken with each stakeholder department along with the final end users—the people and the civil society. By involving all stakeholders from the initial stages of this assessment will help reducing or demystifying complexities.

No, the risks cannot be weighed and ranked in general sense. However, weighting and ranking is carried out in order to prioritize multitude of hazards, vulnerabilities and risks when the available resources and funds are limited. This is a way of incorporating the human element into the system before we take steps to allocate financial resources. For example, if we need to find a disaster risk reduction activity and allocate limited funds, we need to find out an activity which will influence the lives of more beneficiaries or an activity whose benefits will last longer, etc.

Though poverty leads to vulnerability, poverty cannot be equated with vulnerability. Due to rapid urbanization and scarcity of land, poor people are pushed to settle in vulnerable areas. This can be avoided through a multi-pronged approach by taking into account the physical, social and economical aspects of communities. The key is risk sensitive land use planning of urban areas which will ensure that communities are located in safe places and guide the development in a way that will make them capable of withstanding disasters, building back efficiently. Further, the land use zonation, building bye-laws, building codes and standards should be implementable from the point of view of poor and vulnerable sections. Low cost technologies, built materials, etc., should help poor construct safe houses.

28.2 Global Good Practices

The Republic of Moldova, Romania, and Ukraine have, under the UNECE Danube Delta Project, jointly developed a map of hazard sources, including oil terminals in the Delta region, and associated NATECH risk indicators, leading to an improved joint understanding of disaster risk. On this basis, they have strengthened **disaster preparedness**, by developing a joint contingency plan and testing it through the first ever trilateral exercise, in a **multi-stakeholder approach**, involving national and local authorities, industry and other relevant stakeholders.

The Estonian Rescue Board ensures **integration across sectors** and respective **inter-institutional cooperation** between authorities in charge of industrial safety, land-use planning, and environmental assessment. It requires that spatial plans and building design documentation be submitted along with the siting plans for approval and considers environmental impact assessments when deciding whether the accident hazards would increase through the new site or expansion, and accordingly, whether to approve the proposal.

The Implementation Guide also includes references to more specific, cross-cutting or sectoral guidelines developed by the different organizations who have worked together in its development. At UNECE we have, for example, produced Guidance on land-use planning, the siting of hazardous activities and related safety aspects, to foster policy integration among relevant sectors and showcase further good practices. We also encourage government, operators and experts to use UNECE's sector-specific safety guidelines, which are applicable worldwide, to manage technological hazards and reduce related risks, such as the Safety Guidelines and Good Practices for Tailings Management Facilities. Such guidelines, if properly applied, can help prevent accidents, such as the most recent tragic tailings dam collapse, in Brumadinho in Brazil, which took more than 300 lives. The most recently developed UNECE safety guidelines and good practices for the management and retention of firefighting water close the loop back to 1986: They aim at preventing the accidental pollution of soil and water, and the reoccurrence of accidents with severe trans-boundary effects like the one in Schweizerhalle, Switzerland.

Since 1986 a lot has happened: the focus has shifted from the management of disasters to the more proactive management of risks - in a multi-hazard, multi-stakeholder and fully integrated fashion. Man-made/technological hazards are in the limelight, in addition to natural hazards, and the interdependencies between the two are increasingly seen. I am proud to be able to raise awareness by the wider DRR community of existing good practices, guidance and approaches to technological disaster risk management under the UNECE Industrial Accidents Convention, and to continue to foster cross-sectoral and trans-boundary cooperation – essential to prevent technological disasters, and to limit their consequences within and across borders.

28.3 Multi Hazard DRR Approach: Regional Context

Regional disaster risk reduction is one of the essential tools to minimize losses caused by natural hazards across country borders. Due to geological, social, cultural and political resemblance, different countries from one region often share similar characteristics of disaster risks, which make regional disaster risk reduction possible and efficient. South Asian regions face a range of common natural hazards, including cyclones, droughts, earthquakes, floods, landslides and tsunamis. Demographic changes, rapid urbanization, environmental degradation and climate fluctuations have further increased people's exposure to natural hazards, resulting in frequent and severe disasters and compounding the impact of complex emergencies. In addition, declining socio-economic conditions of some populations have increased vulnerabilities to hazards in the region.

Governments in south Asia are investing in disaster risk reduction and improving their countries' response capacities from the community to the national level. Local governments, international and regional organizations, non-governmental organizations and communities may be engaged, to develop effective strategies—tailored to local contexts and the needs of populations—to reduce disaster-related risks.

At the regional level, multi hazard disaster risk reduction approach would bring several changes. First, it would improve preparedness, mitigate severe impacts of disasters and strengthen disaster response mechanisms at the regional level. Second, the regional disaster management system can be strengthened to reduce risks and to improve recovery management at all levels. Then, support can be sought in hazard identification, analysis of existing capacities, monitoring, early warning and early action through partnerships and joint programming. Fourth, new programmes, policies, institutional arrangement legislation, human resource and capacity developments can be executed in view of pre-planning and risk reduction. Also, regional platforms, forums and coordination mechanisms can be established among various stakeholders for exchange of knowledge, information and expertise. Finally, development of common hazard risk management plans amongst the affected countries at regional level can be supported such as integrating a coastal risk mitigation plan.

At the community level, it is necessary to empower local authorities and communities to reduce risks including through resource incentive and decision making responsibilities. It is important to reduce vulnerabilities and increase capacities of vulnerable communities to cope with, prevent, or minimize loss and damage to life, property, and the environment with a view to hasten recovery. Community participation in pre planning, search and rescue, relief, response and post-disaster recovery needs to be enhanced through prioritizing the most vulnerable groups and localities.

28.4 National Context

With the enactment of the Disaster Management Act, 2005, India has already taken a significant step towards paradigm shift, from the erstwhile relief-centric response to a proactive prevention, mitigation and preparedness-driven approach for conserving developmental gains and also to minimize losses of life, livelihoods and property. Now, we are not following the knee-jerk model for responding to a disaster, rather we are moving towards prevention, mitigation and preparedness. You can very well understand this with the latest cyclone Fani in Odisha. With the help of advancement and progress in our early warning facilities, India is now able to get precise early warning related to cyclones and associated phenomena well in advance. This has helped in evacuating people likely to be affected to safer places. And, you can see the results in the significant reduction in lives lost.

We have not only been able to minimize deaths, but are also focusing on resilient infrastructure. The National Disaster Management Plan emphasizes on mitigation and prevention by clearly spelling out responsibilities of all stakeholders for specific mitigation activities. India is also championing a coalition for disaster resilient infrastructure which will be multi-hazard oriented and will help in minimizing loss to property and infrastructure. The coalition was in fact launched during the UN Climate Summit in New York in September 2014. This global coalition for disaster resilient infrastructure would address concerns that are common to developing and developed countries, small and large economies, countries at early and advanced stages of infrastructure development and countries that have moderate or high disaster risk. Few concrete initiatives work at the intersection of the Sendai Framework, Sustainable Development Goals and Climate Change Adaptation and focus on disaster resilient infrastructure.

28.5 Partnerships

National partnerships, for instance through government and multi-donor consortiums; agencies and institutions with specific technical interests. Through these partnerships, ADB will seek to foster a long-term commitment to IDRM, helping to ensure that the benefits of ADB's IDRM investments are sustainable and to leverage these.

Disaster Risk Reduction There is need to invest in local, national, subregional, and regional disaster and climate risk reduction initiatives, building on to strengthen IDRM capacity. It will support both structural and nonstructural, and both standalone and embedded, actions. These will include the incorporation of cost-effective measures to strengthen resilience in engineering design and actions to strengthen disaster resilience. Further, it is also important to explore the underlying causes of vulnerability so that the causal roots can be addressed. Encourages a more comprehensive DRM approach, addressing hazards, community needs (social, economic, environmental), and vulnerabilities.

It will encourage "no regrets" strategies and approaches, pursuing IDRM measures that are justified on the basis of current economic, social, and environmental costs, benefits, and levels and forms of disaster risk but that also support future disaster resilience, without requiring any certainty of knowledge about the frequency or intensity of future hazard events. In addition, ADB will seek to avoid the creation of new risk in its other development investments.

28.5.1 Post-Disaster Assistance

Explore measures to enhance the quality and scope of its post-disaster assistance, supporting a more timely and cost-effective government led response, reducing the need for reprogramming of resources, and including specific measures to address the immediate and long-term needs of women and girls. Actual levels of assistance provided in any single year will depend on the timing, intensity, and location of individual natural hazard events government-led post-disaster damage and loss assessments, strengthening opportunities to ensure that the nature and levels of ADB assistance are carefully aligned with wider response programs and address priority reconstruction needs. Training will be provided to maintain disaster response skills, including PDNA capabilities.

28.5.2 *Disaster Risk Financing Instruments*

The development of DRF instruments and wider DRF strategies for households, businesses, and governments, enhancing the public and private financial management of residual disaster risk.

This support will be designed to (i) help facilitate timely recovery and reconstruction efforts, complementing ADB's post-disaster assistance instruments by providing additional injections of liquidity in the aftermath of a disaster; (ii) encourage the development of optimal bundles of DRF instruments, reflecting the opportunity costs associated with various options as well as considerations relating to the scale of funding required and timeliness of fund disbursement; (iii) help spread the public and private costs of recovery and reconstruction over time; (iv) incentivize investments in DRR through risk-based premium pricing and similar mechanisms; and (v) encourage income-enhancing, rather than risk-averting, livelihood decisions.

Governments can establish comprehensive DRF mechanisms for the rapid post-disaster rehabilitation of federal and state infrastructure, supporting effective post-disaster intervention and stimulating greater investment in DRR.

28.6 Stakeholder Engagement

Successful IDRM requires coordination and collaboration across a wide range of public and private stakeholders; across a wide range of sectors and disciplines; across countries facing common disaster risk, such as shared earthquake fault lines or river systems; and across countries linked through international trade and supply chains.

28.6.1 *Partnerships*

National partnerships, for instance through government and multi-donor consortiums; agencies and institutions with specific technical interests. Through these partnerships, ADB will seek to foster a long-term commitment to IDRM, helping to ensure that the benefits of ADB's IDRM investments are sustainable and to leverage these investments through subsequent government-supported replication and scaling up and more general enhanced government interest in IDRM joint capacity development initiatives, such as the joint development and implementation of standardized methodologies and tools for financial disaster risk assessment and management.

Private sector engagement in integrated disaster risk management. Promote greater private sector engagement in innovative IDRM solutions and encourage and support strengthened disaster resilience of the private sector. Greater private sector engagement in IDRM is essential to help meet the funding gap for DRM, to share and spread post-disaster relief, recovery, and reconstruction costs, and to ensure that all infrastructure constructed by the private sector is disaster resilient. The private sector will play a significant role in meeting the region's huge and increasing infrastructure investment and financial institution needs over the next few decades and a vision of disaster resilience is unrealistic without private sector engagement.

Overall enabling environment for public–private partnerships (PPPs) in IDRM. It will encourage the development of DRF products, continuing some preliminary work already under way. To financial institutions in support of post-disaster housing reconstruction and capital for affected businesses. There are also potential opportunities for greater private sector engagement in other aspects of DRM, such as in technological innovation to enhance the disaster resilience of infrastructure and in income-generating investments that could indirectly reduce disaster risk.

Building a resilient society with appropriate coping mechanisms is the basic principle behind any disaster risk reduction process. Risk assessment allows for the determination of the acceptable level of risk, defined as the level of losses that is acceptable without destroying lives, national economy or personal finances. Once the current and acceptable levels of risk are determined, disaster risk reduction plans and strategies could be revised or developed so that they have the measurable goal of reducing the current risk to acceptable levels. Risk reduction involves—hazard, vulnerability and capacity assessment.

There are two levels of risk assessments—national and local. The first is a strategic risk assessment that supports the design of national disaster risk management strategies, policy and regulations, disaster risk management programming and budget allocation. The second—local risk assessment is an operational risk assessment for disaster risk reduction action planning, contingency planning, pre-disaster recovery planning and urban planning.

The disaster risk reduction process systematically involves different stakeholders in risk reduction planning including local government and other competent authorities like district disaster management authorities, state disaster management authorities; NGO networks; volunteer groups; financing institutions; private/business; media; Red Cross society; hospitals and fire fighting and other services; academic community; and, individual households. Timeliness is the essence with each stakeholder having their responsibilities in different phases—such as disaster prevention, preparedness, response, relief and mitigation.

The various stakeholders who could be part of risk reduction planning are government officials who work for the town or city, such as local geologists, engineers, land-use planners, etc.; academic and research institutions that can provide technical expertise; low-cost staff/ volunteers / NGOs / civil society who can provide up-to-date data on the locality and training resources; local and international non-governmental organizations can offer expertise, as well as local knowledge particularly if they have locally worked for a long period; and community-based organizations including religious, gender and youth-based groups around particular interests, such as environment and social improvement.

Multi-stakeholder workshops and other consultations forums are also some ways of getting inputs from stakeholders. Disaster risk reduction planning is a priority-setting and partnership-building exercise to coordinate the efforts of multiple agencies and levels of government and society. This means the process needs to be inclusive and participatory, and the local planning authority would benefit from identifying and engaging stakeholders in the risk assessment process.

As mentioned earlier, the process of MHRA needs a multi-stakeholder participation, right from the first stage-identifying the hazards. The understanding of MHRA cannot be a top down and a classroom oriented process, but a co-learning experience. When the stakeholders are involved from the first stage, they understand, own and feel the process and accept the outcomes and outputs.

Identifying the stakeholders is the initial step to involve them in the process. Care is taken that all categories of stakeholders—line departments, civil society, teachers from local schools, students, non-governmental organizations, rescue agencies, representatives from emergency services, media representatives, senior citizens, etc. are involved. For instance, senior members of society can offer invaluable information about past events and how good practices evolved in the area.

28.7 Way Forward

Public awareness and education is an essential component of disaster risk reduction. NDMA carries out awareness and education campaigns (do's and don'ts for various disasters) involving various media to reach out to the masses. We use simplified content consisting of text and graphical images in our social media posts. NDMA's social media handles, especially Twitter, are active round the clock. To make our awareness campaigns truly inclusive, we have recently incorporated sign language in our audio-video content on different disasters. We also encourage line departments of central, state and local governments to widely disseminate the do's and don'ts in their local languages.

Also, NDMA regularly conducts multi-state mock exercises to build resilience to different disasters. These exercises help enhance the coping ability of the administrative machinery and help them respond faster, better and in a planned manner during and after a disaster. The National Disaster Response Force (NDRF) regularly conducts community preparedness programmes across the country.

References

Carreño, Martha L, Omar D, Cardona & Alex H Barbat (2012) New methodology for urban seismic risk assessment from a holistic perspective. Bulletin of earthquake engineering 10:547–565.

Fatima Amin Experienced Professional skilled in Emergency Management, Public Speaking, Cyberlaw, Microsoft PowerPoint, and Leadership. Strong professional graduated from University of Kashmir. I am passionate research scholar, enthusiastic and budding disaster management professional who views the world through lenses of disaster risk reduction (DRR) and efforts related to building resilient communities. She is currently working with NIDM as a Young Professional for the ECDRM Division.

Kopal Verma is currently associated with the Ministry of External Affairs, as a consultant and had worked with NIDM as a junior consultant at the ECDRM division at NIDM, Govt. of India.

Pritha Acharya has a background in environmental sciences. She has worked in the field of sustainability, climate change adaptation and biodiversity conservation while working with IUCN India. She is currently working as a research associate and manages the CAP-RES (DST) funded project.

Chapter 29
Anticipatory Adaptation Planning: An Inherent Vulnerability Approach to Climate Change and Disaster Resilience

R. Shukla, Kamna Sachdeva, and P. K. Joshi

Abstract Mitigation measures for Climate Change and disaster risk impacts cannot be impeccable, thus adaptation actions are imperative for building societal resilience to unforeseen and unavoidable impacts. In the present era of extreme events, which is the causal agent of enormous toll on sustainable humanitarian and development planning, a shift from reactive to more active anticipatory planning is essential to foster resilience within communities. Although the need of anticipatory actions is highlighted in several global agreements, however, the dominant conceptualization of adaptation within policy circles at regional and local levels remains overly simplistic, with limited attention to its links with the concept of anticipation. Since, assessments of vulnerability are required to develop adaptation planning, in this article we argue the need of inherent vulnerability approaches for anticipatory adaptation planning for responding to the impacts of Climate Change and climate-induced disasters. Usage of contextual inherent vulnerability approach extends the anticipatory adaptation planning to not only anticipating future risk through current scenarios, but also to identify locations that will be more acutely affected as a result of existing structural vulnerabilities. We propose an inherent vulnerability framework based on the intrinsic social and biophysical to understand the spatial dynamics of vulnerability for the Himalayan state of Uttarakhand to strengthen adaptations response and build resilience to future climatic and disaster risks.

Keywords Anticipatory · Adaptation · Inherent vulnerability · Adaptation planning

---

R. Shukla
Centre of Excellence in Disaster Mitigation and Management, IIT Roorkee, Roorkee, India

K. Sachdeva
Delhi Skill and Entrepreneurship University, New Delhi, India

P. K. Joshi (✉)
Special Centre for Disaster Research, School of Environmental Sciences, Jawaharlal Nehru University, New Delhi, India
e-mail: pkjoshi@mail.jnu.ac.in

A. K. Gupta et al. (eds.), *Disaster Risk and Management Under Climate Change*, Disaster Resilience and Green Growth,
https://doi.org/10.1007/978-981-99-4105-6_29

29.1 Introduction

Contemporary societies are increasingly facing several complex environmental stressors such as Climate Change and increased disasters generating multi-scale and multi-dimensional catastrophic risks (DeLeo 2017). The losses by the natural disasters grew to the annual total damage amounting to US$140 billion for the period of 2005–2014 (Hewitt 2017). Climate Change has been proven to further increase the intensity and frequency of natural disasters like floods, storms, heat waves and droughts (Hirabayashi et al. 2013; Dai 2013).

In the recent report by IPCC on 1.5° warming, the global target is to limit the warming to 1.5° in order to avoid detrimental consequences on human and natural systems. Even under a zero emissions, future warming is determined by the physical inertia of the climate system and the residual greenhouse gas climate forcing. These global efforts to reduce GHG emissions, a leading cause of global warming, are laudable. However, regardless of the success of these efforts some magnitude of changes remains constant and cannot be prevented even by the most ambitious emission reduction strategies. Therefore, a more comprehensive consideration of adaptation as a response to the impact of Climate Change and disasters is imperative to minimize its lethal effects on society and economy. One of the most complex and challenging issues for planning for the mitigation and responses to impacts of changes in climate is the high degree of uncertainty associated with the spatial and temporal distribution associated with it. Since certain aspect of future cannot be forecasted therefore a degree of anticipation in adaptation planning approaches become essential in order to build the capacity of the societies for taking sustainable responses to future unknowns. As Hannah (2010) describes, the emerging global challenges have led to the need of fetishization of actionism, forcing that the policy makers not only need to think about future but also make policies that prevent and prepare for future challenges. Further, research highlights that anticipatory actions are needed not only at global level but also at regional and local scales through multi-scale networks that inform adaptive institutions, strategy formation, decision making and societal resilience (Boyd et al. 2015). A recent report by the United Nations on humanitarian assistance to global challenges called for adopting ‘a more anticipatory approach’ instead of the usual reactive approaches.

The IPCC report defines Climate Change Adaptation (CCA) (in social systems) as “the process of adjustment to actual or expected climate and its effects, in order to moderate harm or exploit beneficial opportunities” and Disaster Risk Reduction (DRR) as “a policy goal or objective, and the strategic and instrumental measure employed for anticipating future disaster risk, reducing existing exposure, hazard, or vulnerability, and improving resilience”. The concept and practice of CCA and DRR systematically analyse the causal factors of heightened vulnerability of people and property to wisely overcome the root causes of vulnerability and enhance the preparedness and adaptive capacity of people to cope with adverse events. Figure 29.1 presents the conceptual framework of impacts of Climate Change on disaster risks. Analysis of vulnerability forms the starting point of facilitating

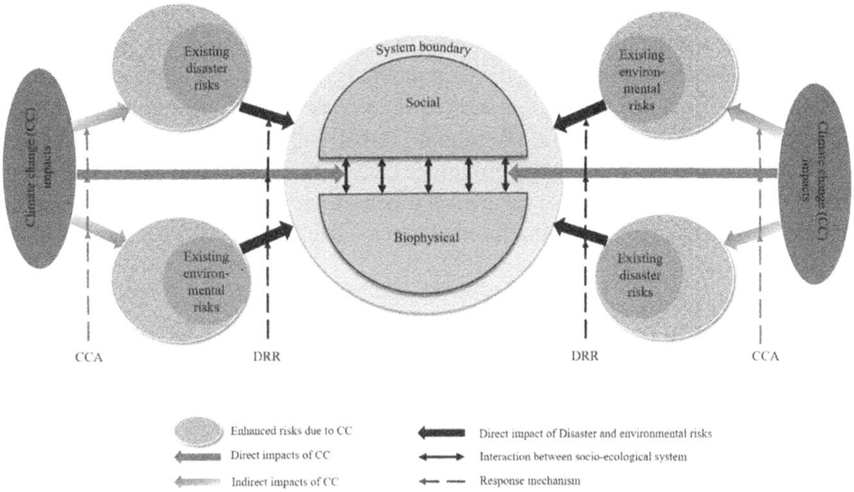

Fig. 29.1 Conceptual framework of impacts of Climate Change on existing disaster and environmental risks, which adversely affect the social and ecological systems. Figure also shows the entry point for CCA and DDR efforts

adaptation actions that aim to limit the impacts of any stressors (Kelly and Adger 2000; Adger 2006). Additionally, vulnerability forms a policy-relevant framework for identifying areas of specific intervention to improve the capacity of the people or any physical system to adapt to stressors.

Consistent with this introduction, the chapter builds on the argument, supporting the need for anticipatory adaptation responses to deal with the emerging Climate Change and disaster risks. The chapter discusses the concept of anticipation, anticipatory adaptation, and vulnerability assessment approaches. Based on the insights gained from these concepts we propose a framework based on the concept of inherent vulnerability for anticipatory adaptation planning. The proposed framework is applied for understanding the main dimension and sub-dimensions of vulnerability of agriculture communities living in the Indian Himalayas. We conclude that anticipatory adaptation planning approaches that are hazard-generic, address the structural inequalities, and compel preventive action in the face of uncertain future disasters and climate risks, assist in developing and mainstreaming action-oriented foresighted policymaking.

29.2 Adaptation

Adaptation is a trans-disciplinary field that involves concepts and understandings from a diversity of areas such as social and natural sciences. Adaptation focuses on improving human well-being through action-oriented processes. Adaptation efforts are a high priority for all the countries, and the issue pertains relevant at local, national and international levels. Although efforts by research communities are growing to understand better and develop methods to inform and support adaptation decision making noted efforts that are required to deal with long-time uncertainty, and diverse knowledge system. (Wise et al. 2014). Furthermore, adaptation actions should also explicitly address the proximate causes of vulnerability emanating from the broader social and political regimes and help in the transformation of prevailing governance mechanism for being more effective (Pahl-Wostl 2007; O'Brien, 2012).

29.2.1 Defining Adaptation

IPCC defines adaptation as an 'adjustment in natural or human systems in response to actual or expected climatic stimuli or their effects, which moderates harm or exploits beneficial opportunities.' The IPCC definition though leaves out space for numerous interpretations, yet mainly highlights that adaptation is not just about protecting against negative impacts but also capturing the socioeconomic benefits that may come along. Adaptation actions have been classified into multiple forms and levels based on the timings of action, and degree of spontaneity (Smit and Wandel 2006). Related to the timing of response to a stressor, adaptation actions can be classified as anticipatory (proactive), concurrent (during), and reactive (responsive). Based on the intent, actions are classified to be either autonomous or planned. On the basis of the spatial scope and extent of the responses can vary from local to regional and to global scale. Lastly, the methods of adaptation intervention vary from technological provisions, behavioral attitudes, financial support, institutional facilitation, and information sharing (Smit and Wandel 2006).

29.2.2 Concept of Anticipation

The concept of Anticipation is widely used within numerous fields such as psychology, future studies, resilience and more recently in governance studies. Hence, several disciplinary definitions exist for anticipation. However, the commonality between all these definitions is the aspect of foresightedness i.e. dealing with the paradoxical question of 'thinking the unthinkable' or 'to act for the unknown'. Few terms that appear recurrently across the several definitions reviewed by Boyd et al. (2015) are future, possibilities, predictions, scenarios and imagination. Broadly, anticipation can be defined as a feature of human behavior, which attempts to understand, imagine and benefit from the future. Nonetheless, research into anticipation has not kept pace with the growing demands of anticipatory practices and policies to deal with the emerging problems.

29.2.3 Anticipatory Adaptation

Anticipatory actions seek to reduce and prevent the effects of stressors (by responding prior to the occurrence of it) through a cushioning effect, and facilitate recovery. The planning process is iterative and allows for planned strategies to not only address multiple stressors but also allows them to be adjusted to developmental goals. Anticipatory approaches do not wait for a climate-induced or disasters event to happen but rather such an approach focuses on ex-ante measures to reduce harmful impacts (Ford & Smit, 2004). The earlier understanding and narratives of anticipatory adaptation highlighted that anticipatory efforts require foreknowledge of Climate Change and its impacts. Scenario planning and predictive models of future climate formed the main basis of anticipatory actions, which were mainly limited to technological solutions to the anticipated Climate Change (Adger 2006). However, these widespread narratives are being challenged due to the growing uncertainty in environmental conditions and identification of critical link between structural vulnerability and adaptive capacity (Shinn 2017). With scholars identifying the role of existing vulnerability, deeply rooted in social, political and economic relationships, in adaptation success, a shift in focus is noted within anticipatory approaches and anticipatory policy making with greater emphasis on identifying the opportunities for a forward-looking learning process to tackle the intersecting developmental and uncertain climatic challenges, rather than learning from the shock. (DeLeo 2017; Shinn 2017). As conception of vulnerability forms the initial steps in adaptation planning, therefore, understanding how evolving understanding of vulnerability has led to the need to re-conceptualize adaptation-planning approach to address anticipatory planning needs becomes crucial. One such example of anticipatory action could be building sea dykes in response to anticipated sea level rise. Another example is of such foreseeable planning, before the occurrence of climatic risk, are the law that defines the maximum acceptable risks of the flooding. Adaptation measures are initiated as soon as the water level rises to the maximum acceptable risks. The first step to anticipatory adaptation planning is assessment of vulnerability to identify vulnerable locations or/and vulnerable population. The section below defines vulnerability and its assessment approaches in details.

29.3 Vulnerability and Vulnerability Assessment Approaches

Vulnerability broadly refers to as the *potential of harm* (Adger 2006). A bewildering array of terms exists around the vulnerability concept such as sensitivity, fragility, and risk. It is a relative property with a negative connotation attached to it (Adger 2006; Ionescu et al. 2009).Varied definitions of vulnerability exists due to the diversity in understanding this concept across multiple disciplines. The recent IPCC (2014) report defines vulnerability as the ‘propensity or predisposition to be

Table 29.1 Vulnerability definitions given in IPCC reports

Assessment report	Approach	Definition
2nd Assessment report (1995)	Sensitivity/vulnerability approach	Vulnerability defines the extent to which Climate Change may damage or harm a system. The magnitude and rate of Climate Change are important in determining the sensitivity, adaptability, and vulnerability of a system
3rd Assessment report (2001)	Sensitivity/ vulnerability approach	Vulnerability is a function of the character, magnitude, and rate of climate variation to which a system is exposed, its sensitivity, and its adaptive capacity
4th assessment report (2007)	Sensitivity/vulnerability approach	Vulnerability is the degree to which a system is susceptible to and unable to cope with, adverse effects of Climate Change, including climate variability and extremes
5th assessment report (2014)	Risk-Hazard approach	The propensity or predisposition to be adversely affected

affected.' An evolution in the vulnerability definitions can be noted across the consecutive IPCC reports since 1995 (Table 29.1). Climate Change was central to the definition of vulnerability with the definitions explicitly mentioning Climate Change, climate variability and extremes until the fifth Assessment Report (AR5), which is climate-neutral. With the evolution of vulnerability definition, it can be noted that, more recently, Climate Change/ variability/ extremes are not seen as the sole foundation of vulnerability rather vulnerability is understood to be generated through complex, context-specific, and multifaceted factors (Füssel 2007; Sugden et al. 2014).

The aforementioned diversity in understanding the concept of vulnerability obfuscates its operationalization. Quantification of vulnerability is difficult owing to its dynamic, multi-dimensional, cross-scalar, and site-specific nature (Adger 2006).

Vulnerability assessment tools aim to identify the hotspots of vulnerable location or communities which are at the margin of tolerance, and further gain insights into factors that create conditions of heightened vulnerability (Preston et al. 2011).The assessment of vulnerability begins with its conceptualization based on the context and locale rationales (Ciurean et al., 2013; Pamungkas et al. 2014).

Mainly two approaches for conceptualizing vulnerability exist namely 'outcome vulnerability' and 'contextual vulnerability', also referred to as 'end point' and 'starting point' vulnerability approaches, respectively (O'brien et al. 2007). These approaches conceptualize vulnerability differently be either an endogenous characteristic or a residual outcome of an exogenous hazard. The outcome vulnerability approach conceptualizes vulnerability to be an outcome of an external hazard and is based on risk-hazard impact models. In this approach, humans are seen merely as a passive receiver of the impacts of the hazard, failing to recognize the adaptive capacity of the humans to adapt (Piya et al. 2015).

The contextual approach bases its paradigm on the construct that vulnerability is a pre-existing condition that arises purely from deeply rooted social inequality, access to assets, and power dynamics, which make some people more vulnerable than others (Cutter et al. 2003). Such a vulnerability analysis following a contextual approach is also termed as social vulnerability. However, the usage of term 'social' is inappropriate when vulnerability is not merely determined by intrinsic socio-economic characteristics but is equally dependent on physical and ecological characteristics of a region. In such cases, the term 'inherent' vulnerability might be more appropriate (Sharma et al. 2015). The inherent vulnerability approach helps in identifying locations or communities who are closer to the margin of tolerance to changing climatic conditions and assists in reducing the vulnerability of the marginalized.

29.3.1 Inherent Vulnerability Framework for Anticipatory Adaptation Planning

Scholars have highlighted that the pre-existing vulnerabilities crystallize the view of differentiated and disproportionate impacts of Climate Change and disaster risks (Adger 2006; Sugden et al. 2014). In order to effectively respond to these risks, anticipatory adaptation planning actions should aim at reducing the vulnerability of the disproportionately vulnerable populations and enhance their capacity to adapt. Therefore, it is important to understand the inherent socio-political, and biophysical factors that hinder or enhance people's access to resources, thereby influencing their capacities to adapt.

Inherent vulnerability represents an antecedent condition, which measures the predisposition of a system to be affected by any harm shaped by intrinsic social and ecological attributes (Rajesh et al. 2014).

Here we propose framework (Fig. 29.2), based on the IPCC AR5 (2014) risk/vulnerability framework, for assessing the inherent vulnerability of agriculture communities. Inherent vulnerability is expressed as a factor of sensitivity and adaptive capacity for identifying hotspots of vulnerability. The extent to which a system would be affected by any stress is referred as sensitivity, and adaptive capacity is the capacity of the system to adapt to these shocks. The equation below represents our conceptualization of inherent vulnerability:

$$\text{Inherent vulnerability} = \text{f}\left(\text{Sensitivity}_{\text{B,S}}; \text{Adaptive capacity}_{\text{BS}}\right)$$

Where, B stands for biophysical and S stands for social vulnerability.

Social vulnerability is a result of socio-political dynamics and unequal asset distribution within a society. In the proposed framework, social sensitivity is measured using three sub-dimensions namely population pressure, livelihood dependency and presence of marginal population. Sensitivity of a community have been

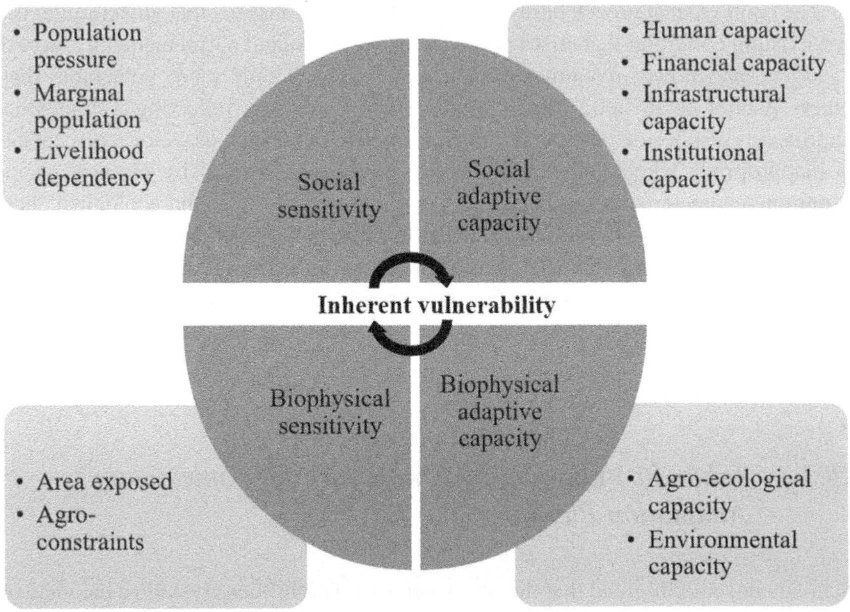

Fig. 29.2 Conceptual framework of inherent vulnerability

identified to get exacerbate due to these conditions. In contrast, the capacity to adjust or cope with existing and unanticipated risks which donates the social adaptive capacity is determined through four sub-dimensions of human capacity, financial capacity, infrastructural capacity and institutional capacity.

Biophysical vulnerability, in the proposed framework, is defined as the vulnerability that arises due to natural resources constraints and its degradation. Itis determined by topographical parameters of physical geography, soil properties, land use, and availability of natural resources. Biophysical sensitivity is due to the heightened agricultural constraint and greater agricultural area exposed to risks. These conditions ameliorate the adverse effects of any stressors. Whereas, biophysical adaptive capacity was determined using environmental capacity and agro-ecological capacity dependent on availability of natural resources (forest and water) and irrigated area form the two sub-dimensions of biophysical adaptive capacity.

29.4 Application of the Proposed Framework: Case Study of Uttarakhand

As prediction of when, where and to what extent emergent climate-induced hazards will manifest in the mountain region is extraordinarily difficult, therefore there is an imperative need to shift from post-hoc to anticipatory planning. Besides, based on

Table 29.2 List of indicators selected under each sub-dimension for assessing social vulnerability

Sub-dimension	Indicator
Social sensitivity	
Agricultural density	Total population (per unit agriculture area)
	Agricultural population (per unit agriculture area)
Livelihood dependency	Main cultivators (% of total population)
	Main agricultural workers (% of total population)
	Marginal agriculture workers (% of total population)
	Marginal cultivators (% of total population)
Marginalized population	Schedule caste (% of total population)
	Schedule tribe (% of total population)
Social adaptive capacity	
Human capacity	Literacy rate (% of literate population)
	Sex- ratio (female per 1000 male)
	Non-agriculture dependent workers (% of total population)
Infrastructure capacity	Availability of road
	Availability of irrigation facilities
	Availability of education facilities
	Availability of communication facilities
	Availability of power supply for agriculture
	Distance from nearest town (km)
Financial capacity	Income (per capita)
	Expenditure (per capita)
Institutional capacity	Availability of self-help groups
	Availability of Agriculture Credit Society
	Availability of Agriculture Marketing Society

the findings of a systematic review by Shukla et al. (2017), it was found that most of the vulnerability studies in Himalayan region do not focus on addressing the underlying factors of chronic vulnerability, that undermine the capacity of the Himalayan communities to undertake adaptation. Therefore, the proposed theoretical framework was applied to assess the inherent vulnerability of the agriculture communities in Indian Himalayan region. We carried out the assessment of inherent vulnerability for the Uttarakhand state of western Himalayas, to identify the hotspots of social, biophysical, and cumulative inherent vulnerability at village level.

Operationalization of the proposed framework was done using an indicator-based approach. Final selection of indicators was done on the basis of the locale-specificities. Indicators were identified for each sub-dimension of social and bio-physical vulnerability as per mountain specific characteristics. Tables 29.2 and 29.3 lists the indicators selected for assessing social and biophysical vulnerability, respectively. The selection of the indicators in any indicators-based approach is limited by the availability of the data, however, remote sensing outputs help in addressing the data gaps. A detailed understanding of each indicator along with the source of data is provided in Shukla et al. (2016b). Data on 36 indicators,

Table 29.3 List of indicators selected under each sub-dimension for assessing social vulnerability

Sub-dimension	Indicator
Biophysical sensitivity	
Agro-constraint	Median elevation (m)
	Mean slope (m)
	Mean aspect (degree)
	Soil erosion (erosivity class)
	Barren land (% of total area (ha))
Area exposed	Net sown area (% of total area (ha))
	Current fallow (% of total area (ha))
	Tree crop (% of total area (ha))
	Culturable wasteland (% of total area (ha))
Biophysical adaptive capacity	
Environmental capacity	High-density forest (% of total forest area (km^2))
	NDVI
	NDWI
Agro-ecological capacity	Net irrigated area (% of agriculture area (ha))
	Forest availability (per unit agriculture area)

characterizing the sensitivity and adaptive capacity was collected from secondary sources for 15,285 villages of Uttarakhand. A weight was assigned to each indicator as per its importance in determining vulnerability. Indicators were weighed using Analytical Hierarchical Processes (AHP). Finally, spatial distribution of the inherent vulnerability, an aggregate of social and biophysical vulnerability, was studied to identify its hotspots.

The results highlight the presence of high biophysical vulnerability (0.82 ± 0.10) and social vulnerability (0.65 ± 0.15). Incidences of high biophysical vulnerability correspond with villages that have a greater extent of cultivated area with least coverage of dense forest. The score of social vulnerability is higher in villages that lack local institutions, infrastructural facilities, and Self Help Groups (SHGs). A plot of tehsils (administrative zones made of a cluster of villages) on the intersectional plane of social and biophysical vulnerability classified them in quarters of combination of social and biophysical vulnerability. About 46% of the tehsils lie in the high biophysical – high social vulnerability quadrant, whereas only 10% of the tehsils lie in the low biophysical – low social vulnerability quadrant.

Overall the state observes high vulnerability with a mean value of 0.66 ± 0.15. The distribution of villages based on the index indicates that 23.6% of the villages exhibit very high, 24.7% high and 26% moderate vulnerability. Out of the remaining 25.7% villages, 18.3% villages are classified under low and merely 7.4% villages under very low vulnerability class. Bhikiyasain, Chaubatakhal, Khalsi, Tehri and Chakrata tehsils have the highest proportion of villages grouped under very high vulnerability class. The statistically significant value of Global Moran's I of 0.42 (Z-score = 79.35, p value < 0.05) denote spatial clustering in inherent vulnerability index within the study region. The results of LISA analysis provide further insight into the clustered pattern of inherent vulnerability by identifying four association types. This includes high-high (HH), high-low (HL), lowhigh (LH) and low-low (LL) associations. The 23.19% of villages in HH clusters identified as vulnerability hotspots in the present study warrant immediate adaptation planning. Furthermore, a distinctive pattern is found in the distribution of vulnerability hotspots i.e. most of the vulnerability hotspots are centered in the middle altitudinal zone of Uttarakhand state. A comprehensive account of the findings of the study is provided in Shukla et al. (2016a, b).

Analysis based on the proposed framework at a fine-scale, like at village level, can potentially help in understanding the underlying patterns of vulnerability that otherwise remains masked in coarse scale assessments (Frazier et al. 2014). Information from similar analysis can robustly inform the policymakers in identifying and prioritizing anticipatory response actions, which will better enable the communities to adapt to multiple yet unknown stressors. Furthermore, practitioners and communities encounter several structural obstacles like lack of awareness, and access to knowledge to adaptation planning (Macchi et al. 2014). Therefore, there is a need of a comprehensive vulnerability assessment that not only provide a base for anticipatory adaptation planning but also address the compounded structural problem of underdevelopment, marginalization, food and livelihood security and socio-economic inequality (Rasul 2014).

29.5 Mainstreaming and Monitoring Anticipatory Adaptation

There is a need for mainstreaming of anticipatory adaptation actions at all levels i.e. national and sub-national levels to foster resilience within communities. Mainstreaming needs to be done within developmental policies to generate co-benefits such as improvements in livelihoods, well-being, and poverty reduction of the vulnerable communities. Further, UNFCC recognizes the need of monitoring of the effectiveness and benefits of adaptation actions to inform the policy makers and integrate 'learning' into 'doing'. The multifaceted and uncertain nature of climatic risks require continuous and long term monitoring to measure and report the extent of progress and achievements of targeted outcomes. However, some Climate Change impacts are seen on longer timescale, making the task of monitoring very challenging. Few of the other challenges that have been identified monitoring and measuring of anticipatory adaption actions are shifting policy norms, selection of suitable set of indicators, and attribution of an effect to specific adaptation intervention. We recommend a parallel analysis should be undertaken to identify the barriers in implementation and monitoring of anticipatory adaptation planning. This will further inform the policy design by evaluating the difficulties in overcoming these barriers in terms of financial, human, and time requirement.

29.6 Conclusion and Way Forward

The premise of anticipatory actions is based on preparedness to respond to potential ways in which future may unfold. Anticipatory adaptation helps in responding complementary to Climate Change and DRM approaches to enhance resilience from reducing risks and exposure and vulnerability. It is prudent to strategically develop a vulnerability assessment approach that focuses on identification of the most vulnerability communities based on the inherent social and biophysical conditions, which will simultaneously benefit the DRR and CCA efforts. The inherent vulnerability framework, proposed in the chapter, conceptualizes vulnerability to be pre-existing condition through which impacts of stressors are mediated resulting in differential vulnerability. The application of proposed framework helps in identification of underlying drivers that exacerbate vulnerability and constrain adaptation actions at a multiple scales. The proposed framework is applicable to other regions as well; however, the selected indicators are specific to mountain agriculture. Further, challenges remain in monitoring and measuring anticipatory actions due to the uncertain nature of future risks. We advocate the use of such approaches in conjunction with participative community planning could minimize the challenges for multi-scaled Climate Change and related disaster risk in order to enhance the resilience of communities. Anticipatory actions based on such approaches will help to identify "no-regret" policy options, reduce the perpetuating structural inequalities, and enable policy makers at all levels to adopt suitable adaptation strategies based on both at micro and macro level contextual needs.

References

Adger WN (2006) Vulnerability. Glob Environ Chang 16(3):268–281. https://doi.org/10.1016/j.gloenvcha.2006.02.006

Boyd E, Nykvist B, Borgström S, Stacewicz IA (2015) Anticipatory governance for social-ecological resilience. Ambio 44(Suppl 1):149–161. https://doi.org/10.1007/s13280-014-0604-x

Ciurean RL, Schroter D, Glade T (2013) Conceptual frameworks of vulnerability assessments for natural disasters reduction. InTech. https://doi.org/10.5772/55538

Cutter SL, Boruff BJ, Shirley WL (2003) Social vulnerability to environmental hazards*. Soc Sci Q 84(2):242–261. https://doi.org/10.1111/1540-6237.8402002

Dai A (2013) Increasing drought under global warming in observations and models. Nat Clim Chang 3(1):52–58. https://doi.org/10.1038/nclimate1633

DeLeo RA (2017) Anticipatory policymaking in global venues: policy change, adaptation, and the UNFCCC. Futures 92:39–47. https://doi.org/10.1016/j.futures.2016.09.001

Ford JD, Smit B (2004) A framework for assessing the vulnerability of communities in the Canadian Arctic to risks associated with climate change. Arctic 57(4):389–400. http://www.jstor.org/stable/40512642

Frazier TG, Thompson CM, Dezzani RJ (2014) A framework for the development of the SERV model: a spatially explicit resilience-vulnerability model. Appl Geogr 51:158–172. https://doi.org/10.1016/j.apgeog.2014.04.004

Füssel H-M (2007) Vulnerability: a generally applicable conceptual framework for Climate Change research. Glob Environ Chang 17(2):155–167. https://doi.org/10.1016/j.gloenvcha.2006.05.002
Hannah M (2010) (Mis)adventures in Rumsfeld space. GeoJournal 75:397–406
Hewitt K (2017) Identifying emerging issues in disaster risk reduction, migration, climate change and sustainable development. In: *Identifying emerging issues in disaster risk reduction, migration, climate change and sustainable development*, 35–51. https://doi.org/10.1007/978-3-319-33880-4
Hirabayashi Y, Mahendran R, Koirala S, Konoshima L, Yamazaki D, Watanabe S, Kim H, Kanae S (2013) Global flood risk under Climate Change. Nat Clim Chang 3(9):816–821. https://doi.org/10.1038/nclimate1911
Ionescu C, Klein RT, Hinkel J, Kavi Kumar KS, Klein R (2009) Towards a formal framework of vulnerability to climate change. Environ Model Assessment 14(1):1–16. https://doi.org/10.1007/s10666-008-9179-x
IPCC (2014) Climate Change 2014: Synthesis Report. Contribution of Working Groups I, II and III to the Fifth Assessment Report of the Intergovernmental Panel on Climate Change. IPCC, Geneva, 151 pp
Kelly PM, Adger WN (2000) Theory and practice in assessing vulnerability to climate change and facilitating adaptation. Clim Chang 47(4):325–352. https://doi.org/10.1023/A:1005627828199
Macchi M, Gurung AM, Hoermann B (2014) Community perceptions and responses to climate variability and change in the Himalayas. Clim Dev 7(5):37–41. https://doi.org/10.1080/17565529.2014.966046
O'Brien K (2012) Global environmental change II: from adaptation to deliberate transformation. Prog Hum Geogr 36(5):667–676. https://doi.org/10.1177/0309132511425767
O'brien K, Eriksen S, Nygaard LP, Schjolden A (2007) Why different interpretations of vulnerability matter in climate change discourses. Clim Pol 7(1):73–88. https://doi.org/10.1080/14693062.2007.9685639
Pahl-Wostl C (2007) Transitions towards adaptive management of water facing climate and global change. Water Resour Manag 21(1):49–62. https://doi.org/10.1007/s11269-006-9040-4
Pamungkas A, Bekessy SA, Lane R (2014) Vulnerability modelling to improve assessment process on community vulnerability. Procedia Soc Behav Sci 135:159–166. https://doi.org/10.1016/j.sbspro.2014.07.341
Piya L, Joshi NP, Maharjan KL (2015) Vulnerability of Chepang households to climate change and extremes in the Mid-Hills of Nepal. Clim Chang 1–17. https://doi.org/10.1007/s10584-015-1572-2
Preston BL, Yuen EJ, Westaway RM (2011) Putting vulnerability to climate change on the map: a review of approaches, benefits, and risks. Sustain Sci 6(2):177–202. https://doi.org/10.1007/s11625-011-0129-1
Rajesh S, Jain S, Sharma P, Bhahuguna R (2014) Assessment of inherent vulnerability of rural communities to environmental hazards in kimsar region of Uttarakhand, India. Environ Dev 12: 16–36. https://doi.org/10.1016/j.envdev.2014.06.003
Rasul G (2014) Food, water, and energy security in South Asia: a nexus perspective from the Hindu Kush Himalayan region. Environ Sci Policy 39:35–48. https://doi.org/10.1016/j.envsci.2014.01.010
Sharma J, Chaturvedi RK, Bala G, Ravindranath NH (2015) Assessing "inherent vulnerability " of forests : a methodological approach and a case study from Western Ghats , India. In: *Mitigiation and adapation strategies for global change*, 573–590. https://doi.org/10.1007/s11027-013-9508-5
Shinn JE (2017) Toward anticipatory adaptation: transforming social-ecological vulnerabilities in the Okavango Delta, Botswana. Geogr J 184(2):1–13. https://doi.org/10.1111/geoj.12244
Shukla R, Sachdeva K, Joshi PK (2016a) An indicator-based approach to assess village-level social and biophysical vulnerability of agriculture communities in Uttarakhand, India. J Mt Sci 13(2): 2260–2271. https://doi.org/10.1007/s11629-016-4058-4

Shukla R, Sachdeva K, Joshi PK (2016b) Inherent vulnerability of agricultural communities in Himalaya: a village-level hotspot analysis in the Uttarakhand state of India. Appl Geogr 74:182–198. https://doi.org/10.1016/j.apgeog.2016.07.013

Shukla R, Sachdeva K, Joshi PK (2017) Demystifying vulnerability assessment of agriculture communities in the Himalayas: a systematic review. Nat Hazards 91(1):409–429. https://doi.org/10.1007/s11069-017-3120-z

Smit B, Wandel J (2006) Adaptation, adaptive capacity and vulnerability. Glob Environ Chang 16(3):282–292. https://doi.org/10.1016/j.gloenvcha.2006.03.008

Sugden F, Maskey N, Clement F, Ramesh V, Philip A, Rai A (2014) Agrarian stress and climate change in the eastern Gangetic Plains : gendered vulnerability in a stratified social formation. Glob Environ Chang 29:258–269. https://doi.org/10.1016/j.gloenvcha.2014.10.008

Wise RM, Fazey I, Stafford Smith M, Park SE, Eakin HC, Archer Van Garderen ERM, Campbell B (2014) Reconceptualising adaptation to climate change as part of pathways of change and response. Glob Environ Chang 28:325–336. https://doi.org/10.1016/j.gloenvcha.2013.12.002

R Shukla is an Assistant Professor at IIT Roorkee. She persued her Ph.D. from TERI School of Advanced Studies, New Delhi and pursued her post-doctoral research from Potsdam, Germany. Her research focuses on assessing differential vulnerability to climate change impacts and understanding barriers to climate change adaptation for agriculture communities in the Indian Himalayan region.

Kamna Sachdeva researches food security, pollution, and climate change impacts. She has a special interest in carbonaceous aerosols and its linkages with climate change. She has been an Associate Professor and the HoD of the Department of Energy and Environment, TERI-SAS. She is currently working as a Professor at the Delhi Skill and Entrepreneurship University, New Delhi.

P. K. Joshi is a Professor at School of Environmental Science at the Jawaharlal Nehru University, New Delhi. His research interests include the application of Remote Sensing and GIS for various disciplines such as forests, agriculture and in urban systems, LULC characterisation, and climate change studies.

Chapter 30
Adaptive Planning for Resilience and Sustainability: Lessons from India - Project CAP-RES and Network

Anil Kumar Gupta and Pritha Acharya

Abstract Sustainability is understood in terms of continuous and futuristic availability of resources with inter-generational equity over the changing scenarios. This implicitly requires the ability to manage and cope with risks of uncertainties and crisis due to extremes and disasters. Environmental resources of nature, cities, industry and communities that dwell with overlap are critical for future of life on earth. Concepts of resource efficiency and circular economy are new integration enabling the convergence and synergies among climate change mitigation (GHG emission reduction and green economy), adaptation and resilience, and disaster risk reduction. The new era of integrated resilience and sustainability has emergence over the lessons of SDG's journey along SFDRR and Paris Climate Agreement, thus, assimilating the new understandings and lessons of recent times towards futuristic sustainability vision. There are lateral aspects integral to sustainability too like health, occupational safety, food safety, livelihood, inclusiveness, etc. and, therefore, holistic perspectives are needed from adaptive governance—policy-planning, localization and participatory actions at all levels. The CAP-RES (Climate Adaptive Planning for Resilience and Sustainability, in Multi-hazard Environment) under the National Mission for Strategic Knowledge on Climate Change (part of the National Action Plan on Climate Change) implemented as Centre for Excellence on Climate Resilience enabled showcasing of a strategic, ambitious and reliable approach of umbrella programme on resilience building towards sustainable future. The latest major endeavor of CAP-RES aimed at futuristic vision (100 years of India's independence—Vision 2047) through "first Resilience and Sustainability Summit RESSUMMIT2047" which was a unique assembly of high level experts on policy-planning, S&T and innovations, practitioners, academics, researchers and NGOs, with a range of partners covering international, UN and wider range of

A. K. Gupta
ECDRM Division, International Cooperation, CAP-RES DST Project, National Institute of Disaster Management (Ministry of Home Affairs), New Delhi, India

P. Acharya (✉)
ECDRM Division, CAP-RES DST Project, Centre for Excellence on Climate Resilience, National Institute of Disaster Management (Ministry of Home Affairs), New Delhi, India

A. K. Gupta et al. (eds.), *Disaster Risk and Management Under Climate Change*, Disaster Resilience and Green Growth,
https://doi.org/10.1007/978-981-99-4105-6_30

organizations. The envisaged evolution is to a global network platform promoted by India for "Resilience and Sustainability Knowledge and Capacities RES-KAP" with a strong – reliable – sustainable network led platform of institutions, agencies, experts, on relevant fields, focused on India. This also comes up from the backdrop of first National Adaptation Communication to UNFCCC in the year 2022.

Keywords Adaptive planning · Climate change · Knowledge network

30.1 Contexts of Resilience—DRR to Sustainability

The manifestation of climate change is witnessed through the increased severity and frequency of extreme events like floods, heatwaves, droughts, forest fires, etc. This also stands as testimony to the fact that climate change is altering the global landscapes of disaster. India exhibits a diverse array of geographies with regional-specific climate-related hazard complexes which makes it among the world's most disaster-prone countries. The current trends of environmental degradation (unsustainable practices, habitat change and loss, pollution, overexploitation of resources, etc.) infrastructure and urban growth and migration of people, and climate change in the backdrop pose additional risk and complexity by increasing the vulnerability to disasters. Given the changing climatic trends, new and emergent disasters are set to happen and also pose challenges for communities particularly which are more vulnerable and lack the capacity to cope with climate-induced disasters.

Adaptation strategies like Disaster Risk Reduction (DRR) are key to addressing the amalgamation of factors contributing to community vulnerability. As disaster management in India is undergoing multiple paradigm shifts from a relief and response-centric approach to a mitigation and preparedness-centric approach to a hazard and vulnerability-centric approach (Gupta et al. 2016) and, thus, in the context of changing climate the key facets of DRR include; (a) understanding the hazard processes and events, (b) reducing the underlying causes of vulnerability, (c) effective planning and disaster forewarning and (d) mainstreaming measures of disaster mitigation into developmental activities at all levels (Bindal et al. 2021a, b, c). The Sendai Framework for Disaster Risk Reduction (SFDRR), Sustainable Development Goals (SDGs) and the Prime Minister's 10 Point Agenda on DRR necessitate the need for an integrated and holistic understanding of disaster risk management in planning processes. Environment-Disaster linkages which are multi-layered, complex and critical to holistic DRR, bring sustainability and resource efficiency at the forefront of resilience framework from the earlier ritualistic ways.

The existing gaps in understanding and proficiency commemorates the need for capacity building by improving the knowledge base of the key players and stakeholders in the sector for achieving climate action goals. Along the similar lines, the present chapter is incited from the "Climate Adaptive Planning for Resilience and Sustainability in Multi-Hazard Environment (CAP-RES)", which aimed to promote capacity building through an interwoven network of knowledge and training support for varied stakeholders and players across sectors. This includes strong systems of case studies, research, policy analysis and advocacy, and strategic dialogues towards enabling an effective policy environment towards realistic localization of resilience agenda.

30.2 Methods and Approaches

The CAP-RES is an umbrella project which also envisaged to support Science-Policy-Practice- Interface based on capacity development for addressing disaster risks from climate and extreme events perspectives. This consortium project also aimed to establish a network platform through; (a) implementing a science-policy-practice framework, (b) understanding the gaps and needs in capacity for addressing disaster risks from climate change and extreme events, and (c) capacity enhancing of key professionals, students, policymakers, researchers through knowledge sharing platforms and policy advocacy (NIDM 2018, Climate Adaptive Planning for Resilience and Sustainable Development in Multi- hazard Environment (CAP-RES)).

CAP-RES is spread across primarily an array of five thematic areas (Fig. 30.1) and follows a '3 M' approach which ensures that the needs of capacity development should follow an integrated framework involving multiple stakeholders, multiple levels and multiple hazards (Fig. 30.2). By doing so CAP-RES focused on a whole range of actors and widening the capacity development scope. Figures 30.3 and 30.4 elaborated on the CAP-RES Contexts, Aspects and Capacity Interventions.

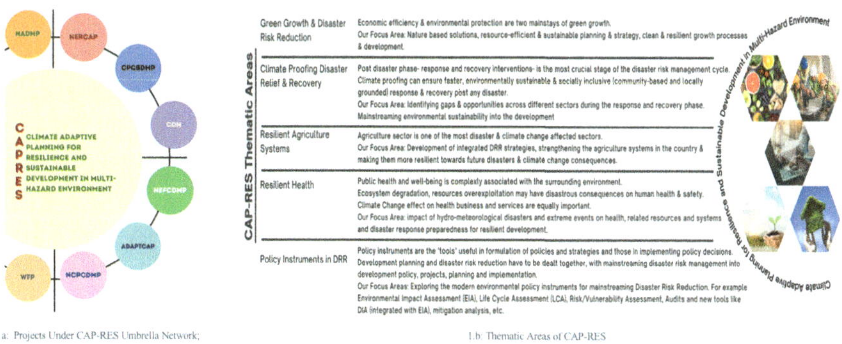

Fig. 30.1 CAP-RES: Umbrella network

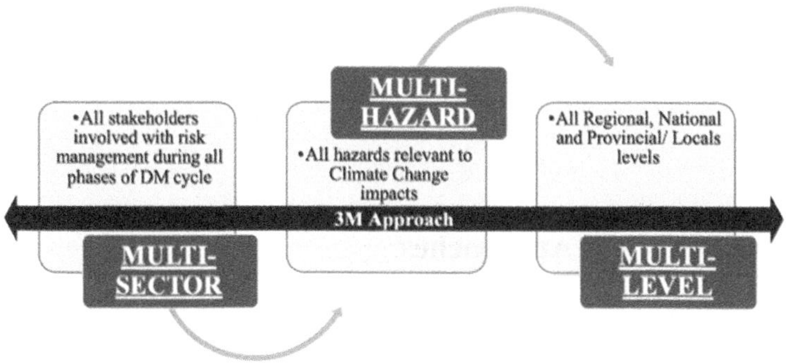

Fig. 30.2 The '3 M' Approach followed by CAP-RES

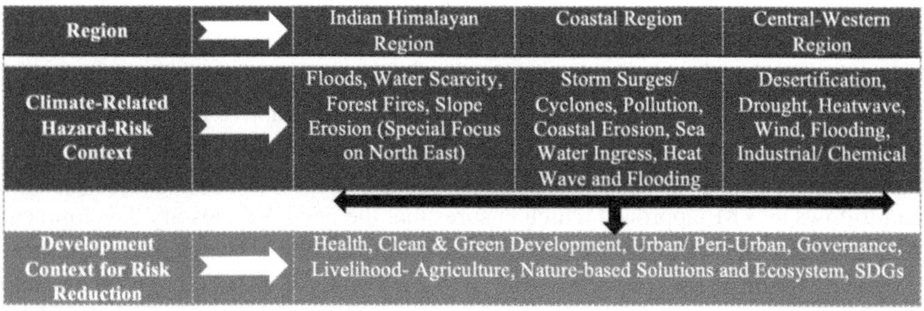

Fig. 30.3 Regional contexts and specific problem context

Fig. 30.4 Focus on the capacity interventions

30.3 State of Art

CAP-RES provides a platform for knowledge and experience sharing via well-documented case studies and best practices (Bindal et al. 2021a, b, c). Further, it established a roadmap for addressing the climate change induces risks and vulnerabilities by adding value to the present understanding of the context, approaches and policies related to DRR. Figure 30.5 enlists the objectives and summarises the capacity development activities and actions undertaken by CAP-RES.

Box 30.1 Resilience and Sustainability Summit: Vision 2047 January 17–192,023 Vigyan Bhawan, New Delhi

(continued)

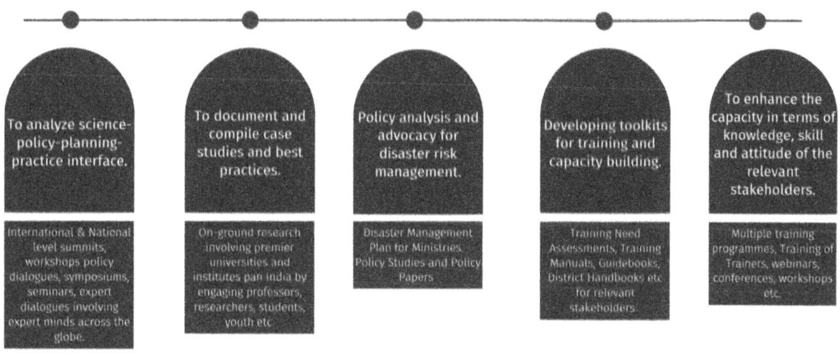

Fig. 30.5 Working scheme of CAP-RES

Box 30.1 (continued)

Since its inception, CAP-RES emerged as a network of networks platform with the active involvement of national and international organizations including national and state governments, Bilateral, United Nations, NGOs, CSOs, Academic and Research institutions and Universities as partners. Under this network, CAP-RES has been associated with more than 60 such organizations and institutions covering both national and international levels. The detailed coverage of activities with specific reference to adaptive planning for resilience and sustainability has been enlisted in the following table (Table 30.1) (1):

30.4 Road Ahead

Sustainability integrated resilience to climate change, disasters and anthropogenic risks in multi-hazard environment covering cities, infrastructure, industry and resources, are the new vision evolved based on recent understanding built on the past experiences as well as future perspectives. In responding to the impacts of climate change and its manifestations, two primary approaches are being most talked about: (a) mitigation and (b) adaptation (NASA 2023). On one hand where adaptation ensures adjusting to the whole gamut of changing climatic scenarios and 'adapting to life' with the goal to reduce climate challenges and risks. Mitigation, on the other hand, deals with reducing climate change so that its manifestations can be reduced. Effective DRR strategies also place mitigation-adaptation synergies at their core. While both these approaches may work at individuals scale and levels, adaptation and mitigation often have consequences for one another. Thus, understanding their linkages and synergies is beneficial and further widens the scope to combat the changing climatic scenario, but at the same time, it is important to understand and realize the trade-offs as well (UNFCCC 2022). While we realize the resources are limited and conventional approaches to their management are inadequate to ensure sustainability over present to future times, the convergence of resource sustainability along with environmental concerns with adaptive and resilience.

Linkages between mitigation and adaptation have been proposed and addressed under various international development agenda including the Paris Climate Agreement, Convention on Biological Diversity, United Nations Convention to Combat Desertification, Kyoto Protocol, Food and Agricultural Organizations, etc. (UNFCCC 2022). The inculcation of mitigation and adaptation approaches for effective DRR measures have also been highlighted by the Sendai Framework for Disaster Risk Reduction and Sustainable Development Goals (Gupta et al. 2016). In addition, the Paris Agreement also provides scope for integrating and establishing the mitigation and adaptation synergies for DRR through Nationally Determined Contributions. The present context of disaster management focuses on the localization and mainstreaming of DRR strategies in policy planning and the national adaptation plan is one way to achieve this target (Bindal et al. 2021a, b, c).

Table 30.1 Action-oriented initiatives for capacity development for adaptive planning for resilience and sustainable development in multi-hazard environment

Thematic coverage	Governance, policy & finance	Sustainability & resources	Climate adaptation & risk mitigation	NBS & Eco-DRR	Resilience – cross cutting & laterals
Activities undertaken					
Knowledge platform					
National Health Programme on Climate Change and Health Sector Resilience (NIDM, GIZ, and WHO 2019)			✓		✓
Sectoral Consultative Workshop on National Agriculture Disaster Management Plan (NIDM 2019a)	✓	✓	✓		
UNCCD COP 14: Special Event Reducing Drought Risk to Improve Land Water Resilience, mainstreaming eco-DRR pathways and tools (NIDM 2019b)		✓	✓	✓	
CAP-RES-22 International Symposium Disaster Resilience and Green Growth for Sustainable Development (Bindal et al. 2021a)	✓	✓	✓	✓	✓
Climate Change, Disaster and Health Epidemiology and Response Preparedness (NIDM and VMMC 2020)			✓		✓
Science Technology Research-Policy-Practice Interface for Climate risk management-Implementing Prime Ministers Agenda 10 on Disaster Risk	✓		✓	✓	✓

(continued)

Table 30.1 (continued)

Thematic coverage	Governance, policy & finance	Sustainability & resources	Climate adaptation & risk mitigation	NBS & Eco-DRR	Resilience – cross cutting & laterals
Management- Item 05 Leveraging Science & Technology (Bindal et al. 2021b)					
Sustainable Future and Climate Change: Insights from Adaptation Gap Report 2020	✓	✓	✓	✓	
High Level Policy Dialogues on Localizing Climate Resilience Agenda: Vision 2050 & 2100 (Gupta and Gupta 2022).	✓	✓	✓		✓
High Level Policy Round Table on Science Technology & Innovation Policy for Disaster Risk Reduction (NIDM 2022)	✓		✓		✓
National Consultation Workshop on Wetland Conservation and Nature Based Solution for Reducing Water Related Stress (NIDM and WISA 2022)		✓	✓	✓	✓
National Consultation Workshop on Wetlands Conservation as Nature-based Solution (NbS) for Disaster Risk		✓	✓	✓	
Resilience and Sustainability Summit: Vision 2047 Science-Policy-Planning-Practice Interface for DRR in changing Climate	✓	✓	✓		✓
Tools and manuals					

(continued)

Table 30.1 (continued)

Thematic coverage	Governance, policy & finance	Sustainability & resources	Climate adaptation & risk mitigation	NBS & Eco-DRR	Resilience – cross cutting & laterals
Climate and Disaster Resilient Agriculture Management (Gupta et al. 2021a)		✓	✓		✓
Handbook for District Administration on Disaster Risk Management in Agriculture Sector (Gupta et al. 2021b)	✓	✓	✓		✓
Health Adaptation and Resilience to Climate Risks (Gupta et al. 2021c)		✓	✓		✓
Crisis Management Plan to Deal with Sudden Contamination of Water Bodies		✓			
Disaster, Risk and Resilience in Chemical and Petrochemical Sector		✓	✓		✓
Capacity Building for Environment and Forest Sector		✓	✓	✓	✓
Climate Change and Extreme Events Risk Reduction (Choldol et al. 2022)			✓		✓
Climate Proofing Disaster Relief and Recovery (Gupta et al. 2022a)	✓	✓	✓		✓
Developing Disaster-Risk Resilience in Cities (Gupta et al. 2019b)	✓		✓		✓
Peri-Urban Ecosystems and Urban Resilience (Gupta et al. 2021f)		✓	✓	✓	✓
Nature-based Solution for Reducing Disaster Risks: A Guidebook for	✓		✓	✓	

(continued)

Table 30.1 (continued)

Thematic coverage	Governance, policy & finance	Sustainability & resources	Climate adaptation & risk mitigation	NBS & Eco-DRR	Resilience – cross cutting & laterals
District Disaster Management Planning (Gupta et al. 2022b)					
Green Growth and Resource Efficiency for DRR: Towards Green Economy and Resilient Development (Gupta et al. 2022f)	✓	✓	✓	✓	✓
Policy plan documents					
National Agriculture Disaster Management Plan (MoAFW 2020)	✓		✓		✓
National Health Adaptation Plan for Disaster Related Health Issues (NIDM 2019a, b, c)	✓		✓		✓
National Chemicals and Petrochemicals Disaster Management Plan	✓	✓			✓
National Environment, Forest and Climate Disaster Management Plan	✓	✓	✓	✓	✓
Disaster Management Plan for the Ministry of Ayush	✓	✓			✓
Research case studies					
Agriculture Resilience Assessment Over Central India Under Climate change (Goyal et al. 2021)			✓		✓
Climatic Variations and Agriculture Landscape: A study on Policies and Practices for Resilience (Nautiyal et al. 2021)	✓	✓	✓		

(continued)

Table 30.1 (continued)

Thematic coverage	Governance, policy & finance	Sustainability & resources	Climate adaptation & risk mitigation	NBS & Eco-DRR	Resilience – cross cutting & laterals
Integrated Farming Systems for Food Security in Indian Himalayas under Climate Change (Tiwari et al. 2021)		✓	✓	✓	✓
Socio-Ecological Resilience to Cyclone Vulnerability: A Case Study of Coastal Odisha		✓	✓		✓
Multi-Hazard Disaster Risk & Resilience at City Levels (NIDM 2021)			✓		✓
Institutional Mechanisms for Evaluating Human Deaths from Disaster (Gupta et al. 2022c)	✓		✓		✓
Landscape Approach and Role in Disaster Management: Moving beyond Conventional Concepts Towards Implementation Frameworks (Agrawal et al. 2022a)	✓	✓	✓	✓	✓
Water Crisis in India Cities: A Systemic Failure or Wrath of Changing Climate (Gupta et al. 2022d)		✓	✓		✓
Underground Urbanism: Re-imagining the role of underground spaces for India's urban future (Agrawal et al. 2022b)		✓			✓
Corporate Carbon Management and ESG Reporting: Investments by Technology Industry	✓				✓

(continued)

Table 30.1 (continued)

Thematic coverage	Governance, policy & finance	Sustainability & resources	Climate adaptation & risk mitigation	NBS & Eco-DRR	Resilience – cross cutting & laterals
in Green growth and Climate actions (Agrawal et al. 2022c)					
Disaster Damages and Losses: Study of Underlying Causes Case Study on Cyclones Tauktae and Yaas (Gupta et al. 2022e)		✓	✓		
Mapping Climatic and Biological Disasters in India (Gupta et al. 2021l)		✓	✓		✓
Climate Risk Management Framework for India: Addressing Loss and Damage (NIDM and GIZ 2019)	✓		✓		✓
Key Biodiversity Areas: Conservation-based on threats and ecosystem services using GIS- based modelling in Central India			✓	✓	✓
Socio-Ecological Resilience to Cyclone Vulnerability—A Case Study of Coastal Odisha		✓	✓		✓
Policy papers/thematic papers					
Roadmap of Resilient Agriculture in India (Gupta et al. 2019a)	✓	✓	✓		✓
Temperature Related Disasters (Heat & Cold Wave) Implications for Health Adaptation and			✓		✓

(continued)

Table 30.1 (continued)

Thematic coverage	Governance, policy & finance	Sustainability & resources	Climate adaptation & risk mitigation	NBS & Eco-DRR	Resilience – cross cutting & laterals
Resilience (Gupta et al. 2021g)					
Air Pollution and Public Health Emergencies (Gupta et al. 2021h)		✓	✓		✓
Water Related Disaster (Flood & Drought) Implications for Health Adaptation (Gupta et al. 2021i)			✓		✓
Environmental Services for Health Protection in Disasters and Emergencies (Gupta et al. 2021j)			✓	✓	✓
Climate Related Adaptation and Resilience in India (Gupta et al. 2021k)		✓	✓		✓
National Chemicals and Petrochemicals Disaster Management Plan: Framework and Roadmap		✓	✓		✓
National Environment, Forest and Climate Disaster Management Plan: Framework and Roadmap		✓	✓	✓	
Green Growth Benefits for Climate and Disaster Resilience (Gupta et al. 2019c)	✓	✓	✓	✓	✓
Articles/chapters					
Climate Change Implication and Adaptation for River Systems (Amin and Gupta 2022)		✓	✓		✓
Impact of COVID-19 on Agro-Food		✓			✓

(continued)

Table 30.1 (continued)

Thematic coverage	Governance, policy & finance	Sustainability & resources	Climate adaptation & risk mitigation	NBS & Eco-DRR	Resilience – cross cutting & laterals
Industry and Transitions towards Food Security (Poonia et al. 2020)					
Understanding Public Health Interventions: Isolation, Quarantine, Social Distancing (Madan and Gupta 2020)					✓
Opportunities and Advances to Mainstream Nature-Based Solutions in Disaster Risk Management and Climate Strategy (Dhyani et al. 2020a)				✓	
Introduction to Virus Outbreaks (Sood et al. 2020)					✓
Environmental Governance for Resilience & SDGs (Gupta et al. 2020)	✓				
Disaster Governance and Legal Systems in India (Gupta 2018)	✓				✓
An evaluation of climate resilient agricultural practices in India: a narrative synthesis of literature (Goswami et al. 2023)	✓		✓		
Public Policy in Environment and Sustainability Strategies: Global & National Scenario (Baidya and Gupta 2022)	✓				✓
India's Health Adaptation Plan: Strategic Tool for Minimizing Disaster	✓		✓		✓

(continued)

Table 30.1 (continued)

Thematic coverage	Governance, policy & finance	Sustainability & resources	Climate adaptation & risk mitigation	NBS & Eco-DRR	Resilience – cross cutting & laterals
Related Losses and Damage (Barwal et al. 2022a)					
Addressing Hydro-climatic Risks Through Sectoral Planning: A Case of National Agriculture Disaster Management Plan (NADMP) (Hodam et al. 2022)	✓	✓	✓		✓
Drought as a Disaster and Its Characterization over Central India (Poonia et al. 2022)		✓	✓	✓	
Climate Finance at International and National Level: Needs, Drivers and Sources (Padhan et al. 2022)	✓		✓		✓
Disaster Preparedness and Emergency Response for Air Pollution and Related Health Extremes (Barwal et al. 2022b)		✓	✓		✓
An Introduction to Extremes in Atmospheric Processes and Phenomena: Assessment, Impacts and Mitigation (Saxena et al. 2022a)			✓		✓
Protection of Mangroves as Core Component for Implementing the NbS (Bindal et al. 2021c)		✓	✓	✓	
Ecosystem-Based Approaches and Policy Perspective from	✓	✓	✓	✓	

(continued)

Table 30.1 (continued)

Thematic coverage	Governance, policy & finance	Sustainability & resources	Climate adaptation & risk mitigation	NBS & Eco-DRR	Resilience – cross cutting & laterals
India (Bhardwaj and Gupta 2021)					
Health Sector Resilience to Climate-Related Disasters: India's Experience of Health Adaptation Planning (Gupta et al. 2021d)		✓	✓		✓
Enhancing Epidemic Resilience: Planning and Institutional Resilience (Bindal et al. 2020)	✓	✓	✓		✓
Lessons from Trans-Domain Assessment of COVID 19 Outbreak (Barwal et al. 2020)		✓			✓
Multi-hazard risk management during pandemic (Chondol et al. 2020)		✓	✓		✓
Nature-Based Solution Entry Points Through Sectoral Policies, Strategic Instruments and Business Continuity (Bhardwaj et al. 2020)	✓	✓	✓		
New Pathways for NbS to Realise and Achieve SDGs and Post 2015 Targets: Transformative Approaches in Resilience Building (Acharya et al. 2020)	✓		✓	✓	
Other strategic contribution					
Health Adaptation and Resilience to Climate Change and Related Disasters: A Compendium of			✓		✓

(continued)

Table 30.1 (continued)

Thematic coverage	Governance, policy & finance	Sustainability & resources	Climate adaptation & risk mitigation	NBS & Eco-DRR	Resilience – cross cutting & laterals
Case Studies (Gupta et al. 2021c)					
Peri-Urban Ecosystems and Urban Resilience: Knowledge Compendium of Case Studies (Gupta et al. 2021e, f)		✓	✓	✓	✓
Book Series on Environmental Sustainability, Green growth and Disaster Resilience		✓	✓	✓	✓
Book on Climate Change and Environmental Impacts: Past, Present and Future Perspective (Phartiyal et al. 2022)		✓	✓	✓	✓
Book on Hydro-Meteorological Extremes and Disasters (Goyal et al. 2022)	✓	✓	✓		✓
Book on Extremes in Atmospheric Processes and Phenomenon: Assessment, Impacts and Mitigation (Saxena et al. 2022b)			✓		✓
Book on Integrated Risk of Pandemic: Covid-19 Impacts, Resilience and Recommendations. (Goyal and Gupta 2020)		✓			✓
Book on Nature-based Solutions for Resilient Ecosystems and Societies (Dhyani et al. 2020b)	✓			✓	✓

With diverse experiences of pilots, projects/programmes, case studies and dealing with extremes and disasters in different eco-geographical settings and under SDG priorities aligned to resilience and climate change mitigation, an umbrella institution of international significance and global eminence, through the network programme can be a lead initiative in India. CAP-RES strategies and approaches pave the way to planning and implementation of such an ambitious plan. The network approach for sustainability and resilience integration based endeavours would help in avoiding duplicacy and in reaping benefits of overlaps, synergies, and co-benefits among DRR, green-growth and resilient development approaches. The adaptation plan at national and regional scales would largely contribute to DRR in more proactive ways. The key concerns of Loss and Damage (L&D), a key decision of COP-27, is a clear advocacy on this line.

Acknowledgement The CAP-RES has been designed and implemented with the vision supported by Dr. Akhilesh Gupta, then Head of Climate change Programme of DST-GOI and presently Secretary SERB, DST-GOI and Senior Advisor. We are also grateful to Dr. Nisha Mendiratta, Advisor-DST, Dr. Susheela Negi, Scientist-DST, Dr. Sweta Baidya, Ms., Shweta Bhardwaj, Dr. Sanayanbi Hodam, Ms. Richa Srivastava, Dr. Anjali Barwal, Ms. Atisha Sood, Ms. Fatima Amin, Dr. Uzma Parveen, Dr. Kopal Verma, for their support in all the endeavors of CAP-RES. We are also thankful to all our strategic and programme partners including GIZ, UNDP, CDKN, GEAG, GGGI, WHO, UNEP, IUCN, Wetlands International, ICIMOD, IIT Indore, IIT Roorkee, Sphere India, IGES, for all their support and cooperation.

References

Acharya P, Gupta AK, Dhyani S, Karki M (2020) New pathways for NbS to realise and achieve SDGs and post 2015 targets: transformative approaches in resilience building. Nature-based Solutions for Resilient Ecosystems and Societies, 435–455

Agrawal M, Friedmann SJ, Palacios L, Gupta AK (2022c) Corporate carbon management and ESG reporting: investments by Technology industry in green growth and climate actions. NIDM, New Delhi

Agrawal M, Gupta AK, Acharya P (2022a) Landscape approach and role in disaster management: moving beyond conventional concepts towards implementation frameworks. NIDM, New Delhi

Agrawal M, Zurita MM, Paraskevopoulou C, Admiraal H, Cornaro A, Gupta AK (2022b) Underground urbanism: re-imagining the role of underground spaces for India's urban future. NIDM, New Delhi

Amin F, Gupta AK (2022) Climate change implication and adaptation for river systems. In River dynamics and flood hazards: studies on risk and mitigation (pp. 497–506). Singapore: Springer Nature Singapore

Baidya S, Gupta AK (2022) Public policy in environment and sustainability strategies: global & national scenario. In Hydro-meteorological extremes and disasters (pp. 17–32). Singapore: Springer Nature Singapore

Barwal A, Bindal S, Gupta AK (2022b) Disaster preparedness and emergency response for air pollution and related health extremes. In Extremes in Atmospheric Processes and Phenomenon: Assessment, Impacts and Mitigation (pp. 329–347). Singapore: Springer Nature Singapore

Barwal A, Sood A, Gupta AK (2022a) India's Health Adaptation Plan: strategic tool for minimizing disaster related losses and damage. In Hydro-meteorological extremes and disasters (pp. 315–328). Singapore: Springer Nature Singapore

Barwal A, Sood A, Gupta AK, Goyal MK (2020) Lessons from trans-domain assessment of COVID 19 outbreak. Integrated Risk of Pandemic: Covid-19 Impacts, Resilience and Recommendations, 481–496

Bhardwaj S, Gupta AK (2021) Ecosystem-based approaches and policy perspective from India. Ecosystem-based disaster and climate resilience: integration of blue-green infrastructure in sustainable development, 101–125. In: Ecosystem-Based Disaster and Climate Resilience: Integration of Blue-Green Infrastructure in Sustainable Development (Disaster and Risk Research: GADRI Book Series). Springer Nature

Bhardwaj S, Gupta AK, Dhyani S, Thummarukudy M (2020) Nature-based solution entry points through sectoral policies, strategic instruments and business continuity. Nature-based Solutions for Resilient Ecosystems and Societies, 409–433

Bindal M, Gupta A, Gupta AK, Acharya P, Bhardwaj S, Hodam S, … Chondol T (2021a) CAP-RES-22 International Symposium DISASTER RESILIENCE AND GREEN GROWTH FOR SUSTAINABLE DEVELOPMENT. CAPRES-22 International Symposium (p. 130). New Delhi: NIDM

Bindal M, Gupta A, Gupta AK, Acharya P, Bhardwaj S, Hodam S … Bindal S (2021b) Science & Technology Research-Policy-Practice Interface for Climate Risk Management IMPLEMENTING PRIME MINISTER'S AGENDA 10 ON DISASTER RISK MANAGEMENT – ITEM 05 LEVERAGING SCIENCE & TECHNOLOGY. Science & Technology Research-Policy-Practice Interface for Climate Risk Management (p. 96). New Delhi: NIDM

Bindal MK, Gupta AK, Baidya S, Acharya P, Bhardwaj S (2021c) Environmental sustainability, green growth and disaster resilience knowledge compendium. NIDM, New Delhi

Bindal S, Acharya P, Gupta AK, Kishore J (2020) Enhancing epidemic resilience: planning and institutional resilience. Integrated Risk of Pandemic: Covid-19 Impacts, Resilience and Recommendations, 463–480

Choldol T, Gupta T, Amin F, Bindal MK (2022) Climate change and extreme events risk reduction. NIDM, New Delhi

Chondol T, Bhardwaj S, Panda AK, Gupta AK (2020) Multi-hazard risk management during pandemic. Integrated Risk of pandemic: Covid-19 impacts, resilience and recommendations, 445–461. In: Integrated Risk of Pandemic: Covid-19 Impacts, Resilience and Recommendations. Springer Nature

Dhyani S, Karki M, Gupta AK (2020a) Opportunities and advances to mainstream nature-based solutions in disaster risk management and climate strategy. Nature-based solutions for resilient ecosystems and societies, 1–26. In: Nature-based Solutions for Resilient Ecosystems and Societies. Springer Nature

Dhyani S, Gupta AK, Karki M (2020b) Nature-based solutions for resilient ecosystems and societies. Springer, New Delhi

Goswami M, Gupta AK, Kishan R, Baidya S, Khan YI, Prakash S et al (2023) An evaluation of climate resilient agricultural practices in India: a narrative synthesis of literature. Environ Sustain:1–17

Goyal MK, Gupta AK (2020) Integrated risk of pandemic: Covid-19 impacts, resilience and recommendations. Springer, New Delhi

Goyal MK, Gupta AK, Gupta A (2022) Hydro-meteorological extremes and disasters. Springer, New Delhi

Goyal MK, Poonia V, Kumar N, Jha S, Gupta AK, Acharya P (2021) Agriculture resilience assessment over Central India under climate change. NIDM, New Delhi

Gupta A, Gupta AK (2022) Localizing climate resilience agenda: Vision 2050 & 2100 (LOCC-RES 2100). National Institute of Disaster Management, Delhi. Page 67

Gupta A, Srivastava R, Hodam S, Singh S, Chary R (2021a) Climate resilientt agriculture management: training manual. NIDM, New Delhi

Gupta AK (2018) Disaster governance and legal systems in India. Disaster Risk Governance in India and Cross Cutting Issues, 39–60

Gupta AK, Acharya P, Baidya S, Agrawal M, Bansal S (2022f) Training manual on green growth and resource efficiency for DRR: towards green economy and resilient development. NIDM, New Delhi
Gupta AK, Barwal A, Madan A, Sood A, Kishore J (2021i) Water related disaster (floods & drought); implication for health adaptation & resilience. NIDM, New Delhi
Gupta AK, Barwal A, Sood A (2021k) Climate related health adaptation and resilience in India: northern region: policy paper. NIDM, New Delhi
Gupta AK, Bindal MK, Barwal A, Sood A (2021h) Air pollution and public health emergencies. NIDM, New Delhi
Gupta AK, Bindal MK, Sethia R, Chaudhary V, Baidya S, Mahajan V (2022e) Disaster damages and losses: study of underlying causes. NIDM, New Delhi
Gupta AK, Bindal MK, Sood A, Barwal A (2021j) Environmental services for health protection in disaster ans emergencies. NIDM, New Delhi
Gupta AK, Chopde S, Baidya S, Bindal MK (2022c) Institutional mechanism for evaluating human deaths from disaster. NIDM, New Delhi
Gupta AK, Chopde S, Nair SS, Singh S, Bindal S (2021l) Mapping climatic and biological disasters in India study of spatial & temporal patterns and lessons for strengthening resilience. NIDM & GIZ, New Delhi
Gupta AK, Chopde S, Singh S, Wajih S, Katyal S (2016) Prime minister's AgendA 10: India's disaster risk management roadmap to climate resilient and sustainable development. NIDM, New Delhi
Gupta AK, Dhyani S, Nair SS, Singh S (2020) Environmental governance for resilience & SDGs. Disaster Management for 2030 Agenda of the SDG, 253–276
Gupta AK, Gotmare S, Nair U, Babu S (2019c) Green growth benefits for climate and disaster resilience: concerns for urban and infrastructure systems. Position paper. NIDM, New Delhi
Gupta AK, Hodam S, Srivastava R, Bhardwaj S, Chary R, Prabhakar M, Sehgal VK (2019a) Roadmap of resilient agriculture in India. National Institute of Disaster Management, Delhi, India, p 19
Gupta AK, Jafar E, Bhardwaj S, Tyagi N, Mahajan V, Sethia R et al (2022a) Climate proofing disaster relief and recovery. NIDM, New Delhi
Gupta AK, Kumar R, Verma D, Acharya P, Awasthi S, Sharma N, Prasad S (2022b) Nature-based solutions for reducing disaster risks a guidebook for district disaster management planning. Wetlands International South Asia, New Delhi, New Delhi. and National Institute of Disaster Management, New Delhi
Gupta AK, Madan A, Acharya P (2022d) Water crisis in Indian cities: a systemic failure or wrath of changing climate. NIDM, New Delhi
Gupta AK, Mani N, Sarkar BB, Singh S, Katyal S (2019b) Developing disaster risk resilient cities. New Delhi: Gorakhpur Environmental Action Group (GEAG), Gorakhpur (UP), National Institute of Disaster Management, New Delhi, India and United Nations Children's Fund (UNICEF)
Gupta AK, Singh S, Agarwal M, Mani N, Wajih SA (2021f) Peri-urban ecosystems and urban resilience: training modules, instructions and reference materials. NIDM, New Delhi
Gupta AK, Singh S, Agarwal M, Nivedita M, Wajih SA (2021e) Peri-urban ecosystems and urban resilience: training modules, instructions and reference materials. Gorakhpur Environmental Action Group, Gorakhpur (U.P.) India, Climate Development and Knowledge Network, Cape Town South Africa, Pages 100
Gupta AK, Sood A, Barwal A (2021d) Health sector resilience to climate-related disasters: India's experience of national health adaptation planning. In Multidimensional Approaches to Impacts of Changing Environment on Human Health (pp. 375–387). CRC Press
Gupta AK, Sood A, Barwal A, Acharya P, Chondol T, Madan A et al (2021c) Health adaptation and resilience to climate risks. NIDM, New Delhi
Gupta AK, Sood A, Chondol T, Sethi S, Barwal A, Kishore J (2021g) Temperature related disasters: heat & cold wave. Implecations for health adaptation & resilience. NIDM, New Delhi

Gupta AK, Srivastava R, Hodam S, Singh S (2021b) Handbook for district administration on disaster risk management in agriculture sector. NIDM, New Delhi
Hodam S, Srivastava R, Gupta AK (2022) Addressing hydro-climatic risks through sectoral planning: a case of national agriculture disaster management plan (NADMP). In Hydro-meteorological extremes and disasters (pp. 301–314). Singapore: Springer Nature Singapore
Madan A, Gupta AK (2020) Understanding public health interventions: isolation, quarantine, social distancing. integrated risk of pandemic: covid-19 impacts, resilience and recommendations, 333–344
MoAFW (2020, November). NATIONAL AGRICULTURE DISASTER MANAGEMENT PLAN NOVEMBER 2020. Retrieved from National Agriculture Disaster Management Plan : NATIONAL AGRICULTURE DISASTER MANAGEMENT PLAN NOVEMBER 2020
NASA. (2023, February 14). Responding to climate change. Retrieved from Global Climate Change: https://climate.nasa.gov/solutions/adaptation-mitigation/
Nautiyal S, Goswami M, Khan IY, Prakash S, Kishan R, Gupta MK, Baidya S (2021) Climatic variations and agricultural landscape: a study on policies and practices for resilience. NIDM, New Delhi
NIDM (2018) Climate adaptive planning for resilience and sustainable development in multi-hazard environment (CAP-RES). DST, New Delhi
NIDM (2019a) Sectoral consultative workshop on national agriculture disaster management plan. NIDM, New Delhi
NIDM (2019b) UNCCD COP 14: special event reducing drought risk to improve land water resilience, mainstreaming eco-DRR pathways and tools. NIDM, New Delhi
NIDM (2019c, December 10) National health adaptation plan for climate change related disasters. Retrieved from NCDC: https://ncdc.gov.in/WriteReadData/l892s/5979424540l632477225.pdf
NIDM (2021) Multi-hazard disaster risk and resilience: practical learning and step-by-step guide to improve disaster resilience at city levels. NIDM, New Delhi
NIDM (2022) High level policy round table on science technology & innovation policy for disaster risk reduction. NIDM, New Delhi
NIDM, GIZ (2019) Climate risk management framework for India: addressing loss and damage. NIDM & GIZ, New Delhi
NIDM, GIZ, WHO (2019) National Programme on "Climate Change & Health Sector Resilience". NIDM, New Delhi
NIDM, VMMC (2020) Climate change, disaster and health: epidemiology and response preparedness. NIDM, New Delhi
NIDM, & WISA (2022, April 28) Wetland conservation: a nature-based solution for reducing water-mediated disaster risks. Retrieved from Wetlands International : https://south-asia.wetlands.org/news/wetland-conservation-a-nature-based-solution-for-reducing-water-mediated-disaster-risks/
Padhan N, Islary M, Gupta AK (2022) Climate finance at international and national level: needs, drivers and sources. In Hydro-Meteorological Extremes and Disasters (pp. 33–44). Springer Nature Singapore, Singapore
Phartiyal B, Mohan R, Chakrabort S, Gupta AK (2022) Climate change and environmental impacts: past, present and future perspective. Springer, New Delhi
Poonia V, Goyal MK, Gupta AK, Gupta A (2022) Drought as a disaster and its characterization over Central India. In Hydro-meteorological extremes and disasters (pp. 243–263). Springer Nature Singapore, Singapore
Poonia V, Goyal MK, Madramootoo CA, Gupta AK, Saraswat C (2020) Impact of COVID-19 on agro-food industry and transitions towards food security. Integrated risk of pandemic: covid-19 impacts, resilience and recommendations, 255–273
Saxena P, Shukla A, Gupta AK (2022a) An introduction to extremes in atmospheric processes and phenomena: assessment, impacts and mitigation. In Extremes in atmospheric processes and phenomenon: assessment, impacts and mitigation (pp. 1–8). Springer Nature Singapore, Singapore

Saxena P, Shukla A, Gupta AK (2022b) Extremes in atmospheric processes and phenomenon: assessment, impacts and mitigation. Springer, New Delho

Sood A, Barwal A, Gupta AK, Goyal MK (2020) Introduction to virus outbreaks. Integrated risk of pandemic: covid-19 impacts, resilience and recommendations, 3–20. In: Integrated Risk of Pandemic: Covid-19 Impacts, Resilience and Recommendations. Springer Nature.

Tiwari PC, Joshi B, Gupta AK, Bhardwaj S (2021) Development of integrated framing for improving food security in Indian Himalayas under climate change: a case illustration of upper Kosi catchment in Kumaun Himalaya, Uttarakhand. NIDM, New Delhi

UNFCCC (2022) Information paper on linkages between adaptation and mitigation. Bonn: United Nations Framework Convention on Climate Change, the Kyoto Protocol and the Paris Agreement

Anil Kumar Gupta Professor of Policy & Strategies, Integrated Resilience. Head of the ECDRM Division, International Cooperation, Programme Director - Centre for Excellence on Climate Resilience, Nodal Officer for Central Ministries at National Institute of Disaster Management, (Govt. of India), New Delhi. He is also a member of the expert Team for WMO Climate Statement and Core Group Member of IUCN-Council of Ecosystem Management.

Pritha Acharya has a background in environmental sciences. She has worked in the field of sustainability, climate change adaptation and biodiversity conservation while working with IUCN India. She is currently working as a research associate and manages the CAP-RES (DST) funded project.

Chapter 31
Disaster Risk Reduction through Climate Adaptive Development: Strategies and Road Ahead

Anil K. Gupta, Shweta Bhardwaj, Manish K. Goyal, and Akhilesh Gupta

Abstract India is vulnerable to natural hazards particularly drought, flood and cyclones. Increase in frequency and intensity of disasters due to climate change are of serious concern for Disaster Management, Climate Change Adaptation and Climate Resilience. Climate Resilient Development aims to build the capacities of vulnerable communities to withstand the serious ecological, economic, and social challenges posed by climate change. To address this many policies and programs have to be seen all through the climate lens. The need for this has been well spelt out in India's National Disaster Management Plan (National Disaster Management Authority, Governmnet of India, 2016) and India's National Action Plan for Climate Change (2008). Internationally, Sustainable Development Goals, Paris Climate Agreement (2015) and Sendai Framework for Disaster Risk Reduction-SFDRR (2015–2030) have formulated global and universal guidelines and targets for action towards sustainable development by synchronizing every tier of governments and agencies of very country. Potential of integrating DRR and CCA has also been recognised in this study. The combined approach has multiple benefits including a lessening of duplicate efforts (considering the case if the two subjects are taken separately), optimization of resource utilization, and development of more resilient communities.

A. K. Gupta (✉)
ECDRM Division, International Cooperation, CAP-RES DST Project, National Institute of Disaster Management (Ministry of Home Affairs), New Delhi, India

S. Bhardwaj
Ernst and Young, New Delhi, India

M. K. Goyal
Dean of Infrastructure, Department of Civil Engineering, IIT Indore, Indore, India

A. Gupta
Department of Science and Technology, Science and Engineering Research Board, S&TI Policy Coordination, Climate Change Programme, Government of India, New Delhi, India

A. K. Gupta et al. (eds.), *Disaster Risk and Management Under Climate Change*,
Disaster Resilience and Green Growth,
https://doi.org/10.1007/978-981-99-4105-6_31

Keywords Climateresilience · Sendai Framework for Disaster Risk Reduction · National Disaster Management Plan · Sustainable Development Goals · Systematic approach

31.1 Introduction

India saw around 2,300 deaths in the year 2017 with 22.5 million people affected (majorly by flood and storm) all across the country (Natural Disasters in 2017: Lower Mortality, higher cost 2018) . It has taken the lives of many and has destroyed property and resources. During the period of 1901–2007, maximum and minimum temperature of India spiked up at the rate of 0.71 and 0.27°C every century (Prajapati et al. 2015). Currently, climate variability has become a major risk to the economic and social development. Majority of researches show that India is one of the most vulnerable country to Climate Change. Millions of population, dense economic activities are one of the leading reasons for positioning India as one of the most vulnerable country to Climate Change. Degrading Himalayan ecosystem, erratic rainfall, delayed monsoon, increased flooding in certain pockets are threatening over-all security of human well-being: food security, bio diversity security, livelihood security etc.

The long-term impact of Climate Change on development sectors necessitates development of policies, practices and schemes to include the information and understanding of Climate Change with an aim to respond quickly to changing climate in long as well as short term period. It signals to mainstream Climate Change resilience in the current and future planning of all development sectors and to address the vulnerabilities and livelihood security of millions of people. For this, not only plans should be in line with Climate Change, but the existing policies and plans should also be revamped. Increasing urbanization and population make the world increasingly unsafe, disaster-prone and insecure for living beings. Hence, to reduce Climate Change and disaster risks, mainstreaming them into development policies, plans, programs and projects emerge as the only best possible strategy to protect lives, livelihoods and assets of human being.

The need of investing in resilient development that can substantially minimize the damage and loss from disasters is widely accepted. Sendai Framework for Disaster Risk Reduction (SFDRR) and Paris Climate Agreement recognizesthis and identifies the underlying causes of disaster risk for which mainstreaming the climate resilience plans into development plans was called for. India subscribes to SFDRR and is committed to building the country and community resilience in pursuit of SFDRR goals.

Ministry of Home Affairs, Government of India, handles all disaster management related initiatives at the national level, other than drought and Climate Change, which are dealt separately by the Ministry of Agriculture. This gap in government departments needs to address due to the cross-cutting nature of disaster impacts that affect human lives, livelihoods, habitat, health, education, and infrastructure altogether. It is never be enough that the disasters are dealt by one line department. As the aftermath of disasters and Climate Change does not select which aspect of development will be damaged, government departments need not to wait for other exclusively disaster-related departments to come up with climate-smart plans. In the end what is left for the system is to mainstream plans which are eco-friendly and climate-friendly in economic, social and cultural development programs of the country.

This chapter investigates the initiatives taken by world and India particularly for Climate Change by looking at current schemes/plans for gaps in convergence, integrating mechanism of Climate Resilience and Disaster Risk Management into climate responsive development plans (at national and sub national level) identifying planning support tools and the roles of stakeholder; research institute and NGOs in implementing climate resilient development at national and sub national level. This chapter also summarizes the benefits of integrating/ mainstreaming Climate Change Adaptation and Disaster Risk Management in the development framework. This integrated approach of CCA and DRR has the potential of:

- Making communities more resilient to the devastating effects of Climate Change and natural disasters,
- Reducing climate-related losses,
- Optimization of natural resources
- Increasing the effectiveness of both (CCA and DRR) (ProAct Network 2008)

31.2 Climate Resilience through Developmental Plans

The economic and social costs (worldwide) to adapt to climate continue to mount year after year. Sometimes the economic costs surpass the GDP (Gross Domestic Product) of a few countries, especially developing and under-developed countries. A scientific journal Nature Climate Change reported that India is one of the top three carbon emitters in the world that pay heaviest cost of $86 on emitting each extra tonne of CO_2, after United States (Ricke et al. 2018). India already bears the economic loss of $206 billion owing to its 2.4 billion metric tonnes of Carbon dioxide (Staff, D. T. 2018). On top of it, every year $9-10 billion are wasted due to extreme events caused by Climate Change. Reports have suggested natural disasters have caused India, economic losses of about $80 Billion during the last 20 years (NDTV 2018). Climate Change poses a threat to the agriculture sector of the country on which half of the population of the country relies for livelihood. Depletion of crop productivity and crop production due to extreme events such as drought, flood, cyclone, dust storm and hailstorm have been widely acclaimed to be the deadly impact of Climate Change. The impact of Climate Change is systemic, impairing every aspect of living being on this planet.

Global Climate Index of 2018 claims India to be the sixth most vulnerable country struggling with weather-related extreme events (Global Climate Risk Index 2018). This claim amplified in its accuracy when IPCC's latest special report;

Global Warming of 1.5°C claimed that countries like India would witness substantial sea level rise, drought, flood and heat waves. Above claims by different internationally recognized reports gives a wake-up alarm for India and the world to not only develop, but also implement climate resilience plans in order to save the planet from irreversible damage. Policy makers, scientific fraternity and communities need to work together to formulate strategies to mainstream CCA and DDR into the development planning process for climate resilience development (Gupta et al. 2016).

Climate-resilient developments are integrated to some extent at the national level through India's commitment to SFDRR and National Disaster Management Plan 2016 and other ministerial-level programs. However, more inclusive and action-oriented plans need to develop by integrating sectorial departments such as Water Supply, Health, Agriculture, Rural Development and Urban Development undertake activities that influence climate and disaster resilience. The co-ordination gap between state level, district level and national plans and projects need to addressed through horizontally and vertically coordination to save India and the lives that reside in it.

31.3 Journey from Disaster Management to Climate Resilience Development

During 1970's the widely accepted global approach for a natural disaster was exclusively disaster response. A permanent disaster management approach came into existence in 1980's which dealt with all aspects of the disaster management cycle: preparedness, response, recovery and reconstruction. After realization of insufficiency of 'only disaster response', continuous efforts have been done for the past twenty-five years to address the issues of consistent onset of disasters. The importance of DRR was first recognized in the Millennium Development Goals 2010. Millennium Development Goals (MDGs) identified that disaster risks are on the rise globally due to increasing vulnerability to natural hazards.

The Hyogo Framework for Action (HFA), 2005–2015 recognized that DRR is across cutting issue to development which should be achieved internationally on agreed developmental goals, including goals of MDGs. Member states of the United Nations Conference on Sustainable Development (Rio+20), 2012, developed a set of Sustainable Development Goals; built upon the foundations ofRio+20 Outcome Document- The Future We Want- has a section on DRR which sets out a substantial foundation for consideration on post-2015 framework to guide nations after the expiry of HFA 2005–2015.

31.4 Universal International Frameworks for Climate Resilience Development

31.4.1 Sendai Framework for Disaster Risk Reduction (SFDRR)

SFDRR was adopted at the Third UN World Conference in Sendai, Japan, on March 18, 2015. It is the successor instrument of HFA 2005–2015 and is founded on elements which ensure continuity with the work done by states and other stakeholders under HFA. It is a non-binding voluntary agreement for 15 years which recognizes the responsibility of the state for DRR and shared responsibility with local governments, the private sector and other stakeholders. This is an improvised version of HFA, by identifying the gaps, good lessons learned and future challenges. Key features of SFDRR are:

- Shifting focus from Disaster Management to disaster risk management
- Emphasis on the need of strengthening the disaster risk governance by placing governments at the centre of DRR
- Wider approach of DRR, focusing on both natural and man-made hazards and related environmental, technological and biological hazards and risks
- Recognition to the implementation of integrated environmental and natural resource management techniques

31.4.2 Paris Climate Agreements

The Paris Agreement on Climate Change was adopted at the UN Climate Change Conference held in December 2015. The agreement was adopted by 195 countries and called for a commitment to work together to safeguard the planet, promote sustainable human development and build more resilient and equitable world for all. The agreement will bring about a process of maintaining a balance between mitigation and adaptation. In future approaches of climate adaptation, the loss and damage caused by disasters will be better incorporated to accomplish the goal of climate resilience. The preamble of agreement mentions about the adoption of SFDRR. National ratification by governments and implementation of national climate action plans are the step towards implementing the Paris Agreement.

31.4.3 Implementing Sendai Framework in India

India is a signatory to SFDRR and strives to comply with it. The country is making efforts to achieve the global targets by making advancement in the entire disaster management cycle by following the recommendations in the Sendai Framework and by adopting internationally accepted best practices. National Disaster Management Plan NDMP, 2016 is the first ever plan at the national level for disaster management. India has aligned the national plan with SFDRR's guidelines and actions for disaster risk reduction.

31.5 Linking Climate Change Adaption and Disaster Risk Reduction

Linkage of CCA and DRR is generally considered as a win-win strategy. Integration of DRR and CCA has the ability to ease the burden of development, by minimizing the redundancies, and this combined approach also makes use of the optimum resource utilization. The following figure summarizes the differences and the convergence areas of DRR and CCA.

There are multiple areas where these two approaches were reflected with common goals. Some are as following (Fig. 31.1):

- Both CCA and DRR have the same objective, i.e. to develop hazard resilient communities
- Environmental management measures benefit both
- Both should be integrated into disaster management operations (including relief, recovery and development plans) to develop resilient communities

Source	**Differences**		**Convergence Areas**
	DRR	**CCA**	
Origin	Disaster Event	Scientific Theory	• DRR increasingly forward-looking • Existing climate variability is an entry point for CCA
Risk Level	Existing risks	New risks	• DRR increasingly forward-looking • Existing climate variability is an entry point for CCA
Knowledge	Traditional/ indigenous knowledge at community level is a basis for resilience	Traditional/ indigenous knowledge at community level may be insufficient for resilience against types and scales of risk yet to be exerienced	Selected examples where integration of scientific knowledge and traditional knowledge for DRR provides learning opportunities
Recognition	Political and widespread recognition often quite weak	Political and widespread recognition increasingly strong	None, other than that climate-related disaster events are now more likely to be analysed and debated with reference to climate change
Structural measures	Designed for safety levels modelled on current and historical evidence	Designed for safety levels modelled on current and historical evidence and predicted changes	• DRR increasingly forward-looking • Existing climate variability is an entry point for CCA

Fig. 31.1 Differences and the convergence areas of DRR and CCA (ProAct Network 2008)

31.6 Tools for Integrating Climate Resilient Plans into Developmental Schemes/Projects

There are several tools and methods through which DRR and CCA measures have been integrated to reduce the risk of natural disasters. These are:

31.6.1 Legal Policy Framework for Disaster Management

National Disaster Management Authority has been constituted for laying down policy and guidelines for Disaster Management in India. Similarly, SDMAs (State Disaster Management Authority) and DDMAs (District Disaster Management Authorities) have been constituted for laying down guidelines for Disaster Management in state and district level. NDMA, SDMA and DDMA have issued several guidelines many a time.

31.6.2 Environmental and Natural Resource Laws in DRR and Integrating CCA

Regulatory provisions related to environment and natural resources- water, land, agriculture, forests, wildlife, habitats, ecosystems; procedures and planning – Environmental clearance, EIA, audit, risk analysis, land-use and zoning, emergency preparedness; and environmental services-drinking water, sanitation, waste management, preventive-health, including climate mitigation and adaptation etc. play an important role in addressing hazards, reducing underlying causes of vulnerability and enhancing capacity, and thereby, relate to Disaster Risk Reduction (Gupta et al. 2016).

31.6.3 Inclusion of DRR into Developmental Schemes and Projects

For mitigating climatic hazards and minimizing the impacts of natural disasters and for improving livelihoods and overall well-being of the people, central and state Governments have implemented a number of schemes, whose activities are facilitated further by the involvement of PRIs, NGOs etc. Some of these important national level programs are discussed in the Table 31.1. These plans and projects can be used as entry points for mainstreaming climate change concerns into development plans in national and sub-national level.

Table 31.1 A brief description of some of the Schemes and Programs where climate and DRR can be integrated

Schemes	Details
Swacch Bharat Mission	It is a national campaign of the Government of India, to clean the streets, roads and infrastructure of the country. It has two arms: SBM (urban) and SBM (rural)
Atal Mission for Rejuvenation of Urban Transformation	It was launched in 2015 to provide basic services to household, i.e. water supply, sewerage, urban transport
Smart Cities Mission	It is the new initiative of the present government that aims to drive the economic growth and quality of life of city people by enabling local area development through the use of technology, information, infrastructure development and services
Urban and Regional Development Plans Formulation and Implementation	It has provisions for rainwater harvesting, conservation of urban water bodies, water supply system, waste water management system, energy efficiency, strategic plan for new and renewable energy efficiency, strategic plan for new and renewable energy, alternate sources of energy to meet the city demand
National Mission on Sustainable Habitat	It is one of the Twelve Missions of National Action Plan on Climate Change (NAPCC) launched in 2010, that takes into account the climate change component and vulnerabilities which is likely to affect infrastructure related to water, sanitation, energy, transport etc.
Affordable Housing in Partnership	GoI approved the scheme of Affordable Housing in Partnership (AHP) as part of Rajiv AwasYojana (RAY) on 3rd September 2013 to increase affordable housing stock, as part of the preventive strategy.
National Rural Health Mission (NRHM)	It was launched by Ministry of Health and Family Welfare in 2005 with an objective to provide support to the health care system of rural areas of 18 states through the provision of physical infrastructure, human resources, equipment, emergent transport, drugs, diagnostics and other support. It provides managing, funding and institutional support to all the selected states to all the facilities starting from sub-centre, public health centres, community health centres, sub-district and district hospitals
Sarva Shiksha Abhiyan	The SSA is Government of India's flagship program for achieving Universalization of Elementary Education (UEE) in a time bound manner, making free and compulsory Education to the children of 6-14 years age group. SSA is being implemented in partnership with State Governments to cover the entire country and address the needs of millions of children.
Pradhan Mantri Gram Sadak Yojana (PMGSY)	PMSGY is a fully funded Centrally sponsored Scheme to provide all-weather road connectivity in rural areas of the country.

(continued)

Table 31.1 (continued)

Schemes	Details
National Rural Livelihood Project (NRLP)	The NRLP is implemented is implemented in 13 high poverty states accounting for about 90 per cent of the rural poor in the country. Intensive livelihood investments would be made by the NRLP in 107 districts and 422 blocks 13 states (Assam, Bihar, Chhattisgarh, Jharkhand, Gujarat, Maharashtra, Madhya Pradesh, Orissa, Rajasthan, Uttar Pradesh, West Bengal, Karnataka and Tamil Nadu)
Special Package for Drought Mitigation Strategies	GoI in 2009 approved a special package for implementing drought mitigation strategies in Bundelkhand region at a cost of Rs 7266 crore comprising Rs 3506 crore for Uttar Pradesh and Rs 3760 crore for Madhya Pradesh, to be implemented throughout 3 years starting 2009–10. It is envisaged to give additional central assistance (ACA) to the tune of Rs 3450 crore for implementation of the package. In continuation of the special Package during the 12th Plan period (2012–17) a financial outlay of Rs 4400 crore was approved under the Backwards Regions Grant Fund (BRGF). The project objectives are to restore ecological balance by harnessing, conserving and developing natural resources like soil, water and forest and improve the ecosystem by checking soil erosion and deforestation
National Watershed Development Project for Rainfed Areas (NWDPRA)	The project aims at in-situ moisture conservation primarily through vegetative measures to conserve rainwater control soil erosion and generate the green cover both on arable and non-arable lands. The scheme is implemented at the field level by an interdisciplinary team of members from line departments of state government and the beneficiaries of the watershed
Pradhan Mantri Krishi Sinchai Yojana	PMKSY is an amalgamating ongoing schemes viz. Accelerated Irrigation Benefit Program (AIBP) of the Ministry of Water Resources, River Development & Ganga Rejuvenation (MoWR,RD&GR), Integrated Watershed Management Program (IWMP) of Department of Land Resources (DoLR) and the On-Farm Water Management (OFWM) of Department of Agriculture and Cooperation (DAC). It mainly undertakes rain water conservation, construction of farm pond, water harvesting structures, small check dams and contour bunding etc. MoWR, RD &GR, undertake various measures for creation of assured irrigation source, construction of diversion canals, field channels, water diversion/lift irrigation, including the development of water distribution systems

(continued)

Table 31.1 (continued)

Schemes	Details
Mid-Day Meal	The Mid-Day Meal Scheme directs state governments to provide cooked mid-day meals in all government and government assisted primary schools
Paramparagat Krishi Vikas Yojana	The government has launched Paramparagat Krishi Vikas Yojana in order to address the critical importance of soil and water for improving agricultural production. The government would support and improve organic farming practices prevalent in India. Following the cluster approach mode of farming, at least 50 farmers would form a group having 50 acres of land to implement organic farming
Rashtria Krishi Vikas Yojana (RKVY)	Launched in 2007, provides 'additional central assistance' to Central government and state schemes related to agriculture. Region-specific agriculture research and preparation of district agriculture plans, taking into account the local needs and conditions
Integrated District Approach (IDA)	It was launched in late 2004 and early 2005 in 17 districts across 14 states in the country (key strategies outlined by UNICEF India) . It focusses on promoting community action and the integrated delivery of services by establishing horizontal linkage between the communities and the line agencies on the one hand and establishing an interface between the communities and the line agencies on the other to ensure responsive, relevant and convergent delivery of services
Integrated Child Development Services	This project strives to provide supplementary nutrition, health care and pre-school education to children below the age of six

Source: Gupta et al. (2016)

31.6.4 Integration of CCA-DRR within Policy-Planning

Certain important environment-policy instruments (instruments that are useful in the formulation of policy and/or implementation of planning) directly useful in developing and implementing CCA and DRR are discussed in the table below (Table 31.2):

Table 31.2 Environment policy-instrument and their role in CCA_DRR

Instrument	Brief description	Role in CCA-DRR integration
Strategic Environmental Assessment (SEA)	EIA of policies, plan and programs	Mainstreaming CCA-DRR towards sustainable development with the ecosystem approach, climate-risk mitigation and post-conflict recovery context
Environmental Impact Assessment (EIA)	Regional EIA, Country EIA, Cumulative EIA, Carrying capacity based Planning Process	The anticipation of hazards, risk hotspots, vulnerability-spatial contexts; Projected mitigation and capacities; Residual risks for emergency response/plan
Life Cycle Assessment (LCA)	Environmental impacts during different stages of life-cycle of a material or a major project	Prediction and forecasting of changing patterns of hazards and risk profiles over time
Ecological footprint	Human demand for natural resources and ecosystem services bearing to regeneration capacity	The anticipation of ecosystem fragility or biotic pressure on land and water resources that lead to hazards and aggravate disaster risks
Environmental Legislation	Policy Statements, Acts & Rules, Ordinances, Notifications, Standards and Codes, Treaties	Provides legal support for reducing hazard precursors, vulnerability causes; offers capacity and recovery potentials etc.
Auditing/Environmental Management System (EMS)	Environment audit, Water balance audit, Safety Health audit, Eco-auditing	Impact of a strategy or activities of an organization/facility, person or business on environment leading to hazards, vulnerability or mitigation, and related data/documentation
Cess/Levees	Charges for natural resource exploitation, environmental services-water and clean-up, etc.	Reduces pressure on landscape and ecosystems; facilitates conservation-reduces hazard intensities, susceptibility and improves response resources
Natural Resource Accounting (NRA)	Transformation of data on environmental features for use in economic decisions	Assessment of prevailing and anticipation of vulnerability; resilience and recovery potentials
Eco-labelling/Eco-mark	Public information on eco-friendly production and product	Promoting people's contribution and concern for reducing hazards in nature and disaster prevention
Environmental Taxes	Polluter pays principle payments to curb the ill-effects on the environment	Curbing environmental precursors of hazards and vulnerability; financing mitigation and sustainability

Source Gupta et al. (2016)

31.6.5 Integration of Climate Resilience Plans into Development Plans

31.6.5.1 Mechanism of Inclusive Climate Resilience Development Plans

Policy Level

At policy level means climate resilient plans are developed by integrating Climate Change Adaptation practices into DRR policies and vice versa. Adaption and Mitigation practices need to be embedded into each strata of the policy framework of state and national government. The framework should be comprehensive, simple and engaging for all the stakeholders involved in the process. It should identify the impacts of Climate Change in a way that vulnerability assessment, hazard risk assessment should be thoroughly studied on a regional way in order toameliorate communities resilience to wide array consequences of Climate Change. This may be reflected in the national development strategy for sectors such as agriculture, forestry, water resources, transportation etc. The foremost step for this is to assess the interconnection of different sectors that directly and indirectly affect the ecosystem and climate. Awareness of the linkage of different sectors/departments of government system will pave the way for outlining the budgeting of Climate resilient practices and plans into fiscal policies of the state as well as central government so that the fund could be utilized by vulnerable communities. The policy framework should take into account the severity, frequency and latest trends of natural disasters and social and economic practices that lead to these disasters at the local and national level. The inclusion of these parameters at the policy level will assist the stakeholders in analysing the key concern areas to mitigate disaster risk, estimating the amount of funds required to mitigate these parameters, reducing the losses due to disaster risk by taking appropriate actions.

Planning Level

The development planning at state, national and local (village) levels is the crucial next step for putting policies into practice. For the start logical strategies are needed to be prepared for climate resilient development for short, medium and long-term. At the local planning level two key entry points exist for integrating CCA and DRR:

- Integration of CCA into DDMPs
- Integration of CCA into decentralized planning processes on DRR by Panchayati Raj Institutions (PRI) bodies and elected community representatives, including VDMPs.
- The effective integration of climate resilient plans into development planning at district and village levels requires capacity building of local authorities. This includes:

- Dialogue workshops and training programs with local and district authorities on climate change and participatory planning for DRR and CCA. Workshops and Shared Learning Dialogues (SLDs) with local authorities and community have been successful for creating awareness among stakeholders on climate resilient development.
- Development of Local Adaptation and Mitigation Guides (LAMGs) for integration of climate resilient plans into community level plans.
- Continuous support for individual meetings, provided by teams and local adaptation specialists leading to improved planning.

Implementation Level

It is the successive level of measures taken in national and state level policies in a top-down process and the starting point for planning and policy development from the bottom-up approach. The bottom-up approach of implementation of policies projects CCA and DRR in the practices of communities, founded on the needs and interests of the local community.

Community-based adaptation measures, worldwide have been successful in demonstrating the resilience building at local levels. It is essential to enable national and local actors - policy makers and planners – to better understand how to support communities which are increasingly at risk of disasters. This should be facilitated by the growing recognition among the different communities of the added value of DRM, DRR and CAA, in particular in a context of tight financial resources (Fig. 31.2).

31.7 Multi-Hazard Scenarios and Sustainable Development Goals

With the manifestation of climate change, disaster risks are exacerbating due to the increase in frequency and intensity of climatic-induced extreme events. Hence, the likelihood of occurrence of floods, droughts, and other extreme weather events (heat waves/cold waves) is increasing day by day. With this increase in frequency of disaster events, the world is witnessing emergence of multi-hazard risk scenarios where possibility of two or more hazards coinciding in a same place and time have also relatively increased. Such multi-hazard risk scenarios may arise either due to concurrence of two or more extreme events or due to cascading effect of any disaster, where a primary event triggers a secondary event(s). Multiple disasters at a same point in time can result into complex emergency and requires advancement of disaster risk reduction efforts based on multi-hazard risk approach. A very recent example of such multi-hazard risk scenario have been witnessed during global pandemic COVID-19 where number of disasters including flood, cyclones,

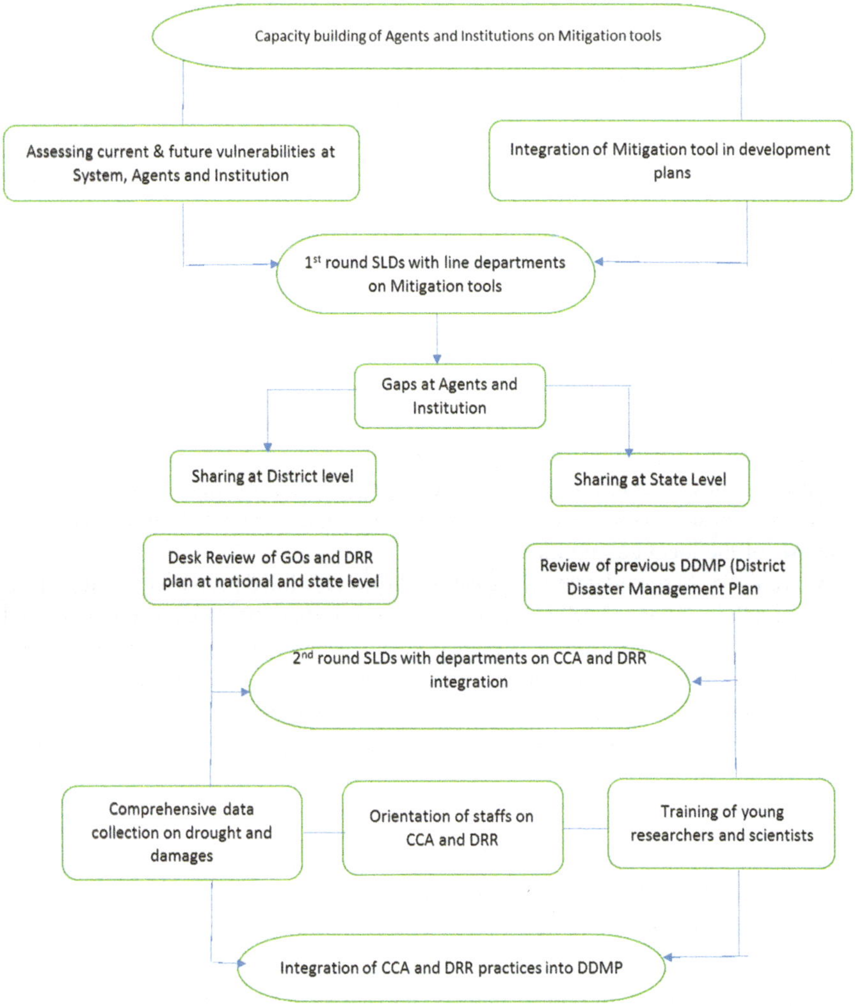

Fig. 31.2 Outline of integration of CCA and DRR practices in District Disaster Management Plan (Source: Gupta et al. 2016)

industrial disasters, earthquakes etc. occurred during the pandemic in several countries. In such a scenario, a Multi-hazard approach is vital for effective and sustainable response and recovery. There is a growing recognition of the fact that climate change adaption and mitigation constitute an important subset within disaster risk reduction as well as sustainable development. Reducing vulnerabilities and risks towards climate change and disaster risks must build and expand on the existing efforts towards sustainable development. Therefore, it becomes necessary to bring on board the climate change, disaster risk reduction and sustainable development discourse together into policy-planning-practice at all the levels.

At present, some of the initial assessments on impacts of the global pandemic COVID-19 indicate enomorous loss of live and livelihoods and also points towards the possibility of decades' old developmental progress standing to be reversed. It is estimated that COVID-19 would lead world to the sharp economic contraction, potentially the largest one since the Great Depression (1929–1930) which would deepen poverty, food insecurity and hunger across the world (United Nations 2020). The closure school, colleges and educational institutions as a preventive measure to contain the spread of pandemic is believed to bring in the hidden and unprecedented education emergency, where millions of children are at the risk of dropping out of schools and education systems forever and which might inturn push them into life-long deficits by exposing them to child labour, early marriage, gender-based violence and abuse. All this would lead to widening and reinforcement of existing social and economic inequalities in the society (Save the Children 2020). This global pandemic has clearly highlighted the developmental gaps and deficits in the global systems in terms of weak-inadequate healthcare systems, lack of access to basic services, exposure to poverty etc. The need to contain and address the impacts of COVID-19 have lead countries to reset their priorities and relocation of resources for combating the pandemic and it will definitely have far reaching long-term implications on the future prospects advancement of SGDs. The impacts of Covid-19 are directly hampering realisation of Sustainable Development Goal (SDG) including Goal 1(No Poverty), Goal 2(Zero Hunger), Goal 3 (Good Health and Well-Being), Goal 4 (Quality Education), Goal 6 (Clean Water and Sanitation) and Goal 10 (reduced inequality) and also is slowly leading to falling apart of SDG 8 (Decent Work and Economic Growth) and SDG 9 (Industry, Innovation and Infrastructure). The close interlinkages between different SGDs allow the negative impacts of pandemic to easily cascade across other SDGs as well. However, the recovery efforts needed for bringing back societies to pre-pandemic situation would require systemic identification of strategic entry points across targeted actions and these actions cannot be de-linked from SDGs and their themes, for ensuring the long-term sustainability of these efforts. Therefore, response and recovery efforts while addressing immediate challenges must also reinforce SDGs, since concerns addressed through SDGs remain central for building resilience of individuals and communities towards any unprecedented shocks and events such as COVID-19 pandemic or any future disaster or climate induced events. Here, it also important to understand that any sustainable recovery and development would be possible only through environmentally sensitive pathways supported by effective plans and policies; and such an approach also provides a window of opportunity for advancing sustainable development while addressing the climate change concerns.

31.8 Conclusion and Way Forward

Climate-resilient pathways should consist of future trajectories of development that combine adaptation and mitigation in the context of sustainable implementation. CCA and DRR integration have shown better resilience and sustainability. In this chapter the capacities of national, state and district level were understood in implementing climate resilience development plans coupled with departmental level disaster management plan in order to achieve sustainable development. Several programs are already underway in many parts of India, often in partnership with local financial institutions and grassroots non-governmental organizations that are working with local communities on project implementation for sustainable livelihoods and infrastructure development across the country.

Importance of integrating, CCA and DRR have also been realized as climate change and disasters share similar consequences. In addition to National Action Plan for Climate Change (NAPCC), India have already started implementing State Action Plan for Climate Change (SAPCC) for ensuring that the regional development priorities and the national objectives for addressing climate change concerns are well aligned with each other. SAPCC helps in creating an enabling environment for implementation of NAPCC while addressing the state specific issues and builds already existing state level developmental policies, plans and programmes. However, SAPCC still lacks effective implementation on grounds due to lack of institutional and human capacities, mainstreaming across line departments, and inadequate financial and technical support. There are measures that India can take to promote sustainable and inclusive growth particularly in a less carbon-intensive and strong infrastructure manner. Many of these actions will have to be designed and implemented by industry and civil society organizations. Recognizing the important role that non-state actors can play in shaping India's response to climate change, the Government of India should take steps to make this an inclusive and consultative process and should invite the participation of all communities, non-governmental organizations, industry and other relevant stakeholders.

References

Global Climate Risk Index. (2018) Germanwatch

Gupta AK, Singh S, Katyal S, Chopde S, Wajih SA, Kumar A (2016) Climate resilient and disaster safe development. Process Framework Training Manual

National Disaster Management Plan (2016) National Disaster Management Authority, Governmnet of India.

Natural Disasters in 2017: Lower Mortality, higher cost. (2018, March). Cred Crunch. Brussels, Belgium: Centre for Research on the Epidemiology of Disasters (CRED), Research Institute Health & Society (IRSS), Université catholique de Louvain.

NDTV (2018, October 12). India Lost $79.5 billion due to natural disasters in 20 years, says United Nations. Retrieved from NDTV.com: https://www.ndtv.com/india-news/india-lost-79-5-billion-due-to-natural-disasters-in-20-years-says-united-nations-1930867

Prajapati RK, Singh SR, Gangwar RK (2015) Effect of climate change on plant diseases in Bundelkhand zone leading to changing in cropping pattern. Res Educ Dev Soc (REGD):120–138

ProAct Network (2008) Proact network climate change adaptation and disaster risk reduction policy paper 2008. ProAct

Ricke K, Drouet L, Caldeira K, Tavoni M (2018, September 25) Country-level social cost of carbon. nature climate change, 895-900. Retrieved from livemint: https://www.livemint.com/Science/AphhJrsCVyNACWGX3sxfEJ/India-among-top-countries-to-incur-max-loss-from-climate-cha.html

Save the Children (2020) Save our education. Save the Children, London

Staff, D. T. (2018, September 28). US, China and India: Top carbon emitters to face the biggest economic losses. Retrieved from Down To Earth : https://www.downtoearth.org.in/news/climate-change/us-china-and-india-top-carbon-emitters-to-face-the-biggest-economic-losses-61747

United Nations (2020) Achieving the SDGs through the COVID-19 response and recovery. United Nations

Anil K Gupta Professor of Policy & Strategies, Integrated Resilience. Head of the ECDRM Division, International Cooperation, Programme Director - Centre for Excellence on Climate Resilience, Nodal Officer for Central Ministries at National Institute of Disaster Management, (Govt. of India), New Delhi. He is also a member of the expert Team for WMO Climate Statement and Core Group Member of IUCN-Council of Ecosystem Management.

Shweta Bhardwaj has done her master's in Disaster Management from TISS, Mumbai. Presently, she has also workes with NIDM in CAP-RES (DST) funded project and looks after the Climate Proofing component of the project. She has worked extensively in the field of community organization and social development, environment issues, disaster risk and vulnerabilities and disaster governance. She is currently associated with Ernst and Young (EY).

Manish K Goyal has about 10 years of professional experience in consulting, research and project management for engineering projects in India and abroad. I am actively involved in project management, feasibility studies and climate change impact assessment studies and have initiated several efforts to enhance quality. He is currently associated with IIT Indore as Professor and Dean of Infrastructure development.

Akhilesh Gupta is the Secretary SERB and Senior Advisor at Department of Science & Technology (DST) Govt. of India coordinating two National Missions on Climate Change under National Action Plan on Climate Change, and Head of Science - Technology Policy. He is known for coordination of establishng the National Research Foundation of India. He has published over 185 research papers in various National & International Journals & conference proceedings. Dr. Gupta was a member of the National Coordination Team which drafted India's National Action Plan on Climate Change in 2008. He has been president of Indian Meteorological Society & Association of Agro meteorologists.

SPRINGER NATURE

GPSR Compliance

The European Union's (EU) General Product Safety Regulation (GPSR) is a set of rules that requires consumer products to be safe and our obligations to ensure this.

If you have any concerns about our products, you can contact us on ProductSafety@springernature.com

In case Publisher is established outside the EU, the EU authorized representative is:

Springer Nature Customer Service Center GmbH
Europaplatz 3
69115 Heidelberg, Germany

The manufacturer's authorised representative in the EU is Springer Nature Customer Service Centre GmbH, Europaplatz 3, 69115 Heidelberg, Germany. If you have any concerns regarding our products, please contact ProductSafety@springernature.com

Printed and bound by CPI Group (UK) Ltd, Croydon, CR0 4YY
30/09/2026
02225318-0020